A TREATISE
ON THE
THEORY OF SCREWS

Originally published by Cambridge University Press in 1900, *A Treatise on the Theory of Screws* is the definitive reference on screw theory. It gives a very complete geometrical treatment of the problems of small movements in rigid dynamics. In recent years the theory of screws has emerged as a novel mathematical resource for addressing complex engineering problems, with important applications to robotics, multibody dynamics, mechanical design, computational kinematics, and hybrid automatic control.

The author, Sir Robert Stawell Ball, was born in Dublin in 1840 and studied at Trinity College, Dublin. When the Royal College of Science was founded in Dublin in 1867, Ball became the first professor of applied mathematics and mechanism. In 1874 he was appointed Royal Astronomer of Ireland, and in 1892 he assumed the Lowndean Chair of Astronomy and Geometry and the Directorship of the University Observatory at Cambridge, where he remained until his death in 1913.

ERRATUM SLIP

Please note that pages xxxv-xliv, Appendix by E. T. Whittaker, are part of the Foreword and should follow pp. xxxi.

We regret this error.

A TREATISE
ON THE
THEORY OF SCREWS

Sir Robert Stawell Ball (1840–1913)

A TREATISE
ON THE
THEORY OF SCREWS

ROBERT STAWELL BALL

CAMBRIDGE UNIVERSITY PRESS
Cambridge, New York, Melbourne, Madrid, Cape Town, Singapore,
São Paulo, Delhi, Dubai, Tokyo, Mexico City

Cambridge University Press
The Edinburgh Building, Cambridge CB2 8RU, UK

Published in the United States of America by Cambridge University Press, New York

www.cambridge.org
Information on this title: www.cambridge.org/9780521636506

First published 1990
First paperback edition 1998

A catalogue record for this publication is available from the British Library

ISBN 978-0-521-63650-6 Paperback

CONTENTS

		PAGE
	FOREWORD	xxi
	PREFACE	xxxiii
	INTRODUCTION	1

CHAPTER I.

TWISTS AND WRENCHES.

§§		
1.	Definition of the word Pitch	6
2.	Definition of the word Screw	7
3.	Definition of the word Twist	7
4.	A Geometrical Investigation	8
5.	The canonical form of a Small Displacement	9
6.	Instantaneous Screws	10
7.	Definition of the word Wrench	10
8.	Restrictions	11

CHAPTER II.

THE CYLINDROID.

9.	Introduction	15
10.	The Virtual Coefficient	17
11.	Symmetry of the Virtual Coefficient	18
12.	Composition of Twists and Wrenches	18
13.	The Cylindroid	19
14.	General Property of the Cylindroid	21
15.	Particular Cases	22
16.	Cylindroid with One Screw of Infinite Pitch	22
17.	Form of the Cylindroid in general	24
18.	The Pitch Conic	24
19.	Summary	24

CHAPTER III.

RECIPROCAL SCREWS.

§§ PAGE

20. Reciprocal Screws . . . 26
21. Particular Instances . . . 26
22. Screw Reciprocal to Cylindroid . . . 26
23. Reciprocal Cone . . . 27
24. Locus of a Screw Reciprocal to Four Screws . . . 29
25. Screw Reciprocal to Five Screws . . . 30
26. Screw upon a Cylindroid Reciprocal to a Given Screw . . . 30
27. Properties of the Cylindroid . . . 30

CHAPTER IV.

SCREW CO-ORDINATES.

28. Introduction . . . 31
29. Intensities of the Components . . . 32
30. The Intensity of the Resultant . . . 33
31. Co-reciprocal Screws . . . 33
32. Co-ordinates of a Wrench . . . 34
33. The Work done . . . 34
34. Screw Co-ordinates . . . 34
35. Identical Relation . . . 35
36. Calculation of Co-ordinates . . . 35
37. The Virtual Coefficient . . . 36
38. The Pitch . . . 36
39. Screw Reciprocal to Five Screws . . . 36
40. Co-ordinates of a Screw on a Cylindroid . . . 37
41. The Canonical Co-reciprocals . . . 38
42. An Expression for the Virtual Coefficient . . . 38
43. Equations of a Screw . . . 38
44. A Screw of Infinite Pitch . . . 39
45. Indeterminate Screw . . . 40
46. A Screw at Infinity . . . 41
47. Screws on One Axis . . . 41
48. Transformation of Screw Co-ordinates . . . 42
49. Principal Screws on a Cylindroid . . . 43

CHAPTER V.

THE REPRESENTATION OF THE CYLINDROID BY A CIRCLE.

50. A Plane Representation . . . 45
51. The Axis of Pitch . . . 46
52. The Distance between Two Screws . . . 47

§§ PAGE
53. The Angle between Two Screws 48
54. The Triangle of Twists 49
55. Decomposition of Twists and Wrenches 50
56. Composition of Twists and Wrenches 50
57. Screw Co-ordinates 51
58. Reciprocal Screws 51
59. Another Representation of the Pitch 52
60. Pitches of Reciprocal Screws 53
61. The Virtual Coefficient 54
62. Another Investigation of the Virtual Coefficient 55
63. Application of Screw Co-ordinates 57
64. Properties of the Virtual Coefficient 59
65. Another Construction for the Pitch 59
66. Screws of Zero Pitch 60
67. A Special Case 60
68. A Tangential Section of the Cylindroid 60

CHAPTER VI.

THE EQUILIBRIUM OF A RIGID BODY.

69. A Screw System 62
70. Constraints 63
71. Screw Reciprocal to a System 63
72. The Reciprocal Screw System 63
73. Equilibrium 64
74. Reaction of Constraints 64
75. Parameters of a Screw System 65
76. Applications of Co-ordinates 65
77. Remark on Systems of Linear Equations 67

CHAPTER VII.

THE PRINCIPAL SCREWS OF INERTIA.

78. Introduction 69
79. Screws of Reference 70
80. Impulsive Screws and Instantaneous Screws 71
81. Conjugate Screws of Inertia 71
82. The Determination of the Impulsive Screw 72
83. System of Conjugate Screws of Inertia 72
84. Principal Screws of Inertia 73
85. An Algebraical Lemma 75
86. Another Investigation of the Principal Screws of Inertia 76
87. Enumeration of Constants 78
88. Kinetic Energy 79

§§ PAGE
89. Expression for Kinetic Energy . . . 80
90. Twist Velocity acquired by an Impulsive Wrench . . . 81
91. Kinetic Energy acquired by an Impulsive Wrench . . . 82
92. Formula for a Free Body . . . 82
93. Lemma . . . 83
94. Euler's Theorem . . . 83
95. Co-ordinates in a Screw System . . . 83
96. The Reduced Wrench . . . 84
97. Co-ordinates of Impulsive and Instantaneous Screws . . . 85

CHAPTER VIII.

THE POTENTIAL.

98. The Potential . . . 87
99. The Wrench evoked by Displacement . . . 88
100. Conjugate Screws of the Potential . . . 89
101. Principal Screws of the Potential . . . 90
102. Co-ordinates of the Wrench evoked by a Twist . . . 91
103. Form of the Potential . . . 92

CHAPTER IX.

HARMONIC SCREWS.

104. Definition of an Harmonic Screw . . . 94
105. Equations of Motion . . . 96
106. Discussion of the Results . . . 99
107. Remark on Harmonic Screws . . . 100

CHAPTER X.

FREEDOM OF THE FIRST ORDER.

108. Introduction . . . 101
109. Screw System of the First Order . . . 101
110. The Reciprocal Screw System . . . 102
111. Equilibrium . . . 103
112. Particular Case . . . 104
113. Impulsive Forces . . . 104
114. Small Oscillations . . . 105
115. Property of Harmonic Screws . . . 106

CHAPTER XI.

FREEDOM OF THE SECOND ORDER.

§§		PAGE
116.	The Screw System of the Second Order	107
117.	Applications of Screw Co-ordinates	107
118.	Relation between Two Cylindroids	108
119.	Co-ordinates of Three Screws on a Cylindroid	108
120.	Screws on One Line	109
121.	Displacement of a Point	110
122.	Properties of the Pitch Conic	110
123.	Equilibrium of a Body with Freedom of the Second Order	111
124.	Particular Cases	112
125.	The Impulsive Cylindroid and the Instantaneous Cylindroid	112
126.	Reaction of Constraints	114
127.	Principal Screws of Inertia	114
128.	The Ellipse of Inertia	114
129.	The Ellipse of the Potential	116
130.	Harmonic Screws	118
131.	Exceptional Case	118
132.	Reaction of Constraints	119

CHAPTER XII.

PLANE REPRESENTATION OF DYNAMICAL PROBLEMS CONCERNING A BODY WITH TWO DEGREES OF FREEDOM.

§§		PAGE
133.	The Kinetic Energy	120
134.	Body with Two Degrees of Freedom	120
135.	Conjugate Screws of Inertia	124
136.	Impulsive Screws and Instantaneous Screws	125
137.	Two Homographic Systems	126
138.	The Homographic Axis	127
139.	Determination of the Homographic Axis	128
140.	Construction for Instantaneous Screws	128
141.	Twist Velocity acquired by an Impulse	129
142.	Another Construction for the Twist Velocity	129
143.	Twist Velocities on the Principal Screws	131
144.	Another Investigation of the Twist Velocity acquired by an Impulse	131
145.	A Special Case	133
146.	Another Construction for the Twist Velocity acquired by an Impulse	134
147.	Constrained Motion	136
148.	Energy acquired by an Impulse	137
149.	Euler's Theorem	138
150.	To determine a Screw that will acquire a given Twist Velocity under a given Impulse	138
151.	Principal Screws of the Potential	140

§§ PAGE
152. Work done by a Twist 141
153. Law of Distribution of v_a 142
154. Conjugate Screws of Potential 142
155. Determination of the Wrench evoked by a Twist 143
156. Harmonic Screws 143
157. Small Oscillations in general 144
158. Conclusion 144

CHAPTER XIII.

The Geometry of the Cylindroid.

159. Another Investigation of the Cylindroid 146
160. Equation to Plane Section of Cylindroid 152
161. Chord joining Two Screws of Equal Pitch 155
162. Parabola 157
163. Chord joining Two Points 160
164. Reciprocal Screws 161
165. Application to the Plane Section 163
166. The Central Section of the Cylindroid 166
167. Section Parallel to the Nodal Line 167
168. Relation between Two Conjugate Screws of Inertia 168

CHAPTER XIV.

Freedom of the Third Order.

169. Introduction 170
170. Screw System of the Third Order 170
171. The Reciprocal Screw System 171
172. Distribution of the Screws 171
173. The Pitch Quadric 172
174. The Family of Quadrics 173
175. Construction of a Three-system from Three given Screws 175
176. Screws through a Given Point 176
177. Locus of the feet of perpendiculars on the generators 178
178. Screws of the Three-System parallel to a Plane 179
179. Determination of a Cylindroid 180
180. Miscellaneous Remarks 182
181. Virtual Coefficients 183
182. Four Screws of the Screw System 184
183. Geometrical Notes 184
184. Cartesian Equation of the Three-System 184
185. Equilibrium of Four Forces applied to a Rigid Body 186
186. The Ellipsoid of Inertia 187
187. The Principal Screws of Inertia 188
188. Lemma 189

§§		PAGE
189.	Relation between the Impulsive Screw and the Instantaneous Screw	189
190.	Kinetic Energy acquired by an Impulse	189
191.	Reaction of the Constraints	191
192.	Impulsive Screw is Indeterminate	191
193.	Quadric of the Potential	192
194.	The Principal Screws of the Potential	192
195.	Wrench evoked by Displacement	193
196.	Harmonic Screws	193
197.	Oscillations of a Rigid Body about a Fixed Point	194

CHAPTER XV.

The Plane Representation of Freedom of the Third Order.

198.	A Fundamental Consideration	197
199.	The Plane Representation	198
200.	The Cylindroid	199
201.	The Screws of the Three-System	200
202.	Imaginary Screws	201
203.	Relation of the Four Planes to the Quadrics	202
204.	The Pitch Conics	204
205.	The Angle between Two Screws	204
206.	Screws at Right Angles	206
207.	Reciprocal Screws	206
208.	The Principal Screws of the System	207
209.	Expression for the Pitch	208
210.	Intersecting Screws in a Three-System	212
211.	Application to Dynamics	214

CHAPTER XVI.

Freedom of the Fourth Order.

212.	Screw System of the Fourth Order	218
213.	Equilibrium with Freedom of the Fourth Order	219
214.	Screws of Stationary Pitch	221
215.	Applications of the Two-System	224
216.	Application to the Three-System	226
217.	Principal Pitches of the Reciprocal Cylindroid	227
218.	Equations to the Screw in a Four-System	229
219.	Impulsive Screws and Instantaneous Screws	229
220.	Principal Screws of Inertia in the Four-System	230
221.	Application of Euler's Theorem	231
222.	General Remarks	232
223.	Quadratic n-systems	233
224.	Properties of a Quadratic Two-System	234

§§ PAGE
225. The Quadratic Systems of Higher Orders 235
226. Polar Screws 238
227. Dynamical Application of Polar Screws 241
228. On the Degrees of certain Surfaces 242

CHAPTER XVII.

FREEDOM OF THE FIFTH ORDER.

229. Screw Reciprocal to Five Screws 246
230. Six Screws Reciprocal to One Screw 247
231. Four Screws of a Five-System on every Quadric 250
232. Impulsive Screws and Instantaneous Screws 251
233. Analytical Method 252
234. Principal Screws of Inertia 252
235. The Limits of the Roots 253
236. The Pectenoid 254

CHAPTER XVIII.

FREEDOM OF THE SIXTH ORDER.

237. Introduction 258
238. Impulsive Screws 258
239. Theorem 259
240. Theorem 260
241. Principal Axis 260
242. Harmonic Screws 261

CHAPTER XIX.

HOMOGRAPHIC SCREW SYSTEMS.

243. Introduction 262
244. On Plane Homographic Systems 262
245. Homographic Screw Systems 263
246. Relations among the Co-ordinates 263
247. The Double Screws 264
248. The Seven Pairs 264
249. Homographic n-systems 265
250. Analogy to Anharmonic Ratio 266
251. A Physical Correspondence 267
252. Impulsive and Instantaneous Systems 267
253. Special type of Homography 268
254. Reduction to a Canonical Form 269
255. Correspondence of a Screw and a System 270

§§ PAGE
256. Correspondence of m and n Systems 271
257. Screws common to the Two Systems 271
258. Corresponding Screws defined by Equations 272
259. Generalization of Anharmonic Ratio 273

CHAPTER XX.

EMANANTS AND PITCH INVARIANTS.

260. The Dyname 274
261. Emanants 275
262. Angle between Two Screws 276
263. Screws at Right Angles 276
264. Conditions that Three Screws shall be parallel to a Plane 277
265. Screws on the same Axis 277
266. A General Expression for the Virtual Coefficient 278
267. Analogy to Orthogonal Transformation 280
268. Property of the Pitches of Six Co-reciprocals 282
269. Property of the Pitches of n Co-reciprocals 285
270. Theorem as to Signs 285
271. Identical Formulæ in a Co-reciprocal System 286
272. Three Pitches Positive and Three Negative 287
273. Linear Pitch Invariant Functions 287
274. A Pitch Invariant 289
275. Geometrical meaning 290
276. Screws at Infinity 291
277. Expression for the Pitch 292
278. A System of Emanants which are Pitch Invariants 294

CHAPTER XXI.

DEVELOPMENTS OF THE DYNAMICAL THEORY.

279. Expression for the Kinetic Energy 296
280. Expression for the Twist Velocity 297
281. Conditions to be fulfilled by Two Pairs of Impulsive and Instantaneous Screws 298
282. Conjugate Screws of Inertia 299
283. A Fundamental Theorem 300
284. Case of a Constrained Rigid Body 303
285. Another Proof 304
286. Twist Velocity acquired by an Impulse 305
287. System with Two Degrees of Freedom 306
288. A Geometrical Proof 306
289. Construction of Chiastic Homography on the Cylindroid 307
290. Homographic Systems on Two Cylindroids 307
291. Case of Normal Cylindroids 308
292. General Conditions of Chiastic Homography 309

§§ PAGE
293. Origin of the Formulæ of § 281 . . . 310
294. Exception to be noted . . . 312
295. Impulsive and Instantaneous Cylindroids . . . 312
296. An Exceptional Case . . . 314
297. Another Extreme Case . . . 316
298. Three Pairs of Correspondents . . . 317
299. Cylindroid Reduced to a Plane . . . 319
300. A difficulty removed . . . 320
301. Two Geometrical Theorems . . . 320

CHAPTER XXII.

THE GEOMETRICAL THEORY.

302. Prelimary . . . 322
303. One Pair of Impulsive and Instantaneous Screws . . . 323
304. An Important Exception . . . 325
305. Two Pairs of Impulsive and Instantaneous Screws . . . 325
306. A System of Rigid Bodies . . . 326
307. The Geometrical Theory of Three Pairs of Screws . . . 330
308. Another Method . . . 332
309. Unconstrained Motion in system of Second Order . . . 332
310. Analogous Problem in a Three-system . . . 334
311. Fundamental Problem with Free Body . . . 336
312. Freedom of the First or Second Order . . . 338
313. Freedom of the Third Order . . . 339
314. General Case . . . 339
315. Freedom of the Fifth Order . . . 340
316. Principal Screws of Inertia of Constrained Body . . . 341
317. Third and Higher Systems . . . 342
318. Correlation of Two Systems of the Third Order . . . 344
319. A Property of Reciprocal Screw Systems . . . 347
320. Systems of the Fourth Order . . . 348
321. Systems of the Fifth Order . . . 350
322. Summary . . . 350
323. Two Rigid Bodies . . . 351

CHAPTER XXIII.

VARIOUS EXERCISES.

324. The Co-ordinates of a Rigid Body . . . 355
325. A Differential Equation satisfied by the Kinetic Energy . . . 356
326. Co-ordinates of Impulsive Screw in terms of the Intantaneous Screw . . . 356
327. Another Proof of Article 303 . . . 357
328. A more general Theorem . . . 357

§§ PAGE
329. Two Three-Systems . . . 357
330. Construction of Homographic Correspondents . . . 358
331. Geometrical Solution of the same Problem . . . 359
332. Co-reciprocal Correspondents in Two Three-systems . . . 360
333. Impulsive and Instantaneous Cylindroids . . . 361
334. The Double Correspondents on Two Cylindroids . . . 363
335. A Property of Co-reciprocals . . . 364
336. Instantaneous Screw of Zero Pitch . . . 365
337. Calculation of a Pitch Quadric . . . 365

CHAPTER XXIV.

THE THEORY OF SCREW-CHAINS.

338. Introduction . . . 367
339. The Graphic and Metric Elements . . . 368
340. The Intermediate Screw . . . 368
341. The Definition of a Screw-chain . . . 369
342. Freedom of the First Order . . . 369
343. Freedom of the Second Order . . . 370
344. Homography of Screw Systems . . . 374
345. Freedom of the Third Order . . . 375
346. Freedom of the Fourth Order . . . 377
347. Freedom of the Fifth Order . . . 378
348. Application of Parallel Projections . . . 379
349. Properties of this correspondence . . . 383
350. Freedom of the Fifth Order . . . 384
351. Freedom of the Sixth Order . . . 386
352. Freedom of the Seventh Order . . . 386
353. Freedom of the Eighth and Higher Orders . . . 388
354. Reciprocal Screw-chains . . . 388
355. Twists on $6m+1$ Screw-chains . . . 390
356. Impulsive Screw-chains and Instantaneous Screw-chains . . . 392
357. The principal Screw-chains of Inertia . . . 394
358. Conjugate Screw-chains of Inertia . . . 396
359. Harmonic Screw-chains . . . 397

CHAPTER XXV.

THE THEORY OF PERMANENT SCREWS.

360. Introduction . . . 399
361. Different Properties of a Principal Axis . . . 400
362. A Property of the Kinetic Energy of a System . . . 401
363. The Identical Equation in Screw-chain Co-ordinates . . . 403
364. The Converse Theorem . . . 404
365. Transformation of the Vanishing Emanant . . . 405

§§ PAGE
366. The General Equations of Motion with Screw-chain Co-ordinates . . . 405
367. Generalization of the Eulerian Equations . . . 406
368. The Restraining Wrench-chain . . . 407
369. Physical meaning of the Vanishing Emanant . . . 408
370. A Displacement without change of Energy . . . 408
371. The Accelerating Screw-chain . . . 409
372. Another Proof . . . 409
373. Accelerating Screw-chain and Instantaneous Screw-Chain . . . 410
374. Permanent Screw-chains . . . 410
375. Conditions of a Permanent Screw-chain . . . 411
376. Another identical Equation . . . 412
377. Different Screws on the same Axis . . . 414
378. Co-ordinates of the Restraining Wrench for a Free Rigid Body . . . 414
379. Limitation to the position of the Restraining Screw . . . 416
380. A Verification . . . 416
381. A Particular Case . . . 417
382. Remark on the General Case . . . 418
383. Two Degrees of Freedom . . . 419
384. Calculation of T . . . 420
385. Another Method . . . 420
386. The Permanent Screw . . . 421
387. Geometrical Investigation . . . 422
388. Another Method . . . 423
389. Three Degrees of Freedom . . . 426
390. Geometrical Construction for the Permanent Screws . . . 427
391. Calculation of Permanent Screws in a Three-system . . . 428
392. Case of Two Degrees of Freedom . . . 430
393. Freedom of the Fourth Order . . . 431
394. Freedom of the Fifth and Sixth Orders . . . 432
395. Summary . . . 432

CHAPTER XXVI.

AN INTRODUCTION TO THE THEORY OF SCREWS IN NON-EUCLIDIAN SPACE.

396. Introduction . . . 433
397. Preliminary notions . . . 433
398. The Intervene . . . 434
399. First Group of Axioms of the Content . . . 435
400. Determination of the Function expressing the Intervene between Two Objects on a Given Range . . . 435
401. Another Process . . . 441
402. On the Infinite Objects in an Extent . . . 442
403. On the Periodic Term in the Complete Expression of the Intervene . . . 443
404. Intervenes on Different Ranges in a Content . . . 444
405. Another Investigation of the possibility of Equally Graduated Ranges . . . 446

§§		PAGE
406.	On the Infinite Objects in the Content	447
407.	The Departure	448
408.	Second Group of Axioms of the Content	448
409.	The Form of the Departure Function	449
410.	On the Arrangement of the Infinite Ranges	449
411.	Relations between Departure and Intervene	450
412.	The Eleventh Axiom of the Content	451
413.	Representation of Objects by Points in Space	453
414.	Poles and Polars	454
415.	On the Homographic Transformation of the Content	454
416.	Deduction of the Equations of Transformation	455
417.	On the Character of a Homographic Transformation which Conserves Intervene	456
418.	The Geometrical Meaning of this Symmetric Function	461
419.	On the Intervene through which each Object is Conveyed	464
420.	The Orthogonal Transformation	465
421.	Quadrics unaltered by the Orthogonal Transformation	466
422.	Proof that U and V have Four Common Generators	467
423.	Verification of the Invariance of Intervene	468
424.	Application of the Theory of Emanants	469
425.	The Vector in Orthogonal Co-ordinates	470
426.	Parallel Vectors	472
427.	The Composition of Vectors	473
428.	Geometrical proof that Two Homonymous Vectors compound into One Homonymous Vector	475
429.	Geometrical proof of the Law of Permutability of Heteronymous Vectors	476
430.	Determination of the Two Heteronymous Vectors equivalent to any given Motor	476
431.	The Pitch of a Motor	478
432.	Property of Right and Left Vectors	478
433.	The Conception of Force in Non-Euclidian Space	480
434.	Neutrality of Heteronymous Vectors	480
	APPENDIX I. Notes on various points	483
	II. A Dynamical Parable	496
	BIBLIOGRAPHICAL NOTES	510
	INDEX	540

FOREWORD.

SIR Robert Stawell Ball's classic book, *A Treatise on the Theory of Screws* (Cambridge University Press, 1900), is the definitive reference on screw theory. It presents a unique approach to mechanics that weds rotational and linear quantities into a single geometrical element, the screw. The timing of this new edition is highly pertinent because so much of the material has important recent applications in robotics, multi-body dynamics, mechanical design, computational geometry, and hybrid automatic control. Consequently, the treatise has re-emerged as a novel mathematical resource for addressing complex engineering problems. Constrained rigid body dynamics is treated in a completely general fashion, but rather than relying on mere abstractions, the author is a master at describing concrete geometric interpretations in three-dimensional, Euclidean space.

This Foreword is intended to provide the reader with a concise history of the author, the gradual disappearance of the theory, its re-emergence some sixty years after publication of the treatise, and a brief guide of the treatise to facilitate study by the present-day reader. Also new to this edition is a portrait of the author and a reprint of "A *Catalogue Raisonné* of Sir Robert Ball's Mathematical Papers," by E. T. Whittaker, both of which appeared in the posthumous autobiography *Reminiscences and Letters of Sir Robert Ball* (ed. by W. V. Ball, Cassell and Co., 1915).

Sir Robert Stawell Ball was born in Dublin on July 1, 1840. After some years at a preparatory school in Dublin, he was sent in 1851 to Dr. Brindley's school at Tarvin, near Chester, where he received his early training in mathematics. In October 1857, he was entered as a student at Trinity College, Dublin. He soon showed his aptitude for mathematics and won numerous prizes. In 1860 he obtaind a scholarship and the Lloyd Exhibition. In 1861 he was Gold Medalist in Mathematics, first Gold Medalist in Experimental and Natural Sciences, and University Student.

In 1867 The Royal College of Science was founded in Dublin, and Ball became the first Professor of Applied Mathematics and Mechanism. In 1870

Ball read a paper before the Royal Irish Academy "On the Small Oscillations of a Rigid Body about a Fixed Point under the Action of any Forces. . . ." This was the first of his many memoirs on the theory of screws.

The following anecdote from the autobiography is attributed to Sir Joseph Larmor and well illustrates the lively geometrical environment that nourished Ball's developments.

> In the early days of the "Theory of Screws" Ball had made the acquaintance of W. K. Clifford at meetings of the British Association. They were drawn together by a common interest in geometrical forms, including non-Euclidean geometry; and both had a play of humour and frolic which made them for some years main upholders of the lighter side of the activities of that scientific body. About the same time he had inoculated his friend Professor J. D. Everett with the "Geometry of Screws," to the great advantage of that study. I can remember an oracular announcement which he made to Everett to the effect that "The Theory of Screws" is now all done with; it is quite obsolete; it is all going over into non-Euclidean space." I also remember him recounting, in a more serious vein, how, at a meeting of the British Association in the 'seventies, there turned up a young geometer from the University of Erlangen, Felix Klein, already a leader in the German mathematical world. Klein told them about Plücker's linear complex, and certain recent developments in the fascinating field of geometrical relations which it involved, a field in which Klein had first shown his own genius. Ball described how he and Clifford captured Klein after the meeting, and sat up half the night exchanging ideas, the interview culminating with the impatient and admiring complaint that there was positively nothing they could tell him that Klein did not seem to know about already.

Ball was appointed successor to Brünnow as Royal Astronomer of Ireland and Andrews Professor of Astronomy in the University of Dublin in 1874. Brünnow himself was the successor of the celebrated mathematician Sir William Rowan Hamilton. Following Ball was C. J. Joly and then former student E. T. Whittaker.

In 1876 Ball published *The Theory of Screws: A Study in the Dynamics of a Rigid Body,* which, together with his series of memoirs, formed the basis of the treatise. H. Gravelius translated the work to German in 1889, updated it with newer memoirs, and added original material for a 619-page volume, *Theoretische Mechanik starrer Systeme. . . .* In the same year I. Zanchevsky published a 131-page volume in Russian, *Theory of Screws and Its Application to Mechanics,* noting the "lack of books on the Theory of Screws both in Russia and abroad. . . ." In 1897 Gabriel Koenigs included an author's note "*Sur la théorie du la vis de M. Ball*" (pp. 451–457) in his publication *Lecons de Cinématique* (Librarie Scientifique, A. Herman, Paris).

In 1892, Sir Robert Ball succeeded Prof. Adams in the Lowndean Chair of

Astronomy and Geometry and the Directorship of the University Observatory at Cambridge, where he spent his remaining years.

The advent of *A Treatise on the Theory of Screws* was hailed enthusiastically by Felix Klein, who in 1902 wrote a great 229-page article "Zur Schraubentheorie von Sir Robert Ball" (*Zeitschrift für Mathematik und Physik,* vol. 47). In fact Klein praised the *anschaulichkeit* "clarity" and the elementary character of *The Theory of Screws,* not withstanding that a need for further deeper investigations on this were necessary (as Ball himself had said).

In 1908 H. E. Timerding published *Geometrie der Kräfte* (Teubner, Leipzig) in which he devoted pages 132–146 to *Die Ballschen Scrauben* (Ball's Screws).

It appears that the last reference in German made to Ball is that of Felix Klein in 1926 in his *Höhere Geometrie* (Springer Verlag, reprinted by Chelsea Publishing Company, New York, 1949). On page 68, Klein refers to the treatise and to *The Theory of Screws,* 1876, where Klein emphasizes the importance of the "Nullsystem" for rigid body mechanics, which goes back to A. F. Möbius's *Lehrbuch der Statik* (Leipzig, 1837).

The Theory of Screws sank into obscurity in the early 20th century in English-speaking communities. The reasons are speculative, but clearly no individual appears to have carried on the mantle of Ball's work. During his tenure as Royal Astronomer of Ireland, Ball's duties were chiefly running the observatory and astronomical endeavors; screws were his avocation. At Cambridge he continued astronomical duties but also offered lectures on screws, a good class consisting of only two or three serious students. He was a renowned lecturer, regularly traveling about giving talks on popular astronomy for a fee. However, C. J. Joly appears to have been a pivotal individual. As a successor to both Hamilton and Ball, he tried to make their works better known through his own publications in quaternions, screw theory, and their combination. Upon Joly's untimely death in 1906 Ball wrote the following:

> Joly was my intimate correspondent for ten years or so, and I have a wonderful series of his letters almost all complete (over 100). They mostly relate to mathematics. It is a grievous blow to me. I prized his friendship and warm intellecutual sympathy more than I can express, and I often thought (selfishly, perhaps) that as he was twenty-four years my junior, I might look forward to enjoying our friendly intercourse during the rest of my life. As is so often the case, I am now regretting many lost opportunities.
>
> I have derived so much benefit in every way from his knowledge and sympathy. We were engaged, in a way, almost alone among English mathmeticians, in the pursuit of Quaternions and the Theory of Screws, and for many months back I had been most diligently studying his writings so as to bring myself up to the point where our correspondence and inter-

> change of views might be still more fruitful. We had as our special object to make the wonderful life work of Sir William Hamilton better known, and now alas! . . . .

However, in the German-speaking community it seems most likely that the "*Motorrechnung*" (Motor Calculus) of R. Von Mises (*Zeitschr. für Angew, Math u. Mech.,* 1924) assisted the tentative survival of screw theory. The "*Motorrechnung*" was later entered in the influential *Handbuch der Physik* (ed. by H. Geiger and K. Scheel, Springer Verlag, Berlin, 1928). There seems little doubt that the German community was also greatly influenced by the in-depth investigations of Eduard Study, *Geometrie der Dynamen* (Teubner, Leipzig, 1903), which also had a significant impact.

It was not until 1964 that screw theory re-emerged in the English-speaking community, when Emeritus Professor K. H. Hunt (Monash University) and Emeritus Associate Professor J. R. Phillips (University of Sydney) rediscovered the cylindroid and subsequently Ball's treatise. Hunt was to write a book entitled *Kinematic Geometry of Mechanisms* (Clarendon Press, Oxford, 1978, 1990), and in 1984 and 1990 Phillips published two volumes entitled *Freedom of Machinery* (Cambridge University Press). All three texts complement Ball's book.

In parallel with the rediscovery by Hunt and Phillips, F. M. Dimentberg published, in Moscow, an excellent analytical treatise on screws, which was published and then translated to English in 1968 by the Foreign Technology Division of the United States Air Force Systems Command. The title was *The Screw Calculus and Its Applications in Mechanics.* His first reference in the Introduction is to Ball's treatise. However, he states in the Foreword that

> . . . The method of screws made its appearance as a method of mechanics during the Seventies of the last century. The screw calculus proper was formulated in its definitive form during the Nineties based upon the ideas of W. K. Clifford, A. P. Kotel'nikov and E. Study, and is a generalization of vector calulus.

A major objective that Dimentberg had in writing his treatise was to popularize the screw calculus among specialists in mechanics:

> Despite the long time that has elapsed since the origin of screw calculus, there is still only a select group of persons to whom it is familiar owing to the lack of necessary literature on the problem.

The influence of the works of E. Study and R. Von Mises on the Russian-speaking community is evident:

> In compilation of the book, the work of A. P. Kotel'nikov and D. N. Zeylonger was referred to most frequently, followed by the papers of R. Ball, I. Zanchevskiy, E. Study, R. Von Mises, S. G. Kislitsyn and other authors.

Further details of the contributions of these and other authors can be found in the Dimentberg introduction.

Following is a brief guide that aims to assist the present-day reader in studying the treatise.

The treatise essentially consists of four main parts: Fundamental Theory, Survey of Screw Systems, Further Developments, and Extensions of the Theory. The first two parts are largely reproduced from the 1876 edition and augmented with additional material; the most notable additions are the plane representations for screw systems of the second and third orders. Additionally there is an appendix of notes, the delightful address "A Dynamic Parable," and a chronologically arranged annotated bibliography.

The first part of the treatise introduces the fundamental concepts and the scope of the theory. The same topics are taken in much the same order in the later parts, some in greater depth than others.

The Introduction commences with Chasles's theorem on reducing a displacement to a rotation and a parallel translation. It is proved with a minimal use of equations, rather developing the argument using synthetic geometry. Descriptions involving infinitely distant and imaginary elements may seem just as remote and complex to the present-day reader but, nevertheless, are often revealing. However, the precision of the author's most finely honed tools to shape a vision for the geometry of mechanics is well illustrated. The synthetic development of Poinsot's theorem is simple and uses ordinary elements. It reduces a system of forces to a single force and a parallel couple.

Chapter I introduces the fundamental notions of pitch, screw, twist, and wrench. The author imposes two important restrictions on the work: i) all motions are small, and ii) systems are conservative. The main consequence of small motion is to linearize the investigation of mechanics for the purpose of geometrical insight. Limiting the study to conservative systems greatly simplifies many properties of deformable systems. While these restrictions hold throughout the treatise, they are not, per se, restrictions of general screw theory. However, no restrictions are placed on the number and types of constraints.

Chapter II presents the cylindroid, a ruled surface of paramount importance; any two screws and their composition (addition) always lie on this third-order surface or one of its special forms. Ball's virtual coefficient is fundamental to the entire work: for a unit wrench applied to a body undergoing a unit twist, it is a bilinear form for one-half the work done; for a screw it is the quadratic form for its pitch. The pitch conic is a planar mapping of the directions and pitches of the screws on a cylindroid. It is used in later chapters to investigate geometric properties.

Chapter III defines a pair of reciprocal screws by the vanishing of the virtual coefficient; a twist and wrench are reciprocal if they produce no work. For a constrained rigid body, the twist freedoms are reciprocal to the wrench constraints. Reciprocity is the key to investigating the dualistic geometric properties of mechanics.

Chapter IV develops the author's screw coordinates used in the remainder of the treatise. A given screw is expressed as a linear combination of six basis screws that are co-reciprocal, that is, each basis screw is reciprocal to the other five. The scalar multipliers of the basis screws are the screw coordinates. Most often the basis is a canonical set of co-reciprocals; these are three pairs of screws along the axes of a Cartesian frame, where each pair has pitches of equal magnitude an opposite sign. Ball's screw coordinates simplify many relations, but others can be made more difficult. Plücker coordinates for screws are preferred in most current works; the basis is not co-reciprocal but composed of three zero-pitch screws (rotations or forces) along the coordinate axes and three "infinite" screws (translations or couples) parallel to the axes. The latter may be regarded either as screws of infinite pitch and indefinite positions or as screws of indefinite pitch and infinitely distant positions.

Chapter V maps the screws of a cylindroid to the points of a circle. This is closely analogous to the Mohr circle representation of normal and shear stresses at a point for all directions in a plane, but now the variables are the pitches and distances from the cylindroid center to the screw axes, the axes in all directions are parallel to the central plane. The principal stresses, for example, correspond to the principal screws. By this compact and invaluable device, many properties of a second-order screw system are easily expressed and readily understood.

Chapter VI examines the equilibrium of a rigid body. The motion freedoms afforded by a screw of the nth order, or n-system, are equivalently represented by the reciprocal (6-n)-system of constraints. The number of parameters required to specify an n-system is determined.

Chapter VII initiates the kinetics investigations with the principal screws of inertia. For a free body these are a set of six canonical co-reciprocal screws, two along each of Euler's principal axes of inertia, the pitchs being plus and minus the respective radii of gyration. For a rigid body at rest with n degrees of freedom, there are n principal screws of inertia, and an impulsive wrench on one of them causes a twist about the same screw. The principal screws of inertia are usually selected as a basis for screw coordinates. They simplify kinetic expressions because they are co-reciprocal and conjugate, the latter term indicating orthogonality with respect to the inertial bilinear form. Though readily applied when the constraints are general, difficulties with the principal screws of inertia arise for some simple constraints like rotation about a point. Here the author devises a limiting argument, where the pitch of the impulsive wrench goes to zero as its magnitude becomes unbounded. In this way a very great force (zero-pitch) induces a rotation (zero-pitch) about the same principal screw of inertia. A similar difficulty occurs for the existence of a so-called reduced wrench, which is along a screw of the n-system that defines the freedom of the body that imparts the same instantaneous twist as a generally specified wrench.

Chapter VIII parallels the inertia development for the equilibrium of a rigid body in a potential field but covers the material more concisely. There are n principal screws of the potential that are co-reciprocal and conjugate with respect to the potential function Hessian (e.g., local stiffness).

Chapter IX studies the free vibrations of an inertial body in a potential field. A body excited so that it starts to oscillate about one of the n harmonic screws continues to do so. Harmonic screws are conjugate screws with respect to both the inertia and the potential.

The second part of the treatise surveys screw systems for freedoms from one to six. The discussion for each system generally follows the format of the previous part: kinematics, equilibrium and reciprocal systems, impulsive dynamics, and oscillations. There are additional chapters that expand on the freedoms of the second and third orders.

Chapter X outlines screw systems for freedoms of the first order. Its minimal order leads to simple porperties, many of which have been already presented.

Chapter XI applies screw coordinates to the geometry of screws on cylindroids. The ellipse of inertia is a mapping of generalized energy-equivalent radii of gyration from screws on a cylindroid to points on a plane. The ellipse of the potential is analogously defined for potential energy. The formulation of these ellipses is similar to that of the pitch conic. The major and minor axes of these three conics correspond respectively to the maximum and minimum values of kinetic energy, potential energy, and pitch.

Chapter XII returns to the plane representation of the cylindroid for dynamical systems. The mapping induced by inertia is thoroughly investigated for geometric properties. Four geometric constructions are presented to determine the twist velocity acquired from an impulsive wrench. For the special, but commonly found, second-order systems that contain at least one infinite-pitch screw (i.e., admits a translational motion or couple) the "geometrical theory then returns merely a vestige of its interest."

Chapter XIII uses planar sections of the cylindroid to develop its geometrical properties further. A general plane cuts the surface in a cubic curve. Chords derived from pairs of reciprocal screws, and pairs of screws of equal pitch, are shown to envelop a hyperbola and a parabola respectively. Their properties are related to the cubic and the cylindroid geometry. Details are given for the physical construction of a cylindroid model.

Chapter XIV explores the freedom of the third order, this being characterized by the reciprocal system being the third order too. All screws of a chosen pitch generate one family of axes on an elliptic hyperboloid of one sheet. Together, the distinct hyperboloids form two concentric sets about either of two principal screw axes that intersect orthogonally; one transitional hyperboloid degenerates to a pair of planes and belongs to both sets. Inertia and potential are mapped to ellipsoids. The pitch quadric is the hyperboloid of zero-pitch screws.

Chapter XV elegantly employs the three screw coordinates for a system of the third order as homogeneous coordinates of points on a plane. A point corresponds to one screw, a straight line to a two-system, and the plane of points to the entire three-system. The family of hyperboloids, each having its characteristic pitch, maps to a pencil of conics. The three principal screws are the vertices of a triangle that is self-conjugate with respect to the pencil. The planar representation illustrates the geometry of the three-system very effectively, as does the circle representation for the two-system.

Chapter XVI discusses freedom of the fourth order. The system is reciprocal to a two-system on a cylindroid, and this is used to derive several properties. The four screw coordinates may be interpreted as the homogeneous coordinates of a point in space. A line corresponds to a cylindroid, a plane to a three-system, and screws of certain chosen pitch to a quadric surface. This mapping is lightly developed, in contrast to the detailed description of the mapping for the three-system. General quadratic screw systems of order n are introduced using one quadratic equation and $(6-n)-1$ linear equations; their geometric properties are discussed.

Chapter XVII investigates the freedom of the fifth order, chiefly by identifying its lower-order subsystems. Also important is the reciprocal system consisting of a single screw. An interesting fourth-order surface Ball calls a pectenoid is used to describe the distribution of pitch for screws passing through a given point. This is generally analogous to the pitch conic of a two-system. Screws of a specified pitch that pass through a point constitute a planar pencil of axes.

Chapter XVIII fits the freedom of the sixth order within the framework of the survey. This corresponds to a free body and the fundamental properties receive a brief treatment.

The third part of the treatise largely studies properties between two screw systems of the same order that are related by a linear transformation.

Chapter XIX considers linear, one-to-one transformations of screws referred to as homographies. A linear transformation may generally be deduced from seven corresponding screw pairs; for twists and wrenches with given magnitudes, only six pairs are required. Determinant ratios, analogous to the anharmonic ratio (cross ratio) in projective geometry, are developed for eight screws and remain invariant under the one-to-one transformation. Reciprocity, however, is not preserved as an invariant. Transformations that are not one-to-one are also touched upon.

Chapter XX expresses the first and second emanates of kinetic and potential energy as the related bilinear and quadratic forms in screw coordinates respectively. Pitch invariants are relations among pitches that remain constant under various transformations. They are useful in the investigation of principal screws and co-reciprocal screws. For example, every set of six co-reciprocal screws has the sum of the inverse pitches vanish; the expression for this sum is a pitch invariant for transformations that preserve reciprocity.

Chapter XXI relates the products of virtual coefficients between three impulsive wrenches and the resulting instantaneous twists to a fundamental equation that holds for systems of all orders. This equation is used to investigate transformation properties between nth-order screw systems of impulsive wrenches and instantaneous twists. Particular detail is given to second-order systems and their special cases. Chiastic homography is a generalization of the transformation by inertia; it represents a symmetric linear transformation between twists and wrenches.

Chapter XXII expands the dynamical investigation of the previous chapter into transformations between screws systems but relies on geometric methods. It contains a general solution that the author eventually discovered after some twenty-five years of work, namely, the geometric determination of the instantaneous twist from the impulsive wrench.

Chapter XXIII contains various exercises and miscellaneous results. Some are alternate proofs of previously presented material and others further augment developed properties.

The fourth part of the treatise contains various extensions of the theory to multiple bodies, permanent screws, and non-Euclidean spaces.

Chapter XXIV generalizes the fundamental theory from a single mass to a mass-chain. This is a collection of masses that are arbitrarily constrained. The movement of a mass-chain is a twist on a screw-chain; an action on a mass-chain is a wrench on a screw-chain. Considering motion, a screw-chain for 1 bodies consists of the 1 screws that describe the motion of the bodies and the $1-1$ screws that describe the relative motion between the bodies. For a screw-chain with freedom of the first order, a twist on a screw-chain is represented by the $1-1$ relative screws and a twist on any single screw, often selected as the "first" screw. The relative screws are termed the intermediate screws. This device extends the basic theory to mass-chains. For example, two twists on two screw-chains compound into a twist on what could be called a cylindroid-chain, although the author does not use this term. It would consist of a series of 1 cylindroids corresponding to the 1 masses, and a series of $1-1$ cylindroids corresponding to the $1-1$ relative screws. Similar to the format of the earlier parts on the treatise, there is a survey of screw-chains. These are detailed up to the freedom of the seventh order and then higher-order systems are considered more generally. Screw-chains are then extended to reciprocal screws, impulsive and instantaneous screws, principal screws of inertia, conjugate screws, and harmonic screws.

Chapter XXV further develops the apparatus of screw-chains with application to permanent screws. A constrained body twisting about a permanent screw continues this motion without an externally applied wrench. More generally, to include constraints that may change with movement of the body, a constrained body twisting about a permanent screw has an instantaneous zero acceleration in absence of an externally applied wrench. The reaction of the constraint is called the restraining wrench. For freedoms of the first and second orders there is one permanent screw; for the third, fourth, and fifth orders there are three permanent screws; and for the sixth order they consist of all screws along the principal axes of the body. Permanent screws are distinct from the n principal screws of inertia of an n-system. For an impulsive wrench applied along a principal screw of inertia, a quiescent body acquires an instantaneous twist about the same screw. For particular constraints, the permanent screws and principal screws of inertia become identical.

Chapter XXVI is an original and axiomatic introduction to non-Euclidean spaces. The development uses homogeneous point coordinates and only briefly touches on screw quantities. A displacement that preserves distances is called a motor and can be represented by the composition of a left and right vector. For rigid bodies in Euclidean space these coalesce into what are called free vectors or infinite screws (i.e., translations). It is shown that the virtual moment vanishes for every left and right vector pair. To develop an expression for pitch, a small-displacement motor is represented by the composition of small rotations about a pair of polar lines. With establishment of these fundamental relations, the author points the reader toward "a region that still awaits a more complete investigation."

The remainder of the treatise is made up of two appendices and bibliographic notes.

Appendix I adds supplemental notes to eight of the treatise's articles. Three of these notes make reference to the author's colleague, Prof. C. J. Joly.

Appendix II presents "A Dynamic Parable," the author's Presidential Address to the Mathematical and Physical Section of the British Association in Manchester in 1887. The humorous myth chronicles a committee investigating the dynamics of a rigid body subjected to bewildering constraints. Their discoveries parallel the presentation of the fundamental theory in the first part of the treatise in the guise of a comical parody pitting the status quo methods of the establishment against the innovations of a young Mr. Helix.

Biographical notes at the end give an annotated history of screw theory through principal works. Included are twelve of the author's sixteen memoirs; the remaining four appearing after the publication of the treatise.

The authors of this Foreword wish to express their appreciation to: Emeritus Prof. K. Wohlhart, Technische Universität Graz, Austria; Prof. J. M. R. Martinez, Instituto Tecnológico de Celaya, Mexico; Prof. J. Hervé, École Centrale Paris, France; Prof. Emeritus M. Keler, F. H. Munich, Germany; and Prof. J. Hoschek, Technische Universität, Darmstadt, Germany, for their contributions to the history; and especially to Emeritus Prof. K. H. Hunt, Monash University, Australia, for his considerable contributions throughout.

The authors are also grateful to Cambridge University Press for re-publishing *A Treatise on the Theory of Screws* almost 100 years after its first appearance. It is as if many years of work – a few decades – of a man has unjustifiably lain dormant and unappreciated for a century. We confidently predict a wide and growing interest in the book.

H. LIPKIN
GEORGIA INSTITUTE OF TECHNOLOGY

J. DUFFY
UNIVERSITY OF FLORIDA

PREFACE.

ABOUT thirty years ago I commenced to develop the consequences of certain important geometrical and dynamical discoveries properly associated with the illustrious names of Poinsot and Chasles, Hamilton and Klein. The result of my labours I have ventured to designate as "The Theory of Screws."

As the theory became unfolded I communicated the results in a long series of memoirs read chiefly before the Royal Irish Academy. To this learned body I tender my grateful thanks for the continual kindness with which they have encouraged this work.

I published in 1876 a small volume entitled *The Theory of Screws: A Study in the Dynamics of a Rigid Body.* This contained an account of the subject so far as it was then known.

But in a few years great advances were made, the geometrical theories were much extended, and the Theory of Screw-chains opened up a wide field of exploration. The volume just referred to became quite out of date.

A comprehensive account of the subject as it stood in 1886 was given in the German work *Theoretische Mechanik starrer Systeme: Auf Grund der Methoden und Arbeiten und mit einem Vorworte von Sir Robert S. Ball, herausgegeben von Harry Gravelius,* Berlin, 1889. This work was largely a translation of the volume of 1876 supplemented by the subsequent memoirs, and Dr. Gravelius made some further additions.

The theory was still advancing, so that in a few years this considerable volume ceased to present an adequate view of the subject. For example, the Theory of Permanent Screws, which forms perhaps one of the most instructive developments, was not communicated to the Royal Irish Academy until 1890. The twelfth and latest memoir of the series containing the solution of an important problem, which had been under consideration for twenty-five years, did not appear until 1898.

It therefore seemed that the time had now come when an attempt should be made to set forth the Theory of Screws as it stands at present. The present work is the result. I have endeavoured to include in these pages every essential part of the Theory as contained in the twelve memoirs and many other papers. But the whole subject has been revised and rearranged, and indeed largely rewritten, many of the earlier parts have been recast with improvements derived from later research, and I should also add that I have found it necessary to introduce much that has not been previously published.

The pleasant duty remains of expressing my thanks for the help that I have received from friends in preparing this book. I have received most useful aid from Prof. W. Burnside, Mr. A. Y. G. Campbell, Mr. G. Chawner, Mr. A. W. Panton, Mr. H. W. Richmond, Mr. R. Russell, and Dr. G. Johnstone Stoney. In the labour of revising the press I have been aided by Mr. A. Berry, Mr. A. N. Whitehead, and lastly by Prof. C. J. Joly, who it will be seen has contributed several valuable notes.

Finally, I must express my hearty thanks to the Cambridge University Press for the liberality with which they undertook the publication of this book and for the willing consent with which they have met all my wishes.

ROBERT S. BALL.

OBSERVATORY,
CAMBRIDGE, 17 *May* 1900.

APPENDIX.

A *CATALOGUE RAISONNÉ* OF SIR ROBERT BALL'S MATHEMATICAL PAPERS

BY

E. T. WHITTAKER

[WHEN sending me the following *catalogue raisonné* of my father's books and mathematical papers, Professor Whittaker wrote:

> Here is my attempt to sketch the development of Sir Robert's mathematical researches. . . . You must use your judgment as to whether it is suitable for publication in the "Reminiscences." On my part it is simply a slight tribute of affection to one whom I loved more than any other of my teachers.

The reader who refers to p. 152, *ante,* will find that the esteem and affection of pupil for teacher was warmly reciprocated.]

It is interesting to trace the steps by which Ball was led to the mathematical discoveries associated with his name; especially as his earliest papers gave no indication of the direction in which his energies were ultimately to be concentrated.

The first of them, written in 1865, when he was a candidate for Fellowship in Trinity, is purely algebraical,* and is of interest chiefly as showing that he already possessed that unifying and co-ordinating power that made him, in later life, the prince of expositors. Taking in succession the methods of solving biquadratic equations given by Ferrari, Simpson, Euler, Descartes, Lagrange, and Cayley, he shows their substantial identity: each of them depends essentially on the same reducing cubic. This subject was carried farther in a second paper,† in which the methods of Tchirnhausen and others

*"Note on the Algebraic Solution of Biquadratic Equations," *Quart Journ. Math.* 7 (1866), pp. 6–9.

†"Notes on Biquadratic Equations (Second Part)," *Quart. Journ. Math.* 7 (1866), pp. 358–69.

were examined; but his election in 1867 to the Chair of Applied Mathematics and Mechanics in the Royal College of Science at Dublin gave a new turn to his thoughts, and his next papers were on wholly different topics.

The first of them,* a slight and isolated note on a phenomenon of the nature of a mirage he observed from the deck of a steamboat, is of interest chiefly as reflecting his keen power of observation and love of applying scientific principles to the happenings around him. It was followed by a paper† of a purely experimental character on vortex-rings. A year previously, W. Thomson (Lord Kelvin) had suggested that the atoms of matter might be constituted of vortex-rings in a perfect fluid, and that the mutual interactions of atoms might be illustrated by the behaviour of smoke-rings. Ball now described some beautiful experiments on the passage of a smoke-ring through a column of smoke. The subject was pursued in later papers‡; but it was not as an experimental physicist that he was destined to achieve his real reputation.

The teaching work of his Chair led him to devise many new lecture-experiments and examinations of mechanical efficiency, which formed the subject of communications in 1869 and 1870§; and to this period belongs also his earliest paper on an astronomical subject.‖ But the needs of his class led him also to work in the theoretical science of dynamics, and it was here that he found a thoroughly congenial topic and developed the ideas on which his greatest work was later constructed.

His first contribution to theoretical dynamics was a new proof¶ of Lagrange's formula connecting the tension and curvature of a membrane subjected to fluid pressure; his second,** written in April 1869, which may be regarded as the germ of his principal series of research, was a discussion of the small oscillations of a particle on any surface acted on by any forces. In a paper written many years later,†† Ball gave an account of the way in which he was led to this investigation. In the spring of 1869 he happened to attend a lecture at the Royal Dublin Society, given by Dr. Johnstone Stoney, in the course of which the lecturer exhibited and explained the progression of the apse in the elliptic path of the bob of a conical pendulum. Interested

*"On an Optical Phenomenon (Mirage)," *Phil. Mag.* 35 (1868), p. 404.

†"On Vortex-rings in Air," *Phil. Mag.* 36 (1868), p. 12.

‡"Account of Experiments upon the Resistance of Air to the Motion of Vortex-rings," *B. A. Rep.* 41 (1871), pp. 26–9; *Phil. Mag.* 42 (1871), p. 208. "Account of Experiments upon the Retardation Experienced by Vortex-rings of Air when Moving through Air," *Irish Acad. Trans.* 25 (1872), pp. 135, 155; *Irish Acad. Proc.* I. (1873–4), p. 113.

§"Lecture-experiments to Illustrate the Laws of Motion," *Phil. Mag.* 37 (1869), pp. 332–9. "Account of Experiments upon the Mechanical Efficiency of Different Forms of Pulley-blocks," *Dub. Soc. Journ.* 6 (1870), pp. 70–5.

‖"On Nebulæ," *Dub. Soc. Journ.* 5 (1870), p. 339.

¶"Note on an Elementary Proof of a Theorem of Lagrange's," *Phil. Mag.* 39 (1870), pp. 107–8.

**"A Problem in Mechanics," *Quart. Journ. Math.* 10 (1870), pp. 220–8.

††*Trans. R. I. A.* 31 (1897), p. 185.

in the exposition, Ball began immediately to work at the mathematical theory of the subject, in the endeavour to understand it more fully: and in this way he arrived at the results embodied in his *Quarterly Journal* memoir.

This was followed by a communication made to the British Association at its Liverpool meeting in 1870,* in which the work was extended to include the small oscillations of rigid bodies, and in which the characteristic features of his theory begin to appear.

There can be little doubt that this theory was originally suggested to his mind by considering the problem of small oscillations in the light of the kinematical theories of the French geometers. Louis Poinsot (1777–1859), Augustin Louis Cauchy (1789–1857), and Michel Chasles (1793–1880) had shown† how the various possible displacements and motions of rigid bodies can be analysed mathematically into certain elementary types. The most important of these types is the "screw-displacement," which is simply the motion of a nut on an ordinary screw; the nut moves forward and at the same time turns around, the amount of the forward motion bears a definite proportion to the amount of rotation. Poinsot had proved that a rigid body can be transferred from one position in space to any other position in space by a "twist" about a certain screw; and that any system of forces acting on a rigid body can be compounded into a "wrench" about a certain screw.

In the British Association communication of 1870, Ball extended this circle of ideas to obtain a complete theory of the small oscillations of a rigid body in terms of the screws. It was shown that the movement of a free rigid body when making small oscillations is compounded of six normal movements, each consisting of a to-and-fro vibration about a *normal screw,* the position, pitch, and period of which depends on the forces; and that if a rigid body has k degrees of freedom, its motion is compounded of vibrations about k normal screws.

In the Theory of Screws which was now taking shape in his hands, Ball regarded a screw as consisting simply of a straight line (the axis of the screw) with which a parameter (the pitch of the screw) is associated. This constitutes a common basis for the study of twists, which are kinematical displacements, and of wrenches, which are systems of mechanical forces. Now with every screw so defined we can associate a certain linear complex of lines,‡ namely, the null lines§ of the wrenches belonging to the screw; and,

*"The Small Oscillations of a Particle and of a Rigid Body," *B. A. Rep.* 40 (1870), pp. 10–12; *Quart. Journ. Math.* 11 (1871), pp. 206–9.

†The matter had been discussed still earlier in a little-known work of G. Mozzi, published at Naples in 1763.

‡That is, a triply-infinite set of lines in space, between whose line co-ordinates a linear relation exists.

§That is, the lines about which the moment of the system is zero.

conversely, Ball's screw is completely defined by this linear complex. It will be evident from this that the Theory of Screws is closely connected with the Theory of Linear Complexes in Line-Geometry*; and, as a matter of history, many of the theorems discovered by Ball were discovered independently by workers in the field of line-geometry. As an example of this we may mention the theory of reciprocal screws. Ball termed two screws "reciprocal" when a wrench acting on either does no work as the body is twisted about the other, and he established many properties of reciprocal screws – e.g., that a screw can be determined to be reciprocal to five given screws. Klein, working at about the same time† on the theory of linear complexes, defined complexes "in involution" by a condition that is really equivalent to Ball's condition of reciprocity, and obtained results regarding six "complexes reciprocally in involution," which are precisely the same as his.

At the period of these earlier investigations Ball was not aware of the work of the line-geometers. He first heard of them from Prof. Felix Klein at the Bradford meeting of the British Association in 1873.

At the British Association meeting of the following year (1871), which was held at Edinburgh, Ball described the cubic surface, of fundamental importance in the theory of screws, to which, at Cayley's suggestion, he gave the name *cylindroid.*‡ The composition of two displacements of a rigid body about two given screws gives a result that could have been produced by displacement about a single screw: the locus of this single screw is the cylindroid. Its fundamental property is that, if any three screws of the surface are taken, and if a body is displaced by being screwed along each of the screws through a small angle proportional to the sine of the angle between the remaining screws, the body, after its last displacement, will occupy the same position that it did before the first.§

The theorem of the cylindroid includes as particular cases the well-known rules for the composition of two displacements parallel to given lines, or of two small rotations about intersecting axes.

From this time commences his long series of communications to the

*The conception of a geometry in which the element of space is taken to be the line instead of the point is due to Plücker; it had been suggested by him as far back as 1846, but its chief development dates only from the publication of his "Neue Geometrie des Raumes" in 1868–9.

†Klein's paper was published a few months before Ball's. Cf. *Math. Ann.* 2, pp. 204, 368.

‡"Exhibition and Description of a Model of a Conoidal Cubic Surface called the 'Cylindroid,' which is presented in the theory of the Geometrical Freedom of a Rigid Body," *Brit. Ass. Rep.* 41 (1871), pp. 8–9; *Phil. Mag.* 42 (1871), pp. 181–3.

§The cylindroid, like the reciprocal screws, had been anticipated in some degree by the line-geometers. Cf. Plücker's "Neue Geometrie des Raumes" (1868–9), p. 97; see also Battaglini, *Napoli Rend.* 8 (1869), p. 87.

Royal Irish Academy on the theory of screws and related problems,* and a number of memoirs on the same subject in various English and Continental journals.† The first of his twelve principal memoirs was published in the Irish Academy's Transactions in 1872‡ and consisted chiefly of a systematic recount of the discoveries that had been presented in a fragmentary way in the shorter papers. The second of the twelve great memoirs appeared§ in the *Philosophical Transactions* of 1874; the chief feature of this paper is the general theorem that a rigid body has as many principal screws of inertia as it has degrees of freedom. This is a generalisation of the well-known property of the principal axes of a rigid body rotating around a fixed point. The third memoir of the series appeared in 1875‖; in this he introduced screw co-ordinates, which play a considerable part in the theory. They may be described as an adaptation for dynamical purposes of Klein's co-ordinates of linear complex referred to six fundamental complexes, of which each pair are in involution.

The results of the first three memoirs were incorporated in a volume Ball published in 1876 under the title, "The Theory of Screws: A Study in the Dynamics of a Rigid Body."¶

In the midst of these investigations he continued his work in applied mechanics, publishing his well-known books** on the subject in 1871 and 1873, and two original papers†† on it in the latter year, while in 1872 we find

*The earliest was "On the Small Oscillations of a Rigid Body about a Fixed Point under the Action of any Forces, and more particulary when Gravity is the only Force Acting," *Irish Acad. Trans.* 24 (1871), pp. 593–628; *Irish Acad. Proc.* I. (1873), pp. 11–13.

†"On a Geometrical Solution of the following Problem: 'A Quiescent Rigid Body Possessing Three Degrees of Freedom Receives an Impulse, Determine the Instantaneous Screw about which the Body Commences to Twist,'" *Brit. Ass. Rep.* 43 (1873), pp. 26–7. "Contributions to the Theory of Screws," *Brit. Ass. Rep.* 43 (1873), pp. 27–8. "On a Screw-complex of the Second Order," *ibid.* 45 (1875), p. 10. "The Theory of Screw: A study in the Dynamics of Rigid Body," *Math. Ann.* 9 (1876), pp. 541–53. "On the Principal of Screws of Inertia of a Free or Constrained Rigid Body," *Phil. Mag.* 6 (1878), p. 274–80.

‡"The Theory of Screws: A Geometrical Study of the Kinematics, Equilibrium, and Small Oscillations of a Rigid Body," *Irish Acad. Trans.* 25 (1872), pp. 157–215; abstract in *Quart. Journ. Math.* 12 (1873), pp. 41–7, and *Irish Acad. Proc.* I. (1873–4), pp. 233–8.

§"Researches in the Dynamics of a Rigid Body by the Aid of the Theory of Screws," *Roy. Soc. Proc.* 21 (1873), pp. 383–6; *Phil. Trans.* 164 (1874), pp. 15–40.

‖"Screw Co-ordinates and their Application to Problems in the Dynamics of a Rigid Body," *Irish Acad. Proc.* I. (1873–4), pp. 552–3; *Irish Acad. Trans.* 25 (1875), pp. 295–327.

¶Dublin; Hodges, 194 pp.

**"Experimental Mechanics," 352 pp. (1871). "Elementary Lessons on Applied Mechanics," 143 pp. (1873).

††"Notes on Applied Mechanics: 1.—Parallel Motion. 2.—The Contact of Curves." *Irish Acad. Proc.* I. (1873–4), pp. 243–5; *Quart Journ. Math.* 12 (1873), pp. 112–4. "Notes on Applied Mechanics: 3.—Of the Theory of Long Pillars. 4.—Note on a Hydrodynamical Theorem Due to Prof. Stokes." *Irish Acad. Proc.* I. (1873–4), pp. 491–3.

two papers on an astronomical subject.[*] Astronomy was, indeed, to claim much of his life henceforward, for in 1874 he exchanged the Professorship of Mechanics in the Royal College of Science for the office of Royal Astronomer of Ireland. The dignity of this position, enhanced by his election to the Royal Society in the previous year, gave him, at the early age of thirty-four, a leading position among Irish men of science. That his merits as a lecturer were already recognised appears from the published abstract, of 1871,[†] of a discourse to the Royal Dublin Society.

Ball's devotion to his new profession was soon made evident by a profusion of books[‡] and original papers.[§] But his interest in the screws in no way abated, and, indeed, developed in several new directions. In 1880 he communicated[‖] to the British Association an extension of the theory to the kinematics of a rigid body in non-Euclidian space. The most general displacement of a rigid body is a rotation about an axis combined with a rotation about the polar axis with regard to the absolute: Ball found geometrically the joint effect of two small displacements. At the meeting of the following year he showed[¶] that the properties of a system of surfaces arising in the theory of screws are only the "survivals" of a more interesting geometrical system in non-Euclidian space. The result was to give a complete geometrical theory of the statics and kinematics of a rigid body with three degrees of freedom in non-Euclidian space. The non-Euclidian researches were presented in detail in the fifth of his great series of mem-

[*] "On the Orbit of the Binary Star ξ Ursæ Majoris," *Monthly Notices R. A. S.* 32 (1872), pp. 336–9. "On a New Approximation to the Orbit of the Binary Star ξ Ursæ Majoris," *Irish Acad. Proc.* I. (1873–4), pp. 316–28.

[†] "On Energy," *Dublin Soc. Journ.* 6 (1875), pp. 187–9.

[‡] *Astronomy,* 166 pp. (1877). *Elements of Astronomy,* 459 pp. (1880). *The Story of the Heavens,* 530 pp. (1886). Time and Tide (1889). In this period he also wrote the article "Gravitaion" for the *Encyclopædia Britannica,* and *Mechanics,* 167 pp. (1877), which was translated into Italian and published *sub tit. Mechanica,* Ball-Benetti (Ulrico Hoepli; Milano) (1880).

[§] "Observations of the Minor Planets (9) (13) (75) (97) (111) with the Transit-circle at Dublin," *Monthly Notices* 37 (1877), pp. 14–5; "On the Annual Parallax of the Star P. III. 242," *Monthly Notices* 41 (1881), pp. 36–42. "Further Researches on the Annual Parallax of 61 Cygni," *Monthly Notices* 41 (1881), pp. 162–6. "On a Simple Approximate Method of Calculating the Effect of Refraction upon the Distance and Position-angle of two Adjacent Stars," *ibid.* pp. 445–7. "Researches on the Annual Parllax of the Star Groombridge 1618," *Copernicus* 1 (1881), pp. 16–22. "Determination of the Annual Parallax of 6 Cygni $\beta = \Sigma$ 2486," *Copernicus* 2 (1882), pp. 159–63. "On the Method of Regulating a Clock Intended to Show Correct Mean Time," *Irish Acad. Proc.* 3 (1883), pp. 66–8. "Researches on the Parallax of 61 (A) Cygni made at Dunsink," *ibid.* pp. 209–12. "Observations in Search of Stars with a Large Annual Parallax," *ibid.* pp. 215–26. "Speculations on the Source of Meteorites," *ibid.* pp. 227–30. "Researches on Annual Parallax made at Dunsink," *ibid.* pp. 355–67; and the "Dunsink Observations and Researches," published separately.

[‖] "Notes on Non-Euclidian Geometry," *B. A. Rep.* (1880), pp. 476–7.

[¶] "On the Elucidation of a Question in Kinemetics by the Aid of Non Euclidian Space," *B. A. Rep.* (1881), pp. 535–6.

oirs,* with which may be associated another paper† of a somewhat later date. The fourth memoir‡ of the series, which had been read to the Royal Irish Academy somewhat earlier in the year, contained an extension of the theory of screws (which had hitherto been occupied with the dynamics of *one* rigid body) to the case of *any connected system* of rigid bodies. In this memoir Ball introduced the notion of *screw-chains*, which bear to such systems the same relation that a screw bears to a single rigid body. By means of this conception he obtained general systems a theory similar to that of the principal screws of inertia of a single rigid body.

Three papers on Lagrange's equations in dynamics and their application to the theory of screws appeared§ before the fifth memoir, and between the fifth and sixth memoirs Ball published a number of shorter papers dealing with the screws. In the first of these‖ he considered a rigid body with freedom of the third order, and studied a representation of the screws in the system by means of three homgeneous co-ordinates of a point in a plane. In the second¶ he discussed the cases where there is a (1,1) correspondence between two systems of screws, and developed for these cases a general theory resembling the theory of homographic correspondence in geometry. In the third** he applied the algebraical theory of emanants to screw co-ordinate transformations. In the fourth†† he showed how certain problems in a system with two degrees of freedom could be studied by means of a merely plane construction. In the fifth‡‡ and the seventh§§ he returned to the subject of displacements in elliptic space; and in the sixth‖‖ he investigated the properties

*"Certain Problems in the Dynamics of a Rigid System Moving in Elliptic Space," *Irish Acad. Trans.* 28 (1881), pp. 159–84.

†"On the Theory of the Content," *Trans. R. I. A.* 29 (1889), pp. 123–82.

‡"Extension of the Theory of Screws to the Dynamics of any Material System," *Irish Acad. Trans.*, 28 (1881), pp. 99–136. Cf. *B. A. Rep.* (1881), pp. 547–8.

§"Note on a Transformation of Lagrange's Equation of Motion in Generalised Co-ordinates, which is Convenient in Physical Astronomy," *Monthly Notices* 37 (1877), pp. 265–8. "On an Elementary Proof of Lagrange's Equations of Motion in Generalised Co-ordinates," *Irish Acad. Proc.* 2 (1877), pp. 463–4. "Note on the Application of Lagrange's Equations of Motion to Problems in the Dynamics of a Rigid Body," *Irish Acad. Proc.* 3 (1879), pp. 213–4.

‖"Preliminary Note on the Plane Representation of Certain Problems in the Dynamics of a Rigid Body," *Irish Acad. Proc.* 3 (1881), p. 428–34.

¶"On Homographic Screw Systems," *Irish Acad. Proc.* 3 (1881), pp. 435–46.

**"Contributions to the Theory of Screws," *Irish Acad. Proc.* 3 (1882), pp. 661–9.

††"On a Plane Representation of Certain Dynamical Problems in the Theory of a Rigid Body," *Irish Acad. Proc.* 4 (1883), pp. 29–37.

‡‡"Notes on the Kinematics and Dynamics of a Rigid System in Elliptic Space," *Proc. R. I. A.* 4 (1884), pp. 252–8.

§§"Note on the Character of the Linear Transformation which Corresponds to the Displacement of a Rigid System in Elliptic Space," *Proc. R. I. A.* 4 (1885), pp. 532–7.

‖‖"Note on a Geometrical Method of Investigating the Dynamical Properties of the Cylindroid," *Proc. R. I. A.* 4 (1885), pp. 518–22.

of the cylindroid by means of the theory of reciprocal screws. Four other short papers* written at this time (1886–88) belonged to the astronomical side of his activity.

The principal series of memoirs was continued in 1886 by the sixth of the sequence.† In this, Ball showed how the dynamical problems connected with the cylindroid can be solved by elementary plane geometry. The screws on the cylindroid are represented by points on the circumference of a circle, the angle between two screws is the angle their corresponding points subtend at the circumference, and the shortest distance of any two screws is the projection of the corresponding chord on a fixed ray in the plane of the circle.

The seventh memoir, which appeared‡ in the following year, was a study of the cylindroid regarded as a conoidal cubic with one nodal line and three right lines in the plane at infinity: the treatment is geometrical rather than dynamical.

The eighth memoir,§ was a development of the first of the shorter papers already mentioned. It was published in 1889, in which year appeared also a German edition of the "Theory of Screws," by Herr Harry Gravelius.‖ This edition contained an account of all the memoirs down to and including the eighth, with original investigations by the German editor.

The ninth memoir, of date 1890, was entitled¶ "The Theory of Permanent Screws," and contains the generalisation to screws of the well-known property of the principal axes of a single rigid body, that if the body be once set in rotation about one of these axes, it will continue to rotate about it.

The tenth paper** was published two years after Ball's translatin to Cambridge, in 1892, and is devoted to the theory of certain invariantive expressions in the co-ordinates of screws and to the notion of *chiastic homography,* which is a type of homography possessed by impulsive and instantaneous systems.

*"Notes on Laplace's Analytical Theory of the Perturbations of Jupiter's Satellites," *Proc. R. I. A.* 4 (1886), pp. 557–67. "Observations of Nova Andromedæ made at Dunsink," *ibid.* p. 641. "Note on the Astronomical Theory of the Great Ice Age," *ibid.* pp. 642–4. "On the Harmonic Tidal Constituents of the Port of Dublin," *ibid.* pp. 190–1 (1888).

†"Dynamics of Modern Geometry: A New Chapter in the Theory of Screws," *Cunningham Memoirs of the R. I. A.* No. IV., pp. 1–44 (1886)

‡"On the Plane Sections of the Cylindroid," *Trans. R. I. A.* 29 (1887), pp. 1–32.

§"How Plane Geometry Illustrates General Problems in the Dynamics of a Rigid Body with Three Degrees of Freedom," *Trans. R. I. A.* 29 (1889), pp. 247–84.

‖"Theoretische Mechanik Starrer Systeme." Berlin; Reimer (1889).

¶*Trans. R. I. A.* 29 (1890), pp. 613–52.

**"The Theory of Pitch Invariants and the Theory of Chiastic Homography," *Trans. R. I. A.* 30 (1894), pp. 559–86.

The eleventh memoir of the series* is occupied with the relations between impulsive screws and instantaneous screws; and the twelfth,† which appeared in the same year (1897), contains the long-sought-for geometrical method of finding the instantaneous screw from the impulsive screw, which was necessary for the complete geometrical method in dynamics.

This twelfth memoir was the last of the formal series. But we have still to notice the number of shorter papers that appeared in the intervals of publication of the later members and also some memoirs of considerable length that appeared in the last twelve years of Ball's life.

Among the shorter papers, a notable place must be given to the brilliant Presidential Address‡ to Section A of the British Association at its Manchester meeting in 1887. This, which occupies ten pages of print, is really a complete popular exposition of the Theory of Screws cast into the form of a parable. The aspects in which the methods of space-geometry are superior to the methods of Cartesian analysis are most strikingly brought out. The address, which was immediately translated into Italian by Professor Vivanti, and later into Hungarian by Dr. Seydler, ranks as on of the best of Ball's minor writings. A number of other short papers§ on related topics appeared in the succeeding period.

A new "Treatise on the Theory of Screws," to replace the one published in 1876, was published by Cambridge University Press in 1900. It forms a handsome imperial octavo volume of 544 pages. Four memoirs‖ of considerable extent on the subject appeared in the decade following its publication; in the last of them, written in his seventieth year, Ball developed the connection, which Joly had pointed out, between the Theory of Screws and the quarterion theory of the linear vector function.

*"Further Development of the Relations between Impulsive Screws and Instantaneous Screws," *Trans. R. I. A.* 31 (1897), pp. 99–144.

†"Concluding Memoir on the Theory of Screws, with a Summary of the Twelve Memoirs," *Trans. R. I. A.* 31 (1897), pp. 145–96.

‡"A Dynamical Parable," *B. A. Rep.* (1887), pp. 568–79. Certain extracts from this will be found at p. 242, *ante*.

§"Note on a Determinant in the Theory of Screws," *Proc. R. I. A.* (3) 1 (1890), pp. 375–8. "On a Geometrical Illustration of a Dynamical Theorem," *B. A. Rep.* (1891), p. 566. "Note on a General Theorem in Dynamics," *B. A. Rep.* (1894), p. 561. "On a Form of the Differential Equations of Dynamics," *Austral. Assoc. Rep.* 6 (1895), pp. 215–7. "Note on Geometrical Mechanics," *Camb. Proc.* 8 (1895), pp. 240–1. "Amendment to the Twelfth and Concluding Memoir on the Theory of Screws," *Proc. R. I. A.* (3) 4 (1898), pp. 667–8. "Note on a Point in Theoretical Dynamics," *Camb. Proc.* 9 (1898), pp. 193–5.

‖"On Further Developments of the Geometrical Theory of Six Screws," *Trans. R. I. A.* 31 (1901), pp. 473–540. "On the Reflection of Screw-systems and Allied Questions," *ibid.* 32 (1903), pp. 101–54. "Some Extensions of the Theory of Screws," *ibid.* pp. 299–366. "Contributions to the Theory of Screws," *Proc. R. I. A.* 28 (1910), pp. 16–68.

It must be remembered that during almost the whole of his working life Ball occupied the position of director of an astronomical observatory, and that much of his attention was given to observational research that appeared in the official publications of the observatory. In addition to these, and to the mathematical papers cited here, he produced in the latter part of his life various isolated astronomical papers,* some popular expositions† of the subject, and a substantial treatise‡ on spherical astronomy, which was his last considerable work.

On a consideration of his mathematical research alone, with which this note is primarily concerned, there can be no doubt that Ball will be regarded by posterity as one of the two or three greatest British mathematicians of his generation.

*"On the Cause of an Ice Age," *B. A. Rep.* (1891), pp. 645–7. "Relative Positions of 223 Stars in the Cluster χ Persei as Determined Photographically" (with A. A. Rambaut), *Trans. R. I. A.* 30 (1893), pp. 231–76. "Note on Mr. A. Y. G. Campbell's Paper 'On the Variation of Uncanonical Arbitrary Constants,'" *Monthly Notices* 57 (1897), pp. 118, 131.

†"Starland" (1889). "The Cause of an Ice Age" (1892). "In the High Heavens" (1893). "The Story of the Sun" (1893). "Great Astronomers" (1895). "The Earth's Beginning" (1901). "A Primer of Astronomy" (1904). "A Popular Guide to the Heavens" (1905). "In Starry Realms" (1906).

‡"A Treatise on Spherical Astronomy." Cambridge; 1908.

THE THEORY OF SCREWS.

INTRODUCTION.

THE Theory of Screws is founded upon two celebrated theorems. One relates to the displacement of a rigid body. The other relates to the forces which act on a rigid body. Various proofs of these theorems are well known to the mathematical student. The following method of considering them may be found a suitable introduction to the present volume.

ON THE REDUCTION OF THE DISPLACEMENT OF A RIGID BODY TO ITS SIMPLEST FORM.

Two positions of a rigid body being given, there is an infinite variety of movements by which the body can be transferred from one of these positions to the other. It has been discovered by Chasles that among these movements there is one of unparalleled simplicity. He has shown that a free rigid body can be moved from any one specified position to any other specified position by a movement consisting of a rotation around a straight line accompanied by a translation parallel to the straight line.

Regarding the rigid body as an aggregation of points its change of place amounts to a transference of each point P to a new point Q. The initial and the final positions of the body being given each point P corresponds to one Q, and each Q to one P. If the coordinates of P be given then those of Q will be determined, and vice versâ. If we represent P by its quadriplanar coordinates x_1, x_2, x_3, x_4, then the quadriplanar coordinates y_1, y_2, y_3, y_4 of Q must be uniquely determined. There must, therefore, be equations connecting these coordinates, and as the correspondence is essentially of the one-to-one type these equations must be linear. We shall, therefore, write them in the form

$$\begin{aligned}
y_1 &= (11)\,x_1 + (12)\,x_2 + (13)\,x_3 + (14)\,x_4,\\
y_2 &= (21)\,x_1 + (22)\,x_2 + (23)\,x_3 + (24)\,x_4,\\
y_3 &= (31)\,x_1 + (32)\,x_2 + (33)\,x_3 + (34)\,x_4,\\
y_4 &= (41)\,x_1 + (42)\,x_2 + (43)\,x_3 + (44)\,x_4.
\end{aligned}$$

If we make $y_1 = \rho x_1$, $y_2 = \rho x_2$, $y_3 = \rho x_3$, $y_4 = \rho x_4$ we can eliminate x_1, x_2, x_3, x_4 and obtain the following biquadratic for ρ,

$$\begin{vmatrix} (11)-\rho, & (12) & (13) & (14) \\ (21) & (22)-\rho, & (23) & (24) \\ (31) & (32) & (33)-\rho, & (34) \\ (41) & (42) & (43) & (44)-\rho \end{vmatrix} = 0.$$

The four roots of this indicate *four* double points, i.e. points which remain unaltered. But these points are not necessarily distinct or real.

What we have written down is of course the general homographic transformation of the points in space. For the displacement of a rigid system is a homographic transformation of all its points, but it is a very special kind of homographic transformation, as will be made apparent when we consider what has befallen the four double points.

In the first place, since the distance between every two points before the transformation is the same as their distance after the transformation it follows that every point in the plane at infinity before any finite transformation must be in the same plane afterwards. Hence the plane at infinity remains in the same position. Further, a sphere before *this* transformation is still a sphere after it. But it is well known that all spheres intersect the plane at infinity in the same imaginary circle Ω. Hence we see not only that the plane at infinity must remain unaltered by the transformation but that a certain imaginary circle in that plane is also unaltered.

A system of points P_1, P_2, P_3, &c. on this circle Ω will, therefore, have as their correspondent points Q_1, Q_2, Q_3, &c. also on Ω. As all anharmonic ratios are unaltered by a linear transformation it follows that the systems P_1, P_2, P_3, &c. and Q_1, Q_2, Q_3, &c. are homographic. There will, therefore, be two double points of this homography, O_1 and O_2, and these will be the same after the transformation as they were before. They are, therefore, two of the four double points of which we were in search.

It should be remarked that the points O_1 and O_2 cannot coincide, for if they coincide in O, then O must be the double point corresponding to a repeated root of the biquadratic for ρ. But such a root is real. Hence O must be real. But every point on Ω is imaginary. Hence this case is impossible.

As Ω is unaltered and O_1 and O_2 are fixed, the tangents at O_1 and O_2 are fixed, and so is therefore T, the intersection of these tangents; this is accordingly the third of the four points wanted. It lies in the plane at infinity, but is a real point. The ray O_1O_2 is also real; it is the vanishing line of the planes perpendicular to the parallel rays, of which T is the vanishing point.

In general in any homographic transformation there cannot be four distinct double points in a plane, unless every point of the plane is a double point. For suppose P_1, P_2, P_3, P_4 were four distinct coplanar double points and that any other point R had a correspondent R'. Draw the conic through P_1, P_2, P_3, P_4, R. Then R' must lie on this conic because the anharmonic ratios $R(P_1, P_2, P_3, P_4)$ and $R'(P_1, P_2, P_3, P_4)$ are equal. We have also $P_1(P_2, P_3, P_4, R)$ and $P_1(P_2, P_3, P_4, R')$ equal, but this is impossible if R and R' be distinct. R is therefore a double point.

In the case of the displaced rigid body suppose there is a fourth distinct double point in the plane at infinity. Each ray connected with the body will then have one double point at infinity, so that after the transformation the ray must again pass through the same point, i.e. the transformed position of each ray must be parallel to its original position. This is a special form of displacement. It is merely a translation of the whole rigid system in which every ray moves parallel to itself.

In the more general type of displacement there can therefore be no double point distinct from T, O_1, O_2 and lying in the plane at infinity. Nor can there be *in general* another double point at a finite position T'. For if so, then the ray TT' is unaltered in position, and any finite point T'' on the ray TT' will be also unaltered, since this homographic transformation does not alter distances. Hence every point on TT' is a double point. Here again we must have fallen on a special case where the double points instead of being only four have become infinitely numerous. In this case every point on a particular ray has become a double point. The change of the body from one position to the other could therefore be effected by simple rotation around this ray.

There must however be four double points even in the most general case. Not one of these is to be finite, and in the plane at infinity not more than three are to be distinct. The fourth double point must be in the plane at infinity, and there it must coincide with either O_1, O_2 or T. Thus we learn that the most general displacement of a rigid system is a homographic transformation of all its points with the condition that two of its double points are on the imaginary circle Ω in the plane at infinity, while the pole of their chord gives a third. Of these three one, we shall presently see which one, is to be regarded as formed of two coincident double points.

All rays through T are parallel rays, and hence we learn that in the general displacement of a rigid body there is one real parallel system of rays each of which L is transformed into a parallel ray L'. Let A be any plane perpendicular to this parallel system. Let L and L' cut A in the points R and R'. Then as L and L' move, R and R' are corresponding points in two plane homographic systems. Any two such systems in a plane will of course have three double points. The special feature of this homographic transformation is that every circle is transformed into a circle. Each circle passes through the two circular points at infinity in its plane. These two points in A are therefore two of the double points of the plane homographic transformation. There remains one real point X in A which is common to the two systems. The normal S to A drawn through X is therefore the one and only ray which the homographic transformation does not alter.

This shows that in the most general change of a rigid system from one position to another there is one real ray unaltered. Hence every point on S before the transformation is also on S afterwards. There must therefore be two double points distinct or coincident on S. But we have already proved that in the general case there is no finite double point. Hence S must have two coincident double points at T. Thus we learn that in the general transformation of a system which is equivalent to the displacement of a rigid body, there is one real point at infinity which is the result of two coinciding double points, and the polar of this point with respect to the imaginary circle on the plane at infinity cuts that circle in the two other double points.

The displacement of the rigid body can thus be produced either by rotating the body around S or by translating the body parallel to S, or by a combination of such movements. We are therefore led to the fundamental theorem discovered by Chasles.

Any given displacement of a rigid body can be effected by a rotation about an axis combined with a translation parallel to that axis.

Of much importance is the fact that this method of procedure is in general unique. It is easily seen that there is only one axis by rotation about which, and translation parallel to which, the body can be brought from one given position to another given position. Suppose there were two axes P and Q, which possessed this property, then by the movement about P, all the points of the body originally on the line P continue thereon; but it cannot be true for any other line that all the points of the body originally on that line continue thereon after the displacement. Yet this would have to be true for Q, if by rotation around Q and translation parallel thereto, the desired change could be effected. We thus see that the displacement of a rigid body can be made to assume an extremely simple form, in which *no arbitrary element* is involved.

ON THE REDUCTION OF A SYSTEM OF FORCES APPLIED TO A RIGID BODY TO ITS SIMPLEST FORM.

It has been discovered by Poinsot that any system of forces which act upon a rigid body can be replaced by a single force, and a couple in a plane perpendicular to the force. Thus a *force, and a couple in a plane perpendicular to the force, constitute an adequate representation of any system of forces applied to a rigid body.*

It is easily seen that all the forces acting upon a rigid body may, by transference to an arbitrary origin, be compounded into a force acting at the origin, and a couple. Wherever the origin be taken, the magnitude and direction of the force are both manifestly invariable; but this is not the case either with the moment of the couple or the direction of its axis.

The origin, however, can always be so selected that the plane of the couple shall be perpendicular to the direction of the force. For at *any* origin the couple can be resolved into two couples, one in a plane containing the force, and the other in the plane perpendicular to the force. The first component can be compounded with the force, the effect being merely to transfer the force to a parallel position; thus the entire system is reduced to a force, and a couple in a plane perpendicular to that force.

It is very important to observe that there is *only one* straight line which possesses the property that a force along this line, and a couple in a plane perpendicular to the line, is equivalent to the given system of forces. Suppose two lines possessed the property, then if the force and couple belonging to one were reversed, they must destroy the force and couple belonging to the other. But the two straight lines must be parallel, since each must be parallel to the resultant of all the forces supposed to act at a point, and the forces acting along these must be equal and opposite. The two forces would therefore form a couple in a plane perpendicular to that of the couple which is found by compounding the two original couples. We should then have two couples in perpendicular planes destroying each other, which is manifestly impossible.

We thus see that any system of forces applied to a rigid body can be made to assume an extremely simple form, in which *no arbitrary element* is involved.

These two principles being established we are able to commence the Theory of Screws.

CHAPTER I.

TWISTS AND WRENCHES.

1. Definition of the word Pitch.

The direct problem offered by the Dynamics of a Rigid Body may be thus stated. To determine at any instant the position of a rigid body subjected to certain constraints and acted upon by certain forces. We may first inquire as to the manner in which the solution of this problem ought to be presented. Adopting one position of the body as a standard of reference, a complete solution of the problem ought to provide the means of deriving the position at any epoch from the standard position. We are thus led to inquire into the most convenient method of specifying one position of a body with respect to another.

To make our course plain let us consider the case of a mathematical point. To define the position of the point P with reference to a standard point A, there can be no more simple method than to indicate the straight line along which it would be necessary for a particle to travel from A in order to arrive at P, as well as the length of the journey. There is a more general method of defining the position of a rigid body with reference to a certain standard position. We can have a movement prescribed by which the body can be brought from the standard position to the sought position. It was shown in the Introduction that there is one simple movement which will always answer. A certain axis can be found, such that if the body be rotated around this axis through a determinate angle, and translated parallel to the axis for a determinate distance, the desired movement will be effected.

It will simplify the conception of the movement to suppose, that at each epoch of the time occupied in the operations producing the change of position, the angle of rotation bears to the final angle of rotation, the same ratio which the corresponding translation bears to the final translation. Under these circumstances the motion of the body is precisely the same as if it were attached to the nut of a uniform screw (in the ordinary sense of the word), which had an appropriate position in space, and an appropriate number of threads to the inch.

In the Theory of Screws the word *pitch* is employed in a particular sense that must be carefully noted. We define the *pitch of a screw* to be the rectilinear distance through which the nut is translated parallel to the axis of the screw, while the nut is rotated through the angular unit of circular measure. The pitch is thus a linear magnitude. It follows from this definition that the rectilinear distance parallel to the axis of the screw through which the nut moves when rotated through a given angle is simply the product of the pitch of the screw and the circular measure of the angle.

2. Definition of the word Screw.

It is a fundamental principle of the theory developed in these pages that the dynamical significance of screws is precisely analogous to their kinematical significance. It is, therefore, essential that in the formal definition of the particular sense the word screw is to bear in this volume no prominence can be assigned to kinematical terms or conceptions unless it can be equally given to dynamical terms and conceptions. This condition is fulfilled by excluding both Kinematics and Dynamics and constituting the screw as the geometrical entity thus described.

A screw is a straight line with which a definite linear magnitude termed the pitch is associated.

We shall often denote a screw by a symbol, and then usually by a small Greek letter. With reference to these symbols, a caution may be necessary. If, for example, a screw be denoted by α, then α is not an ordinary algebraic quantity. It is a symbol which denotes all that is included in the conception of a screw, and requires five quantities for its specification; of these four are required to determine the position of the straight line, and the pitch must be specified by a fifth. It will often be convenient to denote the pitch by a symbol, derived from the symbol employed to denote the screw to which the pitch belongs. The pitch of a screw is accordingly represented by appending to the letter p a suffix denoting the screw. For example, p_α denotes the pitch of α and is an ordinary algebraical quantity.

3. Definition of the word Twist.

We have next to define the use to be made of the word *twist*.

A body is said to receive a twist about a screw when it is rotated uniformly about the screw, while it is translated uniformly parallel to the screw, through a distance equal to the product of the pitch and the circular measure of the angle of rotation.

4. A Geometrical Investigation.

We can now demonstrate that whenever a body admits of an indefinitely small movement of a continuous nature it must be capable of executing that particular kind of movement denoted by a twist about a screw.

Let A_1 be a standard position of the body, and let P be any marked point of the body initially at P_1. As the body is displaced continuously to a neighbouring position, P will generally pursue a certain trajectory which, as the motion is small, may be identified with its tangent on which P_n is a point adjacent to P_1. In travelling from P_1 to P_n, P passes through the several positions, $P_2, \ldots P_{n-1}$. In a similar manner every other point, Q_1, of the rigid body will pass through a series of positions, Q_2, &c., to Q_n. We thus have the points of the body initially at P_1, Q_1, R_1, respectively, and each moves along a straight line through the successive systems of positions P_2, Q_2, R_2, &c., on to the final position P_n, Q_n, R_n. We may thus think of the consecutive positions occupied by the body A_2, A_3, &c., as defined by the groups of points P_1, Q_1, R_1 and P_2, Q_2, R_2, &c. We have now to show that if the body be twisted by a continuous screw motion direct from A_1 to A_n, it will pass through the series of positions A_2, A_3, &c. It must be remembered that this is hardly to be regarded as an obvious consequence. From the initial position A_1 to the final position A_n, the number of routes are generally infinitely various, but when these situations are contiguous, it is always possible to pass by a twist about a screw from A_1 to A_n viâ the positions A_2, $A_3 \ldots A_{n-1}$.

Suppose the body be carried direct by a twist about a screw from the position A_1 to the position A_n. Since this motion is infinitely small, each point of the body will be carried along a straight line, and as P_1 is to be conveyed to P_n, this straight line can be no other than the line P_1P_n. In its progress P_1 will have reached the position P_2, and when it is there the points Q_1, R_1 will each have advanced to certain positions along the lines Q_1Q_n and R_1R_n, respectively. But the points reached by Q_1 and R_1 can be no other than the points Q_2 and R_2, respectively. To prove this we shall take the case where P_1, Q_1, R_1 are collinear. Suppose that when P_1 has advanced to P_2, Q_1 shall not have reached Q_2, but shall be at the intermediate point Q_0. (Fig. 1.) Then the line P_1Q_1 will have moved to P_2Q_0, and as R_1 can only be conveyed along R_1R_2, while at the same time it must lie along P_2Q_0, it follows that the lines P_2Q_0 and R_1R_2 must intersect at the point R_0, and consequently all the lines in this figure lie in a plane. Further, P_2Q_2 and P_2Q_0 are each equal to P_1Q_1, as the body is rigid, and so also P_2R_0 and P_2R_2 are equal to P_1R_1. Hence it follows that Q_1Q_0 and R_1R_0 are parallel, and consequently all the points on the line $P_1Q_1R_1$ are displaced in parallel directions. It would hence follow that the motion of every point in the body was in a parallel direction, and that consequently the entire movement was simply a translation. But even in this case it would be impossible for the points Q_0 and R_0 to be distinct from Q_2 and R_2, because,

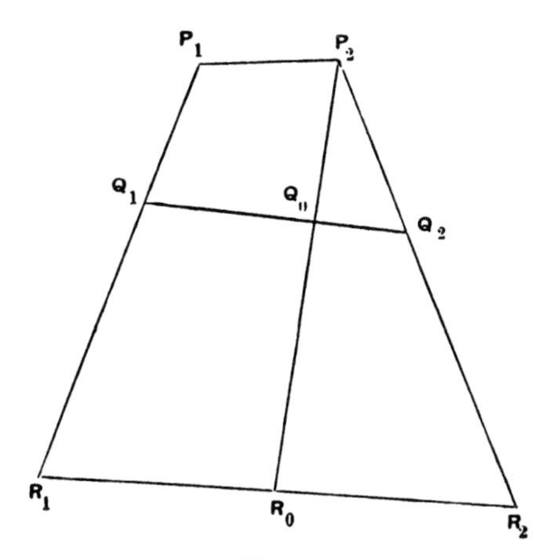

Fig. 1.

when a body is translated so that all its points move in parallel lines, it is impossible, if the body be rigid, for the distances traversed by each point not to be all equal. We have thus demonstrated that if a body is free to move from a position A_1 to an adjacent position A_n by an infinitely small but continuous movement, it is also free to move through the series of positions A_2, A_3, &c., by which it would be conveyed from A_1 to A_n by a twist.

We may also state the matter in a somewhat different manner, as follows:—It would be impossible to devise a system of constraints which would permit a body to be moved continuously from A_1 to A_n, and would at the same time prohibit the body from twisting about the screw which directly conducts from A_1 to A_n. Of course this would not be true except in the case where the motion is infinitely small. The connexion of this result with the present investigation is now obvious. When A is the standard position of the body, and B an adjacent position into which it can be moved, then the body is free to twist about the screw defined by A and B.

5. **The canonical form of a small displacement.**

In the Theory of Screws we are only concerned with the small displacements of a system, and hence we can lay down the following fundamental statement.

The canonical form to which the displacement of a rigid body can be reduced is a twist about a screw.

If a body receive several twists in succession, then the position finally attained could have been reached in a single twist, which is called the *resultant twist*.

Although we have described the twist as a compound movement, yet in the present method of studying mechanics it is essential to consider the twist as one homogeneous quantity. Nor is there anything unnatural in such a supposition. Everyone will admit that the relation between two positions of a point is most simply presented by associating the purely metric element of length with the purely geometrical conception of a directed straight line. In like manner the relation between two positions of a rigid body can be most simply presented by associating a purely metric element with the purely geometrical conception of a screw, which is merely a straight line, with direction, *situation, and pitch.*

It thus appears that a twist bears the same relation to a rigid body which the ordinary vector bears to a point. Each just expresses what is necessary to express the transference of the corresponding object from one given position to another*.

6. Instantaneous Screws.

Whatever be the movement of a rigid body, it is at every instant twisting about a screw. For the movement of the body when passing from one position to another position indefinitely adjacent, is indistinguishable from the twist about an appropriately chosen screw by which the same displacement could be effected. The screw about which the body is twisting at any instant is termed the *instantaneous screw.*

7. Definition of the word Wrench.

It has been explained in the Introduction that a system of forces acting upon a rigid body may be generally expressed by a certain force and a couple whose plane is perpendicular to the force. We now employ the word *wrench,* to denote a force *and* a couple in a plane perpendicular to the force. The quotient obtained by dividing the moment of the couple by the force is a linear magnitude. Everything, therefore, which could be specified about a wrench is determined (if the force be given in magnitude), when the position of a straight line is assigned as the direction of the force, and a linear magnitude is assigned as the quotient just referred to. Remembering the definition of a screw (§ 2), we may use the phrase, *wrench on a screw,* meaning thereby, a force directed along the screw and a couple in a plane perpendicular to the screw, the moment of the couple being equal to the product of the force and the pitch of the screw. Hence we may state that

The canonical form to which a system of forces acting on a rigid body can be reduced is a wrench on a screw.

* Compare M. René de Saussure, *American Journal of Mathematics*, Vol. XVIII. No. 4, p. 337.

If a rigid body be acted upon by several wrenches, then these wrenches could be replaced by one wrench which is called the *resultant wrench*.

A twist about a screw α requires six algebraic quantities for its complete specification, and of these, five are required to specify the screw α. The sixth quantity, which is called the AMPLITUDE OF THE TWIST, and is denoted by α', expresses the *angle of that rotation* which, when united with a translation, constitutes the entire twist.

The *distance of the translation* is the product of the amplitude of the twist and the pitch of the screw, or in symbols, $\alpha' p_\alpha$. The sign of the pitch expresses the sense of the translation corresponding to a given rotation.

If the pitch be zero, the twist reduces to a pure rotation around α. If the pitch be infinite, then a finite twist is not possible except the amplitude be zero, in which case the twist reduces to a pure translation parallel to α.

A *wrench on* a screw α requires six algebraic quantities for its complete specification, and of these, five are required to specify the screw α. The sixth quantity, which is called the INTENSITY OF THE WRENCH, and is denoted by α'', expresses the *magnitude of that force* which, when united with a couple, constitutes the entire wrench.

The *moment of the couple* is the product of the intensity of the wrench and the pitch of the screw, or in symbols, $\alpha'' p_\alpha$. The sign of the pitch expresses the direction of the moment corresponding to a given force.

If the pitch be zero, the wrench reduces to a pure force along α. If the pitch be infinite, then a finite wrench is not possible except the intensity be zero, in which case the wrench reduces to a couple in a plane perpendicular to α.

In the case of a *twisting motion* about a screw α the *rate* at which the amplitude of the twist changes is called the TWIST VELOCITY and is denoted by $\dot{\alpha}$.

8. Restrictions.

It is first necessary to point out the restrictions which we shall impose upon the forces. The rigid body M, whose motion we are considering, is presumed *to be acted upon by the same forces whenever it occupies the same position.* The forces which we shall assume are to be such as form what is known as a *conservative system.* Forces such as those due to a resisting medium are excluded, because such forces do not depend merely on the position of the body, but on the manner in which the body is moving through that position. The same consideration excludes friction which depends on the direction in which the body is moving through the position under consideration.

But the condition that the forces shall be defined, when the position is given, is still not sufficiently precise. We might include, in this restricted group, forces which could have no existence in nature. We shall, therefore, add the condition that *the system is to be one in which the continual creation of energy is impossible.*

An important consequence of this restriction is stated as follows:—The quantity of energy necessary to compel the body M to move from the position A to the position B, is independent of the route by which the change has been effected.

Let L and M be two such routes, and suppose that less energy was required to make the change from A to B viâ L than viâ M. Make the change viâ L, with the expenditure of a certain quantity of energy, and then allow the body to return viâ M. Now, since at every stage of the route M the forces acting on the body are the same whichever way the body be moving, it follows, that in returning from B to A viâ M, the forces will give out exactly as much energy as would have been required to compel the body to move from A to B viâ M; but by hypothesis this exceeds the energy necessary to make the change viâ L, and hence, on the return of the body to A, there is a clear gain of a quantity of energy, while the position of the body and the forces are the same as at first. By successive repetitions of the process an indefinite quantity of energy could be created from nothing. This being contrary to experience, compels us to admit that the quantity of energy necessary to force the body from A to B is independent of the route followed.

It follows that the amount of work done in a number of twists against a wrench is equal to the work that would be done in the resultant twist.

For, the work done in producing a given change of position is independent of the route.

We may calculate *the work done in a twist against a wrench* by determining the amount of work done against three forces which are equivalent to the wrench, in consequence of the movements of their points of application which are caused by the twist.

We shall assume the two lemmas—1st. The work done in the displacement of a rigid body against a force is the same at whatever point in its line of application the force acts. 2nd. The work done in the displacement of a point against a number of forces acting at that point, is equal to the work done in the same displacement against the resultant force.

The theorem to be proved is as follows:—The amount of work done in a given twist against a number of wrenches, is equal to the work done in the same twist against the resultant wrench.

Let n wrenches, which consist of $3n$ forces acting at $A_1, \ldots A_{3n}$, compound into one wrench, of which the three forces act at P, Q, R. The force at A_k may generally be decomposed into three forces along PA_k, QA_k, RA_k. By the 2nd lemma the amount of work (W) done against the $3n$ original forces, equals the amount of work done against the $9n$ components. It, therefore, appears from the 1st lemma, that W will still be the amount of work done against the $9n$ components, of which $3n$ act at P, $3n$ at Q, $3n$ at R. Finally, by the 2nd lemma, W will also be the amount of work done by the original twist against the three resultants formed by compounding each group at P, Q, R. But these resultants constitute the resultant wrench, whence the theorem has been proved.

We thus obtain the following theorem, which we shall find of great service throughout this book.

If a series of twists $A_1, \ldots A_m$, would compound into one twist A, and a series of wrenches $B_1, \ldots B_n$, would compound into one wrench B, then the energy that would be expended or gained when the rigid body performs the twist A, under the influence of the wrench B, is equal to the algebraic sum of the mn quantities of energy that would be expended or gained when the body performs severally each twist $A_1, \ldots A_m$ under the influence of each wrench $B_1, \ldots B_n$.

We have now explained the conceptions, and the language in which the solution of any problem in the Dynamics of a rigid body may be presented. A complete solution of such a problem must provide us, at each epoch, with a screw, by a twist about which of an amplitude also to be specified, the body can be brought from a standard position to the position occupied at the epoch in question. It will also be of much interest to know the instantaneous screw about which the body is twisting at each epoch, as well as its twist velocity. Nor can we regard the solution as quite complete, unless we also have a clear conception of the screw on which all the forces acting on the body constitute a wrench of which we should also know the intensity.

There is one special feature which characterises that portion of Dynamics which is discussed in the present treatise. We shall impose no restrictions on the form of the rigid body, and but little on the character of the constraints by which its movements are limited, or on the forces to which the rigid body is submitted. The restriction which we do make is that *the body, while the object of examination, remains in, or indefinitely adjacent to, its original position.*

As a consequence of this restriction, we here make the remark that *the amplitude of a twist is henceforth to be regarded as a small quantity.*

If it be objected, that with so great a restriction as that just referred to, only a limited field of inquiry remains, the answer is as follows:—A perfectly general investigation could yield but a slender harvest of interesting or valuable results. All the problems of Physical importance are *special* cases of the general question. Thus, a special character in the constraints has produced the celebrated problem of the rotation of a rigid body about a fixed point. To vindicate our particular restriction it seems only necessary to remark, that the restricted inquiry still includes the theory of Equilibrium, of Impulsive Forces, and of Small Oscillations.

Whatever novelty may be found in the following pages will, it is believed, be largely due to the circumstance that, with the important exception referred to, all the conditions of each problem are of absolute generality.

CHAPTER II.

THE CYLINDROID.

9. Introduction.

Let α and β be any two screws which we shall suppose to be fixed both in position and in pitch. Let a body receive a twist of amplitude α' about α, followed by a twist of amplitude β' about β. The position attained could have been arrived at by a single twist about some third screw ρ with an amplitude ρ'. We are always to remember that the amplitudes of the twists are infinitely small quantities. With this assumption the *order* in which the twists about α and β are imparted will be immaterial in so far as the resulting displacements are concerned. The position attained is the same whether α' follows β' or β' follows α'.

Any change in α' or in β' will of course generally entail a change both in the pitch and in the position of ρ. It might thus seem that ρ depended upon *two* parameters, and that consequently the different positions of ρ would form a doubly infinite series, known in linear geometry as a congruence. But this is not the case, for we prove that ρ depends only upon the *ratio* of α' to β' and is thus only singly infinite.

Take any point P and let h_α be the perpendicular distance from P to α, while p_α is as usual the pitch of the screw; then the point P is transferred by the twist about α through the distance

$$\sqrt{p_\alpha^2 + h_\alpha^2}\,\alpha'.$$

The twist about β conveys P to a distance

$$\sqrt{p_\beta^2 + h_\beta^2}\,\beta'.$$

The resultant of these two displacements conveys P in a direction which depends upon the *ratio* of α' to β', and not upon their absolute magnitudes.

Let P and Q be two points on ρ, then the resultant displacement will convey P and Q to points P' and Q' respectively which are also on the axis of ρ. Suppose that α' and β' be varied while their ratio is preserved P and Q will then be transferred to P'' and Q'' while by the property just proved P, P', P'' will be collinear and so will Q, Q', Q''. It therefore follows that as P, P', Q, Q' are collinear so will P, Q, P'', Q'' be collinear. The line PQ will therefore be displaced upon itself for every pair of values α' and β' which retain the same ratio. The position of the resultant screw is thus not altered by any changes of α' and β', which preserves their ratio.

Let ω be the angle between α and β. We take the case of a point P at an infinite distance on the common perpendicular to α and β. This point is displaced through a distance equal to

$$h\sqrt{\alpha'^2 + \beta'^2 + 2\alpha'\beta'\cos\omega},$$

where h stands for the infinite perpendicular distance from P to α or to β. This displacement of P is normal to ρ which itself intersects at right angles the common perpendicular to α and β. As the perpendicular distance from P to ρ can only differ by a finite quantity from h

$$h\rho' = h\sqrt{\alpha'^2 + \beta'^2 + 2\alpha'\beta'\cos\omega},$$

or

$$\rho' = \sqrt{\alpha'^2 + \beta'^2 + 2\alpha'\beta'\cos\omega}.$$

This determines the amplitude of the resulting twist which is, it may be noted, independent of the pitches.

Let ϕ be the angle between the directions in which a point Q on ρ is displaced by the twists about α and β, then the square of the displacement of Q will be

$$(p_\alpha^2 + h_\alpha^2)\,\alpha'^2 + (p_\beta^2 + h_\beta^2)\,\beta'^2 + 2\sqrt{p_\alpha^2 + h_\alpha^2}\sqrt{p_\beta^2 + h_\beta^2}\,\alpha'\beta'\cos\phi\,;$$

but this may also be written

$$p_\rho^2(\alpha'^2 + \beta'^2 + 2\alpha'\beta'\cos\omega),$$

whence we see that p_ρ depends only on the ratio of α' to β'.

The pitch and the position of ρ thus depend on the single numerical parameter expressing the ratio of α' and β'. As this parameter varies so will ρ vary, and it must in successive positions coincide with the several generators of a certain ruled surface. Two of these generators will be the situations of α and of β corresponding to the extreme values of zero and infinity respectively, which in the progress of its variation the parameter will assume.

We shall next ascertain the laws according to which twists (and wrenches) must be compounded together, that is to say, we shall determine the single screw, one twist (or wrench) about which will produce the same effect on the body as two or more given twists (or wrenches) about two or more given screws. It will be found to be a fundamental point of the present theory that the rules for the composition of twists and of wrenches are identical*.

10. The Virtual Coefficient.

Suppose a rigid body be acted upon by a wrench on a screw β, of which the intensity is β''. Let the body receive a twist of small amplitude α' around a screw α. It is proposed to find an expression for the energy required to effect the displacement.

Let d be the shortest distance between α and β, and let θ be the angle between α and β. Take α as the axis of x, the common perpendicular to α and β as the axis of z, and a line perpendicular to x and z for y. If we resolve the wrench on β into forces X, Y, Z, parallel to the axes, and couples of moments L, M, N, in planes perpendicular to the axes we shall have

$$X = \beta'' \cos O ; \ Y = \beta'' \sin O ; \ Z = 0 ;$$

$$L = \beta'' p_\beta \cos O - \beta'' d \sin O ; \ M = \beta'' p_\beta \sin O + \beta'' d \cos O ;$$

$$N = 0.$$

We thus replace the given wrench by four wrenches, viz., two forces and two couples, and we replace the given twist by two twists, viz., one rotation and one translation. The work done by the given twist against the given wrench must equal the sum of the eight quantities of work done by each of the two component twists against each of the four component wrenches. Six of these quantities are zero. In fact a rotation through the angle α' around the axis of x can do work only against L, the amount being

$$\alpha' \beta'' (p_\beta \cos O - d \sin O).$$

The translation $p_\alpha \alpha'$ parallel to the axis of x can do work only against X, the amount being

$$\alpha' \beta'' p_\alpha \cos O.$$

Thus the total quantity of work done is

$$\alpha' \beta'' \{(p_\alpha + p_\beta) \cos O - d \sin O\}.$$

The expression

$$\tfrac{1}{2} [(p_\alpha + p_\beta) \cos O - d \sin O]$$

is of great importance in the present theory†. It is called the *virtual*

* That the analogy between the composition of forces and of rotations can be deduced from the general principle of virtual velocities has been proved by Rodrigues (*Liouville's Journal*, t. 5, 1840, p. 436).

† The theory of screws has many points of connexion with certain geometrical researches on the linear complex, by Plücker and Klein. Thus the latter has shown (*Mathematische Annalen*, Band II., p. 368 (1869)), that if p_α and p_β be each the "Hauptparameter" of a linear complex, and if

$$(p_\alpha + p_\beta) \cos O - d \sin O = 0,$$

where d and O relate to the principal axes of the complexes, then the two complexes possess a special relation and are said to be in "involution."

coefficient of the two screws α and β, and may be denoted by the symbol

$$\varpi_{\alpha\beta}.$$

11. Symmetry of the Virtual Coefficient.

An obvious property of the virtual coefficient is of great importance. If the two screws α and β be interchanged, the virtual coefficient remains unaltered. The identity of the laws of composition of twists and wrenches can be deduced from this circumstance*, and also the Theory of Reciprocal Screws which will be developed in Chap. III.

12. Composition of Twists and Wrenches.

Suppose three twists about three screws α, β, γ, possess the property that the body after the last twist has the same position which it had before the first: then the amplitudes of the twists, as well as the geometrical relations of the screws, must satisfy certain conditions. The particular nature of these conditions does not concern us at present, although it will be fully developed hereafter.

We may at all events *conceive* the following method of ascertaining these conditions :—

Since the three twists neutralize it follows that the total energy expended in making those twists against a wrench, on any screw η, must be zero, whence

$$\alpha'\varpi_{\alpha\eta} + \beta'\varpi_{\beta\eta} + \gamma'\varpi_{\gamma\eta} = 0.$$

This equation is one of an indefinite number (of which six can be shown to be independent) obtained by choosing different screws for η. From each group of three equations the amplitudes can be eliminated, and four of the equations thus obtained will involve all the purely geometrical conditions as to direction, situation, and pitch, which must be fulfilled by the screws when three twists can neutralize each other.

But now suppose that three wrenches equilibrate on the three screws α, β, γ. Then the total energy expended in a twist about *any* screw η against the three wrenches must be zero, whence

$$\alpha''\varpi_{\alpha\eta} + \beta''\varpi_{\beta\eta} + \gamma''\varpi_{\gamma\eta} = 0.$$

An indefinite number of similar equations, one in fact for every screw η, must be also satisfied.

By comparing this system of equations with that previously obtained, it is obvious that the geometrical conditions imposed on the screws α, β, γ, in

* This pregnant remark, or what is equivalent thereto, is due to Klein (*Math. Ann.*, Vol. IV. p. 413 (1871)).

the two cases are identical. The amplitudes of the three twists which neutralise are, therefore, proportional to the intensities of the three wrenches which equilibrate.

When three twists (or wrenches) neutralise, then a twist (or wrench) equal and opposite to one of them must be the resultant of the other two. Hence it follows that the laws for the *composition* of twists and of wrenches must be identical.

13. The Cylindroid.

We next proceed to study the composition of twists and wrenches, and we select twists for this purpose, though wrenches would have been equally convenient.

A body receives twists about three screws; under what conditions will the body, after the last twist, resume the same position which it had before the first?

The problem may also be stated thus:—It is required to ascertain the single screw, a twist about which would produce the same effect as any two given twists. We shall first examine a special case, and from it we shall deduce the general solution.

Take, as axes of x and y, two screws α, β, intersecting at right angles, whose pitches are p_α and p_β. Let a body receive twists about these screws of amplitudes $\theta' \cos l$ and $\theta' \sin l$. The translations parallel to the coordinate axes are $p_\alpha\theta' \cos l$ and $p_\beta\theta' \sin l$. Hence the axis of the resultant twist makes an angle l with the axis of x; and the two translations may be resolved into two components, of which $\theta' (p_\alpha \cos^2 l + p_\beta \sin^2 l)$ is parallel to the axis of the resultant twist, while $\theta' \sin l \cos l \,(p_\alpha - p_\beta)$ is perpendicular to the same line. The latter component has the effect of transferring the resultant axis of the rotations to a distance $\sin l \cos l \,(p_\alpha - p_\beta)$, the axis moving parallel to itself in a plane perpendicular to that which contains α and β. The two original twists about α and β are therefore compounded into a single twist of amplitude θ' about a screw θ whose pitch is

$$p_\alpha \cos^2 l + p_\beta \sin^2 l.$$

The position of the screw θ is defined by the equations

$$y = x \tan l,$$

$$z = (p_\alpha - p_\beta) \sin l \cos l.$$

Eliminating l we have the equation

$$z\,(x^2 + y^2) - (p_\alpha - p_\beta)\, xy = 0.$$

The conoidal cubic surface represented by this equation has been called the cylindroid*.

Each generating line of the surface is conceived to be the residence of a screw, the pitch of which is determined by the expression

$$p_\alpha \cos^2 l + p_\beta \sin^2 l.$$

When a cylindroid is said to contain a screw, it is not only meant that the screw is one of the generators of the surface, but that the pitch of the screw is identical with the pitch appropriate to the generator with which the screw coincides.

We shall first show that it is impossible for more than one cylindroid to contain a given pair of screws θ and ϕ. For suppose that two cylindroids A and B could be so drawn. Then twists about θ and ϕ will compound into a twist on the cylindroid A and also on the cylindroid B (§ 14). Therefore the several screws on A would have to be identical with the screws on B, i.e. the two surfaces could not be different. That *one* cylindroid can always be drawn through a given pair of screws is proved as follows.

Let the two given screws be θ and ϕ, the length of their common perpendicular be h, and the angle between the two screws be A; we shall show that by a proper choice of the origin, the axes, and the constants p_α and p_β, a cylindroid can be found which contains θ and ϕ.

If l, m be the angles which two screws on a cylindroid make with the axis of x, and if z_1, z_2 be the values of z, we have the equations of which the last four are deduced from the first six

* This surface has been described by Plücker (*Neue Geometrie des Raumes*, 1868–9, p. 97); he arrives at it as follows:—Let $\Omega=0$, and $\Omega'=0$ be two linear complexes of the first degree, then all the complexes formed by giving μ different values in the expression $\Omega+\mu\Omega'=0$ form a system of which the axes lie on the surface $z(x^2+y^2)-(k^0-k_0)xy=0$. The parameter of any complex of which the axis makes an angle ω with the axis of x is $k=k^0\cos^2\omega+k_0\sin^2\omega$. Plücker also constructed a model of this surface.

Plücker does not appear to have noticed the mechanical and kinematical properties of the cylindroid which make this surface of so much importance in Dynamics; but it is worthy of remark that the distribution of pitch which is presented by physical considerations is exactly the same as the distribution of parameter upon the generators of the surface, which Plücker fully discussed.

The first application of the cylindroid to Dynamics was made by Battaglini, who showed that this surface was the locus of the wrench resulting from the composition of forces of varying ratio on two given straight lines (Sulla serie dei sistemi di forze, *Rendic. Acc. di Napoli*, 1869, p. 133). See also the Bibliography at the end of this volume.

The name *cylindroid* was suggested by Professor Cayley in 1871 in reply to a request which I made when, in ignorance of the previous work of both Plücker and Battaglini, I began to study this surface. The word originated in the following construction, which was then communicated by Professor Cayley. Cut the cylinder $x^2+y^2=(p_\beta-p_\alpha)^2$ in an ellipse by the plane $z=x$, and consider the line $x=0$, $y=p_\beta-p_\alpha$. If *any* plane $z=c$ cuts the ellipse in the points A, B and the line in C, then CA, CB are two generating lines of the surface.

$$p_\theta = p_\alpha \cos^2 l + p_\beta \sin^2 l, \qquad z_1 = (p_\alpha - p_\beta) \sin l \cos l,$$

$$p_\phi = p_\alpha \cos^2 m + p_\beta \sin^2 m, \qquad z_2 = (p_\alpha - p_\beta) \sin m \cos m,$$

$$A = l - m, \qquad h = z_1 - z_2,$$

$$p_\alpha - p_\beta = \frac{\sqrt{h^2 + (p_\theta - p_\phi)^2}}{\sin A},$$

$$p_\alpha + p_\beta = p_\theta + p_\phi - h \cot A,$$

$$l = \tfrac{1}{2}\left(A + \tan^{-1}\frac{p_\theta - p_\phi}{h}\right), \quad z_1 = \tfrac{1}{2}(p_\beta - p_\alpha)\cot A + \frac{h}{2},$$

with similar values for m and z_2. It is therefore obvious that the cylindroid is determined, and that the solution is unique.

It will often be convenient to denote by (θ, ϕ) the cylindroid drawn through the two screws θ and ϕ.

On any cylindroid there are in general two but only two screws which like α and β intersect and are at the same time at right angles. These two important screws are often termed the *principal screws* of the surface.

14. General Property of the Cylindroid.

If a body receive twists about three screws on a cylindroid, and if the amplitude of each twist be proportional to the sine of the angle between the two non-corresponding screws, then the body after the last twist will have regained the same position that it held before the first.

The proof of this theorem must, according to (§ 12), involve the proof of the following:—If a body be acted upon by wrenches about three screws on a cylindroid, and if the intensity of each wrench be proportional to the sine of the angle between the two non-corresponding screws, then the three wrenches equilibrate.

The former of these properties of the cylindroid is thus proved:—Take any three screws θ, ϕ, ψ, upon the surface which make angles l, m, n, with the axis of x, and let the body receive twists about these screws of amplitudes θ', ϕ', ψ'. Each of these twists can be decomposed into two twists about the screws α and β which lie along the axes of x and y. The entire effect of the three twists is, therefore, reduced to two rotations around the axes of x and y, and two translations parallel to these axes.

The rotations are through angles equal respectively to

$$\theta' \cos l + \phi' \cos m + \psi' \cos n$$

and

$$\theta' \sin l + \phi' \sin m + \psi' \sin n.$$

The translations are through distances equal to

$$p_\alpha(\theta' \cos l + \phi' \cos m + \psi' \cos n)$$

and

$$p_\beta(\theta' \sin l + \phi' \sin m + \psi' \sin n).$$

These four quantities vanish if

$$\frac{\theta'}{\sin(m-n)} = \frac{\phi'}{\sin(n-l)} = \frac{\psi'}{\sin(l-m)},$$

and hence the fundamental property of the cylindroid has been proved.

The cylindroid affords the means of compounding two twists (or two wrenches) by a rule as simple as that which the parallelogram of force provides for the composition of two intersecting forces. Draw the cylindroid which contains the two screws; select the screw on the cylindroid which makes angles with the given screws whose sines are in the inverse ratio of the amplitudes of the twists (or the intensities of the wrenches); a twist (or wrench) about the screw so determined is the required resultant. The amplitude of the resultant twist (or the intensity of the resultant wrench) is proportional to the diagonal of a parallelogram of which the two sides are parallel to the given screws, and of lengths proportional to the given amplitudes (or intensities).

15. Particular Cases.

If $p_\alpha = p_\beta$ the cylindroid reduces to a plane, and the pitches of all the screws are equal. If all the pitches be zero, then the general property of the cylindroid reduces to the well-known construction for the resultant of two intersecting forces, or of rotations about two intersecting axes. If all the pitches be infinite, the general property reduces to the construction for the composition of two translations or of two couples.

16. Cylindroid with one Screw of Infinite pitch.

Let OP, Fig. 2, be a screw of pitch p about which a body receives a small twist of amplitude ω.

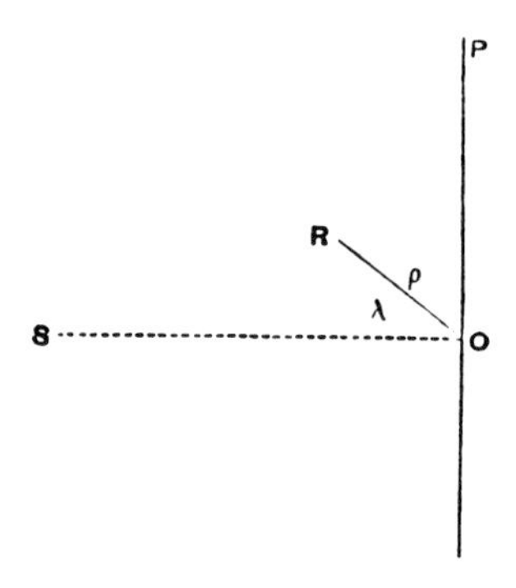

Fig. 2.

Let OR be the direction in which all points of the rigid body are translated through equal distances ρ by a twist about a screw of infinite pitch parallel to OR. It is desired to find the cylindroid determined by these two screws.

In the plane POR draw OS perpendicular to OP and denote $\angle ROS$ by λ.

The translation of length ρ along OR may be resolved into the components $\rho \sin \lambda$ along OP and $\rho \cos \lambda$ along OS.

Erect a normal OT to the plane of POR with a length determined by the condition

$$\omega OT = \rho \cos \lambda.$$

The joint result of the two motions is therefore a twist of amplitude ω about a screw θ through T and parallel to OP.

The pitch p_θ of the screw is given by the equation

$$\omega p_\theta = \omega p + \rho \sin \lambda,$$

whence

$$p_\theta - p = OT \tan \lambda.$$

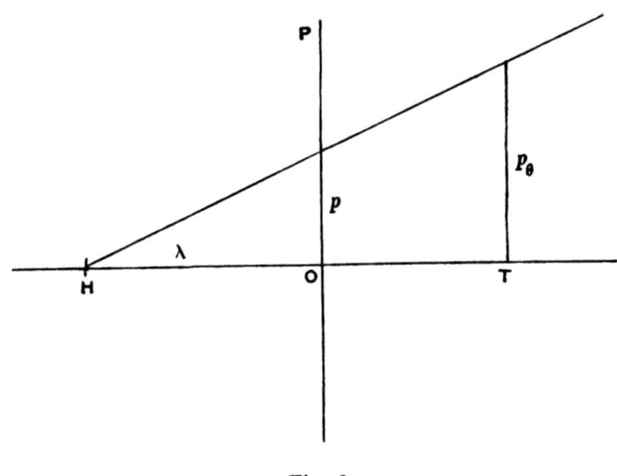

Fig. 3.

In Fig. 3 we show the plane through OP perpendicular to the plane POR in Fig. 2. The ordinate is the pitch of the screw through any point T.

If $p_\theta = 0$ then $OT = -OH$. Thus H is the point through which the one screw of zero pitch on the cylindroid passes, and we have the following theorem:

If one screw on a cylindroid have infinite pitch, then the cylindroid reduces to a plane. The screws on the cylindroid become a system of parallel lines, and the pitch of each screw is proportional to the perpendicular distance from the screw of zero pitch.

17. Form of the Cylindroid in general.

The equation of the surface contains only the single parameter $p_\alpha - p_\beta$, consequently all cylindroids are similar surfaces differing only in absolute magnitude.

The curved portion of the surface is contained between the two parallel planes $z = \pm (p_\alpha - p_\beta)$, but it is to be observed that the nodal line $x=0$, $y=0$, also lies upon the surface.

The intersection of the nodal line of the cylindroid with a plane is a node or a conjugate point upon the curve in which the plane is cut by the cylindroid according as the point does lie or does not lie between the two bounding planes.

18. The Pitch Conic.

It is very useful to have a clear view of the distribution of pitch upon the screws contained on the surface. The equation of the surface involves only the difference of the pitches of the two principal screws and one arbitrary element must be further specified. If, however, two screws be given, then both the surface and the distribution are determined. Any constant added to all the pitches of a certain distribution will give another possible distribution for the same cylindroid.

Let p_θ be the pitch of a screw θ on the cylindroid which makes an angle l with the axis of x; then (§ 13)

$$p_\theta = p_\alpha \cos^2 l + p_\beta \sin^2 l.$$

Draw in the plane x, y, the pitch conic

$$p_\alpha x^2 + p_\beta y^2 = H,$$

where H is any constant; then if r be the radius vector which makes an angle l with the axis of x, we have

$$p_\theta = \frac{H}{r^2},$$

whence the pitch of each screw on a cylindroid is proportional to the inverse square of the parallel diameter of the conic.

This conic is known as the *pitch conic.* By its means the pitches of all the screws on the cylindroid are determined. The asymptotes, real or imaginary, are parallel to the two screws of zero pitch.

19. Summary.

We shall often have occasion to make use of the fundamental principles demonstrated in this chapter, viz.,

That one, but only one, cylindroid can always be drawn so that two of its generators shall coincide with any two given screws α *and* β, *and that when all the generators of the surface become screws by having pitches assigned to them consistent with the law of distribution characteristic of the cylindroid, the pitches assigned to the generators which coincide with* α *and* β *shall be equal to the given pitches of* α *and* β.

Thus the cylindroid must become a familiar conception with the student of the Theory of Screws. A model of this surface is very helpful, and fortunately there can be hardly any surface which is more easy to construct. In the Frontispiece a photograph of such a model is shown, and a plate representing another model of the same surface will be found in Chap. XIII.

We shall develop in Chap. V an extremely simple method by which the screws on a cylindroid are represented by the points on a circle, and every property of the cylindroid which is required in the Theory of Screws can be represented by the corresponding property of points on a circle.

CHAPTER III.

RECIPROCAL SCREWS.

20. Reciprocal Screws.

If a body only free to twist about a screw α be in equilibrium, though acted upon by a wrench on the screw β, then conversely a body only free to twist about the screw β will be in equilibrium, though acted upon by a wrench on the screw α.

The principle of virtual velocities states, that if the body be in equilibrium the work done in a small displacement against the external forces must be zero. That the virtual coefficient should vanish is the necessary and the sufficient condition, or (§ 10)

$$(p_\alpha + p_\beta) \cos \theta - d \sin \theta = 0.$$

The symmetry shows that precisely the same condition is required whether the body be free to twist about α, while the wrench act on β, or vice versâ. *A pair of screws are said to be reciprocal when their virtual coefficient is zero.*

21. Particular Instances.

Parallel or intersecting screws are reciprocal when the sum of their pitches is zero. Screws at right angles are reciprocal either when they intersect, or when one of the pitches is infinite. Two screws of infinite pitch are reciprocal, because a couple could not move a body which was only susceptible of translation. A screw whose pitch is zero or infinite is reciprocal to itself*.

22. Screw Reciprocal to Cylindroid.

If a screw η be reciprocal to two given screws θ and ϕ, then η is reciprocal to every screw on the cylindroid (θ, ϕ).

* See also Professor Everett, F.R.S., *Messenger of Mathematics*, New Series (1874), No. 39.

For a body only free to twist about η would be undisturbed by wrenches on θ and ϕ; but a wrench on *any* screw ψ of the cylindroid can be resolved into wrenches on θ and ϕ; therefore a wrench on ψ cannot disturb a body only free to twist about η; therefore ψ and η are reciprocal. We may say for brevity that η is reciprocal to the cylindroid.

η cuts the cylindroid in three points because the surface is of the third degree, and one screw of the cylindroid passes through each of these three points; these three screws must, of course, be reciprocal to η. But two intersecting screws can only be reciprocal when they are at right angles, or when the sum of their pitches is zero. The pitch of the screw upon the cylindroid which makes an angle l with the axis of x is

$$p_\alpha \cos^2 l + p_\beta \sin^2 l.$$

This is also the pitch of the screw $\pi - l$. There are, therefore, two screws of any given pitch; but there cannot be more than two. It follows that η can at most intersect two screws upon the cylindroid of pitch equal and opposite to its own; and, therefore, η must be perpendicular to the third screw. Hence any screw reciprocal to a cylindroid must intersect one of the generators at right angles. We easily infer, also, that a line intersecting one screw of a cylindroid at right angles must cut the surface again in two points, and the screws passing through these points have equal pitch.

These important results can be otherwise proved as follows. A wrench can always be expressed by a force at *any* point O, and a couple in a plane L through that point but not of course in general normal to the force.

For wrenches on the several screws of a cylindroid, the forces at any point all lie on a plane and the couples all intersect in a ray.

The first part of this statement is obvious since all the screws on the cylindroid are parallel to a plane.

To prove the second it is only necessary to note that any wrench on the cylindroid can be decomposed into forces along the two screws of zero pitch. Their moments will be in the planes drawn through O and the two screws of zero pitch. The transversal across the two screws of zero pitch drawn from O must therefore lie in every plane L.

We hence see that the third screw on the cylindroid which is crossed by such a transversal must be perpendicular to that transversal.

23. Reciprocal Cone.

From any point P perpendiculars can be let fall upon the generators of the cylindroid, and if to these perpendiculars pitches are assigned which are equal in magnitude and opposite in sign to the pitches of the two remaining screws on the cylindroid intersected by the perpendicular, then the perpendiculars form a cone of reciprocal screws.

We shall now prove that this cone is of the second order, and we shall show how it can be constructed.

Let O be the point from which the cone is to be drawn, and through O let a line OT be drawn which is parallel to the nodal line, and, therefore, perpendicular to all the generators. This line will cut the cylindroid in one real point T (Fig. 4), the two other points of intersection coalescing into the infinitely distant point in which OT intersects the nodal line.

Draw a plane through T and through the screw LM which, lying on the cylindroid, has the same pitch as the screw through T. This plane can cut the cylindroid in a conic section only, for the line LM and the conic will then

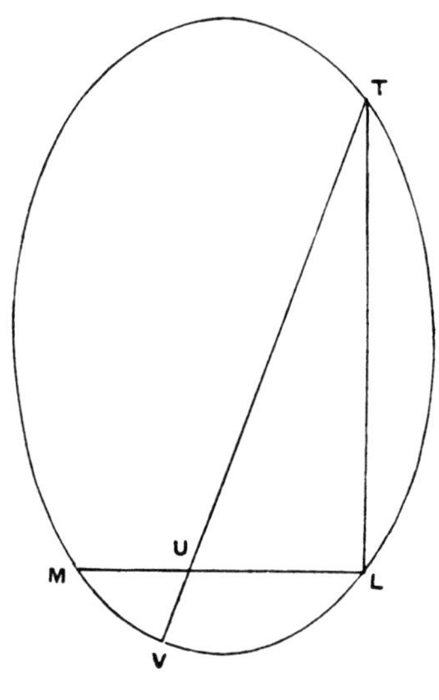

Fig. 4.

make up the curve of the third degree, in which the plane must intersect the surface. Also since the entire cylindroid (or at least its curved portion) is included between two parallel planes (§ 17), it follows that this conic must be an ellipse.

We shall now prove that this ellipse is the locus of the feet of the perpendiculars let fall from O on the generators of the cylindroid. Draw in the plane of the ellipse any line TUV through T; then, since this line intersects two screws of equal pitch in T and U, it must be perpendicular to that generator of the cylindroid which it meets at V. This generator is, therefore, perpendicular to the plane of OT and VT, and, therefore, to the line OV. It follows that V must be the foot of the perpendicular from O on the generator through V, and that, therefore, the cone drawn from O to the ellipse $TLVM$ is the cone required.

We hence deduce the following construction for the cone of reciprocal screws which can be drawn to a cylindroid from any point O.

Draw through O a line parallel to the nodal line of the cylindroid, and let T be the one real point in which this line cuts the surface. Find the second screw LM on the cylindroid which has a pitch equal to the pitch of the screw which passes through T. A plane drawn through the point T and the straight line LM will cut the cylindroid in an ellipse, the various points of which joined to O give the cone required*.

We may further remark that as the plane TLM passes through a generator it must be a tangent plane to the cylindroid at one of the intersections, suppose L, while at the point M the line LM must intersect another generator. It follows (22) that L must be the foot of the perpendicular from T upon LM, and that M must be a point upon the nodal line.

24. Locus of a Screw Reciprocal to Four Screws.

Since a screw is determined by five quantities, it is clear that when the four conditions of reciprocity are fulfilled the screw must generally be confined to one ruled surface. But this surface *can be no other than a cylindroid.* For, suppose three screws λ, μ, ν, which were reciprocal to the four given screws did not lie on the same cylindroid, then *any* screw ϕ on the cylindroid (λ, μ), and *any* screw ψ on the cylindroid (λ, ν) must also fulfil the conditions, and so must also every screw on the cylindroid (ϕ, ψ) (22). We should thus have the screws reciprocal to four given screws, limited not to one surface, as above shown, but to any member of a family of surfaces. The construction of the cylindroid which is the locus of all the screws reciprocal to four given screws, may be effected in the following manner:—

Let α, β, γ, δ be the four screws, of which the pitches are in descending order of magnitude. Draw the cylindroids (α, γ) and (β, δ). If σ be a linear magnitude intermediate between p_β and p_γ, it will be possible to choose two screws of pitch σ on (α, γ), and also two screws of pitch σ on (β, δ). Draw the two transversals which intersect the four screws thus selected; attribute to each of these transversals the pitch $-\sigma$, and denote the screws thus produced by θ, ϕ. Since intersecting screws are reciprocal when the sum of their pitches is zero, it follows that θ and ϕ must be reciprocal to the cylindroids (α, γ) and (β, δ). Hence all the screws on the cylindroid (θ, ϕ) must be reciprocal to α, β, γ, δ, and thus the problem has been solved.

* M. Appell has proved conversely that the cylindroid is the only *conoidal* surface for which the feet of the perpendiculars from any point on the generators form a plane curve. *Revue de Mathématiques Spéciales*, v. 129—30 (1895). More generally we can prove that this property cannot belong to any ruled surface whatever except a cylindroid and of course a cylinder.

25. Screw Reciprocal to Five Screws.

The determination of a screw reciprocal to five given screws must in general admit of only a finite number of solutions, because the number of conditions to be fulfilled is the same as the number of disposable constants. It is very important to observe that this number must be unity. For if *two* screws could be found which fulfilled the necessary conditions, then these conditions would be equally fulfilled by every screw on the cylindroid determined by those screws (§ 22), and therefore the number of solutions of the problem would not be finite.

The construction of the screw whose existence is thus demonstrated, can be effected by the results of the last article. Take any four of the five screws, and draw the reciprocal cylindroid which must contain the required screw. Any other set of four will give a different cylindroid, which also contains the required screw. These cylindroids must therefore intersect in the single screw, which is reciprocal to the five given screws.

26. Screw upon a Cylindroid Reciprocal to a Given Screw.

Let ϵ be the given screw, and let λ, μ, ν, ρ be any four screws reciprocal to the cylindroid; then the single screw η, which is reciprocal to the five screws ϵ, λ, μ, ν, ρ, must lie on the cylindroid because it is reciprocal to λ, μ, ν, ρ, and therefore η is the screw required.

The solution must generally be unique, for if a second screw were reciprocal to ϵ, then the whole cylindroid would be reciprocal to ϵ; but this is not the case unless ϵ fulfil certain conditions (§ 22).

27. Properties of the Cylindroid*.

We enunciate here a few properties of the cylindroid for which the writer is principally indebted to that accomplished geometer the late Dr Casey.

The ellipse in which a tangent plane cuts the cylindroid has a circle for its projection on a plane perpendicular to the nodal line, and the radius of the circle is the minor axis of the ellipse.

The difference of the squares of the axes of the ellipse is constant wherever the tangent plane be situated.

The minor axes of all the ellipses lie in the same plane.

The line joining the points in which the ellipse is cut by two screws of equal pitch on the cylindroid is parallel to the major axis.

The line joining the points in which the ellipse is cut by two intersecting screws on the cylindroid is parallel to the minor axis.

* For some remarkable quaternion investigations into "the close connexion between the theory of linear vector functions and the theory of screws" see Professor C. J. Joly, *Trans. Royal Irish Acad.*, Vol. XXX. Part XVI. (1895), and also *Proc. Royal Irish Acad.*, Third Series, Vol. V. No. 1, p. 73 (1897).

CHAPTER IV.

SCREW CO-ORDINATES.

28. Introduction.

We are accustomed, in ordinary statics, to resolve the forces acting on a rigid body into three forces acting along given directions at a point and three couples in three given planes. In the present theory we are, however, led to regard a force as a wrench on a screw of which the pitch is zero, and a couple as a wrench on a screw of which the pitch is infinite. The ordinary process just referred to is, therefore, only a special case of the more general method of resolution by which the intensities of the six wrenches on six given screws can be determined, so that, when these wrenches are compounded together, they shall constitute a wrench of given intensity on a given screw*.

The problem which has to be solved may be stated in a more symmetrical manner as follows:—

To determine the intensities of the seven wrenches on seven given screws, such that, when these wrenches are applied to a rigid body, which is entirely free to move in every way, they shall equilibrate.

The solution of this problem is identical (12) with that of the problem which may be enunciated as follows:—

To determine the amplitudes of seven small twists about seven given screws, such that, if these twists be applied to a rigid body in succession, the body after the last twist shall have resumed the same position which it occupied before the first.

The problem we have last stated has been limited as usual to the case where the amplitudes of the twists are small quantities, so that the motion of a point by each twist may be regarded as rectilinear. Were it

* If all the pitches be zero, the problem stated above reduces to the determination of the six forces along six given lines which shall be equivalent to a given force. If further, the six lines of reference form the edges of a tetrahedron, we have a problem which has been solved by Möbius, *Crelle's Journal*, t. XVIII. p. 207 (1838).

not for this condition a distinct solution would be required for every variation of the order in which the successive twists were imparted.

If the number of screws were greater than seven, then both problems would be indeterminate; if the number were less than seven, then both problems would be impossible (unless the screws were specially related); the number of screws being seven, the problem of the determination of the ratios of the seven intensities (or amplitudes) has, in general, one solution. We shall solve this for the case of wrenches.

Let the seven screws be α, β, γ, δ, ϵ, ζ, η. Find the screw ψ which is reciprocal to γ, δ, ϵ, ζ, η. Let the seven wrenches act upon a body only free to twist about ψ. The reaction of the constraints which limit the motion of the body will neutralize every wrench on a screw reciprocal to ψ (20). We may, therefore, so far as a body thus circumstanced is concerned, discard all the wrenches except those on α and β. Draw the cylindroid (α, β), and determine thereon the screw ρ which is reciprocal to ψ. The body will not be in equilibrium unless the wrenches about α and β constitute a wrench on ρ, and hence the ratio of the intensities α'' and β'' is determined. By a similar process the ratio of the intensities of the wrenches on any other pair of the seven screws may be determined, and thus the problem has been solved. (See Appendix, note 1.)

29. Intensities of the Components.

Let the six screws of reference be ω_1, &c. ω_6, and let ρ be a given screw on which is a wrench of given intensity ρ''. Let the intensities of the components be ρ_1'', &c. ρ_6'', and let η be any screw. A twist about η must do the same quantity of work acting directly against the wrench on ρ as the sum of the six quantities of work which would be done by the same twist against each of the six components of the wrench on ρ. If $\varpi_{\eta n}$ be the virtual coefficient of η and the nth screw of reference, we have

$$\rho''\varpi_{\eta\rho} = \rho_1''\varpi_{\eta 1} + \text{\&c. } \rho_6''\varpi_{\eta 6}.$$

By taking five other screws in place of η, five more equations are obtained, and from the six equations thus found ρ_1'', &c. ρ_6'' can be determined. This process will be greatly simplified by judicious choice of the six screws of which η is the type. Let η be reciprocal to ω_2, &c. ω_6, then $\varpi_{\eta 2} = 0$, &c. $\varpi_{\eta 6} = 0$, and we have

$$\rho''\varpi_{\eta\rho} = \rho_1''\varpi_{\eta 1}.$$

From this equation ρ_1'' is at once determined, and by five similar equations the intensities of the five remaining components may be likewise found.

Precisely similar is the investigation which determines the amplitudes of the six twists about the six screws of reference into which any given twist may be decomposed.

30. The Intensity of the Resultant may be expressed in terms of the intensities of its components on the six screws of reference.

Let α be any screw of pitch p_α, let p_1, p_2, &c. p_6 be the pitches of the six screws of reference $\omega_1, \omega_2, \ldots \omega_6$; then taking each of the screws of reference in succession, for η in § 29, and remembering that the virtual coefficient of two coincident screws is simply equal to the pitch, we have the following equations:—

$$\alpha''\varpi_{\alpha 1} = \alpha_1''p_1 + \alpha_2''\varpi_{12} + \ldots + \alpha_6''\varpi_{16},$$

$$\ldots\ldots\ldots\ldots\ldots$$

$$\alpha''\varpi_{\alpha 6} = \alpha_1''\varpi_{61} + \ldots + \alpha_5''\varpi_{65} + \alpha_6''p_6.$$

But taking the screw ρ in place of η we have

$$\alpha''p_\alpha = \alpha_1''\varpi_{\alpha 1} + \ldots \alpha_6''\varpi_{\alpha 6}.$$

Substituting for $\varpi_{\alpha 1} \ldots \varpi_{\alpha 6}$ from the former equations, we deduce

$$p_\alpha \alpha''^2 = \Sigma\,(p_1\alpha_1''^2) + 2\Sigma\,(\alpha_1''\alpha_2''\varpi_{12}).$$

This result may recall the well-known expression for the square of a force acting at a point in terms of its components along three axes passing through the point. This expression is of course greatly simplified when the three axes are rectangular, and we shall now show how by a special disposition of the screws of reference, a corresponding simplification can be made in the formula just written.

31. Co-Reciprocal Screws.

We have hitherto chosen the six screws of reference quite arbitrarily; we now proceed in a different manner. Take for ω_1, any screw; for ω_2, any screw reciprocal to ω_1; for ω_3, any screw reciprocal to both ω_1 and ω_2; for ω_4, any screw reciprocal to ω_1, ω_2, ω_3; for ω_5, any screw reciprocal to ω_1, ω_2, ω_3, ω_4; for ω_6, the screw reciprocal to ω_1, ω_2, ω_3, ω_4, ω_5.

A set constructed in this way possesses the property that each pair of screws is reciprocal. Any set of screws not exceeding six, of which each pair is reciprocal, may be called for brevity a set of co-reciprocals*.

Thirty constants determine a set of six screws. If the set be co-reciprocal, fifteen conditions must be fulfilled; we have, therefore, fifteen elements still disposable, so that we are always enabled to select a co-reciprocal set with special appropriateness to the problem under consideration.

* Klein has discussed (*Math. Ann.* Band II. p. 204 (1869)) six linear complexes, of which each pair is in involution. If the axes of these complexes be regarded as screws, of which the "Hauptparameter" are the pitches, then these six screws will be co-reciprocal.

The facilities presented by rectangular axes for questions connected with the dynamics of a particle have perhaps their analogues in the conveniences which arise from the use of co-reciprocal sets of screws in the present theory.

If the six screws of reference be co-reciprocal, then the formula of the last section assumes the very simple form

$$p_\alpha \alpha''^2 = p_1 \alpha_1''^2 + \ldots + p_6 \alpha_6''^2.$$

32. Co-ordinates of a Wrench.

We shall henceforth usually suppose that the screws of reference are co-reciprocal. We may also speak of the co-ordinates of a wrench*, meaning thereby the *intensities of its six components on the six screws of reference.* So also we may speak of the co-ordinates of a twist, meaning thereby *the amplitudes of its six components about the six screws of reference.*

The co-ordinates of a wrench of intensity α'' on the screw α are denoted by $\alpha_1'', \ldots \alpha_6''$. The co-ordinates of a twist of amplitude α' about α are denoted by $\alpha_1', \ldots \alpha_6'$.

The co-ordinates of a twist-velocity $\dot{\alpha}$ about α are denoted by $\dot{\alpha}_1, \dot{\alpha}_2, \ldots \dot{\alpha}_6$. The actual motion of the body is in this case a translation with velocity $\dot{\alpha} p_\alpha$ parallel to α and a rotation around ρ with the angular velocity $\dot{\alpha}$.

33. The Work done in a twist of amplitude α' about a screw α, by a wrench of intensity β'' on the screw β, can be expressed in terms of the co-ordinates.

Replace the twist and the wrench by their respective components about the co-reciprocals. Then the total work done will be equal to the sum of the thirty-six quantities of work done in each component twist by each component wrench. Since the screws are co-reciprocal, thirty of these quantities disappear, and the remainder have for their sum†

$$2p_1 \alpha_1' \beta_1'' + \ldots + 2p_6 \alpha_6' \beta_6''.$$

34. Screw Co-ordinates.

A wrench on the screw α, of which the intensity is *one unit*, has for its components, on six co-reciprocal screws, wrenches of which the intensities may be said to constitute *the co-ordinates of the screw* α. These co-ordinates may be denoted by $\alpha_1, \ldots \alpha_6$.

* Plücker introduced the conception of the six co-ordinates of a system of forces—*Phil. Trans.*, Vol. CLVI. p. 362 (1866). See also Battaglini, "Sulle dinami in involuzione," *Atti di Napoli* IV., (1869); Zeuthen, *Math. Ann.*, Band I. p. 432 (1869).

† That the work done can be represented by an expression of this type was announced by Klein, *Math. Ann.* Band IV. p. 413 (1871).

When the co-ordinates of a screw are given, the screw itself may be thus determined. Let ϵ be any small quantity. Take a body in the position A, and impart to it successively twists about each of the screws of reference of amplitudes $\epsilon\alpha_1$, $\epsilon\alpha_2$, ... $\epsilon\alpha_6$. Let the position thus attained be B; then the twist which would bring the body directly from A to B is about the required screw α.

35. Identical Relation.

The six co-ordinates of a screw are not independent quantities, but fulfil one relation, the nature of which is suggested by the relation between three direction cosines.

When two twists are compounded by the cylindroid (§ 14), it will be observed that the amplitude of the resultant twist, as well as the direction of its screw, depend solely on the amplitudes of the given twists, and the directions of the given screws, and not at all upon either their pitches or their absolute situations. So also when any number of twists are compounded, the amplitude and direction of the resultant depend only on the amplitudes and directions of the components. We may, therefore, state the following general principle. If n twists neutralize (or n wrenches equilibrate) then a closed polygon of n sides can be drawn, each of the sides of which is proportional to the amplitude of one of the twists (or intensity of one of the wrenches), and parallel to the corresponding screw.

Let a_n, b_n, c_n, be the direction cosines of a line parallel to any screw of reference ω_n, and drawn through a point through which pass three rectangular axes.

Then since a unit wrench on α has components of intensities $\alpha_1, \ldots \alpha_6$, we must have

$$(a_1\alpha_1 + \ldots + a_6\alpha_6)^2 + (b_1\alpha_1 + \ldots + b_6\alpha_6)^2 + (c_1\alpha_1 + \ldots + c_6\alpha_6)^2 = 1,$$

whence
$$\Sigma\alpha_1^2 + 2\Sigma\alpha_1\alpha_2 \cos(12) = 1,$$

if we denote by cos (12) the cosine of the angle between two straight lines parallel to ω_1 and ω_2.

36. Calculation of Co-ordinates.

We may conceive the formation of a table of triple entry from which the virtual coefficient of any pair of screws may be ascertained. The three arguments will be the angle between the two screws, the perpendicular distance, and the sum of the pitches. These arguments having been ascertained by ordinary measurement of lines and angles, the virtual coefficient can be extracted from the tables.

Let α be a screw, of which the co-ordinates are to be determined. The work done by the unit wrench on α in a twist of amplitude ω_1' about the screw ω_1 is

$$2\omega_1'\varpi_{\alpha 1},$$

but this must be equal to the work done in the same twist by a wrench of intensity α_1 on the screw ω_1, whence

$$2p_1\alpha_1\omega_1' = 2\omega_1'\varpi_{\alpha 1},$$

or

$$\alpha_1 = \frac{\varpi_{\alpha 1}}{p_1}.$$

Thus, to compute each co-ordinate α_n, it is only necessary to ascertain from the tables the virtual coefficient of α_1 and ω_n and to divide this quantity by p_n.

37. The Virtual Coefficient of two screws may be expressed with great simplicity by the aid of screw co-ordinates.

The components of a twist of amplitude α' are of amplitudes $\alpha'\alpha_1, \ldots \alpha'\alpha_6$.

The components of a wrench of intensity β'' are of intensities $\beta''\beta_1, \ldots \beta''\beta_6$.

Comparing these expressions with § 32, we see that

$$\alpha_n' = \alpha'\alpha_n, \quad \beta_n'' = \beta''\beta_n,$$

and we find that the expression for the work done in the twist about α, by the wrench on β, is

$$\alpha'\beta''\,[2p_1\alpha_1\beta_1 + \ldots + 2p_6\alpha_6\beta_6].$$

The quantity inside the bracket is twice the virtual coefficient, whence we deduce the important expression

$$\varpi_{\alpha\beta} = \Sigma p_1\alpha_1\beta_1.$$

Since α and β enter symmetrically into this expression, we are again reminded of the reciprocal character of the virtual coefficient.

38. The Pitch of a screw is at once expressed in terms of its co-ordinates, for the virtual coefficient of two coincident screws being equal to the pitch, we have

$$p_\alpha = \Sigma p_1\alpha_1^2.$$

39. Screw Reciprocal to five Screws.

We can determine the co-ordinates of the single screw ρ, which is reciprocal to five given screws, $\alpha, \beta, \gamma, \delta, \epsilon$. (§ 25.)

The quantities $\rho_1, \ldots \rho_6$, must satisfy the condition

$$\Sigma p_1\rho_1\alpha_1 = 0,$$

and four similar equations; hence $p_n \rho_n$ is proportional to the determinant obtained by omitting the nth column from the matrix or:

$$\left|\begin{matrix} \alpha_1, & \alpha_2, & \alpha_3, & \alpha_4, & \alpha_5, & \alpha_6, \\ \beta_1, & \beta_2, & \beta_3, & \beta_4, & \beta_5, & \beta_6, \\ \gamma_1, & \gamma_2, & \gamma_3, & \gamma_4, & \gamma_5, & \gamma_6, \\ \delta_1, & \delta_2, & \delta_3, & \delta_4, & \delta_5, & \delta_6, \\ \epsilon_1, & \epsilon_2, & \epsilon_3, & \epsilon_4, & \epsilon_5, & \epsilon_6, \end{matrix}\right| .$$

and affixing a proper sign. The ratios of $\rho_1, \ldots \rho_6$, being thus found, the actual values are given by § 35.

If there were a sixth screw ζ the evanescence of the determinant which written in the usual notation is $(\alpha_1, \beta_2, \gamma_3, \delta_4, \epsilon_5, \zeta_6)$ would express that the six screws had a common reciprocal. This is an important case in view of future developments.

40. Co-ordinates of a Screw on a Cylindroid.

We may define the screw θ on the cylindroid by the angle l, which it makes with α, one of the two principal screws α and β. Since a wrench of unit intensity on θ has components of intensities $\cos l$ and $\sin l$ on α and β (§ 14), and since each of these components may be resolved into six wrenches on any six co-reciprocal screws, we must have (§ 34)

$$\theta_n = \alpha_n \cos l + \beta_n \sin l.$$

From this expression we can find the pitch of θ: for we have

$$p_\theta = \Sigma p_1 (\alpha_1 \cos l + \beta_1 \sin l)^2,$$

whence expanding and observing that as α and β are reciprocal $\Sigma p_1 \alpha_1 \beta_1 = 0$, and also that $\Sigma p_1 \alpha_1^2 = p_\alpha$ and $\Sigma p_1 \beta_1^2 = p_\beta$, we have the expression already given (§ 18), viz.

$$p_\theta = p_\alpha \cos^2 l + p_\beta \sin^2 l.$$

If two screws, θ and ϕ, upon the cylindroid, are reciprocal, then (m being the defining angle of ϕ),

$$\Sigma p_1 (\alpha_1 \cos l + \beta_1 \sin l)(\alpha_1 \cos m + \beta_1 \sin m) = 0,$$

or
$$p_\alpha \cos l \cos m + p_\beta \sin l \sin m = 0.$$

Comparing this with § 20, we have the following useful theorem:—

Any two reciprocal screws on a cylindroid are parallel to conjugate diameters of the pitch conic.

Since the sum of the squares of two conjugate diameters in an ellipse is constant, we obtain the important result that *the sum of the reciprocals of the pitches of two reciprocal screws on a cylindroid is constant* *.

* Compare *Octonions*, p. 190, by Alex. McAulay, 1898.

41. The Canonical Co-Reciprocals.

If all the six screws of a co-reciprocal system are to pass through the same point, they must in general constitute a pair of screws of pitches $+a$ and $-a$ on an axis OX, a pair of screws of pitches $+b$ and $-b$ on an axis OY which intersects OX at right angles, and a pair of screws of pitches $+c$ and $-c$ on an axis OZ perpendicular to both OX and OY.

It is convenient to speak of a co-reciprocal system thus arranged as a set of *canonical co-reciprocals.* The three rectangular axes OX, OY, OZ we may refer to as the *associated Cartesian axes.*

If $\alpha_1, \alpha_2, \ldots \alpha_6$ be the six co-ordinates of a screw referred to the canonical co-reciprocals, then the pitch is given in general by the equation

$$p_\alpha = a(\alpha_1^2 - \alpha_2^2) + b(\alpha_3^2 - \alpha_4^2) + c(\alpha_5^2 - \alpha_6^2).$$

It must be remembered that in this formula we assume that the co-ordinates satisfy the condition § 35

$$1 = (\alpha_1 + \alpha_2)^2 + (\alpha_3 + \alpha_4)^2 + (\alpha_5 + \alpha_6)^2.$$

Of course this condition is not necessarily complied with when $\alpha_1, \alpha_2, \ldots$ or some of them are infinite, as they are in the case of a screw of infinite pitch § 44.

In general the direction cosines of the screw α are

$$\alpha_1 + \alpha_2, \quad \alpha_3 + \alpha_4, \quad \alpha_5 + \alpha_6.$$

42. An Expression for the Virtual Coefficient.

Let λ', μ', ν' be the direction cosines of the screw θ (of pitch p_θ) which passes through the point x', y', z'. Let λ'', μ'', ν'' be the direction cosines of the screw α (of pitch p_α) which passes through the point x'', y'', z''. Then it can easily be shown that the virtual coefficient of θ and α is half the expression

$$(p_\theta + p_\phi)(\lambda'\lambda'' + \mu'\mu'' + \nu'\nu'') - \begin{vmatrix} x' - x'', & y' - y'', & z' - z'' \\ \lambda' \quad , & \mu' \quad , & \nu' \\ \lambda'' \quad , & \mu'' \quad , & \nu'' \end{vmatrix}.$$

43. Equations of a Screw.

Given the six co-ordinates $\alpha_1, \alpha_2, \ldots \alpha_6$ of a screw, with reference to a set of six canonical co-reciprocals, it is required to find the equations of that screw with reference to the associated Cartesian axes.

If we take for θ in the expression just written the screw of pitch a in the canonical system, thus making

$$\lambda' = 1; \; \mu' = 0; \; \nu' = 0; \; x' = 0; \; y' = 0; \; z' = 0,$$

we have

$$2a\alpha_1 = (\ a + p_\alpha)\lambda'' - (\mu''z' - \nu''y'),$$

similarly

$$-2a\alpha_2 = (-a + p_\alpha)\lambda'' - (\mu''z' - \nu''y'),$$

we thus find

$$\lambda'' = \alpha_1 + \alpha_2;\ \nu''y' - \mu''z' = a(\alpha_1 - \alpha_2) - p_\alpha(\alpha_1 + \alpha_2).$$

In like manner we obtain two similar pairs of equations for the required equations of the screw α,

$$\left.\begin{aligned}(\alpha_5 + \alpha_6)\,y - (\alpha_3 + \alpha_4)\,z &= a(\alpha_1 - \alpha_2) - p_\alpha(\alpha_1 + \alpha_2),\\ (\alpha_1 + \alpha_2)\,z - (\alpha_5 + \alpha_6)\,x &= b(\alpha_3 - \alpha_4) - p_\alpha(\alpha_3 + \alpha_4),\\ (\alpha_3 + \alpha_4)\,x - (\alpha_1 + \alpha_2)\,y &= c(\alpha_5 - \alpha_6) - p_\alpha(\alpha_5 + \alpha_6).\end{aligned}\right\}\quad\ldots\ldots(\text{i}).$$

The expressions on the right-hand side of these equations are the co-ordinates of the extremity of a vector from the origin of length equal to the perpendicular distance of α from the centre, and normal to the plane containing both α and the origin.

The co-ordinates of the foot of the perpendicular from the origin on the screw α are easily shown to be

$$x = (\alpha_5 - \alpha_6)(\alpha_3 + \alpha_4)\,c - (\alpha_5 + \alpha_6)(\alpha_3 - \alpha_4)\,b,$$

$$y = (\alpha_1 - \alpha_2)(\alpha_5 + \alpha_6)\,a - (\alpha_1 + \alpha_2)(\alpha_5 - \alpha_6)\,c,$$

$$z = (\alpha_3 - \alpha_4)(\alpha_1 + \alpha_2)\,b - (\alpha_3 + \alpha_4)(\alpha_1 - \alpha_2)\,a.$$

44. A Screw of Infinite Pitch.

The conception of the screw co-ordinates as defined in § 41 require special consideration in the case of a screw of infinite pitch. Consider a wrench on such a screw. If the intensity of the wrench be one unit, then the moment of the couple which forms part of the wrench is infinite. As the pitches of the screws of reference or any of those pitches are not in general to be infinite, it follows that the wrench of unit intensity on a screw of infinite pitch must have for its components on one or more of the screws of reference wrenches of infinite intensity.

If therefore $\alpha_1, \alpha_2, \ldots \alpha_6$ be the co-ordinates of a wrench of infinite pitch, it is essential that one or more of the quantities $\alpha_1, \alpha_2, \ldots \alpha_6$ shall be infinite.

In the case where the screws of reference form a canonical system we can obtain the co-ordinates as follows:

$$\alpha_1 = \frac{(p_\alpha + a)\cos(\alpha 1) - d_{\alpha 1}\sin(\alpha 1)}{2a};\quad \alpha_2 = \frac{(p_\alpha - a)\cos(\alpha 1) - d_{\alpha 1}\sin(\alpha 1)}{-2a}.$$

If p_α be indefinitely great with respect to a and $d_{\alpha 1}$, then

$$\alpha_1 = \frac{p_\alpha \cos(\alpha 1)}{2a}; \quad \alpha_2 = -\frac{p_\alpha \cos(\alpha 1)}{2a};$$

$$\alpha_3 = \frac{p_\alpha \cos(\alpha 3)}{2b}; \quad \alpha_4 = -\frac{p_\alpha \cos(\alpha 3)}{2b};$$

$$\alpha_5 = \frac{p_\alpha \cos(\alpha 5)}{2c}; \quad \alpha_6 = -\frac{p_\alpha \cos(\alpha 5)}{2c}.$$

If the co-ordinates of a screw not itself at infinity satisfy

$$\alpha_1 + \alpha_2 = 0; \quad \alpha_3 + \alpha_4 = 0; \quad \alpha_5 + \alpha_6 = 0;$$

then we must have

$$p_\alpha = \infty$$

for the equations

$$\alpha_1 = \frac{(p_\alpha + a)\cos(\alpha 1) - d_{\alpha 1}\sin(\alpha 1)}{2a},$$

$$\alpha_2 = \frac{(p_\alpha - a)\cos(\alpha 1) - d_{\alpha 1}\sin(\alpha 1)}{-2a},$$

and two similar pairs could not be otherwise satisfied.

We are not however entitled to assume the converse, i.e. that if the pitch is infinite then the three equations $\alpha_1 + \alpha_2 = 0$, &c. must be satisfied. It will however be true that

$$\frac{\alpha_1}{\alpha_2} = -1; \quad \frac{\alpha_3}{\alpha_4} = -1; \quad \frac{\alpha_5}{\alpha_6} = -1,$$

but some at least of the co-ordinates being infinite, we are in general prevented from replacing these equations by the ordinary linear form.

45. Indeterminate Screw.

It may however be instructive to investigate otherwise the circumstances of a screw α possessing the property that its six co-ordinates $\alpha_1, \alpha_2 \ldots \alpha_6$ are submitted to the three conditions

$$\alpha_1 + \alpha_2 = 0; \quad \alpha_3 + \alpha_4 = 0; \quad \alpha_5 + \alpha_6 = 0.$$

Two distinct cases must be considered. Either the screw α must have some finite points, or it must lie altogether at infinity. The first alternative is now supposed. The second will be discussed in the next article.

If there be any finite points on α then for such points the three left-hand members of the equations in § 43 are all zero. The three right-hand members must also reduce to zero. The only way in which this can be accomplished (for we need not consider the case in which all the co-ordinates are zero) is by making p_α infinite.

The direction of the screw of infinite pitch is indicated by the fact that as a twist about it is a translation with components $a(\alpha_1 - \alpha_2)$, $b(\alpha_3 - \alpha_4)$, $c(\alpha_5 - \alpha_6)$, the screw must be parallel to a ray of which these three quantities are proportional to the direction cosines.

As the three equations to the screw have disappeared, the situation of the screw is indeterminate. This is of course what might be expected, because a couple is equally efficacious in any position in its plane.

46. A Screw at infinity.

If we have

$$\alpha_1 + \alpha_2 = 0; \quad \alpha_3 + \alpha_4 = 0; \quad \alpha_5 + \alpha_6 = 0;$$

then the three equations (i), of § (43) will be satisfied for a screw entirely at infinity, no matter what its pitch may be. From this and the last article we see that the three equations

$$\alpha_1 + \alpha_2 = 0; \quad \alpha_3 + \alpha_4 = 0; \quad \alpha_5 + \alpha_6 = 0$$

may mean either a screw of infinite pitch and indefinite position, or a screw of indefinite pitch lying in the plane at infinity.

47. Screws on one axis.

The co-ordinates being referred to six canonical co-reciprocals, it is required to determine the co-ordinates of the screws of various pitches which lie on the same axis as a given screw α.

We have from § 36

$$\alpha_1 = \frac{(a + p_\alpha)(\alpha_1 + \alpha_2) - d_{\alpha_1} \sin(\alpha 1)}{2a},$$

$$\omega_1 = \frac{(a + p_\omega)(\alpha_1 + \alpha_2) - d_{\alpha_1} \sin(\alpha 1)}{2a},$$

whence

$$\alpha_1 - \omega_1 = \frac{p_\alpha - p_\omega}{2a}(\alpha_1 + \alpha_2).$$

We thus have the useful results

$$\omega_1 = \alpha_1 - \frac{p_\alpha - p_\omega}{2a}(\alpha_1 + \alpha_2); \quad \omega_2 = \alpha_2 + \frac{p_\alpha - p_\omega}{2a}(\alpha_1 + \alpha_2),$$

$$\omega_3 = \alpha_3 - \frac{p_\alpha - p_\omega}{2b}(\alpha_3 + \alpha_4); \quad \omega_4 = \alpha_4 + \frac{p_\alpha - p_\omega}{2b}(\alpha_3 + \alpha_4),$$

$$\omega_5 = \alpha_5 - \frac{p_\alpha - p_\omega}{2c}(\alpha_5 + \alpha_6); \quad \omega_6 = \alpha_6 + \frac{p_\alpha - p_\omega}{2c}(\alpha_5 + \alpha_6).$$

These formulae may be verified by observing that one of the equations (§ 43) defining ω is

$$(\omega_5 + \omega_6)\,y - (\omega_3 + \omega_4)\,z = a(\omega_1 - \omega_2) - p_\omega(\omega_1 + \omega_2).$$

Introducing the values just given for ω this equation becomes

$$(\alpha_5 + \alpha_6)\, y - (\alpha_3 + \alpha_4)\, z = a\,(\alpha_1 - \alpha_2) - p_\alpha\,(\alpha_1 + \alpha_2),$$

as of course it ought to do, for the pitch is immaterial when the question is only as to the situation of the screw.

48. Transformation of Screw-co-ordinates.

Let $\alpha_1 \ldots \alpha_6$ be the co-ordinates of a screw which we shall call ω, with reference to a canonical system of screws of reference with pitches $+a$ and $-a$ on an axis OX; $+b$ and $-b$ on an intersecting perpendicular axis OY, and $+c$ and $-c$ on the intersecting axis OZ which is perpendicular to both OX and OY.

Let x_0, y_0, z_0 be the co-ordinates of any point O' with reference to the associated system of Cartesians.

Draw through O' a system of rectangular axes $O'X'$, $O'Y'$, $O'Z'$ parallel to the original system OX, OY, OZ.

Let a new system of canonical screws of reference be arranged with pitches $+a$ and $-a$ on $O'X'$, $+b$ and $-b$ on $O'Y'$, and $+c$ and $-c$ on $O'Z'$.

Let θ_1, $\theta_2 \ldots \theta_6$ be the co-ordinates of the screw ω with regard to these new screws of reference. It is required to find these quantities in terms of $\alpha_1, \ldots \alpha_6$.

Let x', y', z' be the current co-ordinates of a point on ω referred to the new axes, the co-ordinates of this point with respect to the old axes being x, y, z,

then $$x = x' + x_0; \quad y = y' + y_0; \quad z = z' + z_0.$$

The equations of ω with respect to the new axes are (§ 43),

$$\left.\begin{aligned}(\theta_5 + \theta_6)\, y' - (\theta_3 + \theta_4)\, z' &= a\,(\theta_1 - \theta_2) - p_\omega\,(\theta_1 + \theta_2)\\ (\theta_1 + \theta_2)\, z' - (\theta_5 + \theta_6)\, x' &= b\,(\theta_3 - \theta_4) - p_\omega\,(\theta_3 + \theta_4)\\ (\theta_3 + \theta_4)\, x' - (\theta_1 + \theta_2)\, y' &= c\,(\theta_5 - \theta_6) - p_\omega\,(\theta_5 + \theta_6)\end{aligned}\right\} \quad \ldots\ldots\ldots \text{(i)}.$$

We have also

$$\left.\begin{aligned}(\alpha_5 + \alpha_6)\, y - (\alpha_3 + \alpha_4)\, z &= a\,(\alpha_1 - \alpha_2) - p_\omega\,(\alpha_1 + \alpha_2)\\ (\alpha_1 + \alpha_2)\, z - (\alpha_5 + \alpha_6)\, x &= b\,(\alpha_3 - \alpha_4) - p_\omega\,(\alpha_3 + \alpha_4)\\ (\alpha_3 + \alpha_4)\, x - (\alpha_1 + \alpha_2)\, y &= c\,(\alpha_5 - \alpha_6) - p_\omega\,(\alpha_5 + \alpha_6)\end{aligned}\right\} \quad \ldots\ldots\ldots \text{(ii)}.$$

Remembering that the new axes are parallel to the original axes we have

$$\theta_1 + \theta_2 = \alpha_1 + \alpha_2; \quad \theta_3 + \theta_4 = \alpha_3 + \alpha_4; \quad \theta_5 + \theta_6 = \alpha_5 + \alpha_6 \ldots\ldots\ldots \text{(iii)}.$$

Hence by subtracting the several formulae (i) from the formulae (ii) we obtain

$$\left.\begin{aligned} y_0(\alpha_5+\alpha_6)-z_0(\alpha_3+\alpha_4)&=a(\alpha_1-\alpha_2)-a(\theta_1-\theta_2)\\ z_0(\alpha_1+\alpha_2)-x_0(\alpha_5+\alpha_6)&=b(\alpha_3-\alpha_4)-b(\theta_3-\theta_4)\\ x_0(\alpha_3+\alpha_4)-y_0(\alpha_1+\alpha_2)&=c(\alpha_5-\alpha_6)-c(\theta_5-\theta_6)\end{aligned}\right\}\ \ldots\ldots\ldots \text{(iiii)}.$$

The six equations (iii) and (iiii) determine $\theta_1, \ldots \theta_6$ in terms of $\alpha_1, \ldots \alpha_6$.

49. Principal Screws on a Cylindroid.

If two screws are given we determine as follows the pitches of the two principal screws on the cylindroid which the two given screws define.

Let α and β be the two given screws. Then the co-ordinates of these screws referred to six canonical co-reciprocals are

$$\alpha_1, \ldots \alpha_6 \text{ and } \beta_1, \ldots \beta_6.$$

The co-ordinates of any other screw on the same cylindroid are proportional to

$$\rho\alpha_1+\beta_1,\ \ \rho\alpha_2+\beta_2, \ldots \rho\alpha_6+\beta_6;$$

when ρ is a variable parameter.

The pitch p of the screw so indicated is given by the equation (§ 41)

$$\begin{aligned} a(\rho\alpha_1+\beta_1)^2-a(\rho\alpha_2+\beta_2)^2+b(\rho\alpha_3+\beta_3)^2-b(\rho\alpha_4+\beta_4)^2&\\ +c(\rho\alpha_5+\beta_5)^2-c(\rho\alpha_6+\beta_6)^2&\\ =p\left[\{\rho(\alpha_1+\alpha_2)+\beta_1+\beta_2\}^2+\rho\{(\alpha_3+\alpha_4)+\beta_3+\beta_4\}^2+\rho\{(\alpha_5+\alpha_6)+\beta_5+\beta_6\}^2\right],&\end{aligned}$$

or

$$\rho^2 p_\alpha+2\rho\varpi_{\alpha\beta}+p_\beta=p\{\rho^2+2\rho\cos(\alpha\beta)+1\},$$

or

$$\rho^2(p_\alpha-p)+2\rho\{\varpi_{\alpha\beta}-p\cos(\alpha\beta)\}+p_\beta-p=0.$$

For the principal screws ρ is to be determined so that p shall be a maximum or a minimum (§ 18), whence the equation for p is

$$\{\varpi_{\alpha\beta}-p\cos(\alpha\beta)\}^2=(p_\alpha-p)(p_\beta-p),$$

or

$$p^2\sin^2(\alpha\beta)+p(2\varpi_{\alpha\beta}\cos(\alpha\beta)-p_\alpha-p_\beta)+p_\alpha p_\beta-\varpi^2_{\alpha\beta}=0.$$

The roots of this quadratic are the required values of p.

The quadratic may also receive the form

$$\begin{aligned} 0=(p-p_\alpha)(p-p_\beta)\sin^2(\alpha\beta)+\tfrac{1}{2}d_{\alpha\beta}\sin(\alpha\beta)\cos(\alpha\beta)(p_\alpha+p_\beta-2p)&\\ -\tfrac{1}{4}(p_\alpha-p_\beta)^2\cos^2(\alpha\beta)-\tfrac{1}{4}d^2_{\alpha\beta}\sin^2(\alpha\beta),&\end{aligned}$$

where $d_{\alpha\beta}$ is the shortest distance of α and β.

In this form it is obvious that to increase each of the three quantities p_α, p_β, p by m does not affect the equation. This is of course the well-known property of the cylindroid. (§ 20.)

If the quadratic have two equal roots, then all the screws of the cylindroid form a plane pencil, and all have equal pitches.

The discriminant of the quadratic is

$$4\varpi_{\alpha\beta}^2\cos^2(\alpha\beta)+(p_\alpha+p_\beta)^2-4\varpi_{\alpha\beta}(p_\alpha+p_\beta)\cos(\alpha\beta)-4p_\alpha p_\beta\sin^2(\alpha\beta)+4\varpi_{\alpha\beta}^2\sin^2(\alpha\beta)$$
$$\equiv \{2\varpi_{\alpha\beta}-(p_\alpha+p_\beta)\cos(\alpha\beta)\}^2+(p_\alpha+p_\beta)^2\sin^2(\alpha\beta)-4p_\alpha p_\beta\sin^2(\alpha\beta)$$
$$\equiv (d_{\alpha\beta}^2+(p_\alpha-p_\beta)^2)\sin^2(\alpha\beta).$$

Hence discarding the case when $(\alpha\beta)=0$ we have

$$d_{\alpha\beta}=0,\quad p_\alpha-p_\beta=0,$$

as was to be expected.

CHAPTER V.

THE REPRESENTATION OF THE CYLINDROID BY A CIRCLE*.

50. A Plane Representation.

The essence of the present chapter lies in the geometrical representation of a screw by a point. The series of screws which constitute the cylindroid correspond to, or are represented by, a series of points in a plane. By choosing a particular type of correspondence we can represent the screws of the cylindroid by the points of a circle†. Various problems on the cylindroid can then be studied by the aid of the corresponding circle. We commence with a very simple process for the discovery of the circle. It will in due course appear how this circular representation is suggested by the geometry of the cylindroid itself (§ 68).

It has been shown (§ 13) that the positions of the several screws on the cylindroid may be concisely defined by the intersections of the pairs of planes,

$$y = x \tan \theta,$$

$$z = m \sin 2\theta.$$

In these equations, θ varies in correspondence with the several screws, while m is a parameter expressing the *size* of the cylindroid. In fact, the whole surface, except parts of the nodal line, is contained between two parallel planes, the distance between which is $2m$.

The pitch of the screw corresponding to θ is expressed by

$$p = p_0 + m \cos 2\theta,$$

where p_0 is a constant.

* See papers in *Proceedings of the Royal Irish Academy*, Ser. II. Vol. IV. p. 29 (1883), and the *Cunningham Memoirs of the Royal Irish Academy*, No. 4 (1886).

† I may refer to a paper by Professor Mannheim, in the *Comptes rendus* for 2nd February, 1885, entitled "Représentation plane relative aux déplacements d'une figure de forme invariable assujettie à quatre conditions." Professor Mannheim here shows how the above plane representation might also have been deduced from the instructive geometrical theory which he had brought before the Academy of Sciences on several occasions.

Eliminating θ between the equations for z and p, we obtain

$$(p - p_0)^2 + z^2 = m^2.$$

Let p and z be regarded as the current co-ordinates of a point. Then the locus of this point is the circle which forms the foundation of the plane representation*.

m is, of course, the radius, and p_0 is the distance of the centre from a certain axis. Any point on this circle being given, then its co-ordinates p and z are completely determined. Thus $\sin 2\theta$ and $\cos 2\theta$, and, consequently, $\tan\theta$, are known. We therefore see that the position of a screw and its pitch are completely determined when the corresponding point on the circle is known. To each point of the circle corresponds one screw on the cylindroid. To each screw on the cylindroid corresponds one point on the circle. This may be termed the *representative* circle of the cylindroid.

51. The Axis of Pitch.

Let T (fig. 5) be the origin. Then p_0 is the perpendicular ST from the centre S of the circle to the axis PT. The ordinate AP is the pitch of the

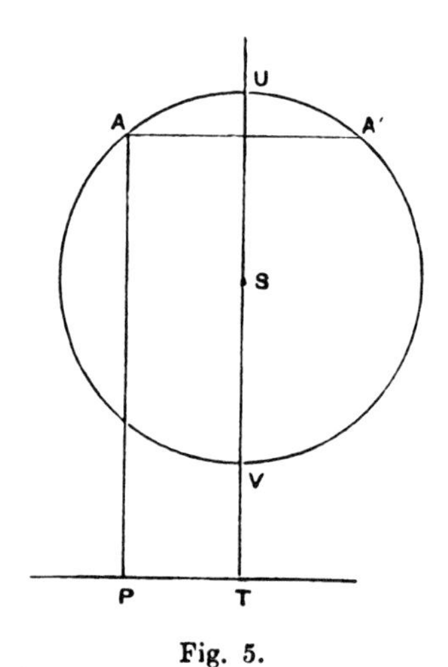

Fig. 5.

* The following elegant construction for the cylindroid is given by Mr T. C. Lewis, *Messenger of Mathematics*, Vol. IX. pp. 1—5, 1879. "Suppose that a point P moves with uniform velocity around a circle while the circle itself rotates uniformly about an axis in its plane with half the angular velocity that P has around the centre. Then the perpendiculars from P on the axis of rotation trace out the cylindroid, while the lengths of those perpendiculars are the pitches of the corresponding screws." This construction is of special interest in connexion with the representation of the cylindroid by a circle discussed in this chapter. The construction of Mr Lewis shows that if the circle rotate around the axis of pitch with half the angular velocity of the point P around the circle, then not only does P represent the screw in this circle but the perpendicular from P on the axis of pitch is the position of the screw itself.

screw, and the line PT may be called the *axis of pitch*. We have, accordingly, the following theorem:—

The pitch of any screw on the cylindroid is equal to the perpendicular let fall on the axis of pitch from the corresponding point on the circle.

A parallel AA' to the axis of pitch cuts the circle in two points, A and A', which have equal pitch. The diameter perpendicular to the pitch axis intersects the circle in the points U, V of maximum and minimum pitch. These points, of course, correspond to the two principal screws on the cylindroid. The two screws of zero pitch are defined by the two real or imaginary points in which the axis of pitch cuts the circle.

A fundamental law of the pitch distribution on the several screws of a cylindroid is simply illustrated by this geometrical representation. The law states that if all the pitches be augmented by a constant addition, the pitches so modified will still be a possible distribution. So far as the cylindroid is concerned, such a change would only mean a transference of the axis of pitch to some other parallel position. The diameter $2m$ merely expresses the size of the cylindroid, and is, of course, independent of the constant part in the expression of the pitch.

52. The Distance between two Screws.

We shall often find it convenient to refer to a screw as simply equivalent to its corresponding point on the circle. Thus, in fig. 6, the two points, A

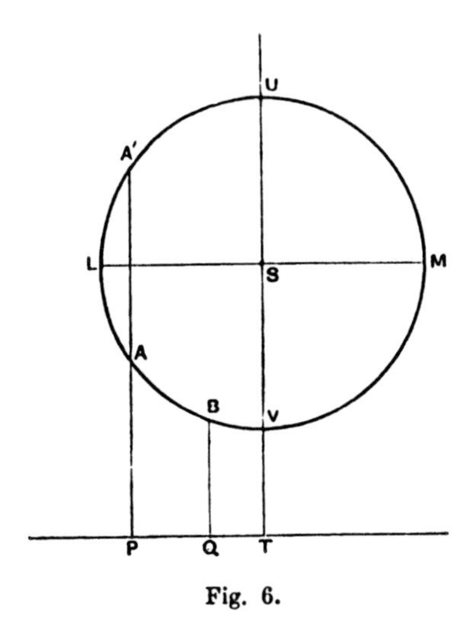

Fig. 6.

and B, may conveniently be called the screws A and B. The propriety of this language will be admitted when it is found that everything about a screw can be ascertained from the position of its corresponding point on the circle.

Let us, for instance, seek the shortest distance between the two screws A and B. Since all screws intersect the nodal axis of the cylindroid at right angles, the required shortest distance is simply the difference between the values of $m \sin 2\theta$ for the two screws: this is, of course, the difference of their abscissae, i.e. the length PQ. Hence we have the following theorem:—

The shortest distance between two screws, A and B, is equal to the projection of the chord AB on the axis of pitch.

We thus see that every screw A on the cylindroid must be intersected by another screw A', and the chord AA' is, of course, perpendicular to the axis of pitch. The ray through S, parallel to the axis of pitch, will give two screws, L and M. These are the bounding screws of the cylindroid, and in each a pair of intersecting screws have become coincident. The two principal screws, U and V, lying on a diameter perpendicular to the axis of pitch, must also intersect.

If all the pitches be reduced by p_0, then the pitch axis passes through the centre of the circle, and the case assumes a simple type. The extremities of a chord perpendicular to the axis of pitch define screws of equal and opposite pitches, and every pair of such screws must intersect. The screws of zero pitch will then be the bounding screws, while the two principal screws will have pitches $+m$ and $-m$, respectively.

53. The Angle between two Screws.

This important function also admits of simple representation by the corresponding circle. Let A, B (fig. 7) denote the two screws; then, if θ and θ' be the angles corresponding to A and B,

$$AST = 2\theta; \quad BST = 2\theta',$$

whence

$$ASB = 2(\theta - \theta').$$

If H be any point on the circle, then

$$AHB = \theta - \theta',$$

and we deduce the following theorem:—

The angle between two screws is equal to the angle subtended in the circle by their chord.

The extremities of a diameter denote a pair of screws at right angles: thus, A', in fig. 7, is the one screw on the cylindroid which is at right angles to A. The principal screws, U and V, are also seen to be at right angles.

The circular representation of the cylindroid is now complete. We see how the pitch of each screw is given, and how the perpendicular distance

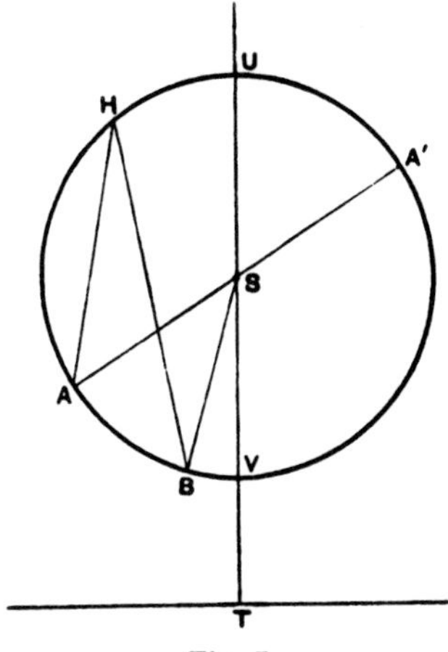

Fig. 7.

and the angle between every pair of screws can be concisely represented. We may therefore study the dynamical and kinematical properties of the cylindroid by its representative circle. We commence by proving a fundamental principle very analogous to an elementary theorem in Statics.

54. The Triangle of Twists.

It has been already shown (§ 14) that any three screws on the cylindroid possess the following property:—

If a body receive twists about three screws, so that the amplitude of each twist is proportional to the sine of the angle between the two non-corresponding screws, the body, after the last twist, will be restored to where it was before the first.

With the circular representation of the cylindroid we transform this theorem into the following:—

If any three screws, A, B, C (Fig. 8), be taken on the circle, and if twists be applied to a body in succession, so that the amplitude of each twist is proportional to the opposite side of the triangle ABC, then the body will be restored by the last twist to the place it had before the first.

From the analogy of wrenches, and of twist velocities to twists, we are also able to enunciate the following theorems:—

If wrenches upon the three screws A, B, C be applied to any rigid body, then these wrenches will equilibrate, provided that the intensity of each is proportional to the opposite side of the triangle.

If twist velocities about the three screws A, B, C animate a rigid body, then these twist velocities will neutralize if they are respectively proportional to the opposite sides of the triangle.

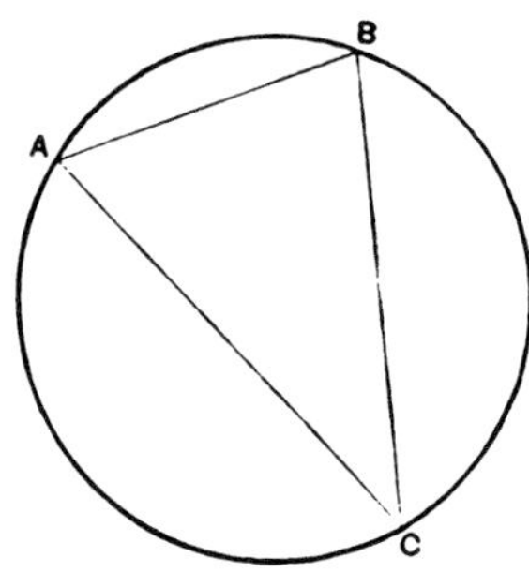

Fig. 8.

55. Decomposition of Twists and Wrenches.

The theorems we have just enunciated lead to simple rules for effecting the composition, or the decomposition of twists or of wrenches. Let a twist on a screw X be given, and let it be required to find the components of this twist on any two given screws A, B, all three, of course, lying on the same cylindroid. Let ω be the amplitude of the twist on X. Then, by the last article, the following triad of twists on the screws X, A, B, respectively, will neutralize:—

$$\omega, \quad \omega\frac{BX}{AB}, \quad \omega\frac{AX}{AB},$$

whence the components on A and B of the twist about X are, so far as magnitudes are concerned,

$$\omega\frac{BX}{AB} \text{ and } \omega\frac{AX}{AB}.$$

A similar proposition holds of course for wrenches.

56. Composition of Twists and Wrenches.

Let two twists, of amplitudes α and β, about the screws A and B, respectively, be applied to a rigid body. It is required to find the single resultant screw X, and the amplitude ω of the resulting twist. Divide AB (Fig. 9) in the point I, so that the segments AI and BI shall be in the inverse ratio of α to β. Bisect the arc AB at H, and draw HI, which will cut the circle in the required point X.

The value of ω is obtained from the equations

$$\frac{\omega}{AB} = \frac{\alpha}{BX} = \frac{\beta}{AX}.$$

If the amplitudes α and β had opposite signs, then the point I should have divided AB externally in the given ratio.

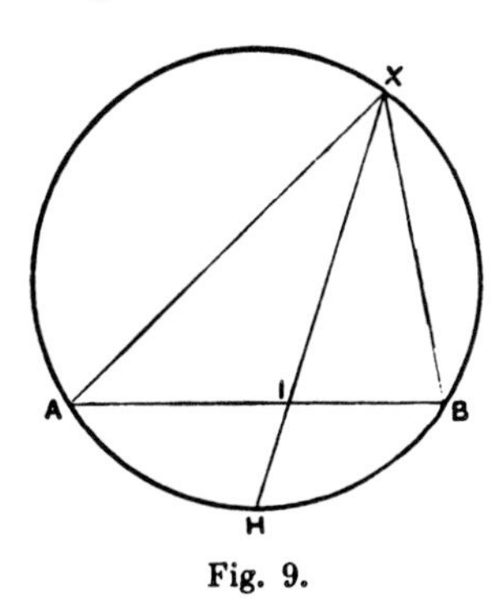

Fig. 9.

57. Screw Co-ordinates.

We have developed in the last chapter the general conception of Screw Co-ordinates. In the case of the cylindroid, the co-ordinates of any screw X, with respect to two standard screws A and B, are found by resolving a wrench of unit intensity on X into its two components on A and B. These components are said to be the co-ordinates of the screw. If we denote the co-ordinates of X by X_1 and X_2, we have

$$X_1 = \frac{BX}{AB}; \quad X_2 = \frac{AX}{AB}.$$

The co ordinates satisfy the identical relation,

$$X_1^2 - 2X_1X_2 \cos \epsilon + X_2^2 = 1,$$

where ϵ denotes the angle between the two screws of reference, that is, the angle subtended by the chord AB.

58. Reciprocal Screws.

Every screw A on the cylindroid has one other reciprocal screw B lying also on the cylindroid (§ 26). Denoting as usual A and B by their corresponding points on the circle, we may enunciate the following theorem:—

The chord joining a pair of reciprocal screws passes through the pole of the axis of pitch.

The condition that two screws shall be reciprocal is

$$(p_\alpha + p_\beta) \cos \theta - d_{\alpha\beta} \sin \theta = 0,$$

where p_α and p_β are the pitches, θ is the angle between the two screws, and $d_{\alpha\beta}$ their shortest distance. It is easy to show that this condition is fulfilled for any two screws A and B (Fig. 10), whose chord passes through O, the pole of the axis of pitch PQ.

Since $SO \,.\, ST = SA^2 = SB^2$, we have $\angle STA = \angle SAB$, and $\angle STB = \angle SBA$, whence $\angle ATB$ is bisected by ST, and therefore

$$\angle ATP = \tfrac{1}{2} \angle ASB = \theta = \angle BTQ.$$

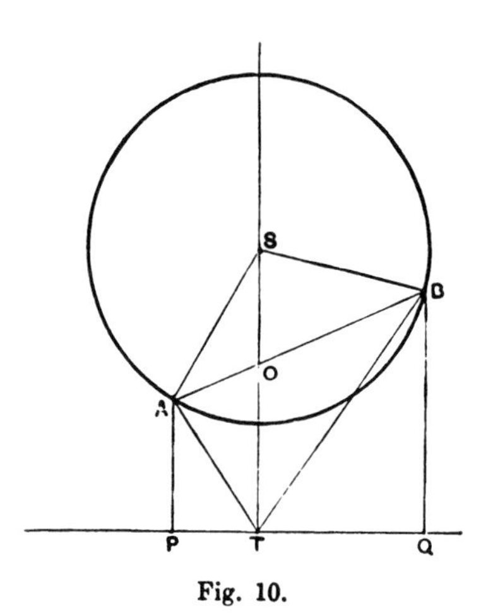

Fig. 10.

It follows that $AP \cos\theta = PT \sin\theta$, since each is equal to the perpendicular from P on AT.

Similarly,

$$BQ \cos\theta = QT \sin\theta;$$

whence $$(AP + BQ) \cos\theta - (PT + QT) \sin\theta = 0,$$

which reduces to

$$(p_\alpha + p_\beta) \cos\theta - d_{\alpha\beta} \sin\theta = 0.$$

The theorem has thus been proved.

We have, therefore, a simple construction for finding the screw B reciprocal to a given screw A. It is only necessary to join A to O, the pole of the axis of pitch, and the point in which this cuts the circle again gives B the required reciprocal screw.

We also notice that the two principal screws of the cylindroid are reciprocal, inasmuch as their chord passes through O.

59. Another Representation of the Pitch.

We can obtain another geometrical expression for the pitch, which will be often more convenient than the perpendicular distance from the point to the axis of pitch.

Let A (Fig. 11) be the point of which the pitch is required. Join AOB, draw AP perpendicular to the axis of pitch PT, and produce AP to intersect BT at E. Then, since O is the pole of PT, the line PT bisects the angle ATE (§ 58), and therefore AE must be bisected at P.

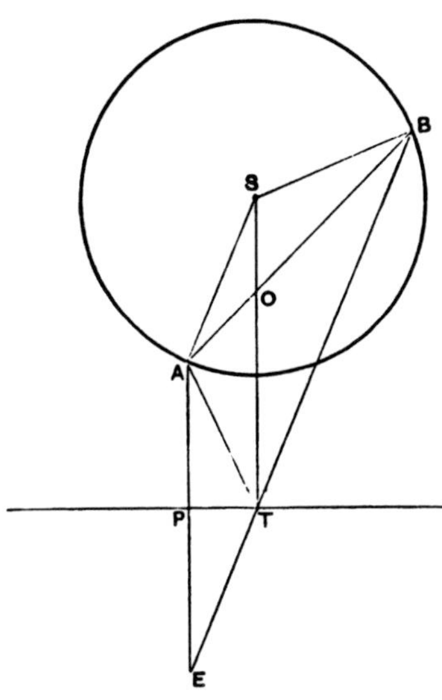

Fig. 11.

From similar triangles,

$$OB : AB :: OT : AE;$$

whence, if p_a be the pitch of A, and, of course, equal to AP, or $\frac{1}{2}AE$,

$$2p_a = \frac{AB \,.\, OT}{OB}.$$

But since the quadrilateral $ASBT$ is inscribable in a circle,

$$OT \,.\, OS = OA \,.\, OB;$$

whence, eliminating OT, we have, finally,

$$p_a = \frac{AO \,.\, AB}{2OS}:$$

as OS is constant, we see that p_a varies as $AO \,.\, AB$, whence the following theorem :—

If AB be any chord passing through O, the pole of the axis of pitch, then the pitch of the screw A is proportional to the product $AO \,.\, AB$.

60. Pitches of Reciprocal Screws.

It is known that the sum of the reciprocals of the pitches of a pair of reciprocal screws on the cylindroid is constant (§ 40). This is also plain from the geometrical representation. For, since the triangles APT and BQT (Fig. 10) are similar, we have

$$AP : BQ :: TP : TQ :: OA : OB;$$

whence O is the centre of gravity of particles of masses $\frac{1}{p_\alpha}$ and $\frac{1}{p_\beta}$ placed at A and B, respectively.

From the known property of the centre of gravity,

$$AP\frac{1}{p_\alpha} + BQ\frac{1}{p_\beta} = OT.\left(\frac{1}{p_\alpha} + \frac{1}{p_\beta}\right);$$

but each of the terms on the left-hand side is unity, whence, as required,

$$\frac{1}{p_\alpha} + \frac{1}{p_\beta} = \frac{2}{OT}.$$

The second mode of representing the pitch also verifies this theorem. For since (§ 59)

$$p_\alpha = \frac{AO.AB}{2OS},$$

$$p_\beta = \frac{BO.BA}{2OS};$$

we have

$$p_\alpha + p_\beta = \frac{AB^2}{2OS}; \quad p_\alpha p_\beta = \frac{AB^2.AO.BO}{4OS^2},$$

from which

$$\frac{1}{p_\alpha} + \frac{1}{p_\beta} = \frac{2.OS}{OA.OB};$$

but $OA.OB$ is constant for every chord through O; and, as OS is constant, it follows that the sum of the reciprocals of the pitches of two reciprocal screws on any cylindroid must be constant.

61. The Virtual Coefficient.

Let A and B (Fig. 12) be the two screws. Let, as usual, O be the pole of the axis of pitch PT. Let O' be the point in which the chord AB intersects OT the perpendicular drawn from O to the axis of pitch, and let $P'T'$ be the polar of O', which is easily shown to be perpendicular to SO. From T let fall the perpendicular TF upon AT', and from O let fall the perpendicular OG upon AB.

As before (§ 58), we have $\angle AT'P' = \angle T'TF = \theta$; also, since

$$\angle SAO' = \angle AT'O', \text{ and } \angle SAO = \angle ATO,$$

we must have $\angle SAO' - \angle SAO = \angle AT'O' - \angle ATO$, or $\angle OAG = \angle TAF$; whence the triangles OAG and TAF are similar, and, consequently,

$$TF = OG.\frac{AT}{AO} = OG.\frac{AS}{OS};$$

but, as in § 58, we have

$$(p_\alpha - TT' + p_\beta - TT')\cos\theta - d_{\alpha\beta}\sin\theta = 0;$$

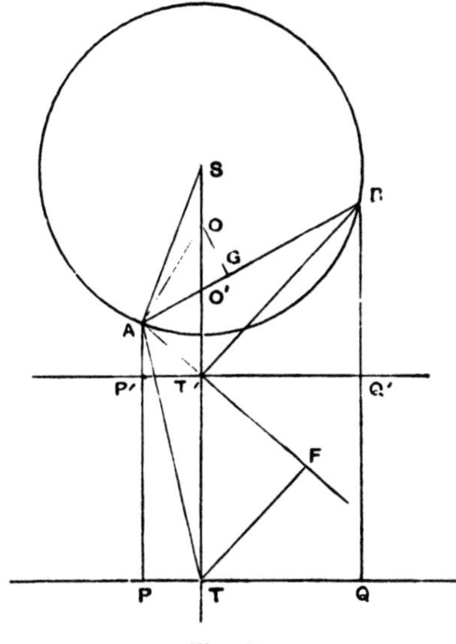

Fig. 12.

whence the virtual coefficient is simply,

$$TT'\cos\theta = OG\frac{AS}{OS},$$

and we have the following theorem :

The virtual coefficient of any pair of screws varies as the perpendicular distance of their chord from the pole of the axis of pitch.

We also notice that the line TF expresses the actual value of the virtual coefficient.

The theorem of course includes, as a particular case, that property of reciprocal screws, which states that their chord passes through the pole of the axis of pitch (§ 58).

62. Another Investigation of the Virtual Coefficient.

It will be instructive to investigate the theorem of the last article by a different part of the theory. We shall commence with a proposition in elementary geometry.

Let ABC (Fig. 13) be a triangle circumscribed by a circle, the lengths of the sides being, as usual, a, b, c. Draw tangents at A, B, C, and thus form the triangle XYZ. It can be readily shown that if masses a^2, b^2, c^2 be placed at A, B, C, their centre of gravity must lie on the three lines AX, BY, CZ. These lines must therefore be concurrent at I, which is the centre of gravity of the three masses.

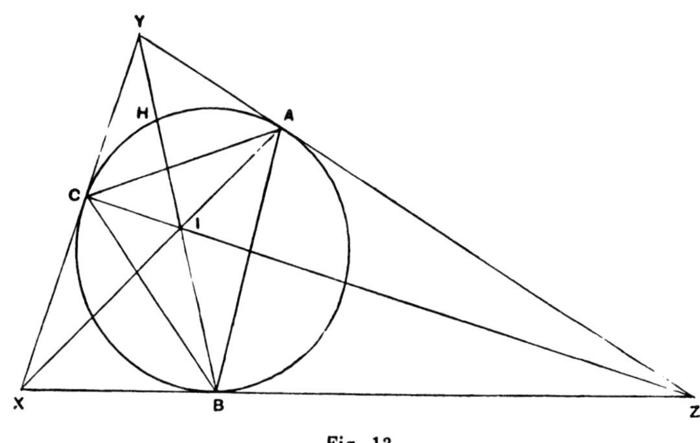

Fig. 13.

Let BY intersect the circle again at H. Then, since AC is the polar of Y, the arc AC is divided harmonically at H and B; consequently the four points A, C, B, H subtend a harmonic pencil at any point on the circle. Let that point be B, then BC, BI, BA, BZ form a harmonic pencil; hence CZ is cut harmonically, and consequently Z must be the centre of gravity of particles, $+a^2$ at A, $+b^2$ at B, and $-c^2$ at C.

Suppose the axis of pitch to be drawn (it is not shown in the figure), and let h be the perpendicular let fall from Z on this axis, also let p_1, p_2, p_3 be the pitches of the screws A, B, C.

Then, by a familiar property of the centre of gravity, we must have

$$p_1a^2 + p_2b^2 - p_3c^2 = (a^2 + b^2 - c^2)\,h = 2abh \cos C.$$

We shall take A, B as the two screws of reference, and if ρ_1 and ρ_2 be the co-ordinates of C with respect to A and B; then, from the principles of screw co-ordinates (§ 30), we have

$$p_3 = p_1\rho_1^2 + p_2\rho_2^2 + 2\varpi_{12}\rho_1\rho_2,$$

where ϖ_{12} is the virtual coefficient of A and B. In the present case we have

$$\rho_1 = \frac{a}{c}; \quad \rho_2 = \frac{b}{c};$$

whence

$$p_1a_2^2 + p_2b^2 - p_3c^2 + 2\varpi_{12}\,ab = 0;$$

and, finally,

$$\varpi_{12} = -h \cos C.$$

The negative sign has no significance for our present purpose, and hence we have the following theorem:

The virtual coefficient of two screws is equal to the cosine of the angle subtended by their chord, multiplied into the perpendicular from the pole of the chord on the axis of pitch.

This is, perhaps, the most concise geometrical expression for the virtual coefficient. It vanishes if the perpendicular becomes zero, for then the chord must pass through the pole of the pitch axis, and the two screws be reciprocal. The cosine enters the expression in order that its evanescence, when $C = 90°$, may provide for the circumstance that the perpendicular is then infinite.

This result is easily shown to be equivalent to that of the last article by the well-known theorem:—

If any two chords be drawn in a circle, then the cosine of the angle subtended by the first chord, multiplied into the perpendicular distance from its pole to the second chord, is equal to the cosine of the angle subtended by the second chord, multiplied into the perpendicular from its pole to the first chord.

It follows that the virtual coefficient must be equal to the perpendicular from the pole of the axis of pitch upon the chord joining the two screws, multiplied into the cosine of the angle in the arc cut off by the axis of pitch. This is the expression of § 61, namely,

$$OG \frac{AS}{OS}.$$

63. Application of Screw Co-ordinates.

It will be useful to show how the geometrical form for the virtual coefficient is derived from the theory of screw co-ordinates. Let α_1, α_2, and β_1, β_2 be the co-ordinates of two screws on the cylindroid; then, if the screws of reference be reciprocal, the virtual coefficient is (§ 37)

$$p_1\alpha_1\beta_1 + p_2\alpha_2\beta_2.$$

Let A, B (Fig. 14) be the screws of reference, and let C and C' be the two screws of which the virtual coefficient is required. Let PQ be the axis of pitch of which O is the pole, then O lies on AB, as the two screws of reference are reciprocal (§ 58).

As AB is divided harmonically at O and H, we have

$$AO : OB :: HA : HB :: AP : BQ :: p_1 : p_2;$$

whence O is the centre of gravity of masses $\frac{1}{p_1}$, $\frac{1}{p_2}$ at A and B, respectively.

If, therefore, AX, BY, OG be perpendiculars on CC', we have, from the principle of the centre of gravity,

$$\frac{1}{p_1} AX + \frac{1}{p_2} BY = \left(\frac{1}{p_1} + \frac{1}{p_2}\right) OG,$$

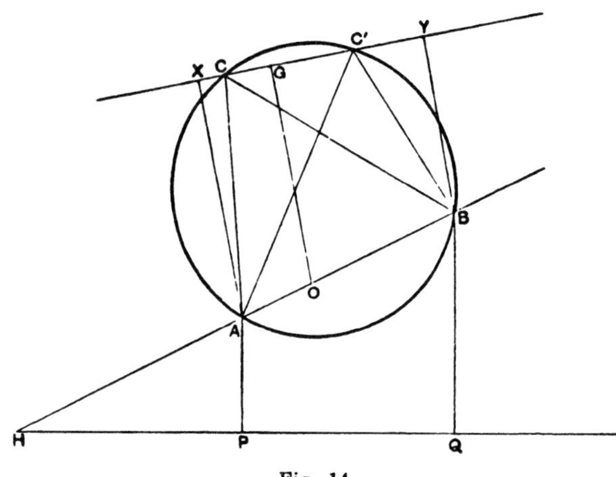

Fig. 14.

or,
$$p_2 AX + p_1 BY = (p_1 + p_2) OG;$$

but, by a well-known property of the circle, if m be the radius,

$$2mAX = AC \, . \, AC'; \quad 2mBY = BC \, . \, BC';$$

whence

$$p_1 BC \, . \, BC' + p_2 AC \, . \, AC' = 2m (p_1 + p_2) OG = m \frac{OG \, . \, AB^2}{OS} \quad (\S\ 60),$$

or
$$p_1 \frac{BC}{AB} \cdot \frac{BC'}{AB} + p_2 \frac{AC}{AB} \cdot \frac{AC'}{AB} = m \frac{OG}{OS}.$$

But, from the expressions for screw co-ordinates (§ 57), this reduces to

$$p_1 \alpha_1 \beta_1 + p_2 \alpha_2 \beta_2 = m \frac{OG}{OS}.$$

The required expression has thus been demonstrated.

We can give another proof of this theorem as follows:—

If the two screws of reference be reciprocal, and if ρ_1 and ρ_2 be the co-ordinates of another screw, then it is known, from the theory of the co-ordinates, that the virtual coefficients of this screw, with respect to the screws of reference, are $p_1\rho_1$ and $p_2\rho_2$, respectively (§ 37).

Thus (Fig. 15) the virtual coefficient of X and A must be (§ 57),

$$p_1 \frac{BX}{AB};$$

but we know (§ 59) $$p_1 = \frac{AO \cdot AB}{2SO};$$

whence the virtual coefficient is

$$\frac{AO \cdot BX}{2SO} = \frac{2m \sin A \cdot AO}{2SO} = m\frac{OG}{OS},$$

as already determined. This is an instructive proof, besides being much shorter than the other methods.

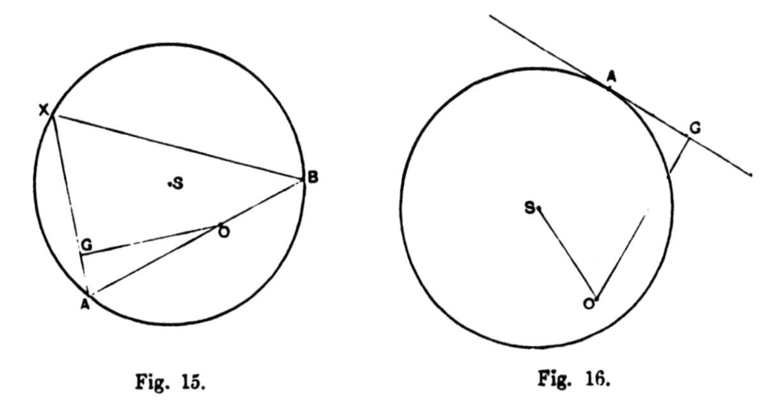

Fig. 15. Fig. 16.

64. Properties of the Virtual Coefficient.

If the virtual coefficient be given then the chord envelopes a circle with its centre at the pole of the axis of pitch.

Two screws can generally be found which have a given virtual coefficient with a given screw.

Let A (Fig. 15) be a given screw, and X a variable screw; then their virtual coefficient is proportional to OG, and therefore to the sine of A, that is, to the length BX. Thus, as X varies, its virtual coefficient with A varies proportionally to the distance of X from the fixed point B.

65. Another Construction for the Pitch.

As the virtual coefficient of two coincident screws is equal to their pitch, we shall obtain another geometrical construction for the pitch by supposing two screws to coalesce. For (in Fig. 16), let AG be the chord joining the two coincident screws, that is the tangent, then, from § 61, we have for the pitch,

$$m\frac{OG}{OS},$$

whence the following theorem :—

The pitch of any screw is proportional to the perpendicular on the tangent at the point let fall from the pole of the axis of pitch.

66. Screws of Zero Pitch.

A screw of zero pitch is reciprocal to itself. The tangent at a point corresponding to a screw of zero pitch, being the chord joining two reciprocal screws, must pass through the pole of the axis of pitch. This is, of course, the same thing as to say that the axis of pitch intersects the circle in two screws, each of which has zero pitch.

67. A Special Case.

We have supposed that the axis of pitch occupies any arbitrary position. Let us now assume that it is a tangent to the representative circle. This specialization of the general case could be produced by augmenting the pitches of all the screws on the cylindroid, by such a constant as shall make one of the two principal screws have zero pitch.

The following properties of the screws on the cylindroid are then obvious:—

1. There is only one screw of zero pitch, O.

2. The pitches of all the other screws have the same sign.

3. The maximum pitch is double the radius.

4. The screw O is reciprocal to every screw on the surface, and this is the only case in which a screw on the cylindroid is reciprocal to every other screw thereon.

68. A Tangential Section of the Cylindroid*.

Let the plane of section be the plane of the paper in Fig. 17, and let the plane contain one of the screws of zero pitch OA. Let OH be the projection of the nodal axis on the plane of the paper. Then OA being perpendicular to the nodal axis must be perpendicular to OH. Let P be the point where the second screw of zero pitch cuts the curve. Then since any ray through P and across AO, meets two screws of equal pitch, it must be perpendicular to the third screw which it also meets on the cylindroid (§ 22). Hence PH is perpendicular to the screw through H, and as the latter lies in the normal plane through OH it follows that the angle at H is a right angle.

Any chord perpendicular to AO must for the same reason intersect two screws of equal pitch, and therefore $APHO$ must be a rectangle.

If tangents be drawn at A and P intersecting at T, then it can be shown that any chord TLM through T cuts the ellipse in points L and M on two reciprocal screws.

* For proofs of theorems in this article see a paper in the *Transactions of the Royal Irish Academy*, Vol. XXIX. pp. 1—32 (1887).

If from any point X a perpendicular XR be let fall on AP, then the pitch of the screw through X is $XR \tan\theta$, where $\sin\theta$ is the eccentricity of the ellipse. Also θ is the angle between the normal to the plane and the nodal axis.

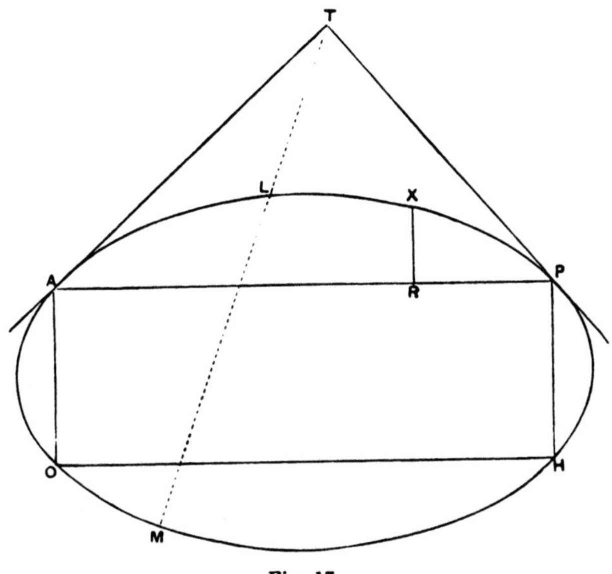

Fig. 17.

Let two circles be described, one with the major axis of the ellipse as diameter and the other with the line joining the two foci as a diameter. Let X_1 be the point in which the ordinate through X meets the first circle and X_2 be the point in which a ray drawn from X_1 to the centre meets the second circle. Then this point X_2 on this inner circle will be exactly the circular representation of the screws on the cylindroid with which this chapter commenced. There is only one such circle, for the distance between the foci is the same for every tangential section, and so is the distance from the centre to the axis of pitch.

CHAPTER VI.

THE EQUILIBRIUM OF A RIGID BODY.

69. A Screw System.

To specify with precision the nature of the freedom enjoyed by a rigid body, it is necessary to ascertain all the screws about which the constraints will permit the body to be twisted. When the attempt has been made for every screw in space, the results will give us all the information conceivable with reference to the freedom of the body, and also with reference to the constraints by which the movement may be hampered.

Suppose that by these trials, n screws $A_1, \dots A_n$ have been found about each of which the body can receive a twist. It is evident, without further trial, that twisting about an infinite number of other screws must also be possible ($n > 1$): for suppose the body receive any n twists about $A_1, \dots A_n$ the position attained could have been reached by a twist about some single screw A. The body can therefore twist about A. Since the amplitudes of the n twists may have any magnitude (each not exceeding an infinitely small quantity), A is merely one of an infinite number of screws, about which twisting must be possible. *All these screws, together with $A_1, \dots A_n$, will in general form what we call a screw system of the nth order.*

If it be found that the body cannot be twisted about any screw which does not belong to the screw system of the nth order, then the body is said to have freedom of the nth order. It is assumed that $A_1, \dots A_n$ are *independent* screws, i.e. not themselves members of a screw system of order lower than n. If this were the case, the screws about which the body could be twisted would only consist of the members of that lower screw system.

Since the amplitudes of the n twists about $A_1, \dots A_n$ are arbitrary, it might be thought that there are n disposable quantities in the selection of a screw S from a screw system of the nth order. It is, however, obvious from § 14 that the determination of the position and pitch of S depends only upon the *ratios* of the amplitudes of the twists about $A_1, \ldots A_n$ and hence *in the selection of a screw from the screw system of the nth order, we have $n-1$ disposable quantities.*

70. Constraints.

An essential feature of a system of constraints consists in the number of independent quantities which are necessary to specify the position of the body when displaced in conformity with the requirements of the constraints. That number which cannot be less than one, nor greater than six, is the order of the freedom. To each of the six orders of freedom a certain type of screw system is appropriate.

The study of the six types of screw system as here defined is a problem of kinematics, but the statical and kinematical properties of screws are so interwoven that we derive great advantages by not attempting to relegate the statics and kinematics to different chapters. We shall not require any detailed examination of the constraints. Every conceivable condition of constraints must have been included when the six screw systems have been discussed in their most general form. Nor does it come within our scope, except on rare occasions, to specialize the enunciation of any problem, further than by mentioning the order of the freedom permitted to the body.

71. Screw Reciprocal to a System.

If a screw X be reciprocal to n independent screws, $A_1, \ldots A_n$, of a screw system of order n, then X is reciprocal to every other screw A which belongs to the same screw system. For, by the property of the screw system, it appears that twists of appropriate amplitudes about $A_1, \ldots A_n$, would compound into a twist about A. It follows (§ 69) that wrenches on $A_1, \ldots A_n$, of appropriate intensities (§ 30) compound into a wrench on A. Suppose these wrenches on $A_1, \ldots A_n$, were applied to a body only free to twist about X, then since X is reciprocal to $A_1, \ldots A_n$, the equilibrium of the body would be undisturbed. The resultant wrench on A must therefore be incapable of moving the body, therefore A and X must be reciprocal.

72. The Reciprocal Screw System.

All the screws which are reciprocal to a screw system P of order n constitute a screw system Q of order $6-n$. This important theorem is thus proved:—

Since only one condition is necessary for a pair of screws to be reciprocal, it follows, from the last section, that if a screw X be reciprocal to P it will fulfil n conditions. The screw X has, therefore, $5-n$ elements still disposable, and consequently ($n < 5$) an infinite number of screws Q can be found which are reciprocal to the screw system P. The theory of reciprocal screws will now prove that Q must really be a screw system of order $6-n$. In the first place it is manifest that Q must be a screw system of some order, for if a body be capable of twisting about even six independent screws, it must be perfectly free. Here, however, if a body were able to twist about the infinite number of screws embodied in Q, it would still not be free, because it would remain in equilibrium, though acted upon by a wrench about any screw of P. It follows that Q can only denote the collection of screws about which a body can twist which has some definite order of freedom. It is easily seen that that number must be $6-n$, for the number of constants disposable in the selection of a screw belonging to a screw system is one less than the order of the system (§ 36). But we have seen that the constants disposable in the selection of X are $5-n$, and, therefore, Q must be a screw system of order $6-n$.

We thus see, that *to any screw system P of order n corresponds a reciprocal screw system Q of order* $6-n$. Every screw of P is reciprocal to all the screws of Q, and vice versâ. This theorem provides us with a definite test as to whether any given screw α is a member of the screw system P. Construct $6-n$ screws of the reciprocal system. If then α be reciprocal to these $6-n$ screws, α must in general belong to P. We thus have $6-n$ conditions to be satisfied by any screw when a member of a screw system of order n.

73. Equilibrium.

If the screw system P expresses the freedom of a rigid body, then the body will remain in equilibrium though acted upon by a wrench on any screw of the reciprocal screw system Q. This is, perhaps, the most general theorem which can be enunciated with respect to the equilibrium of a rigid body. This theorem is thus proved:—Suppose a wrench to act on a screw η belonging to Q. If the body does not continue at rest, let it commence to twist about α. We would thus have a wrench about η disturbing a body which twists about α, but this is impossible, because α and η are reciprocal.

In the same manner it may be shown that a body which is free to twist about all the screws of Q will not be disturbed by a wrench about any screw of P. Thus, of two reciprocal screw systems, each expresses the locus of a wrench which is unable to disturb a body free to twist about any screw of the other.

74. Reaction of Constraints.

It also follows that the reactions of the constraints by which the movements of a body are confined to twists about the screws of a system P can only be wrenches on the reciprocal screw system Q, for the reactions of the constraints are only manifested by the success with which they resist the efforts of certain wrenches to disturb the equilibrium of the body.

75. Parameters of a Screw System.

We next consider the *number* of parameters required to specify a screw system of the nth order often called for brevity an n-system. Since the system is defined when n screws are given, and since five data are required for each screw, it might be thought that $5n$ parameters would be necessary. It must be observed, however, that the given $5n$ data suffice not only for the purpose of defining the screw system but also for pointing out n special screws upon the screw system, and as the pointing out of each screw on the system requires $n-1$ quantities (§ 69), it follows that the number of parameters actually required to define the system is only

$$5n - n(n-1) = n(6-n).$$

This result has a very significant meaning in connexion with the theory of reciprocal screw systems P and Q. Assuming that the order of P is n, the order of Q is $6-n$; but the expression $n(6-n)$ is unaltered by changing n into $6-n$. It follows that the number of parameters necessary to specify a screw system is identical with the number necessary to specify its reciprocal screw system. This remark is chiefly of importance in connexion with the systems of the fourth and fifth orders, which are respectively the reciprocal systems of a cylindroid and a single screw. We are now assured that a collection of all the screws which are reciprocal to an arbitrary cylindroid can be nothing less than a screw system of the fourth order in its most general type, and also, that all the screws in space which are reciprocal to a single screw must form the most general type of a screw system of the fifth order.

76. Applications of Co-ordinates.

If the co-ordinates of a screw satisfy n linear equations, the screw must belong to a screw system of the order $6-n$. Let η be the screw, and let one of the equations be

$$A_1\eta_1 + \ldots + A_6\eta_6 = 0,$$

whence η must be reciprocal to the screw whose co-ordinates are proportional to

$$\frac{A_1}{p_1}, \ldots \frac{A_6}{p_6}, \text{ (§ 37).}$$

It follows that η must be reciprocal to n screws, and therefore belong to a screw system of order $6-n$.

Let α, β, γ, δ be for example four screws about which a body receives twists of amplitudes α', β', γ', δ'. It is required to determine the screw ρ and the amplitude ρ' of a twist about ρ which will produce the same effect as the four given twists. We have seen (§ 37) that the twist about any screw α may be resolved in one way into six twists of amplitudes $\alpha'\alpha_1, \ldots \alpha'\alpha_6$, on the six screws of reference; we must therefore have

$$\rho'\rho_1 = \alpha'\alpha_1 + \beta'\beta_1 + \gamma'\gamma_1 + \delta'\delta_1,$$

$$\ldots\ldots\ldots\ldots\ldots\ldots\ldots\ldots$$

$$\rho'\rho_6 = \alpha'\alpha_6 + \beta'\beta_6 + \gamma'\gamma_6 + \delta'\delta_6,$$

whence ρ' and $\rho_1, \ldots \rho_6$ can be found (§ 35).

A similar process will determine the co-ordinates of the resultant of any number of twists, and it follows from § 12 that the resultant of any number of wrenches is to be found by equations of the same form. In ordinary mechanics, the conditions of equilibrium of any number of forces are six, viz. that each of the three forces, and each of the three couples, to which the system is equivalent shall vanish. In the present theory the conditions are likewise six, viz. that the intensity of each of the six wrenches on the screws of reference to which the given system is equivalent shall be zero.

Any screw will belong to a system of the nth order if it be reciprocal to $6-n$ independent screws; it follows that $6-n$ conditions must be fulfilled when $n+1$ screws belong to a screw system of the nth order.

To determine these conditions we take the case of $n = 3$, though the process is obviously general. Let $\alpha, \beta, \gamma, \delta$ be the four screws, then since twists of amplitudes $\alpha', \beta', \gamma', \delta'$ neutralise, we must have ρ' zero and hence the six equations

$$\alpha'\alpha_1 + \beta'\beta_1 + \gamma'\gamma_1 + \delta'\delta_1 = 0,$$

&c.

$$\alpha'\alpha_6 + \beta'\beta_6 + \gamma'\gamma_6 + \delta'\delta_6 = 0;$$

from any four of these equations the quantities $\alpha', \beta', \gamma', \delta'$ can be eliminated, and the result will be one of the three required conditions.

It is noticeable that the $6-n$ conditions are often presented in the evanescence of a single function, just as the evanescence of the sine of an angle between a pair of straight lines embodies the two conditions necessary that the direction cosines of the lines coincide. The function is suggested by the following considerations:—If $n+2$ screws belong to a screw system of the $(n+1)$th order, twists of appropriate amplitudes about the screws neutralise. The amplitude of the twist about any one screw must be proportional to a function of the co-ordinates of all the other screws. We thus see that the evanescence of one function must afford all that is necessary for $n+1$ screws to belong to a screw system of the nth order.*

* *Philosophical Transactions*, 1874, p. 23.

77. Remark on systems of Linear Equations.

Let a right line be, as usual, represented by the two equations

$$Ax + By + Cz + D = 0,$$
$$A'x + B'y + C'z + D' = 0.$$

There are here six independent constants involved, while a right line is completely defined by four constants. The fact of course is that these two equations not only determine the right line on which our attention is fixed, but they also determine two planes through that line. Four constants are needed for the straight line and one more for each of the planes, so that there are six constants in all.

If we are concerned with the straight line only the intrusion of two superfluous constants is often inconvenient. We can remove them by first eliminating y and then x, thus giving the two equations the form

$$x = Pz + Q,$$
$$y = P'z + Q'.$$

We have here no more than the four constants P, Q, P', Q', which are indispensable for the specification of the straight line.

Of course it may be urged that these equations also represent two planes. No doubt they do, but the equation $z = Px + Q$ is a plane parallel to the axis of y, which is absolutely determined when the straight line is known. The plane $Ax + By + Cz + D = 0$ may represent any one of the pencil of planes which can be drawn through the straight line.

Analogous considerations arise when the screws of an n-system are represented by a series of linear equations. We commence with the case of the two-system, in which of course the screws are limited to the generators of a cylindroid.

Let $\theta_1, \theta_2, \ldots \theta_6$ be the co-ordinates of a screw θ referred to any six screws of reference.

Let these co-ordinates satisfy the four linear equations

$$A_1\theta_1 + A_2\theta_2 + \ldots + A_6\theta_6 = 0,$$
$$B_1\theta_1 + B_2\theta_2 + \ldots + B_6\theta_6 = 0,$$
$$C_1\theta_1 + C_2\theta_2 + \ldots + C_6\theta_6 = 0,$$
$$D_1\theta_1 + D_2\theta_2 + \ldots + D_6\theta_6 = 0,$$

where A_1, A_2, ..., B_1, B_2, ... C_1, C_2 and D_1, D_2, ... are constants.

Then it is a fundamental part of the present Theory that the locus so defined is a cylindroid (§ 76).

But it will be observed that there is here a mass of not fewer than 20 independent constants, while the cylindroid is itself completely defined by eight constants (§ 75). The reason is that these four equations really each specify one screw, i.e. four screws in all, and as each screw needs five constants the presence of 20 constants is accounted for.

But when it is the cylindroid alone that we desire to specify there is no occasion to know these four particular screws. All we want is the system of the fourth order which contains those screws. For the specification of the position of a screw in a four-system three constants are required. Thus the selection of four screws in a given four-system requires 12 constants. These subtracted from 20 leave just so many as are required for the cylindroid.

This is of course the interpretation of the process of solving for $\theta_3, \theta_4, \theta_5, \theta_6$ in terms of θ_1 and θ_2. We get

$$\theta_3 = P\theta_1 + Q\theta_2; \ \theta_4 = P'\theta_1 + Q'\theta_2; \ \theta_5 = P''\theta_1 + Q''\theta_2; \ \theta_6 = P'''\theta_1 + Q'''\theta_2.$$

Thus we find that the constants are now reduced to eight, which just serve to specify the cylindroid.

An instructive case is presented in the case of the three-system. The three linear equations of the most general type contain 15 constants. But a three-system is defined by 9 constants (§ 75). This is illustrated by solving the equations for $\theta_2, \theta_4, \theta_6$ in terms of $\theta_1, \theta_3, \theta_5$, when we have

$$\theta_2 = P\theta_1 + Q\theta_3 + R\theta_5,$$
$$\theta_4 = P'\theta_1 + Q'\theta_3 + R'\theta_5,$$
$$\theta_6 = P''\theta_1 + Q''\theta_3 + R''\theta_5.$$

This symmetrical process is specially convenient when the screws of reference are six canonical co-reciprocals.

The general theory may also be set down. An n-system of screws is defined by $6 - n$ linear equations. These contain $5(6 - n) = 30 - 5n$ constants. We can, however, solve for $6 - n$ of the variables in terms of the remaining n. Thus we get $6 - n$ equations, each of which has n constants, i.e. $n(6 - n)$ in all. This is just the number of constants necessary to specify an n-system. The original number in the equation $30 - 5n$ may be written

$$n(6 - n) + (6 - n)(5 - n).$$

The redundancy of $(6-n)(5-n)$ expresses the number of constants necessary for specifying $6 - n$ screws in a system of the $(6 - n)$th order.

CHAPTER VII.

THE PRINCIPAL SCREWS OF INERTIA*.

78. Introduction.

If a rigid body be free to rotate about a fixed point, then it is well known that an impulsive couple about an axis parallel to one of the principal axes which can be drawn through the point will make the body commence to rotate about that axis. Suppose that there was on one of the principal axes a screw η with a very small pitch, then a twisting motion about η would closely resemble a simple rotation about the corresponding axis. An impulsive wrench on η (i.e. a wrench of great intensity acting for a small time) will reduce to a couple when compounded with the necessary reaction of the fixed point. If we now suppose the pitch of η to be evanescent, we may still assert that an impulsive wrench on η of very great intensity will cause the body, if previously quiescent, to commence to twist about η.

We have stated a familiar property of the principal axes in this indirect manner, for the purpose of showing that it is merely an extreme case for a body with freedom of the third order of the following general theorem:—

If a quiescent rigid body have freedom of the nth order, then n screws can always be found (but not generally more than n), such that if the body receive an impulsive wrench on any one of these screws, the body will commence to twist about the same screw.

These n screws are of great significance in the present method of studying Dynamics, and they may be termed the *principal screws of inertia*. In the present chapter we shall prove the general theorem just stated, while in the chapters on the special orders of freedom we shall show how the principal screws of inertia are to be determined for each case.

* *Philosophical Transactions*, 1874, p. 27.

79. Screws of Reference.

We have now to define the group of six co-reciprocal screws (§ 31) which are peculiarly adapted to serve as the screws of reference in Kinetic investigations. Let O be the centre of inertia of the rigid body, and let OA, OB, OC be the three principal axes through O, while a, b, c are the corresponding radii of gyration. Then two screws along OA, viz. ω_1, ω_2, with pitches $+a$, $-a$; two screws along OB, viz. ω_3, ω_4, with pitches $+b$, $-b$, and two along OC, viz. ω_5, ω_6, with pitches $+c$, $-c$, are the co-reciprocal group which we shall employ. The group thus indicated form of course a set of canonical co-reciprocals (§ 41). For convenience in writing the formulae, we shall often use $p_1, \ldots p_6$, to denote the pitches as before.

We shall first prove that the six screws thus defined are the principal screws of inertia of the rigid body when perfectly free. Let the mass of the body be M, and let a great constant wrench on ω_1 act for a short time e. The intensity of this wrench is ω_1'', and the moment of the couple is $a\omega_1''$. We now consider the effect of the two portions of the wrench separately. The effect of the force ω_1'' is to give the body a velocity of translation parallel to OA and equal to $\frac{e}{M}\omega_1''$. The effect of the couple is to impart an angular velocity $\dot{\omega}_1$ about the axis OA. This angular velocity is easily determined. The effective force which must have acted upon a particle dm at a perpendicular distance r from OA is $\frac{r\dot{\omega}_1}{e}dm$. The sum of the moments of all these forces is $Ma^2\frac{\dot{\omega}_1}{e}$. This quantity is equal to the moment of the given couple so that

$$Ma^2\frac{\dot{\omega}_1}{e} = a\omega_1'',$$

whence

$$\dot{\omega}_1 = \frac{e}{aM}\omega_1''.$$

The total effect of the wrench on ω_1 is, therefore, to give the body a velocity of translation parallel to OA, and equal to $\frac{e}{M}\omega_1''$, and also a velocity of rotation about OA equal to $\frac{e}{aM}\omega_1''$. These movements unite to form a twisting motion about a screw on OA, of which the pitch, found by dividing the velocity of translation by the velocity of rotation, is equal to a. This same quantity is however the pitch of ω_1, and thus it is proved that an impulsive wrench on ω_1 will make the body commence to twist about ω_1. We shall in future represent $e\omega_1''$ by the symbol ω_1''', which is accordingly to express the *intensity of the impulsive wrench.*

80. Impulsive Screws and Instantaneous Screws.

If a free quiescent rigid body receive an impulsive wrench on a screw η, the body will immediately commence to twist about an instantaneous screw α. The co-ordinates of α being given for the six screws of reference just defined, we now seek the coordinates of η.

The impulsive wrench on η of intensity η''' is to be decomposed into components of intensities $\eta'''\eta_1, \ldots \eta'''\eta_6$ on $\omega_1, \ldots \omega_6$. The component on ω_n will generate a twist velocity about ω_n amounting to

$$\frac{1}{M}\frac{\eta'''\eta_n}{p_n},$$

but if $\dot{\alpha}$ be the twist velocity about α which is finally produced, the expression just written must be equal to $\dot{\alpha}\alpha_n$, and hence we have the following useful result:—

If the co-ordinates of the instantaneous screw be proportional to $\alpha_1, \ldots \alpha_6$, then the co-ordinates of the corresponding impulsive screw are proportional to $p_1\alpha_1, \ldots p_6\alpha_6$.

81. Conjugate Screws of Inertia.

Let α be the instantaneous screw about which a quiescent body either free or constrained in any way will commence to twist in consequence of receiving an impulsive wrench on any screw whatever η. Let β be the instantaneous screw in like manner related to another impulsive screw ζ.

We have to prove that if ζ be reciprocal to α then shall η be reciprocal to β.

When the body receives an impulsive wrench on ζ of intensity ζ''' there is generally a simultaneous reaction of the constraints, which takes the form of an impulsive wrench of intensity μ''' on a screw μ. The effect on the body is therefore the same as if the body had been free, but had received an impulsive wrench of which the component wrench on the first screw of reference had the intensity $\zeta'''\zeta_1 + \mu'''\mu_1$. This and the similar quantities will be proportional to the co-ordinates of the impulsive screw which had the body been perfectly free would have β as an instantaneous screw. These latter, as we have shown in § 80, are proportional to $p_1\beta_1, p_2\beta_2 \ldots p_6\beta_6$. Hence it follows that, h being some quantity differing from zero, we have

$$\zeta'''\zeta_1 + \mu'''\mu_1 = hp_1\beta_1,$$
$$\zeta'''\zeta_2 + \mu'''\mu_2 = hp_2\beta_2,$$
$$\cdots\cdots\cdots\cdots\cdots\cdots$$
$$\zeta'''\zeta_6 + \mu'''\mu_6 = hp_6\beta_6.$$

Multiplying the first of these equations by $p_1\alpha_1$, the second by $p_2\alpha_2$, &c. adding the six products and remembering that α and ζ are reciprocal by hypothesis while α and μ are reciprocal by the nature of the reactions of the constraints, we have

$$p_1^2\alpha_1\beta_1 + \ldots + p_6^2\alpha_6\beta_6 = 0.$$

The symmetry of this equation shows that in this case η must be reciprocal to β. Hence we have the following theorem which is of fundamental importance in the subject of the present volume.

If α be the instantaneous screw about which a quiescent rigid body either perfectly free or constrained in any manner whatever commences to twist in consequence of an impulsive wrench on some screw η, and if β be another instantaneous screw, similarly related to an impulsive screw ζ, then whenever ζ is reciprocal to α we shall find that η is reciprocal to β.

When this relation is fulfilled the screws α and β are said to be *conjugate screws of inertia.*

82. The Determination of the Impulsive Screw, corresponding to a given instantaneous screw, is a definite problem when the body is perfectly free. If, however, the body be constrained, we shall show that *any* screw selected from a certain screw system will, in general, fulfil the required condition.

Let $B_1, \ldots B_{6-n}$ be $6 - n$ screws selected from the screw system which is reciprocal to that corresponding to the freedom of the nth order possessed by the rigid body. Let S be the screw about which the body is to twist. Let X be *any* one of the screws, an impulsive wrench about which would make the body twist about S; then any screw Y belonging to the screw system of the $(7 - n)$th order, specified by the screws, $X, B_1, \ldots B_{6-n}$ is an impulsive screw, corresponding to S as an instantaneous screw. For the wrench on Y may be resolved into $7 - n$ wrenches on $X, B_1, \ldots B_{6-n}$; of these, all but the first are instantly destroyed by the reaction of the constraints, so that the wrench on Y is practically equivalent to the wrench on X, which, by hypothesis, will make the body twist about S.

As an example:—if the body had freedom of the fifth order, then an impulsive wrench on *any* screw on a certain cylindroid will make the body commence to twist about a given screw.

As another example:—if a body have freedom of the third order, then the "locus" of an impulsive wrench which would make the body twist about a given screw consists of all the screws in space which are reciprocal to a certain cylindroid.

83. System of Conjugate Screws of Inertia.

We shall now show that from the screw system of the nth order P, which expresses the freedom of the rigid body, generally n screws can be selected so that every pair of them are conjugate screws of inertia (§ 81). Let B_1, &c. B_{6-n} be $(6-n)$ screws defining the reciprocal screw system. Let A_1 be any screw belonging to P. Then in the choice of A_1 we have $n-1$ arbitrary quantities. Let I_1 be any impulsive screw corresponding to A_1 as an instantaneous screw. Choose A_2 reciprocal to $I_1, B_1, \ldots B_{6-n}$, then A_1 and A_2 are conjugate screws, and in the choice of the latter we have $n-2$ arbitrary quantities. Let I_2 be any impulsive screw corresponding to A_2 as an instantaneous screw. Choose A_3 reciprocal to $I_1, I_2, B_1, \ldots B_{6-n}$, and proceed thus until A_n has been attained, then each pair of the group A_1, &c. A_n are conjugate screws of inertia. The number of quantities which remain arbitrary in the choice of such a group amount to

$$n-1+n-2+\ldots+1=\frac{n(n-1)}{2},$$

or exactly half the total number of arbitrary constants disposable in the selection of any n screws from a system of the nth order.

84. Principal Screws of Inertia.

We have now to prove the important theorem in Dynamics which affirms the existence of n principal Screws of Inertia in a rigid body with n degrees of freedom.

The proof that we shall give is, for the sake of convenience, enunciated with respect to the freedom of the third order, but the same method applies to each of the other degrees of freedom.

Let θ be one of the principal screws of inertia, then an impulsive wrench on θ must make the body commence to twist about θ. In the most general case when the body is submitted to constraint, the impulsive wrench on θ will of course be compounded with the reaction on some screw λ of the reciprocal system. The result will be to produce the impulsive wrench which would, if the body had been free, have generated an instantaneous twist velocity about θ.

We thus have the following equations (§ 80) where x and y are unknown:

$$p_1\theta_1 = x\theta_1 + y\lambda_1,$$

$$p_2\theta_2 = x\theta_2 + y\lambda_2,$$

$$\cdots\cdots\cdots\cdots$$

$$p_6\theta_6 = x\theta_6 + y\lambda_6.$$

Let α be one of the screws of the three-system in question. Then since λ must be reciprocal to α we have by multiplying these equations respectively by $p_1\alpha_1, \ldots p_6\alpha_6$ and adding,

$$p_1^2\alpha_1\theta_1 + p_2^2\alpha_2\theta_2 + \ldots + p_6^2\alpha_6\theta_6 = xp_1\alpha_1\theta_1 + xp_2\alpha_2\theta_2 + \ldots + xp_6\alpha_6\theta_6.$$

In like manner if β, γ be two other screws of the three-system,

$$p_1^2\beta_1\theta_1 + p_2^2\beta_2\theta_2 + \ldots + p_6^2\beta_6\theta_6 = xp_1\beta_1\theta_1 + xp_2\beta_2\theta_2 \ldots + xp_6\beta_6\theta_6,$$

$$p_1^2\gamma_1\theta_1 + p_2^2\gamma_2\theta_2 + \ldots + p_6^2\gamma_6\theta_6 = xp_1\gamma_1\theta_1 + xp_2\gamma_2\theta_2 \ldots + xp_6\gamma_6\theta_6.$$

But as θ belongs to the three-system its co-ordinates must satisfy three linear equations. These we may take to be

$$F_1\theta_1 + F_2\theta_2 + \ldots + F_6\theta_6 = 0,$$

$$G_1\theta_1 + G_2\theta_2 + \ldots + G_6\theta_6 = 0,$$

$$H_1\theta_1 + H_2\theta_2 + \ldots + H_6\theta_6 = 0.$$

We have thus six linear equations in the co-ordinates of θ. We can therefore eliminate those co-ordinates, thus obtaining a determinantal equation which gives a cubic for x.

The three roots of this cubic will give accordingly three screws in the three-system which possess the required property.

Thus we demonstrate that in any three-system there are three principal screws of inertia, and a precisely similar proof for each of the six values of n establishes by induction the important theorem that there are n principal screws of inertia in the screw system of the nth order. It is shown in § 86 that all the roots are real.

We shall now prove that the Principal Screws of Inertia are co-reciprocal. Let θ and ϕ be two such screws, corresponding to different roots x', x'' of the equation in x.

Then we have

$$\theta_1 = \frac{y\lambda_1}{p_1 - x'}, \quad \theta_2 = \frac{y\lambda_2}{p_2 - x'}, \quad \ldots \quad \theta_6 = \frac{y\lambda_6}{p_6 - x'}.$$

Let μ be the screw of the reciprocal system on which the impulsive wrench is generated by the impulse given on ϕ.

Then

$$\phi_1 = \frac{y\mu_1}{p_1 - x''}, \quad \phi_2 = \frac{y\mu_2}{p_2 - x''}, \quad \ldots \quad \phi_6 = \frac{y\mu_6}{p_6 - x''}.$$

As μ is reciprocal to θ and λ is reciprocal to ϕ, we have

$$\frac{p_1\lambda_1\mu_1}{p_1 - x'} + \frac{p_2\lambda_2\mu_2}{p_2 - x'} + \ldots + \frac{p_6\lambda_6\mu_6}{p_6 - x'} = 0,$$

$$\frac{p_1\lambda_1\mu_1}{p_1 - x''} + \frac{p_2\lambda_2\mu_2}{p_2 - x''} + \ldots + \frac{p_6\lambda_6\mu_6}{p_6 - x''} = 0.$$

Subtracting these equations and discarding the factor $x' - x''$, we get

$$\frac{p_1\lambda_1\mu_1}{(p_1 - x')(p_1 - x'')} + \frac{p_2\lambda_2\mu_2}{(p_2 - x')(p_2 - x'')} + \ldots + \frac{p_6\lambda_6\mu_6}{(p_6 - x')(p_6 - x'')} = 0,$$

which is of course

$$\Sigma p_1 \theta_1 \phi_1 = 0;$$

whence θ and ϕ are reciprocal, and the same being true for each pair of principal screws of inertia we thus learn that they form a co-reciprocal system.

We can also show that each pair of the Principal Screws of Inertia are Conjugate Screws of Inertia.

It is easy to see that

$$\frac{x'}{x' - x''} \Sigma \frac{p_1 \lambda_1 \mu_1}{p_1 - x'} + \frac{x''}{x'' - x'} \Sigma \frac{p_1 \lambda_1 \mu_1}{p_1 - x''} = \Sigma \frac{p_1^2 \lambda_1 \mu_1}{(p_1 - x')(p_1 - x'')}.$$

As each of the terms on the left-hand side of this equation is zero, the expression on the right-hand is also zero, but this is equivalent to

$$\Sigma p_1^2 \theta_1 \phi_1 = 0;$$

whence we show that θ and ϕ are conjugate screws of inertia and the required theorem has been proved.

85. An algebraical Lemma.

Let U and V be two homogeneous functions of the second degree in n variables. If *either* U or V be of such a character that it could be expressed by linear transformation as the *sum* of n squares, then the discriminant of $U + \lambda V$ when equated to zero gives an equation of the nth degree in λ of which all the roots are real*.

Suppose that V can by linear transformation assume the form

$$x_1^2 + x_2^2 \ldots + x_n^2,$$

and adopt $x_1, x_2 \ldots x_n$ as new variables, so that

$$U = a_{11} x_1^2 + a_{22} x_2^2 + 2a_{12} x_1 x_2 \ldots .$$

The discriminant of $U + \lambda V$ will, when equated to zero, give the equation for λ,

$$\begin{vmatrix} a_{11} + \lambda, & a_{12} & , \ldots a_{1n} \\ a_{21} & , a_{22} + \lambda, & \ldots a_{2n} \\ \vdots & \vdots & \vdots \\ a_{n1} & , a_{n2} & , \ldots a_{nn} + \lambda \end{vmatrix} = 0,$$

and the discriminant being an invariant the roots of this equation will be

* A discussion is found in Zanchevsky, *Theory of Screws and its application to Mechanics*, Odessa, 1889. Mr G. Chawner has most kindly translated the Russian for me.

the same as those of the original equation. The required theorem will therefore be proved if it can be shown that all the roots of this equation are real. That this is so is shown in Salmon's *Modern Higher Algebra*, Lesson VI.*

86. Another investigation of the Principal Screws of Inertia.

The n Principal Screws of Inertia can also be investigated in the following fundamental manner by the help of Lagrange's equations of motion in generalized co-ordinates.

Let $\zeta_1 \ldots \zeta_n$ be the co-ordinates (§ 95) of the impulsive screw. Let $\phi_1, \ldots \phi_n$ be the co-ordinates of the body, then $\dot{\phi}_1, \ldots \dot{\phi}_n$ will be the co-ordinates of the instantaneous screw, and from Lagrange's equations,

$$\frac{d}{dt}\left(\frac{dT}{d\dot{\phi}_1}\right) - \frac{dT}{d\phi_1} = P_1,$$

where T is the kinetic energy and where $P_1 \delta\phi_1$ denotes the work done in a twist $\delta\phi_1$ against the wrench.

If the screws of reference be co-reciprocal and if ζ'' be the intensity of a wrench on ζ, then

$$P_1 = 2p_1\zeta''\zeta_1.$$

As we are considering the action of only an impulsive wrench the effect of which is to generate a finite velocity in an infinitely small time we must have the acceleration infinitely great while the wrench is in action. The term $\frac{dT}{d\phi_1}$ is therefore negligible in comparison with $\frac{d}{dt}\left(\frac{dT}{d\dot{\phi}_1}\right)$ and hence for the impulsive motion†

$$\frac{d}{dt}\left(\frac{dT}{d\dot{\phi}_1}\right) = 2p_1\zeta_1\zeta''.$$

We may regard ζ_1 and ζ'' as both constant during the indefinitely small time e of operation of the impulsive wrench, whence (§ 79)

$$2\zeta_1\,\zeta''' = \frac{1}{p_1}\frac{dT}{d\dot{\phi}_1}.$$

Hence replacing $\dot{\phi}_1, \ldots \dot{\phi}_n$ by $\theta_1, \ldots \theta_n$ we deduce the following (§§ 95, 96).

If T be the kinetic energy of a body with freedom of the nth order, twisting about a screw θ whose co-ordinates referred to any n co-reciprocals belonging to the system expressing the freedom are $\theta_1, \ldots \theta_n$, then the co-ordinates

* See also Williamson and Tarleton's *Dynamics*, 2nd edition, p. 457 (1889), and Routh's *Rigid Dynamics*, Part II, p. 49 (1892).

† Niven, *Messenger of Math.*, May 1867, quoted by Routh, *Rigid Dynamics*, Part I, pp. 327–8.

of an impulsive wrench by which the actual motion of the body could be produced are proportional to

$$\frac{1}{p_1}\frac{dT}{d\theta_1}, \dots \frac{1}{p_n}\frac{dT}{d\theta_n}.$$

The existence of n Principal Screws of Inertia can now be readily deduced, for suppose that

$$-2\lambda\theta_1 = \frac{1}{p_1}\frac{dT}{d\theta_1}, \dots -2\lambda\theta_n = \frac{1}{p_n}\frac{dT}{d\theta_n},$$

where λ is an unknown factor. If then we make

$$T = a_{11}\theta_1^2 + a_{22}\theta_2^2 + 2a_{12}\theta_1\theta_2 \dots$$

we have an equation of the nth degree for λ as follows:

$$\left|\begin{array}{llll} a_{11}+p_1\lambda, & a_{12} & ,\dots & a_{1n} \\ a_{21} & , a_{22}+p_2\lambda, & \dots & a_{2n} \\ \vdots & \vdots & & \vdots \\ a_{n1} & , a_{n2} & ,\dots & a_{nn}+p_n\lambda \end{array}\right| = 0.$$

It is essential to note that T is a function of such a character that by linear transformation it can be expressed as the *sum* of n squares, for suppose it could be expressed as

$$H_1^2 + H_2^2 \dots - H_n^2,$$

it would be possible to find a real screw which made $H_1, H_2, \dots H_{n-1}$ each zero, and then the kinetic energy of the body twisting about that screw would be negative. Of course this is impossible. Hence we deduce from § 85 the important principle that all the Principal Screws of Inertia are real.

If the equation had a repeated root the number of Principal Screws of Inertia is infinite. We take $n=4$, but the argument applies to 3 and 2 also. (There can be no repeated root when n is either 5 or 6. See chaps. XVII. and XVIII.) We can choose variables such that T becomes

$$M\dot{\theta}^2(u_1^2\theta_1^2 + u_2^2\theta_2^2 + u_3^2\theta_3^2 + u_4^2\theta_4^2),$$

and the pitch λ becomes simultaneously

$$p_1\theta_1^2 + p_2\theta_2^2 \dots + p_4\theta_4^2.$$

If therefore the discriminant of $T + \lambda p$, equated to zero, has a pair of equal values for λ, we must have a condition like

$$\frac{u_1^2}{p_1} = \frac{u_2^2}{p_2}.$$

Take any screw of the system for which $\theta_3 = 0$, $\theta_4 = 0$, then

$$T = M\dot{\theta}^2(u_1^2\theta_1^2 + u_2^2\theta_2^2),$$

$$p = p_1\theta_1^2 + p_2\theta_2^2,$$

or

$$T = \frac{M\dot{\theta}^2 u_1^2}{p_1} p.$$

Hence we find that for all screws on the cylindroid represented by $\theta_1, \theta_2, 0, 0$ the energy will vary as the pitch when the twist velocity remains the same. It appears from the representation of the Dynamical problem in chap. XII. that in this case *all* the screws of the cylindroid $\theta_1, \theta_2, 0, 0$ must be principal screws of Inertia. The number of principal screws of inertia is therefore infinite in this case. (See Routh's Theorem, Appendix, Note 2.)

87. Enumeration of Constants.

It is the object of this article to show that there are sufficient constants available to permit us to select from the screw system of the nth order expressing the freedom of a rigid body, one group of n screws, of which every pair are both conjugate and reciprocal, and that these constitute the principal screws of inertia (§ 78).

To prove this, it is sufficient to show that when half the available constants have been disposed of in making the n screws conjugate (§ 81) the other half admit of adjustment so as to make the screws also co-reciprocal. Choose A_1 reciprocal to $B_1, \ldots B_{6-n}$, with $n-1$ arbitrary quantities; A_2 reciprocal to $A_1, B_1, \ldots B_{6-n}$, with $n-2$ arbitrary quantities, and so on, then the total number of arbitrary quantities in the choice of n co-reciprocal screws from a system of the nth order is

$$n-1+n-2\ldots+1=\frac{n(n-1)}{2}.$$

Hence, by suitable disposition of the $n(n-1)$ constants it might be anticipated that we can find at least one group of n screws which are both conjugate and co-reciprocal.

We have now to show that these screws would be the principal screws of inertia (§ 78). We shall state the argument for the freedom of the third order, the argument for any other order being precisely similar.

Let A_1, A_2, A_3 be the three conjugate and co-reciprocal screws which can be selected from a system of the third order. Let B_1, B_2, B_3 be any three screws belonging to the reciprocal screw system. Let R_1, R_2, R_3 be *any* three impulsive screws corresponding respectively to A_1, A_2, A_3 as instantaneous screws.

An impulsive wrench on any screw belonging to the screw system of the 4th order defined by R_1, B_1, B_2, B_3 will make the body twist about A_1 (§ 82), but the screws of such a system are reciprocal to A_2 and A_3; for since A_1 and A_2 are conjugate, R_1 must be reciprocal to A_2 (§ 81), and also to A_3, since A_1 and A_3 are conjugate. It follows from this that an impulsive wrench on any screw reciprocal to A_2 and A_3 will make the body commence to twist about A_1, but A_1 is itself reciprocal to A_2 and A_3, and hence an impulsive wrench on A_1 will make the body commence to twist about A_1. Hence A_1 and also A_2 and A_3 are principal screws of inertia.

We shall now show that with the exception of the n screws here determined, generally no other screw possesses the property. Suppose another screw S were to possess this property. Decompose the wrench on S into n wrenches of intensities S_1'', ... S_n'' on A_1, ... A_n; this must be possible, because if the body is to be capable of twisting about S this screw must belong to the system specified by A_1, ... A_n. The n impulsive wrenches on A_1, ... A_n will produce twisting motions about the same screws, but these twisting motions are to compound into a twisting motion on S. It follows that the component twist velocities $\dot{S}_1$, ... $\dot{S}_n$ must be proportional to the intensities S_1'', ... S_n''. But if this were the case, then every screw of the system would be a principal screw of inertia; for let X be any impulsive screw of the system, and suppose that Y is the corresponding instantaneous screw, the components of X on A_1, ... A_n, have intensities X_1'', ... X_n'', these will generate twist velocities equal to

$$\frac{\dot{S}_1}{S_1''} X_1'', \ldots \frac{\dot{S}_n}{S_n''} X_n'',$$

and these quantities must equal the components of the twist velocity about Y. But the ratios

$$\frac{\dot{S}_1}{S_1''}, \ldots \frac{\dot{S}_n}{S_n''}$$

are all equal, and hence the twist velocities of the components on the screws of reference of the twisting motion about Y must be proportional to the intensities of the components on the same screws of reference of the wrench on X. Remembering that twisting motions and wrenches are compounded by the same rules, it follows that Y and X must be identical.

As it is not generally true that all the screws of the system defining the freedom possess the property enjoyed by a principal screw of inertia, it follows that the number of principal screws of inertia must be generally equal to the order of the freedom.

88. Kinetic Energy.

The twisting motion of a rigid body with freedom of the nth order may be completely specified by the twist velocities of the components of the twisting motion on any n screws of the system defining the freedom. If the screws of reference be a set of conjugate screws of inertia, the expression for the kinetic energy of the body consists of n square terms. This will now be proved.

If a free or constrained rigid body be at rest in a position L, and if the body receive an impulsive wrench, the body will commence to twist about a screw α with a kinetic energy E_α. Let us now suppose that a second impulsive wrench acts upon the body on a screw μ, and that if the body had been at rest in the position L, it would have commenced to twist about a screw β, with a kinetic energy E_β.

We are to consider how the amount of energy acquired by the second impulse is affected by the circumstance that the body is then not at rest in L, but is moving through L in consequence of the former impulse. The amount will in general differ from E_β, for the movement of the body may cause it to do work against the wrench on μ during the short time that it acts, so that not only will the body thus expend some of the kinetic energy which it previously possessed, but the efficiency of the impulsive wrench on μ will be diminished. Under other circumstances the motion through A might be of such a character that the impulsive wrench on μ *acting for a given time* would impart to the body a larger amount of kinetic energy than if the body were at rest. Between these two cases must lie the intermediate one in which the kinetic energy imparted is precisely the same as if the body had been at rest. It is obvious that this will happen if each point of the body at which the forces of the impulsive wrench are applied be moving in a direction perpendicular to the corresponding force, or more generally if the screw α about which the body is twisting be reciprocal to μ. When this is the case α and β must be conjugate screws of inertia (§ 81), and hence we infer the following theorem:—

If the kinetic energy of a body twisting about a screw α with a certain twist velocity be E_α, and if the kinetic energy of the same body twisting about a screw β with a certain twist velocity be E_β, then when the body has a motion compounded of the two twisting movements, its kinetic energy will amount to $E_\alpha + E_\beta$ provided that α and β are conjugate screws of inertia.

Since this result may be extended to any number of conjugate screws of inertia, and since the terms E_α, &c., are essentially positive, the required theorem has been proved.

89. Expression for Kinetic Energy.

If a rigid body have a twisting motion about a screw α, with a twist velocity $\dot{\alpha}$, what is the expression of its kinetic energy in terms of the co-ordinates of α?

We adopt as the unit of force that force which acting upon the unit of mass for the unit of time will give the body a velocity which would carry it over the unit of distance in the unit of time. The unit of energy is the work done by the unit force in moving over the unit distance. If, therefore, a body of mass M have a movement of translation with a velocity v its kinetic energy expressed in these units is $\frac{1}{2}Mv^2$.

The movement is to be decomposed into twisting motions about the screws of reference ω_1, &c. ω_6, the twist velocity of the component on ω_n being $\dot{\alpha}\alpha_n$. One constituent of the twisting motion about ω_m consists of a velocity of translation equal to $\dot{\alpha}p_n\alpha_n$, and on this account the body has a kinetic energy equal to $\frac{1}{2}M\dot{\alpha}^2p_n^2\alpha_n^2$. On account of the rotation around the axis with an angular velocity $\dot{\alpha}\alpha_n$ the body has a kinetic energy equal to

$$\tfrac{1}{2}\dot{\alpha}^2\alpha_n^2\int r^2dm$$

where r denotes the perpendicular from the element dM on ω_m. Remembering that p_m is the radius of gyration this expression also reduces to $\frac{1}{2}M\dot{\alpha}^2p_m^2\alpha_m^2$, and hence the total kinetic energy of the twisting motion about ω_m is

$$M\dot{\alpha}^2p_n^2\alpha_n^2.$$

We see, therefore (§ 88), that the kinetic energy due to the twisting motion about α is

$$M\dot{\alpha}^2(p_1^2\alpha_1^2 + \dots + p_6^2\alpha_6^2).$$

The quantity inside the bracket is the square of a certain *linear magnitude* which is determined by the distribution of the material of the body with respect to the screw α. It will facilitate the kinetic applications of the present theory to employ *the symbol u_α to denote this quantity.* It is then to be understood that the kinetic energy of a body of mass M, animated by a twisting motion about the screw α with a twist velocity $\dot{\alpha}$, is represented by

$$M\dot{\alpha}^2u_\alpha^2.$$

90. Twist Velocity acquired by an Impulsive Wrench.

A body of mass M, which is only free to twist about a screw α, is acted upon by an impulsive wrench of intensity η''' on a screw η. It is required to find the twist velocity $\dot{\alpha}$ which is acquired.

The initial reaction of the constraints is an impulsive wrench of intensity λ''' on a screw λ. Then the body moves as if it were free, but had been acted upon by an impulsive wrench of which the component on ω_m had the intensity

$$\eta'''\eta_n + \lambda'''\lambda_n.$$

This component would generate a velocity of translation parallel to ω_n and equal to $\frac{1}{M}(\eta'''\eta_n + \lambda'''\lambda_n)$. The twist velocity about ω_n produced by this component is found by dividing the velocity of translation by p_n. On the other hand, since the co-ordinates of the screw α are $\alpha_1, \ldots \alpha_6$, the twist velocity about ω_n may also be represented by $\dot{\alpha}\alpha_n$ (§ 34), whence

$$\dot{\alpha}\alpha_n = \frac{1}{M}\frac{\eta'''\eta_n + \lambda'''\lambda_n}{p_n}.$$

If we multiply this equation by $p_n{}^2\alpha_n$, add the six equations found by giving n all values from 1 to 6, and remember that α and λ are reciprocal, we have (§ 39)

$$\dot{\alpha}u_\alpha{}^2 = \frac{1}{M}\eta'''\varpi_{\eta\alpha};$$

whence $\dot{\alpha}$ is determined.

This expression shows that the twist velocity produced by an impulsive wrench on a given rigid body constrained to twist about a given screw, varies directly as the virtual coefficient and the intensity of the impulsive wrench, and inversely as the square of u_α. (See Appendix, Note 3.)

91. The Kinetic Energy acquired by an Impulsive Wrench can be easily found by § 89; for, from the last equation,

$$M\dot{\alpha}^2 u_\alpha{}^2 = \frac{1}{M}\frac{\eta'''^2}{u_\alpha{}^2}\varpi_{\eta\alpha}{}^2;$$

hence the kinetic energy produced by the action of an impulsive wrench on a body constrained to twist about a given screw varies directly as the product of the square of the virtual coefficient of the two screws and the square of the intensity of the impulsive wrench, and inversely as the square of u_α.

92. Formula for a free body.

We shall now express the kinetic energy communicated by the impulsive wrench on η to the body when perfectly free. The component on ω_n of intensity $\eta'''\eta_n$ imparts a kinetic energy equal to

$$\frac{1}{M}\eta'''^2\eta_n{}^2;$$

whence the total kinetic energy is found by adding these six terms.

The difference between the kinetic energy acquired when the body is perfectly free, and when the body is constrained to twist about α, is equal to

$$\frac{1}{M}\frac{\eta'''^2}{u_\alpha{}^2}[u_\alpha{}^2\{\Sigma(\eta)^2\} - \{\Sigma(p_1\alpha_1\eta_1)\}^2].$$

The quantity inside the bracket reduces to the sum of 15 square terms, of which $(p_1\alpha_1\eta_2 - p_2\alpha_2\eta_1)^2$ is a specimen. The entire expression being therefore essentially positive shows that a given impulsive wrench imparts greater energy to a quiescent body when free than to the same quiescent body when constrained to twist about a certain screw.

93. Lemma.

If a group of instantaneous screws belong to a system of the nth order, then the body being quite free the corresponding group of impulsive screws also belong to a system of the nth order; for, suppose that $n+1$ twisting motions about $n+1$ screws neutralise, then the corresponding $n+1$ impulsive wrenches must equilibrate, but this would not be possible unless all the impulsive screws belonged to a screw system of the nth order.

94. Euler's Theorem.

If a free or constrained rigid body is acted upon by an impulsive wrench, the body will commence to move with a larger kinetic energy when it is permitted to select its own instantaneous screw from the screw system P defining the freedom, than it would have acquired, had it been arbitrarily restricted to any other screw of the system.

Let Q be the reciprocal system of the $(6-n)$th order, and let P' be the screw system of the nth order, consisting of those impulsive screws which, if the body were free, would correspond to the screws of P as instantaneous screws.

Let η be any screw on which the body receives an impulsive wrench. Decompose this wrench into components on a system of six screws consisting of any n screws from P', and any $6-n$ screws from Q. The latter are neutralised by the reactions of the constraints, and may be omitted, while the former compound into one wrench on a screw ζ belonging to P'; we may therefore replace the given wrench by a wrench on ζ. If the body were perfectly free, an impulsive wrench on ζ must make the body twist about some screw α on P. In the present case, although the body is not perfectly free, yet it is free so far as twisting about α is concerned, and we may therefore, with reference to this particular impulse about ζ, consider the body as being perfectly free. It follows from § 92 that there would be a loss of energy if the body were compelled to twist about any screw other than α, which is the one it naturally chooses.

95. Co-ordinates in a Screw System.

The co-ordinates of a screw belonging to a specified screw-system can be greatly simplified by taking n co-reciprocal screws belonging to the given screw system as a portion of the six screws of reference. The remaining $6-n$ screws of reference must then belong to the reciprocal screw system. It follows that out of the six co-ordinates $\alpha_1, \ldots \alpha_6$ of a screw α, which belongs to the system, $6-n$ are actually zero. Thus we are enabled to give the more general definition of screw co-ordinates which is now enunciated.

If a wrench, of which the intensity is one unit on a screw α, which belongs to

6—2

a certain screw system of the nth order, be decomposed into n wrenches of intensities $\alpha_1, \ldots \alpha_n$ *on n co-reciprocal screws belonging to the same screw system, then the n quantities* $\alpha_1, \ldots \alpha_n$ *are said to be the co-ordinates of the screw* α. Thus the pitch of α will be represented by $p_1\alpha_1^2 + \ldots + p_n\alpha_n^2$. The virtual coefficient of α and β will be $(p_1\alpha_1\beta_1 + \ldots + p_n\alpha_n\beta_n)$.

We may here remark that in general one screw can be found upon a screw system of the nth order reciprocal to $n-1$ given screws of the same system. For, take $6-n$ screws of the reciprocal screw system, then the required screw is reciprocal to $6-n+n-1=5$ known screws, and is therefore determined (§ 25).

96. The Reduced Wrench.

A wrench which acts upon a constrained rigid body may in general be replaced by a wrench on a screw belonging to the screw system, which defines the freedom of the body.

Take n screws from the screw system of the nth order which defines the freedom, and $6-n$ screws from the reciprocal system. Decompose the given wrench into components on these six screws. The component wrenches on the reciprocal system are neutralized by the reactions of the constraints, and may be discarded, while the remainder must compound into a wrench on the given screw system.

Whenever a given external wrench is replaced by an equivalent wrench upon a screw of the system which defines the freedom of the body, the latter may be termed, for convenience, the *reduced wrench*.

It will be observed, that although the reduced wrench can be determined from the given wrench, that the converse problem is indeterminate ($n < 6$).

We may state this result in a somewhat different manner. A given wrench can in general be resolved into two wrenches—one on a screw of any given system, and the other on a screw of the reciprocal screw system. The former of these is what we denote by the reduced wrench.

This theorem of the reduced wrench ceases to be true in the case when the screw system and the reciprocal screw system have one screw in common. As such a screw must be reciprocal to both systems it follows that all the screws of both systems must be comprised in a single five-system. This is obviously a very special case, but whenever the condition indicated is satisfied it will not be possible to resolve an impulsive wrench into components on the two reciprocal systems, unless it should also happen that the impulsive wrench itself belongs to the five-system*.

* I am indebted to Mr Alex. McAulay for having pointed out in his book on *Octonions*, p. 251, that I had overlooked this exception when enunciating the Theorem of the reduced wrench in the *Theory of Screws* (1876).

97. Co-ordinates of Impulsive and Instantaneous Screws.

Taking as screws of reference the n principal screws of inertia (§ 84), we require to ascertain the relation between the co-ordinates of a reduced impulsive wrench and the co-ordinates of the corresponding instantaneous screw. If the co-ordinates of the reduced impulsive wrench are $\eta_1''', \ldots \eta_n'''$, and those of the twist velocity are $\dot{\alpha}_1, \dot{\alpha}_2, \ldots \dot{\alpha}_n$, then, remembering the property of a principal screw of inertia (§ 78), and denoting by $u_1, \ldots u_n$, the values of the magnitude u (§ 89) for the principal screws of inertia, we have, from § 90,

$$\dot{\alpha}_1 \frac{u_1^2}{p_1} = \frac{1}{M}\eta_1''',$$

whence observing that $\dot{\alpha}_1 = \dot{\alpha}\alpha_1; \ldots \dot{\alpha}_n = \dot{\alpha}\alpha_n$ we deduce the following theorem, which is the generalization of § 80.

If a quiescent rigid body, which has freedom of the nth order, commence to twist about a screw α, of which the co-ordinates, with respect to the principal screws of inertia, are $\alpha_1, \ldots \alpha_n$ and if $p_1, \ldots p_n$ be the pitches, and $u_1, \ldots u_n$ the constants defined, in § 89, of the principal screws of inertia, then the co-ordinates of the reduced impulsive wrench are proportional to

$$\frac{u_1^2}{p_1}\alpha_1, \ldots \frac{u_n^2}{p_n}\alpha_n.$$

Let T denote the kinetic energy of the body of mass M when animated by a twisting motion about the screw α, with a twist velocity $\dot{\alpha}$. Let the twist velocities of the components on *any* n conjugate screws of inertia be denoted by $\dot{\alpha}_1, \dot{\alpha}_2, \ldots \dot{\alpha}_n$. [These screws will not be co-reciprocal unless in the special case where they are the principal screws of inertia.] It follows (§ 88) that the kinetic energy will be the sum of the n several kinetic energies due to each component twisting motion. Hence we have (§ 89)

$$T = Mu_1^2\dot{\alpha}_1^2 + \ldots + Mu_n^2\dot{\alpha}_n^2,$$

and also
$$u_\alpha^2 = u_1^2\alpha_1^2 + \ldots + u_n^2\alpha_n^2.$$

Let $\alpha_1, \ldots \alpha_n$ and $\beta_1, \ldots \beta_n$ be the co-ordinates of any two screws belonging to a screw system of the nth order, referred to any n conjugate screws of inertia, *whether co-reciprocal or not*, belonging to the same screw system, then the condition that α and β should be conjugate screws of inertia is

$$u_1^2\alpha_1\beta_1 + \ldots + u_n^2\alpha_n\beta_n = 0.$$

To prove this, take the case of $n = 4$, and let A, B, C, D be the four screws of reference, and let $A_1, \ldots A_6$ be the co-ordinates of A with respect to the six principal screws of inertia of the body when free (§ 79). The unit wrench on α is to be resolved into four wrenches of intensities $\alpha_1, \ldots \alpha_4$ on A, B, C, D: each of these components is again to be resolved into six wrenches on the screws of reference. The six co-ordinates of α, with respect to the same screws, are therefore

$$A_1\alpha_1 + B_1\alpha_2 + C_1\alpha_3 + D_1\alpha_4,$$

$$\dots\dots\dots\dots\dots\dots$$

$$A_6\alpha_1 + B_6\alpha_2 + C_6\alpha_3 + D_6\alpha_4.$$

We can now express the condition that α and β are conjugate screws of inertia. This condition is (§ 81)

$$\Sigma p_1^2 (A_1\alpha_1 + B_1\alpha_2 + C_1\alpha_3 + D_1\alpha_4)(A_1\beta_1 + B_1\beta_2 + C_1\beta_3 + D_1\beta_4) = 0.$$

Denoting $p_1^2 A_1^2 + \dots + p_6 A_6^2$ by u_1^2, and observing that $\Sigma p_1^2 A_1 B_1$ and similar expressions are zero, we deduce

$$u_1^2 \alpha_1\beta_1 + \dots + u_4^2 \alpha_4\beta_4 = 0.$$

A similar proof may be written down for each of the remaining degrees of freedom.

CHAPTER VIII.

THE POTENTIAL.

98. The Potential.

Suppose a rigid body which possesses freedom of the nth order be submitted to a system of forces. Let the symbol O define a position of the body from which the forces would be unable to disturb it. By a twist of amplitude θ' about a screw θ belonging to the screw system, the body may be displaced from O to an adjacent position P, the energy consumed in making the twist being denoted by the Potential V, and no kinetic energy being supposed to be acquired. The same energy would be required, whatever be the route by which the movement is made from O to P. So far as we are at present concerned V varies only with the changes of the position of P with respect to O. The most natural co-ordinates by which the position P can be specified with respect to O are the co-ordinates of the twist (§ 32) by which the movement from O to P could be effected. In general these co-ordinates will be six in number; but if n of the screws of reference be selected from the screw system defining the freedom of the body, then (§ 95) there will be only n co-ordinates required, and these may be denoted by $\theta_1', \dots \theta_n'$.

The Potential V must therefore depend only upon certain quantities independent of the position and upon the n co-ordinates $\theta_1', \dots \theta_n'$; and since these are small, it will be assumed that V must be capable of development in a series of ascending powers and products of the co-ordinates, whence we may write

$$V = H + H_1\theta_1' + \dots + H_n\theta_n'$$

$$+ \text{ terms of the second and higher orders,}$$

where $H, H_1, \dots H_n$ are constants, in so far as different displacements are concerned.

In the first place, it is manifest that $H = 0$; because if no displacement be made, no energy is consumed. In the second place, $H_1, \dots H_n$ must also be each zero, because the position O is one of equilibrium; and therefore, by the principle of virtual velocities, the work done by small twists about the screws of reference must be zero, as far as the first power of small quantities is concerned. Finally, neglecting all terms above the second order, on account of their minuteness, we see that *the function V, which expresses the potential energy of a small displacement from a position of equilibrium, is generally a homogeneous function of the second degree of the n co-ordinates, by which the displacement is defined.*

99. The Wrench evoked by Displacement.

When the body has been displaced to P, the forces no longer equilibrate. They have now a certain resultant wrench. We propose to determine, by the aid of the function V, the co-ordinates of this wrench, or, more strictly, the co-ordinates of the equivalent reduced wrench (§ 96) upon a screw of the system, by which the freedom of the body is defined.

If, in making the displacement, *work has been done* by the agent which moved the body, then the equilibrium of the body was stable when in the position O, so far as this displacement was concerned. Let the displacement screw be θ, and let a reduced wrench be evoked on a screw η of the system, while the intensities of the components on the screws of reference are η_1'', ... η_n''. Suppose that the body be displaced from P to an excessively close position P', the co-ordinates of P', with respect to O, being (§ 95)

$$\theta_1' + \delta\theta_1', \ldots \theta_n' + \delta\theta_n'.$$

The potential V' of the position P' is

$$V + \frac{dV}{d\theta_1'}\delta\theta_1' + \ldots + \frac{dV}{d\theta_n'}\delta\theta_n',$$

it being understood that $\delta\theta_1'$, ... $\delta\theta_n'$ are infinitely small magnitudes of a higher order than θ_1', ... θ_n'.

The work done in forcing the body to move from P to P' is $V' - V$. This must be equal to the work done in the twists about the screws of reference whose amplitudes are $\delta\theta_1'$, ... $\delta\theta_n'$, by the wrenches on the screws of reference whose intensities are η_1'', ... η_n''. As the screws of reference are co-reciprocal, this work will be equal to (§ 33)

$$+ 2\eta_1''p_1\delta\theta_1' + \ldots + 2\eta_n''p_n\delta\theta_n.$$

Since the expression just written must be equal to $V' - V$ for *every* position P' in the immediate vicinity of P, we must have the coefficients of $\delta\theta_1'$, ... $\delta\theta_n'$ equal in the two expressions, whence we have n equations, of which the first is

$$\eta_1'' = + \frac{1}{2p_1}\frac{dV}{d\theta_1'}.$$

Hence, we deduce the following useful theorem:—

If a free or constrained rigid body be displaced from a position of equilibrium by twists of small amplitudes, $\theta_1', \dots \theta_n'$, about n co-reciprocal screws of reference, and if V denote the work done in producing this movement, then the reduced wrench has, for components on the screws of reference, wrenches of which the intensities are found by dividing twice the pitch of the corresponding reference screw into the differential coefficient of V with respect to the corresponding amplitude, and changing the sign of the quotient.

It is here interesting to notice that the co-ordinates of the reduced impulsive wrench referred to the principal screws of inertia, which would give the body a kinetic energy T on the screw θ, are proportional to

$$\frac{1}{2p_1}\frac{dT}{d\theta_1}, \dots \frac{1}{2p_n}\frac{dT}{d\theta_n} \text{ (§ 97).}$$

100. Conjugate Screws of the Potential.

Suppose that a twist about a screw θ evokes a wrench on a screw η, while a twist about a screw ϕ evokes a wrench on a screw ζ. If θ be reciprocal to ζ, then must ϕ be reciprocal to η. This will now be proved.

The condition that θ and ζ are reciprocal is

$$p_1\theta_1\zeta_1 + \dots + p_n\theta_n\zeta_n = 0;$$

but the intensities (or amplitudes) of the components of a wrench (or twist) are proportional to the co-ordinates of the screw on which the wrench (or twist) acts, whence the last equation may be written

$$p_1\theta_1'\zeta_1'' + \dots + p_n\theta_n'\zeta_n'' = 0;$$

but we have seen (§ 99) that

$$\zeta_1'' = +\frac{1}{2p_1}\frac{dV_\phi}{d\phi_1'}, \dots \zeta_n'' = +\frac{1}{2p_n}\frac{dV_\phi}{d\phi_n'};$$

whence the condition that θ and ζ are reciprocal is

$$\theta_1'\frac{dV_\phi}{d\phi_1'} + \dots + \theta_n'\frac{dV_\phi}{d\phi_n'} = 0.$$

Now, as V_ϕ is an homogeneous function of the second order of the quantities $\phi_1', \dots \phi_n'$, we may write

$$V_\phi = A_{11}\phi_1'^2 + \dots + A_{nn}\phi_n'^2 + 2A_{12}\phi_1'\phi_2' + 2A_{13}\phi_1'\phi_3' + \dots,$$

in which $A_{hk} = A_{kh}$.

Hence we obtain:—

$$\frac{dV_\phi}{d\phi_1'} = 2\{A_{11}\phi_1' + A_{12}\phi_2' + \dots + A_{1n}\phi_n'\}.$$

Introducing these expressions we find, for the condition that θ and ζ should be reciprocal,

$$\theta_1'(A_{11}\phi_1' + \ldots A_{1n}\phi_n') + \ldots + \theta_n'(A_{n1}\phi_1' + \ldots + A_{nn}\phi_n) = 0.$$

This may be written in the form :—

$$A_{11}\theta_1'\phi_1' + \ldots A_{nn}\theta_n'\phi_n' + A_{12}(\theta_1'\phi_2' + \theta_2'\phi_1') + \ldots = 0.$$

But this equation is symmetrical with respect to θ and ϕ, and therefore we should have been led to the same result by expressing the condition that ϕ was reciprocal to η.

When θ and ϕ possess this property, they are said to be *conjugate screws of the potential*, and the condition that they should be so related, expressed in terms of their co-ordinates, is obtained by omitting the accents from the last equation.

If a screw ϕ be reciprocal to η, then ϕ is a conjugate screw of the potential to θ. If we consider the screw θ to be given, we may regard the screw system of the fifth order, which embraces all the screws reciprocal to η, as in a certain sense the locus of ϕ. All the screws conjugate to θ, and which, at the same time, belong to the screw system C by which the freedom of the body is defined, must constitute in themselves a screw system of the $(n-1)$th order. For, besides fulfilling the $6-n$ conditions which define the screw system C, they must also fulfil the condition of being reciprocal to η; but all the screws reciprocal to $7-n$ screws constitute a screw system of the $(n-1)$th order (§ 72).

The reader will be careful to observe the distinction between two conjugate screws *of inertia* (§ 81), and two conjugate screws *of the potential*. Though these pairs possess some useful analogies, yet it should be borne in mind that the former are *purely intrinsic* to the rigid body, inasmuch as they only depend on the distribution of its material, while the latter involve *extrinsic* considerations, arising from the forces to which the body is submitted.

101. Principal Screws of the Potential.

We now prove that in general n screws can be found such that when the body is displaced by a twist about any one of these screws, a reduced wrench is evoked on the same screw. The screws which possess this property are called the *principal screws of the potential*. Let α be a principal screw of the potential, then we must have, § 99:—

$$\alpha_1'' = +\frac{1}{2p_1}\frac{dV_\alpha}{d\alpha_1'},$$

and $(n-1)$ similar equations.

Introducing the value of V_α, and remembering (§ 34) that $\alpha_1'' = \alpha''\alpha_1$ and $\alpha_1' = \alpha'\alpha_1$, we have the following n equations:—

$$\alpha_1\left(A_{11} - \frac{\alpha''}{\alpha'}p_1\right) + \alpha_2 A_{12} + \ldots + \alpha_n A_{1n} = 0,$$

&c., &c.

$$\alpha_1 A_{n1} + \alpha_2 A_{n2} + \ldots + \alpha_n\left(A_{nn} - \frac{\alpha''}{\alpha'}p_n\right) = 0.$$

From these linear equations $\alpha_1, \ldots \alpha_n$ can be eliminated, and we obtain an equation of the nth degree in $\frac{\alpha''}{\alpha'}$. The values of $\frac{\alpha''}{\alpha'}$ substituted successively in the linear equations just written will determine the co-ordinates of the n principal screws of the potential. If the position of equilibrium be one which is stable for all displacements then V_α must under all circumstances be positive. As it can be reduced to the sum of n squares all the roots of this equation will be real (§ 86) and consequently all the n principal screws of the potential will be real.

We can now show that these n screws are co-reciprocal. It is evident, in the first place, that if S be a principal screw of the potential, and if θ be a displacement screw which evokes a wrench on η, the principle of § 100 asserts that, when θ is reciprocal to S, then must also η be reciprocal to S. Let the n principal screws of the potential be denoted by $S_1, \ldots S_n$, and let T_n be that screw of the screw system which is reciprocal to $S_1, \ldots S_{n-1}$ (§ 95), then if the body be displaced by a twist about T_n, the wrench evoked must be on a screw reciprocal to $S_1, \ldots S_{n-1}$; but T_n is the only screw of the screw system possessing this property; therefore a twist about T_n must evoke a wrench on T_n, and therefore T_n must be a principal screw of the potential. But there are only n principal screws of the potential, therefore T_n must coincide with S_n, and therefore S_n must be reciprocal to $S_1, \ldots S_{n-1}$.

102. Co-ordinates of the Wrench evoked by a Twist.

The work done in giving the body a twist of small amplitude α' about a screw α, may be denoted by

$$Fv_\alpha^2\alpha'^2.$$

In fact, remembering that $\alpha'\alpha_1 = \alpha_1', \ldots$, and substituting these values for α_1' in V (§ 100), we deduce the expression:—

$$Fv_\alpha^2 = A_{11}\alpha_1^2 + \ldots + A_{nn}\alpha_n^2 + 2A_{12}\alpha_1\alpha_2 + 2A_{13}\alpha_1\alpha_3 + \ldots$$

where F is independent of α and has for its dimensions a mass divided by the square of a time, and where v_α *is a linear magnitude specially appropriate to each screw* α, *and depending upon the co-ordinates of* α, *and the constants in the function* V (§ 98).

The parameter v_α may be contrasted with the parameter u_α considered in § 89. Each is a linear magnitude, but the latter depends only upon the co-ordinates of α, and the distribution of the material of the rigid body. Both quantities may be contrasted with the pitch p_α, which is also a linear magnitude, but depends on the screw, and neither on the rigid body nor the forces.

If a body receive a twist of small amplitude α' about one of the principal screws of the potential, then the intensity of the wrench evoked on the same screw is (§ 99):—

$$+\frac{1}{2p_\alpha}\frac{dV_\alpha}{d\alpha'};$$

but we have just seen that $V = Fv_\alpha^2\alpha'^2$, whence we have the following theorem:—

If a body which has freedom of the nth order be displaced from a position of equilibrium by a twist about a screw α, of which the co-ordinates with respect to the principal screws of the potential are $\alpha_1, \ldots \alpha_n$, then a reduced wrench (§ 96) is evoked on a screw with co-ordinates proportional to $\frac{v_1^2}{p_1}\alpha_1, \ldots \frac{v_n^2}{p_n}\alpha_n$, where $v_1, \ldots, p_1, \ldots$, are the values of the quantity v, and the pitch p, for the principal screws of the potential.

We can now express with great simplicity the condition that two screws θ and ϕ shall be conjugate screws of the potential. For, if θ be reciprocal to the screw whose co-ordinates are proportional to

$$\frac{v_1^2}{p_1}\phi_1, \ldots \frac{v_n^2}{p_n}\phi_n,$$

we have:—

$$v_1^2\theta_1\phi_1 + \ldots + v_n^2\theta_n\phi_n = 0.$$

The expression for the potential assumes the simple form

$$Fv_1^2\alpha_1'^2 + \ldots + Fv_n^2\alpha_n'^2.$$

If the function V be proportional to the product of the pitch of the displacement screw and the square of the amplitude, then every displacement screw will coincide with the screw about which the wrench is evoked.

103. Form of the Potential.

The n principal screws of the potential form a unique group, inasmuch as they are *co-reciprocal*, as well as being conjugate screws of the potential. They therefore fulfil $n(n-1)$ conditions, being the total number available in the selection of n screws in an n system.

We now show that the expression of the potential will consist of the sum of n square terms, when referred to *any* set of n conjugate screws of the potential.

The energy consumed in giving a body a twist of amplitude θ' from the position of equilibrium O to a new position P, is equal to $Fv_\theta^2\theta'^2$ (§ 102), and η is the screw on which the wrench is evoked. Suppose that from the position P the body receive a twist of amplitude ϕ' about a screw ϕ, it would generally not be correct to assert that the energy consumed in the second twist was proportional to the square of its amplitude. For, during the second twist, either a portion of the energy will be consumed in doing work against the wrench on η, or the energy expended in the second twist will be rendered less, in consequence of the assistance afforded by the wrench on η. If, however, η be reciprocal to ϕ, then the quantity of energy consumed in the twist about ϕ will be unaffected by the presence of a wrench on η. Hence if θ and ϕ be two conjugate screws of the potential, the energy expended in giving the body first a twist of amplitude θ' about θ, and then a twist of amplitude ϕ' about ϕ, is to be represented by

$$Fv_\theta^2\theta'^2 + Fv_\phi^2\phi'^2.$$

By taking a third screw, conjugate to both θ and ϕ, and then a fourth screw conjugate to the remaining three, and so on, we see finally that the potential reduces to the sum of n square terms, where each pair of the screws of reference are conjugate screws of the potential.

CHAPTER IX.

HARMONIC SCREWS.

104. Definition of an Harmonic Screw.

We have seen in § 97 that to each screw θ of a screw system of the nth order corresponds a screw λ, belonging to the same screw system. The relation between θ and λ is determined when the rigid body, and also the screw system which defines its freedom, are completely known. The physical connection between the two screws θ and λ may be thus stated. If an impulsive wrench act on the screw λ, the body, if previously quiescent, will commence to move by twisting about θ.

We have also seen (§ 102) that to each screw θ of a screw system of the nth order corresponds a certain screw η belonging to the same screw system. The relation between θ and η is determined when the rigid body, the forces, and the screw system which defines the freedom, is known. The physical connexion between the two screws θ and η may be thus stated. If the body be displaced from a position of equilibrium by a twist about θ, the evoked wrench, when reduced (§ 96) to the screw system, acts on η.

The rigid body being given in a position of equilibrium, and the forces which act on the body being known, and also the screw system by which the freedom of the body is prescribed, we then have corresponding to each screw θ of the given screw system, two other screws λ and η, which also belong to the same screw system.

Considering the very different physical character of the two systems of correspondence, it will of course usually happen that the two screws λ and η are not identical. But a little reflection will enable us to foresee what we shall afterwards prove, viz., that when θ has been appropriately chosen, then λ and η may coincide. For since $n-1$ arbitrary quantities are disposable in the selection of a screw from a screw system of the nth order (§ 69), it follows that for any two screws (for example λ and η) to coincide, $n-1$ conditions must be fulfilled; but this is precisely the number of arbitrary elements available in the selection of θ. We can thus conceive that for one or more particular screws θ, the two corresponding screws λ and η are identical; and we shall now prove the following important theorem:—

If a rigid body be displaced from a position of equilibrium by a twist about a screw θ, and if the evoked wrench tend to make the body commence to twist about the same screw θ, then if we call θ an harmonic screw (§ 106), we assert that the number of harmonic screws is generally the same as the order of the screw system which defines the freedom of the rigid body.

We shall adopt as the screws of reference the n principal screws of inertia. The impulsive screw, which corresponds to θ as an instantaneous screw, will have for co-ordinates

$$h\frac{u_1^2}{p_1}\theta_1, \ldots h\frac{u_n^2}{p_n}\theta_n \text{ (§ 97)},$$

where h is a certain constant which is determined by making the co-ordinates satisfy the condition (§ 35). If θ be an harmonic screw, then, remembering that the screws of reference are co-reciprocal (§ 87), we must have n equations, of which the first is (§ 102):—

$$h\frac{u_1^2}{p_1}\theta_1''' = +\frac{1}{2p_1}\frac{dV}{d\theta_1'}.$$

Assuming $\frac{h\theta''}{\theta'} = -Ms^2$, where M is the mass of the body, and s an unknown quantity, and developing V, we deduce the n equations:—

$$\theta_1(A_{11} + Ms^2u_1^2) + \theta_2 A_{12} + \ldots + \theta_n A_{1n} = 0,$$

$$\ldots\ldots\ldots\ldots\ldots\ldots\ldots$$

$$\theta_1 A_{n1} + \theta_2 A_{n2} + \ldots + \theta_n(A_{nn} + Ms^2u_n^2) = 0.$$

Eliminating $\theta_1, \ldots \theta_n$, we have an equation of the nth degree for s^2. The n roots of this equation are all real (§ 85), and each one substituted in the set of n equations will determine, by a system of n linear equations, the ratios of the n co-ordinates of one of the harmonic screws.

It is a remarkable property of the n harmonic screws that each pair of them are conjugate screws of inertia, and also conjugate screws of the potential. Let $H_1, \ldots H_{n-1}$, be $n-1$ of the harmonic screws, to which correspond the impulsive screws $S_1, \ldots S_{n-1}$. Also suppose T to be that one screw of the given screw system which is reciprocal to $S_1, \ldots S_{n-1}$ (§ 95), then T must form with each one of the screws $H_1, \ldots H_{n-1}$ a pair of conjugate screws of inertia (§ 81). But, since $S_1, \ldots S_{n-1}$ are the screws on which wrenches are evoked by twists about $H_1, \ldots H_{n-1}$ respectively, it is evident that T must form with each one of these screws $H_1, \ldots H_{n-1}$ a pair of conjugate screws of the potential (§ 100). It follows that the impulsive screw, corresponding to T as the instantaneous screw, must be reciprocal to $H_1, \ldots H_{n-1}$; and also that a twist about T must evoke a wrench on a screw reciprocal to $H_1, \ldots H_{n-1}$. As in general only one screw of the system can be reciprocal to $H_1, \ldots H_{n-1}$, it follows that the impulsive screw, which corresponds to T as an instantaneous screw, must also be the screw on which a wrench is evoked by a twist about T. Hence, T must be a harmonic screw, and as there are only n harmonic screws, it is plain that T must coincide with H_n, and that therefore H_n is a conjugate screw of inertia, as well as a conjugate screw of the potential, to each one of the remaining $n-1$ harmonic screws. Similar reasoning will, of course, apply to each of the harmonic screws taken in succession.

105. Equations of Motion.

We now consider the kinetical problem, which may be thus stated. A free or constrained rigid body, which is acted upon by a system of forces, is displaced by an initial twist of small amplitude, from a position of equilibrium. The body also receives an initial twisting motion, with a small twist velocity, and is then abandoned to the influence of the forces. It is required to ascertain the nature of its subsequent movements.

Let T represent the kinetic energy of the body, in the position of which the co-ordinates, *referred to the principal screws of inertia*, are $\theta_1', \ldots \theta_n'$. Then we have (§ 97):—

$$T = M\left[u_1^2\left(\frac{d\theta_1'}{dt}\right)^2 + \ldots + u_n^2\left(\frac{d\theta_n'}{dt}\right)^2\right]$$

while the potential energy which, as before, we denote by V, is an homogeneous function of the second order of the quantities $\theta_1', \ldots \theta_n'$.

By the use of Lagrange's method of generalized co-ordinates we are enabled to write down at once the n equations of motion in the form:—

$$\frac{d}{dt}\left(\frac{dT}{d\dfrac{d\theta_1'}{dt}}\right) - \frac{dV}{d\theta_1'} = 0.$$

Substituting for T we have:—

$$2Mu_1^2\frac{d^2\theta_1'}{dt^2} - \frac{dV}{d\theta_1'} = 0,$$

with $(n-1)$ similar equations. Finally, introducing the expression for V (§ 100), we obtain n linear differential equations of the second order.

The equations which we require can be otherwise demonstrated as follows.

Suppose the body to be in motion under the influence of the forces, and that at any epoch t the co-ordinates of the twisting motion are

$$\frac{d\theta_1'}{dt}, \ldots, \frac{d\theta_n'}{dt} \ldots,$$

when referred to the principal screws of inertia. Let $\zeta_1'', \ldots \zeta_n''$ be the co-ordinates of a wrench which, had it acted upon the body at rest for the small time e, would have communicated to the body a twisting motion identical with that which the body actually has at the epoch t. The co-ordinates of the impulsive wrench which would, in the time e, have produced from rest the motion which the body actually has at the epoch $t+e$, are:

$$\zeta_1'' + e\frac{d\zeta_1''}{dt}, \ldots \zeta_n'' + e\frac{d\zeta_n''}{dt}.$$

On the other hand, the motion at the epoch $t+e$ may be considered to arise from the influence of the wrench $\zeta_1'', \ldots \zeta_n''$ for the time e, followed by the influence of the evoked wrench for the time e. The final effect of the two wrenches must, by the second law of motion, be the same as if they acted simultaneously for the time e upon the body initially at rest.

The co-ordinates of the evoked wrench being:

$$+\frac{1}{2p_1}\frac{dV}{d\theta_1'}, \ldots +\frac{1}{2p_n}\frac{dV}{d\theta_n'},$$

we therefore have the equation:—

$$\zeta_1'' + e\frac{d\zeta_1''}{dt} = \zeta_1'' + \frac{1}{2p_1}\frac{dV}{d\theta_1'},$$

or

$$e\frac{d\zeta_1''}{dt} = +\frac{1}{2p_1}\frac{dV}{d\theta_1'},$$

and $n-1$ similar equations; but we see from § 97 that

$$e\zeta_1'' = M\frac{u_1^2}{p_1}\frac{d\theta_1'}{dt}.$$

Differentiating this equation with respect to the time, and regarding e as constant, we have

$$e\frac{d\zeta_1''}{dt} = M\frac{u_1^2}{p_1}\frac{d^2\theta_1'}{dt^2};$$

whence

$$2Mu_1^2\frac{d^2\theta_1'}{dt^2} - \frac{dV}{d\theta_1'} = 0,$$

the same equation as that already found by Lagrange's method.

To integrate the equations we assume

$$\theta_1' = f_1\Omega, \ldots \theta_n' = f_n\Omega;$$

where $f_1, \ldots f_n$ are certain constants, which will be determined, and where Ω is an unknown function of the time: introducing also the value of V, given in § 100, we find for the equations of motion:

$$-Mu_1^2 f_1 \frac{d^2\Omega}{dt^2} + (A_{11}f_1 + A_{12}f_2 + \ldots + A_{1n}f_n)\,\Omega = 0,$$

&c.

$$-Mu_n^2 f_n \frac{d^2\Omega}{dt^2} + (A_{n1}f_1 + A_{n2}f_1 + \ldots + A_{nn}f_n)\,\Omega = 0.$$

If the quantity s, and the ratios of the n quantities $f_1, \ldots f_n$, be determined by the n equations:—

$$f_1(A_{11} + Mu_1^2 s^2) + f_2 A_{12} + \ldots + f_n A_{n1} = 0,$$

&c., &c.

$$f_1 A_{n1} + f_2 A_{n2} + \ldots + f_n (A_{nn} + Mu_n^2 s^2) = 0;$$

then the n equations of motion will reduce to the single equation:

$$\frac{d^2\Omega}{dt^2} + s^2\Omega = 0.$$

By eliminating $f_1, \ldots f_n$ from the n equations, we obtain precisely the same equation for s^2 as that which arose (§ 104) in the determination of the n harmonic screws. The values of $f_1, \ldots f_n$, which correspond to any value of s^2, are therefore proportional to the co-ordinates of a harmonic screw.

The equation for Ω gives:

$$\Omega = H \sin(st + c).$$

Let $H_1, \ldots H_n, c_1, \ldots c_n$ be $2n$ arbitrary constants. Let f_{pq} denote the value of f_q, when the root s_p^2 has been substituted in the linear equations. Then by the known theory of linear differential equations*,

$$\theta_1' = f_{11}H_1 \sin(s_1 t + c_1) + \ldots + f_{n1}H_n \sin(s_n t + c_n),$$

......

$$\theta_n' = f_{1n}H_1 \sin(s_1 t + c_1) + \ldots + f_{nn}H_n \sin(s_n t + c_n).$$

In proof of this solution it is sufficient to observe, that the values of $\theta_1', \ldots \theta_n'$ satisfy the given differential equations of motion, while they also contain the requisite number of arbitrary constants.

* Lagrange's Method, Routh, *Rigid Dynamics*, Vol. I., p. 369.

106. Discussion of the Results.

For the position of the body before its displacement to have been one of stable equilibrium, it is manifest that the co-ordinates must not increase indefinitely with the time, and therefore all the values of s^2 must be essentially positive, since otherwise the values of $\theta_1', \ldots \theta_n'$ would contain exponential terms.

The $2n$ arbitrary constants are to be determined by the initial circumstances. The initial displacement is to be resolved into n twists about the n screws of reference (§ 95). This will provide n equations, by making $t = 0$, and substituting for $\theta_1', \ldots \theta_n'$, in the equations just mentioned, the amplitudes of the initial twists. The initial twisting motion is also to be resolved into twisting motions about the n screws of reference. The twist velocities of these components will be the values of $\frac{d\theta_1'}{dt}, \ldots, \frac{d\theta_n'}{dt}$, when $t = 0$; whence we have n more equations to complete the determination of the arbitrary constants.

If the initial circumstances be such that the constants $H_2, \ldots, H_n$ are all zero, then the equations assume a simple form:

$$\theta_1' = f_{11} H_1 \sin(s_1 t + c),$$

$$\ldots\ldots$$

$$\theta_n' = f_{1n} H_1 \sin(s_1 t + c).$$

The interpretation of this result is very remarkable. We see that the co-ordinates of the body are always proportional to $f_{11}, \ldots, f_{1n}$; hence the body can always be brought from the initial position to the position at any time by twisting it about that screw, whose co-ordinates are proportional to $f_{11}, \ldots, f_{1n}$; but, as we have already pointed out, the screw thus defined is a harmonic screw, and hence we have the following theorem:—

If a rigid body occupy a position of stable equilibrium under the influence of a conservative system of forces, then n harmonic screws can be selected from the screw system of the nth order, which defines the freedom of the body, and if the body be displaced from its position of equilibrium by a twist about a harmonic screw, and if it also receive any small initial twist velocity about the same screw, then the body will continue to perform twist oscillations about that harmonic screw, and the amplitude of the twist will be always equal to the arc of a certain circular pendulum, which has an appropriate length, and was appropriately started.

The integrals in their general form prove the following theorem:—

A rigid body is slightly displaced by a twist from a position of stable equilibrium under the influence of a system of forces, and the body receives a small initial twisting motion. The twist, and the twisting motion, may each be resolved into their components on the n harmonic screws: n circular pendulums are to be constructed, each of which is isochronous with one of the harmonic screws. All these pendulums are to be started at the same instant as the rigid body, each with an arc, and an angular velocity equal respectively to the initial amplitude of the twist, and to the twist velocity, which have been assigned to the corresponding harmonic screw. To ascertain where the body would be at any future epoch, it will only be necessary to calculate the arcs of the n pendulums for that epoch, and then give the body twists from its position of equilibrium about the harmonic screws, whose amplitudes are equal to these arcs.

107. Remark on Harmonic Screws.

We may to a certain extent see the actual reason why the body, when once oscillating upon a harmonic screw, will never depart therefrom. The body, when displaced from the position of equilibrium by a twist upon a harmonic screw θ, and then released, is acted upon by the wrench upon a certain screw η, which is evoked by the twist. But the actual effect of an impulsive wrench on η would be to make the body twist about the harmonic screw (§ 104), and as the continued action of the wrench on η is indistinguishable from an infinite succession of infinitely small impulses, we can find in the influence of the forces no cause adequate to change the motion of the body from twisting about the harmonic screw θ.

THE CYLINDROID.

To face p. 150

CHAPTER X.

FREEDOM OF THE FIRST ORDER.

108. Introduction.

In the present chapter we shall apply the principles developed in the preceding chapters to study the Statics and Dynamics of a rigid body which has freedom of the first order. Ensuing chapters will be similarly devoted to the other orders of freedom. We shall in each chapter first ascertain what can be learned as to the kinematics of a rigid body, so far as small displacements are concerned, from merely knowing the *order* of the freedom which is permitted by the constraints. This will conduct us to a knowledge of the special screw system which defines the freedom enjoyed by the body. We shall then be enabled to determine the reciprocal screw system, which involves the theory of equilibrium. The next group of questions will be those which relate to the effect of an impulse upon a quiescent rigid body, free to twist about any screw of the screw system. Finally, we shall discuss the small oscillations of a rigid body in the vicinity of a position of stable equilibrium, under the influence of a given system of forces, the movements of the body being limited as before to the screws of the screw system.

109. Screw System of the First Order.

A body which has freedom of the first order can execute no movement which is not a twist about one definite screw. The position of a body so circumstanced is to be specific by a single datum, viz., the amplitude of the twist about the given screw, by which the body can be brought from a standard position to any other position which it is capable of attaining. As examples of a body which has freedom of the first order, we may refer to the case of a body free to rotate about a fixed axis, but not to slide along it, or of a body free to slide along a fixed axis, but not to rotate around it. In the former case the screw system consists of one screw, whose pitch is zero; in the latter case the screw system consists of one screw, whose pitch is infinite.

110. The Reciprocal Screw System.

The integer which denotes the order of a screw system, and the integer which denotes the order of the reciprocal screw system, will, in all cases, have the number six for their sum (§ 72). Hence a screw system of the first order will have as its reciprocal a screw system of the fifth order.

For a screw θ to belong to a screw system of the fifth order, the necessary and sufficient condition is, that θ be reciprocal to one given screw α. This condition is expressed in the usual form:—

$$(p_\alpha + p_\theta)\cos O - d_{\alpha\theta}\sin O = 0,$$

where O is the angle, and $d_{\alpha\theta}$ the perpendicular distance between the screws θ and α.

We can now show that every straight line in space, when it receives an appropriate pitch, constitutes a screw of a given screw system of the fifth order. For the straight line and α being given, $d_{\alpha\theta}$ and O are determined, and hence the pitch p_θ can be determined by the linear equation just written.

Consider next a point A, and the screw α. Every straight line through A, when furnished with the proper pitch, will be reciprocal to α. Since the number of lines through A is doubly infinite, it follows that a singly infinite number of screws of given pitch can be drawn through A, so as to be reciprocal to α. We shall now prove that all the screws of *the same pitch which pass through A, and are reciprocal to α, lie in a plane.* This we shall first show to be the case for all the screws of zero pitch*, and then we shall deduce the more general theorem.

By a twist of small amplitude about α the point A is moved to an adjacent point B. To effect this movement against a force at A which is perpendicular to AB, no work will be required; hence every line through A, perpendicular to AB, may be regarded as a screw of zero pitch, reciprocal to α.

We must now enunciate a principle which applies to a screw system of any order. We have already referred to it with respect to the cylindroid (§ 18). If all the screws of a screw system be modified by the addition of the same linear magnitude (positive or negative) to the pitch of every screw, then the collection of screws thus modified still form a screw system of the same order. The proof is obvious, for since the virtual co-efficient depends on the *sum* of the pitches, it follows that, if all the pitches of a system be

* This theorem is due to Möbius, who has shown, that, if small rotations about six axes can neutralise, and if five of the axes be given, and a point on the sixth axis, then the sixth axis is limited to a plane. ("Ueber die Zusammensetzung unendlich kleiner Drehungen," *Crelle's Journal*, t. xviii., pp. 189—212.) (Berlin, 1838.)

increased by a certain quantity, and all the pitches of the reciprocal system be diminished by the same quantity, then all the first set of screws thus modified are reciprocal to all the second group as modified. Hence, since a screw-system of the nth order consists of all the screws reciprocal to $6 - n$ screws, it follows that the modified set must still be a screw system.

We shall now apply this principle to prove that all the screws λ of *any* given pitch k, which can be drawn through A, to be reciprocal to α, lie in a plane. Take a screw η, of pitch $p_\alpha + k$, on the same line as α, then we have just shown that all the screws μ, of zero pitch, which can be drawn through the point A, so as to be reciprocal to η, lie in a plane. Since μ and η are reciprocal, the screws on the same straight lines as μ and η will be reciprocal, provided the sum of their pitches is the pitch of η; therefore, a screw λ, of pitch k, on the same straight line as μ, will be reciprocal to the screw α, of pitch p_α; but all the lines μ lie in a plane, therefore all the screws λ lie in the same plane.

Conversely, given a plane and a pitch k, a point A can be determined in that plane, such that all the screws drawn through A in the plane, and possessing the pitch k, are reciprocal to α. To each pitch k_1, k_2, ..., will correspond a point A_1, A_2 ...; and it is worthy of remark, that all the points A_1, A_2 must lie on a right line which intersects α at right angles; for join A_1, A_2, then a screw on the line A_1A_2, which has for pitch either k_1 or k_2, must be reciprocal to α; but this is impossible unless A_1A_2 intersect α at a right angle.

111. Equilibrium.

If a body which has freedom of the first order be in equilibrium, then the necessary and sufficient condition is, that the forces which act upon the body shall constitute a wrench on a screw of the screw system of the fifth order, which is reciprocal to the screw which defines the freedom. We thus see that every straight line in space may be the residence of a screw, a wrench on which is consistent with the equilibrium of the body.

If two wrenches act upon the body, then the condition of equilibrium is, that, when the two wrenches are compounded by the aid of a cylindroid, the single wrench which replaces them shall lie upon that one screw of the cylindroid, which is reciprocal to α (§ 26).

We can express with great facility, by the aid of screw co-ordinates, the condition that wrenches of intensities θ'', ϕ'', on two screws θ, ϕ, shall equilibrate, when applied to a body only free to twist about α.

Adopting any six co-reciprocals as screws of reference, and resolving each of the wrenches on θ and ϕ into its six components on the six screws of reference, we shall have for the intensity of the component of the resultant wrench on ω_n—

$$\theta''\theta_n + \phi''\phi_n.$$

Hence the co-ordinates of the resultant wrench are proportional to

$$\theta''\theta_1 + \phi''\phi_1, \ldots \theta''\theta_6 + \phi''\phi_6.$$

For equilibrium this screw must be reciprocal to α, whence we have

$$p_1\alpha_1(\theta''\theta_1 + \phi''\phi_1) + \ldots + p_6\alpha_6(\theta''\theta_6 + \phi''\phi_6) = 0,$$

or,

$$\theta''\varpi_{\alpha\theta} + \phi''\varpi_{\alpha\phi} = 0.$$

This equation merely expresses that the sum of the works done in a small twist about α against the wrenches on θ and ϕ is zero.

We also perceive that a given wrench may be always replaced by a wrench of appropriate intensity on any other screw, in so far as the effect on a body only free to twist about α is concerned.

It may not be out of place to notice the analogy which the equation just written bears to the simple problem of the determination of the condition that two forces should be unable to disturb the equilibrium of a particle only free to move on a straight line. If P, Q be the two forces, and if l, m be the angles which the forces make with the direction in which the particle can move, then the condition is—

$$P\cos l + Q\cos m = 0.$$

This suggests an analogy between the virtual co-efficient of two screws, and the cosine of the angle between two lines.

112. Particular Case.

If a body having freedom of the first order be in equilibrium under the action of gravity, then the vertical through the centre of inertia must lie in the plane of reciprocal screws of zero pitch, drawn through the centre of inertia.

113. Impulsive Forces.

If an impulsive wrench of intensity η''' act on the screw η, while the body is only permitted to twist about α, then we have seen in § 90 how the twist velocity produced can be found. We shall now determine the impulsive reaction of the constraints. This reaction must be an impulsive wrench of intensity λ''' on a screw λ, which is reciprocal to α. The determination of λ may be effected geometrically in the following manner:—Let μ be the screw, an impulsive wrench on which would, if the body were perfectly free, cause an instantaneous twisting motion about α (§ 80). Draw the cylindroid (η, μ).

Then λ must be that screw on the cylindroid which is reciprocal to α, for a wrench on λ, and the given wrench on η, must compound into a wrench on μ, whence the three screws must be co-cylindroidal*; also λ must be reciprocal to α, so that its position on the cylindroid is known (§ 26). Finally, as the impulsive intensity η''' is given, and as the three screws η, λ, μ are all known, the impulsive intensity λ''' becomes determined (§ 14).

114. Small Oscillations.

We shall now suppose that a rigid body which has freedom of the first order occupies a position of stable equilibrium under the influence of a system of forces. If the body be displaced by a small twist about the screw α which prescribes the freedom, and if it further receive a small initial twist velocity about the same screw, the body will continue to perform small twist oscillations about the screw α. We propose to determine the time of an oscillation.

The kinetic energy of the body, when animated by a twist velocity $\frac{d\alpha'}{dt}$ is $Mu_\alpha^2\left(\frac{d\alpha'}{dt}\right)^2$ (§ 89). The potential energy due to the position attained by giving the body a twist of amplitude α' away from its position of equilibrium, is $Fv_\alpha^2\alpha'^2$ (§ 102). But the sum of the potential and kinetic energies must be constant, whence

$$Mu_\alpha^2\left(\frac{d\alpha'}{dt}\right)^2 + Fv_\alpha^2\alpha'^2 = \text{const.}$$

Differentiating we have

$$\frac{d^2\alpha'}{dt^2} + \frac{Fv_\alpha^2}{Mu_\alpha^2}\alpha' = 0.$$

Integrating this equation we have

$$\alpha' = A\sin\sqrt{\frac{Fv_\alpha^2}{Mu_\alpha^2}}\,t + B\cos\sqrt{\frac{Fv_\alpha^2}{Mu_\alpha^2}}\,t,$$

where A and B are arbitrary constants. The time of one oscillation is therefore

$$2\pi\frac{u_\alpha}{v_\alpha}\sqrt{\frac{M}{F}}.$$

Regarding the rigid body and the forces as given, and comparing *inter se* the periods about different screws α, on which the body might have been constrained to twist, we see from the result just arrived at that the time for each screw α is proportional to $\frac{u_\alpha}{v_\alpha}$.

* We shall often for convenience speak of three screws on the same cylindroid as *co-cylindroidal*.

115. Property of Harmonic Screws.

As the time of vibration is affected by the position of the screw to which the motion is limited, it becomes of interest to consider how a screw is to be chosen so that the time of vibration shall be a maximum or minimum. With slightly increased generality we may state the problem as follows:

Given the potential for every position in the neighbourhood of a position of stable equilibrium, it is required to select from a given screw system the screw or screws on which, if the body be constrained to twist, the time of vibration will be a maximum or minimum, relatively to the time of vibration on the neighbouring screws of the same screw system.

Take the n principal screws of inertia belonging to the screw system, as screws of reference, then we have to determine the n co-ordinates of a screw α by the condition that the function $\frac{u_\alpha}{v_\alpha}$ shall be a maximum or a minimum.

Introducing the value of u_α (§ 97), and of v_α (§ 102), in terms of the co-ordinates, we have to determine the maximum and minimum of the function

$$\frac{A_{11}\alpha_1^2 + \ldots + A_{nn}\alpha_n^2 + 2A_{12}\alpha_1\alpha_2 + 2A_{13}\alpha_1\alpha_3 + \ldots}{u_1^2\alpha_1^2 + \ldots + u_n^2\alpha_n^2} = x.$$

Multiplying this equation by the denominator of the left-hand side, differentiating with respect to each co-ordinate successively, and observing that the differential coefficients of x must be zero, we have the n equations:—

$$(A_{11} - u_1^2x)\,\alpha_1 + A_{12}\alpha_2 \ldots + A_{1n}\alpha_n = 0,$$

&c., &c.

$$A_{n1}\alpha_1 + A_{n2}\alpha_2 \ldots + (A_{nn} - u_n^2x)\,\alpha_n = 0.$$

We hence see that there are n screws belonging to each screw of the nth order on which the time of vibration is a maximum or minimum, and by comparison with § 104 we deduce the interesting result that these n screws are also the harmonic screws.

Taking the screw system of the sixth order, which of course includes every screw in space, we see that if the body be permitted to twist about one of the six harmonic screws the time of vibration will be a maximum or minimum, as compared with the time of vibration on *any* adjacent screw.

If the six harmonic screws were taken as the screws of reference, then u_α^2 and v_α^2 would each consist of the sum of six square terms (§§ 89, 102). If the coefficients in these two expressions were proportional, so that u_α^2 only differed from v_α^2 by a numerical factor, we should then find that every screw in space was an harmonic screw, and that the times of vibrations about all these screws were equal.

CHAPTER XI.

FREEDOM OF THE SECOND ORDER.

116. The Screw System of the Second Order.

When a rigid body is capable of being twisted about two screws θ and ϕ, it is capable of being twisted about every screw on the cylindroid (θ, ϕ) (§ 14). If it also appear that the body cannot be twisted about any screw which does not lie on the cylindroid, then as we know the body has freedom of the second order, and the cylindroid is the screw system of the second order by which the freedom is defined (§ 219).

Eight numerical data are required for determination of a cylindroid (§ 75). We must have four for the specification of the nodal line, two more are required to define the extreme points in which the surface cuts the nodal line, one to assign the direction of one generator, and one to give the pitch of one screw, or the eccentricity of the pitch conic.

Although only eight constants are required to define the cylindroid, yet ten constants must be used in defining two screws θ, ϕ, from which the cylindroid could be constructed. The ten constants not only define the cylindroid, but also point out two special screws upon the surface (§ 77).

117. Applications of Screw Co-ordinates.

We have shown (§ 40) that if α, β be the two screws of a cylindroid, which intersect at right angles, then the co-ordinates of any screw θ, which makes an angle l with the screw α, are:

$$\alpha_1 \cos l + \beta_1 \sin l, \ldots \alpha_6 \cos l + \beta_6 \sin l,$$

reference being made as usual to any set of six co-reciprocals.

In addition to the examples of the use of these co-ordinates already given (§ 40), we may apply them to the determination of that single screw θ upon the cylindroid (α, β), which is reciprocal to a given screw η.

From the condition of reciprocity we must have:

$$p_1\eta_1(\alpha_1 \cos l + \beta_1 \sin l) + \ldots + p_6\eta_6(\alpha_6 \cos l + \beta_6 \sin l) = 0,$$

or,

$$\varpi_{\alpha\eta} \cos l + \varpi_{\beta\eta} \sin l = 0.$$

From this $\tan l$ is deduced, and therefore the screw θ becomes known (§ 26).

In general if $\varpi_{\eta\theta}$ be the virtual coefficient of *any* screw η and a screw θ on the cylindroid, we have

$$\varpi_{\eta\theta} = \varpi_{\alpha\eta} \cos l + \varpi_{\beta\eta} \sin l;$$

whence if on each screw θ a distance be set off from the nodal line equal to the virtual coefficient between η and θ, the points thus found will lie on a right circular cylinder, of which the equation is;

$$x^2 + y^2 = \varpi_{\eta\alpha} x + \varpi_{\beta\eta} y.$$

Thus the screw which has the greatest virtual coefficient with η is at right angles to the screw reciprocal to η, and in general two screws can be found upon the cylindroid which have a given virtual coefficient with any given external screw.

118. Relation between Two Cylindroids.

We may here notice a curious reciprocal relation between two cylindroids, which is manifested when one condition is satisfied. If a screw can be found on one cylindroid, which is reciprocal to a second cylindroid, then conversely a screw can be found on the latter, which is reciprocal to the former. Let the cylindroids be (α, β), and (λ, μ). If a screw can be found on the former, which is reciprocal to the latter, then we have:

$$p_1\lambda_1(\alpha_1 \cos l + \beta_1 \sin l) + \ldots + p_n\lambda_n(\alpha_n \cos l + \beta_n \sin l) = 0,$$

$$p_1\mu_1(\alpha_1 \cos l + \beta_1 \sin l) + \ldots + p_n\mu_n(\alpha_n \cos l + \beta_n \sin l) = 0.$$

Whence eliminating l, we find:—

$$\varpi_{\alpha\lambda}\varpi_{\beta\mu} - \varpi_{\beta\lambda}\varpi_{\alpha\mu} = 0.$$

As this relation is symmetrical with regard to the two cylindroids, the theorem has been proved.

119. Co-ordinates of Three Screws on a Cylindroid.

The co-ordinates of three screws upon a cylindroid are connected by four independent relations. In fact, two screws define the cylindroid, and the third screw must then satisfy four equations of the form (§ 20). These relations can be expressed most symmetrically in the form of six equations, which also involve three other quantities.

Let λ, μ, ν be three screws upon a cylindroid, and let A, B, C denote the angles between μ ν, between ν λ, and between λ μ, respectively. If wrenches of intensities λ'', μ'', ν'', on λ, μ, ν, respectively, are in equilibrium, we must have (§ 14):—

$$\frac{\lambda''}{\sin A} = \frac{\mu''}{\sin B} = \frac{\nu''}{\sin C}.$$

But we have also as a necessary condition that if each wrench be resolved into six component wrenches on six screws of reference, the sum of the intensities of the three components on each screw of reference is zero; whence

$$\lambda_1 \sin A + \mu_1 \sin B + \nu_1 \sin C = 0,$$

$$\cdots\cdots$$

$$\lambda_6 \sin A + \mu_6 \sin B + \nu_6 \sin C = 0.$$

From these equations we deduce the following corollaries:—

The screw of which the co-ordinates are proportional to $a\lambda_1 + b\mu_1, \ldots$ $a\lambda_6 + b\mu_6$, lies on the cylindroid (λ, μ), and makes angles with the screws λ, μ, of which the sines are inversely proportional to a and b.

The two screws, of which the co-ordinates are proportional to

$$a\lambda_1 \pm b\mu_1, \ldots a\lambda_6 \pm b\mu_6,$$

and the two screws λ, μ are respectively parallel to the four rays of a plane harmonic pencil.

120. Screws on One Line.

There is one case in which a body has freedom of the second order that demands special attention. Suppose the two given screws θ, ϕ, about which the body can be twisted, happen to lie on the same straight line, then the cylindroid becomes illusory. If the amplitudes of the two twists be θ', ϕ', then the body will have received a rotation $\theta' + \phi'$, accompanied by a translation $\theta' p_\theta + \phi' p_\phi$. This movement is really identical with a twist on a screw of which the pitch is:

$$\frac{\theta' p_\theta + \phi' p_\phi}{\theta' + \phi'}.$$

Since θ', ϕ' may have any ratio, we see that, under these circumstances, the screw system which defines the freedom consists of all the screws with pitches ranging from $-\infty$ to $+\infty$, which lie along the given line. It follows (§ 47), that the co-ordinates of all the screws about which the body can be twisted are to be found by giving x all the values from $-\infty$ to $+\infty$ in the expressions:

$$\theta_1 + \frac{x}{4p_1}\frac{dR}{d\theta_1}, \ldots \theta_6 + \frac{x}{4p_6}\frac{dR}{d\theta_6},$$

in which $$R = (\theta_1 + \theta_2)^2 + (\theta_3 + \theta_4)^2 + (\theta_5 + \theta_6)^2.$$

121. Displacement of a Point.

Let P be a point, and let α, β be any two screws upon a cylindroid. If a body to which P is attached receive a small twist about α, the point P will be moved to P'. If the body receive a small twist about β, the point P would be moved to P''. Then whatever be the screw γ on the cylindroid about which the body be twisted, the point P will still be displaced in the plane $PP'P''$.

For the twist about γ can be resolved into two twists about α and β, and therefore every displacement of P must be capable of being resolved along PP' and PP''.

Thus through every point P in space a plane can be drawn to which the small movements of P, arising from twists about the screws on a given cylindroid are confined. The simplest construction for this plane is as follows:—Draw through the point P two planes, each containing one of the screws of zero pitch; the intersection of these planes is normal to the required plane through P.

The construction just given would fail if P lay upon one of the screws of zero pitch. The movements of P must then be limited, not to a plane, but to a line. The line is found by drawing a normal to the plane passing through P, and through the other screw of zero pitch.

We thus have the following curious property due to M. Mannheim*, viz., that a point in the rigid body on the line of zero pitch will commence to move in the same direction whatever be the screw on the cylindroid about which the twist is imparted.

This easily appears otherwise. Appropriate twists about any two screws, α and β, can compound into a twist about the screw of zero pitch λ, but the twist about λ cannot disturb a point on λ. Therefore a twist about β must be capable of moving a point originally on λ back to its position before it was disturbed by α. Therefore the twists about β and α must move the point in the same direction.

122. Properties of the Pitch Conic.

Since the pitch of a screw on a cylindroid is proportional to the inverse square of the parallel diameter of the pitch conic (§ 18), the asymptotes must be parallel to the screws of zero pitch; also since a pair of reciprocal screws are parallel to a pair of conjugate diameters (§ 40), it follows that the two screws of zero pitch, and any pair of reciprocal screws, are parallel to the rays of an harmonic pencil. If the pitch conic be an ellipse, there

* *Journal de l'école Polytechnique*, T. xx. cah. 43, pp. 57—122 (1870).

are no real screws of zero pitch. If the pitch conic be a parabola, there is but one screw of zero pitch, and this must be one of the two screws which intersect at right angles.

123. Equilibrium of a Body with Freedom of the Second Order.

We shall now consider more fully the conditions under which a body which has freedom of the second order is in equilibrium. The necessary and sufficient condition is, that the forces which act upon the body shall constitute a wrench upon a screw which is reciprocal to the cylindroid which defines the freedom of the body.

It has been shown (§ 23), that the screws which are reciprocal to a cylindroid exist in such profusion, that through every point in space a cone of the second order can be drawn, of which the entire superficies is made up of such screws. We shall now examine the distribution of pitch upon such a cone.

The pitch of each reciprocal screw is equal in magnitude, and opposite in sign, to the pitches of the two screws of equal pitch, in which it intersects the cylindroid (§ 22). Now, the greatest and least pitches of the screws on the cylindroid are p_α and p_β (§ 18). For the quantity $p_\alpha \cos^2 l + p_\beta \sin^2 l$ is always intermediate between $p_\alpha \cos^2 l + p_\alpha \sin^2 l$ and $p_\beta \cos^2 l + p_\beta \sin^2 l$. Hence it follows that the generators of the cone which meet the cylindroid in *three real* points must have pitches intermediate between p_α and p_β. It is also to be observed that, as only one line can be drawn through the vertex of the cone to intersect any two given screws on the cylindroid, so only one screw of any given pitch can be found on the reciprocal cone.

One screw can be found upon the reciprocal cone of every pitch from $-\infty$ to $+\infty$. The line drawn through the vertex parallel to the nodal line is a generator of the cone to which infinite pitch must be assigned. Setting out from this line around the cone the pitch gradually decreases to zero, then becomes negative, and increases to negative infinity, when we reach the line from which we started. We may here notice that when a screw has infinite pitch, we may regard the infinity as either + or − indifferently. If we conceive distances marked upon each generator of the cone from the vertex, equal to the pitch of that generator, then the parallel to the nodal line drawn from the vertex forms an asymptote to the curve so traced upon the cone. It is manifest that we must admit the cylindroid to possess imaginary screws, whose pitch is nevertheless real.

The reciprocal cone drawn from a point to a cylindroid, is decomposed into two planes, when the point lies upon the cylindroid. The first plane is normal to the generator passing through the point. Every line in this plane must, when it receives the proper pitch, be a reciprocal screw. The second plane is that drawn through the point, and through the other screw on the cylindroid, of equal pitch to that which passes through the point.

We have, therefore, solved in the most general manner the problem of the equilibrium of a rigid body with two degrees of freedom. We have shown that the necessary and sufficient condition is, that the resultant wrench be about a screw reciprocal to the cylindroid expressing the freedom, and we have seen the manner in which the reciprocal screws are distributed through space. We now add a few particular cases.

124. Particular Cases.

A body which has two degrees of freedom is in equilibrium under the action of a force, whenever the line of action of the force intersects both the screws of zero pitch upon the cylindroid.

If a body acted upon by gravity have freedom of the second order, the necessary and sufficient condition of equilibrium is, that the vertical through the centre of inertia shall intersect both of the screws of zero pitch.

A body which has freedom of the second order will be in equilibrium, notwithstanding the action of a couple, provided the axis of the couple be parallel to the nodal line of the cylindroid.

A body which has freedom of the second order will remain in equilibrium, notwithstanding the action of a wrench about a screw of *any* pitch on the nodal line of the cylindroid.

125. The Impulsive Cylindroid and the Instantaneous Cylindroid.

A rigid body M is at rest in a position P, from which it is either partially or entirely free to move. If M receive an impulsive wrench about a screw X_1, it will commence to twist about an instantaneous screw A_1, if, however, the impulsive wrench had been about X_2 or X_3 (M being in either case at rest in the position P) the instantaneous screw would have been A_2, or A_3. Then we have the following theorem:—

If X_1, X_2, X_3 lie upon a cylindroid S (which we may call the impulsive cylindroid), then A_1, A_2, A_3 lie on a cylindroid S' (which we may call the instantaneous cylindroid).

For if the three wrenches have suitable intensities they may equilibrate, since they are co-cylindroidal; when this is the case the three instantaneous twist velocities must, of course, neutralise; but this is only possible if the instantaneous screws be co-cylindroidal (§ 93).

If we draw a pencil of four lines through a point parallel to four generators of a cylindroid, the lines forming the pencil will lie in a plane. We may define the *anharmonic ratio of four generators on a cylindroid* to be the anharmonic ratio of the parallel *pencil*. We shall now prove the following theorem :—

The anharmonic ratio of four screws on the impulsive cylindroid is equal to the anharmonic ratio of the four corresponding screws on the instantaneous cylindroid.

Before commencing the proof we remark that,

If an impulsive wrench of intensity F acting on the screw X be capable of producing the unit of twist velocity about A, then an impulsive wrench of intensity $F\omega$ on X will produce a twist velocity ω about A.

Let X_1, X_2, X_3, X_4 be four screws on the impulsive cylindroid, the intensities of the wrenches appropriate to which are $F_1\omega_1$, $F_2\omega_2$, $F_3\omega_3$, $F_4\omega_4$. Let the four corresponding instantaneous screws be A_1, A_2, A_3, A_4, and the twist velocities be ω_1, ω_2, ω_3, ω_4. Let ϕ_m be the angle on the impulsive cylindroid defining X_m, and let θ_m be the angle on the instantaneous cylindroid defining A_m.

If three impulsive wrenches equilibrate, the corresponding twist velocities neutralize by the second law of motion: hence (§ 14) certain values of ω_1, ω_2, ω_3, ω_4 must satisfy the following equations :—

$$\frac{\omega_1}{\sin(\theta_2-\theta_3)}=\frac{\omega_2}{\sin(\theta_3-\theta_1)}=\frac{\omega_3}{\sin(\theta_1-\theta_2)},$$

$$\frac{F_1\omega_1}{\sin(\phi_2-\phi_3)}=\frac{F_2\omega_2}{\sin(\phi_3-\phi_1)}=\frac{F_3\omega_3}{\sin(\phi_1-\phi_2)},$$

$$\frac{\omega_2}{\sin(\theta_3-\theta_4)}=\frac{\omega_3}{\sin(\theta_4-\theta_2)}=\frac{\omega_4}{\sin(\theta_2-\theta_3)},$$

$$\frac{F_2\omega_2}{\sin(\phi_3-\phi_4)}=\frac{F_3\omega_3}{\sin(\phi_4-\phi_2)}=\frac{F_4\omega_4}{\sin(\phi_2-\phi_3)},$$

whence

$$\frac{\sin(\theta_1-\theta_2)\sin(\theta_3-\theta_4)}{\sin(\theta_3-\theta_1)\sin(\theta_4-\theta_2)}=\frac{\sin(\phi_1-\phi_2)\sin(\phi_3-\phi_4)}{\sin(\phi_3-\phi_1)\sin(\phi_4-\phi_2)},$$

which proves the theorem.

If we are given three screws on the impulsive cylindroid, and the corresponding three screws on the instantaneous cylindroid, the connexion between every other corresponding pair is, therefore, geometrically determined.

126. Reaction of Constraints.

Whatever the constraints may be, their reaction produces an impulsive wrench R_1 upon the body at the moment of action of the impulsive wrench X_1. The two wrenches X_1 and R_1 compound into a third wrench Y_1. If the body were free, Y_1 is the impulsive wrench to which the instantaneous screw A_1 would correspond. Since X_1, X_2, X_3 are co-cylindroidal, A_1, A_2, A_3 must be co-cylindroidal, and therefore also must be Y_1, Y_2, Y_3. The nine wrenches X_1, X_2, X_3, R_1, R_2, R_3, $-Y_1$, $-Y_2$, $-Y_3$ must equilibrate; but if X_1, X_2, X_3 equilibrate, then the twist velocities about A_1, A_2, A_3 must neutralize, and therefore the wrenches about Y_1, Y_2, Y_3 must equilibrate. Hence R_1, R_2, R_3 equilibrate, and are therefore co-cylindroidal.

Following the same line of proof used in the last section, we can show that

If impulsive wrenches on any four co-cylindroidal screws act upon a partially free rigid body, the four corresponding initial reactions of the constraints also constitute wrenches about four co-cylindroidal screws; and, further, the anharmonic ratios of the two groups of four screws are equal.

127. Principal Screws of Inertia.

If a quiescent body with freedom of the second order receive impulsive wrenches on three screws X_1, X_2, X_3 on the cylindroid which expresses the freedom, and if the corresponding instantaneous screws on the same cylindroid be A_1, A_2, A_3, then the relation between any other impulsive screw X on the cylindroid and the corresponding instantaneous screw A is completely defined by the condition that the anharmonic ratio of X, X_1, X_2, X_3 is equal to the anharmonic ratio of A, A_1, A_2, A_3.

If three rays parallel to X_1, X_2, X_3 be drawn from a point, and from the same point three rays parallel to A_1, A_2, A_3, then, all six rays being in the same plane, it is well known that the problem to determine a ray Z such that the anharmonic ratio of Z, A_1, A_2, A_3 is equal to that of Z, X_1, X_2, X_3, admits of two solutions. There are, therefore, two screws on a cylindroid such that an impulsive wrench on one of these screws will cause the body to commence to twist about the same screw.

We have thus arrived by a special process at the two principal screws of inertia possessed by a body which has freedom of the second order. This is, of course, a particular case of the general theorem of § 78. We shall show in the next section how these screws can be determined in another manner.

128. The Ellipse of Inertia.

We have seen (§ 89) that a linear parameter u_α may be conceived appropriate to any screw α of a system, so that when the body is twisting about the screw α with the unit of twist velocity, the kinetic energy is found by multiplying the mass of the body into the square of the line u_α.

We are now going to consider the distribution of this magnitude u_α on the screws of a cylindroid. If we denote by u_1, u_2 the values of u_α for any pair of conjugate screws of inertia on the cylindroid (§ 81), and if by α_1, α_2 we denote the intensities of the components on the two conjugate screws of a wrench of unit intensity on α, we have (§ 97)—

$$u_\alpha^2 = u_1^2\alpha_1^2 + u_2^2\alpha_2^2.$$

From the centre of the cylindroid draw two straight lines parallel to the pair of conjugate screws of inertia, and with these lines as axes of x and y construct the ellipse of which the equation is

$$u_1^2x^2 + u_2^2y^2 = H,$$

where H is any constant. If r be the radius vector in this ellipse, we have (§ 35)

$$\frac{x}{r} = \alpha_1 \text{ and } \frac{y}{r} = \alpha_2;$$

whence by substitution we deduce

$$u_\alpha^2 = \frac{H}{r^2},$$

which proves the following theorem:—

The linear parameter u_α on any screw of the cylindroid is inversely proportional to the parallel diameter of a certain ellipse, and a pair of conjugate screws of inertia on the cylindroid are parallel to a pair of conjugate diameters of the same ellipse. This ellipse may be called the *ellipse of inertia*.

The major and minor axes of the ellipse of inertia are parallel to screws upon the cylindroid, which for a given twist velocity correspond respectively to a maximum and minimum kinetic energy.

An impulsive wrench on a screw η acts upon a quiescent rigid body which has freedom of the second order. It is required to determine the screw θ on the cylindroid expressing the freedom about which the body will commence to twist.

The ellipse of inertia enables us to solve this problem with great facility. Determine that one screw ϕ on the cylindroid which is reciprocal to η (§ 26). Draw a diameter D of the ellipse of inertia parallel to ϕ. Then the required screw θ is simply that screw on the cylindroid which is parallel to the diameter conjugate to D in the ellipse of inertia.

The converse problem, viz., to determine the screw η, an impulsive wrench on which would make the body commence to twist about θ, is indeterminate. Any screw in space which is *reciprocal to* ϕ would fulfil the required condition (§ 136).

We have seen in § 96 that an impulsive wrench on any screw in space may generally be replaced by a precisely equivalent wrench upon the cylindroid which expresses the freedom. We are now going to determine the screw η, on the cylindroid of freedom, an impulsive wrench on which would make the body twist about a *given* screw θ on the same cylindroid. This can be easily determined with the help of the pitch conic; for we have seen (§ 40) that a pair of reciprocal screws on the cylindroid of freedom are parallel to a pair of conjugate diameters of the pitch conic. The construction is therefore as follows:—Find the diameter A which is conjugate, with respect to the *ellipse of inertia*, to the diameter parallel to the given screw θ. Next find the diameter B which is conjugate to the diameter A with respect to the *pitch conic*. The screw on the cylindroid parallel to the line B thus determined is the required screw η.

Two concentric ellipses have one pair of common conjugate diameters. In fact, the four points of intersection form a parallelogram, to the sides of which the pair of common conjugate diameters are parallel. We can now interpret physically the common conjugate diameters of the pitch conic, and the ellipse of inertia. The two screws on the cylindroid parallel to these diameters are conjugate screws of inertia, and they are also reciprocal; they are, therefore, the *principal screws of inertia*, to which we have been already conducted (§ 127).

If the distribution of the material of the body bear certain relations to the arrangement of the constraints, we can easily conceive that the pitch conic and the ellipse of inertia might be both similar and similarly situated. Under these exceptional circumstances it appears that every screw of the cylindroid would possess the property of a principal screw of inertia.

129. The Ellipse of the Potential.

We are now to consider another ellipse, which, though possessing many useful mathematical analogies to the ellipse of inertia, is yet widely different from a physical point of view. We have introduced (§ 102) the conception of the linear magnitude v_α, the square of which is proportional to the work done in effecting a twist of given amplitude about a screw α from a position of stable equilibrium under the influence of a system of forces. We now propose to consider the distribution of the parameter v_α upon the screws of a cylindroid. It appears from § 102 that if v_1, v_2 denote the values of the quantity v_α for each of two conjugate screws of the potential, and if α_1, α_2 denote the intensities of the components on the two conjugate screws of a wrench of unit intensity on a screw α, which also lies upon the cylindroid, then—

$$v_\alpha^2 = v_1^2\alpha_1^2 + v_2^2\alpha_2^2.$$

From the centre of the cylindroid draw two straight lines parallel to the pair of conjugate screws of the potential, and with these lines as axes of x and y construct the ellipse, of which the equation is—

$$v_1^2x^2 + v_2^2y^2 = H,$$

where H is any constant. If r be the radius vector in this ellipse, we have—

$$\frac{x}{r} = \alpha_1 \text{ and } \frac{y}{r} = \alpha_2;$$

whence by substitution we deduce—

$$v_\alpha^2 = \frac{H}{r^2},$$

which proves the following theorem:—

The linear parameter v_α on any screw of the cylindroid is inversely proportional to the parallel diameter of a certain ellipse, and a pair of conjugate screws of the potential are parallel to a pair of conjugate diameters of the same ellipse.

This ellipse may be called the *ellipse of the potential.*

The major and minor axes of the ellipse of the potential are parallel to screws upon the cylindroid, which, for a twist of given amplitude, correspond to a maximum and minimum potential energy.

When the body has to relinquish its original position of equilibrium by the addition of a wrench on a screw η to the forces previously in operation, the twist by which the body may proceed to its new position of equilibrium is about a screw θ, which can be constructed by the ellipse of the potential. Determine the screw ϕ (on the cylindroid of freedom) which is reciprocal to η (§ 26), then ϕ, and the required screw θ, are parallel to a pair of conjugate diameters of the ellipse of the potential.

The common conjugate diameters of the pitch conic, and the ellipse of the potential, are parallel to the two screws on the cylindroid, which we have designated the *principal screws of the potential* (§ 101).

When a body is disturbed from its position of equilibrium by a small wrench upon a principal screw of the potential, then the body could be moved to the new position of equilibrium required in its altered circumstances by a small twist about the same screw.

130. Harmonic Screws.

The common conjugate diameters of the ellipse of inertia, and the ellipse of the potential, are parallel to the two harmonic screws on the cylindroid (§ 104). This is evident, because the pair of screws thus determined are conjugate screws both of inertia and of the potential.

If the body be displaced by a twist about one of the harmonic screws, and be then abandoned to the influence of the forces, the body will continue to perform twist oscillations about that screw.

If the ellipse of inertia, and the ellipse of the potential, be similar, and similarly situated, it follows that every screw on the cylindroid will be a harmonic screw.

131. Exceptional Case.

We have now to consider the modifications which the results we have arrived at undergo when the cylindroid becomes illusory in the case considered (§ 120).

Suppose that ξ and ζ were a pair of conjugate screws of inertia on the straight line about which the body was free to rotate and slide independently. Then taking the six absolute principal screws of inertia as screws of reference, we must have (§ 97)—

$$\Sigma p_1^2 \left(\eta_1 + \frac{p_\xi}{4p_1}\frac{dR}{d\eta_1}\right)\left(\eta_1 + \frac{p_\zeta}{4p_1}\frac{dR}{d\eta_1}\right) = 0,$$

where η denotes the screw of zero pitch on the same straight line.

Expanding this equation, and reducing, we find

$$u_\eta^2 + \frac{1}{4}(p_\xi + p_\zeta)\left(\Sigma p_1\eta_1 \frac{dR}{d\eta_1}\right) + \frac{1}{16} p_\xi p_\zeta \left\{\Sigma \left(\frac{dR}{d\eta_1}\right)^2\right\} = 0.$$

This result can be much simplified. By introducing the condition that as in § 120—

$$R = (\eta_1 + \eta_2)^2 + (\eta_3 + \eta_4)^2 + (\eta_5 + \eta_6)^2,$$

we obtain

$$\Sigma p_1\eta_1 \frac{dR}{d\eta_1} = 2\Sigma p_1\eta_1^2 = 2p_\eta = 0\,; \quad \Sigma \left(\frac{dR}{d\eta_1}\right)^2 = 8.$$

Hence we can prove (§ 133) that in this case the product of the pitches of two conjugate screws of inertia is equal to minus the square of the radius of gyration about the common axis of the screws.

132. Reaction of Constraints.

We shall now consider the following problem:—A body which is free to twist about all the screws of a cylindroid C receives an impulsive wrench on a certain screw η. It is required to find the screw λ, a wrench on which constitutes the impulsive reaction of the constraints. Let C' represent the cylindroid which, if the body were perfectly free, would form the locus of those screws, impulsive wrenches on which correspond to all the screws of C as instantaneous screws. Since a wrench on η, and one on λ, make the body twist about some screw on C, it follows that the cylindroid (η, λ) must have a screw ρ in common with C'. The wrench on λ might be resolved into two, one on η, and the other on ρ, and the latter might be again resolved into two wrenches on any two screws of C'. It therefore follows that λ must belong to the screw system of the third order, which may be defined by η, and by any two screws from C'. Take any three screws reciprocal to this system, and any two screws on C. We have then five screws to which λ is reciprocal, and it is therefore geometrically determined (§ 26).

When λ is found, the cylindroid (η, λ) can be drawn, and thus ρ is determined. The position of ρ on C' will point out the screw on C, about which the body will commence to twist, while the position of ρ on (η, λ), and the known intensity of the wrench on η, will determine the intensity of the wrench on λ.

CHAPTER XII.

PLANE REPRESENTATION OF DYNAMICAL PROBLEMS CONCERNING A BODY WITH TWO DEGREES OF FREEDOM*.

133. The Kinetic Energy.

If a rigid body of mass M twist about a screw θ, with the twist velocity $\dot{\theta}$, then the kinetic energy of the body may be written in the form

$$Mu_\theta^2\dot{\theta}^2,$$

where u_θ is a linear magnitude appropriate to the screw θ (§ 89).

The function u_θ^2 is the arithmetic mean between the square of the radius of gyration and the square of the pitch, for the kinetic energy of the body when twisting about θ is the sum of two parts: one, the kinetic energy of the rotation; the other, of the translation. The energy of the rotation is simply

$$\tfrac{1}{2}M\rho_\theta^2\dot{\theta}^2,$$

this being in accordance with the definition of the radius of gyration ρ_θ. The kinetic energy due to the translation is, of course,

$$\tfrac{1}{2}Mp_\theta^2\dot{\theta}^2,$$

whence the total kinetic energy is

$$\tfrac{1}{2}M\dot{\theta}^2(\rho_\theta^2 + p_\theta^2),$$

and therefore

$$u_\theta^2 = \tfrac{1}{2}(\rho_\theta^2 + p_\theta^2).$$

134. Body with two Degrees of Freedom.

The movements are under these circumstances restricted to twists about the screws of a cylindroid, and we shall now examine the law of distribution

* *Royal Irish Academy, Cunningham Memoirs*, No. 4, p. 19 (1886); see also *Proceedings of the Royal Irish Academy*, 2nd Series, Vol. IV. p. 29 (1883).

of u_θ upon the several screws of the cylindroid (§ 128). The representative circle (§ 50) will give a convenient geometrical construction.

Let θ_1 and θ_2 be the two co-ordinates of θ relatively to any two screws of reference on the cylindroid. Then the components of the twist velocity will be $\dot{\theta}\theta_1$ and $\dot{\theta}\theta_2$. The actual velocity of any point of the body will necessarily be a linear function of these components. The square of the velocity will contain terms in which $\dot{\theta}^2$ is multiplied into θ_1^2, $\theta_1\theta_2$, θ_2^2, respectively. If, then, by integration we obtain the total kinetic energy, it must assume the form

$$M\dot{\theta}^2 (\lambda\theta_1^2 + 2\mu\theta_1\theta_2 + \nu\theta_2^2),$$

whence, from the definition of u_θ,

$$u_\theta^2 = \lambda\theta_1^2 + 2\mu\theta_1\theta_2 + \nu\theta_2^2.$$

The three constants, λ, μ, ν, are the same for all screws on the cylindroid. They are determined by the material disposition of the body relatively to the cylindroid.

We have taken the two screws of reference arbitrarily, but this equation can receive a remarkable simplification when the two screws of reference have been chosen with special appropriateness.

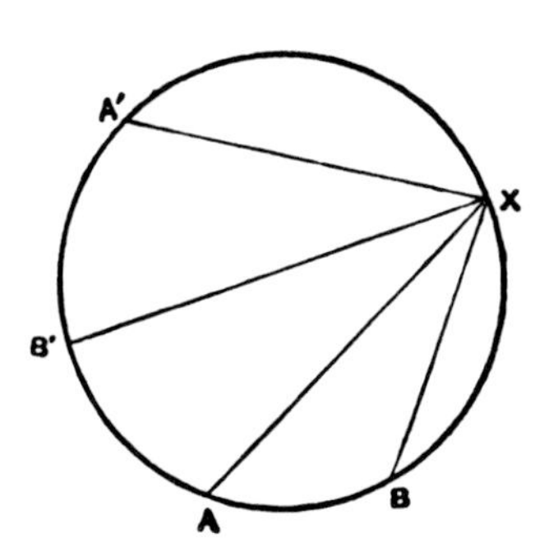

Fig. 18.

Let the lengths AX and BX (fig. 18) be denoted by ρ_1 and ρ_2, and if ϵ be the angle subtended by AB, we have from § 57,

$$\lambda\rho_1^2 + 2\mu\rho_1\rho_2 + \nu\rho_2^2 - u_\theta^2 (\rho_1^2 - 2\rho_1\rho_2 \cos\epsilon + \rho_2^2) = 0.$$

Let us now transform this equation from the screws of reference A, B to another pair of screws A', B'. Let ρ_1', ρ_2' be the distances of X from A', B', respectively; then, from Ptolemy's theorem, we have the following equations:—

$$\rho_1 \,.\, A'B' = \rho_2' \,.\, AA' - \rho_1' \,.\, AB',$$

$$\rho_2 \,.\, A'B' = \rho_2' \,.\, A'B - \rho_1' \,.\, BB'.$$

We thus see that ρ_1 and ρ_2 are linear functions of ρ_1' and ρ_2', the several coefficients $A'B'$, $A'B$, &c., in these two equations being constant. The equation for u_θ^2 is thus to be transformed by a linear substitution for ρ_1 and ρ_2. Of course u_θ, being dependent only upon the position of X, is quite unaffected by the change of the screws of reference. We can therefore apply the well-known principle that the invariant of this binary quantic can only differ by a constant factor from the transformed value. The invariant is

$$(\lambda - u_\theta^2)(\nu - u_\theta^2) - (\mu + u_\theta^2 \cos \epsilon)^2.$$

This must be true for *every* point X, and therefore for all values of u_θ^2. It is necessary that the coefficients of the terms in the expression

$$u_\theta^4 \sin^2 \epsilon - u_\theta^2 (\lambda + \nu + 2\mu \cos \epsilon) + \lambda\nu - \mu^2$$

shall be severally proportional to those in the transformed expression

$$u_\theta^4 \sin^2 \epsilon' - u_\theta^2 (\lambda' + \nu' + 2\mu' \cos \epsilon') + \lambda'\nu' - \mu'^2.$$

We thus obtain the two equations of condition,

$$\frac{\sin^2 \epsilon'}{\sin^2 \epsilon} = \frac{\lambda' + \nu' + 2\mu' \cos \epsilon'}{\lambda + \nu + 2\mu \cos \epsilon} = \frac{\lambda'\nu' - \mu'^2}{\lambda\nu - \mu^2}.$$

The four quantities, λ', μ', ν', ϵ', may now be chosen arbitrarily, subject to these two equations, which are the necessary as well as the sufficient conditions. Indeed it is obvious that there must be but two independent quantities corresponding to the two positions of A' and B'.

We may impose two conditions on the four quantities, and for our present purpose we shall make

$$\lambda' = \nu'; \qquad \mu' = 0.$$

The equations of λ' and ϵ' are then

$$\frac{\sin^2 \epsilon'}{\sin^2 \epsilon} = \frac{2\lambda'}{\lambda + 2\mu \cos \epsilon + \nu} = \frac{\lambda'^2}{\lambda\nu - \mu^2},$$

and we obtain

$$\lambda' = \frac{2(\lambda\nu - \mu^2)}{\lambda + 2\mu \cos \epsilon + \nu},$$

$$\sin^2 \epsilon' = \sin^2 \epsilon \frac{4(\lambda\nu - \mu^2)}{(\lambda + 2\mu \cos \epsilon + \nu)^2};$$

λ' is thus uniquely determined, and the expression for $\sin^2 \epsilon'$ gives for ϵ' four values of the type $\pm \epsilon'$, $\pm (\pi - \epsilon')$. The negative values are meaningless, and the two others are coincident, because the arc which subtends ϵ' on one side subtends $\pi - \epsilon'$ on the other.

There is thus a single pair of screws of reference which permit the expression for u_θ^2 to be exhibited in the canonical form

$$c^2 u_\theta^2 = \lambda' (\rho_1^2 + \rho_2^2).$$

We are now led to a simple geometrical representation for u_θ^2. Let A, B (fig. 19) be the two canonical screws of reference. Bisect AB in O', then

$$\begin{aligned} \rho_1^2 + \rho_2^2 &= BX^2 + AX^2, \\ &= 2AO'^2 + 2XO'^2, \\ &= 2XO' . YO' + 2XO'^2, \\ &= 2XY . XO'. \end{aligned}$$

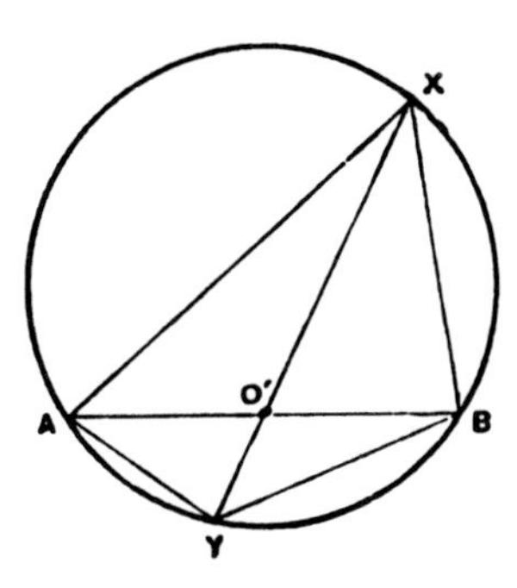

Fig. 19.

It is obvious that the point O' must have a critical importance in the kinetic theory, and its fundamental property, which has just been proved, is expressed in the following theorem:—

If a rigid body be twisting with the unit of twist velocity about any screw X on the cylindroid, then its kinetic energy is proportional to the rectangle $XO' . XY$, where O' is a fixed point.

We are at once reminded of the theorem of § 59, in which a similar law is found for the distribution of pitch, only in this case another point, O, is used instead of the point O'. Both points, O and O', are of much significance in the representative circle. We can easily prove the following theorem, in which we call the polar of O' the *axis of inertia*:—

If a rigid body be twisting with the unit of twist velocity about X, then its kinetic energy is proportional to the perpendicular distance from X to the axis of inertia.

The geometrical construction for the pitch given in § 51 can also be applied to determine u_θ^2. This quantity is therefore proportional to the perpendicular from O' on the tangent at X. It thus appears that the representative circle gives a graphic illustration of the law of distribution of u_θ^2 around the screws on a cylindroid.

The axis of inertia cannot cut the representative circle in real points, for otherwise we should have at either intersection a twist velocity without any kinetic energy. There is no similar restriction to the axis of pitch. We thus see that O' must always lie inside the circle, but that O may be in any part of the plane.

135. Conjugate Screws of Inertia.

We have already made much use of the conception of Conjugate Screws of Inertia. We shall here approach the subject in a manner different from that previously employed.

Let α be a screw about which a rigid body is twisting with a twist velocity $\dot{\alpha}$; let the body be simultaneously animated by a twist velocity $\dot{\beta}$ about a screw β. These two will compound into a twist velocity $\dot{\theta}$ about some screw θ. If the body only had the first twist velocity, its kinetic energy would be $Mu_\alpha^2\dot{\alpha}^2$. If it only had the second, the energy would be $Mu_\beta^2\dot{\beta}^2$. When it has both twist velocities together, the kinetic energy is $Mu_\theta^2\dot{\theta}^2$. Generally it will not be true that the resulting kinetic energy is equal to the sum of the components; but, under a special relation between α and β, we can have this equality; and as shown in § 88 under these circumstances α and β are *conjugate screws of inertia*. The necessary condition is thus expressed:—

$$u_\theta^2\dot{\theta}^2 = u_\alpha^2\dot{\alpha}^2 + u_\beta^2\dot{\beta}^2.$$

We have now to prove the following important theorem:—

Any chord through the pole of the axis of inertia intersects the representative circle in a pair of conjugate screws of inertia.

For we have

$$\dot{\theta}^2 : \dot{\alpha}^2 : \dot{\beta}^2 :: AB^2 : BX^2 : AX^2;$$

but if AB passes through the pole of the axis of inertia, then the centre of gravity of masses $-AB^2$ at X, $+BX^2$ at A, and $+AX^2$ at B, will lie on the axis of inertia; and, accordingly,

$$AB^2u_\theta^2 = BX^2u_\alpha^2 + AX^2u_\beta^2;$$

whence

$$u_\theta^2\dot{\theta}^2 = u_\alpha^2\dot{\alpha}^2 + u_\beta^2\dot{\beta}^2,$$

which proves the theorem.

Or we might have proceeded thus:—From Ptolemy's theorem (fig. 19),

$$AB.XY = AX.BY + AY.BX:$$

multiplying by $AB.XO'$,

$$AB^2.XY.XO' = AX.AB.BY.XO' + AY.XO'.BX.AB;$$

but, from the property of the circle,

$$BY \,.\, XO' = AX \,.\, BO'; \quad AY \,.\, XO' = BX \,.\, AO';$$

whence

$$AB^2 \,.\, XY \,.\, XO' = AX^2 \,.\, AB \,.\, BO' + BX^2 \,.\, AB \,.\, AO',$$

from which we obtain, as before,

$$\dot{\theta}^2 u_\theta^2 = \dot{\alpha}^2 u_\alpha^2 + \dot{\beta}^2 u_\beta^2.$$

136. Impulsive Screws and Instantaneous Screws.

A rigid body having two degrees of freedom lies initially at rest. It is suddenly acted upon by an impulsive wrench of large intensity acting for a short time. The body will, in general, commence to move by twisting about some screw on the cylindroid, and the kinetic problem now to be studied is the following:—Given the impulsive screw, and the intensity of the impulsive wrench, find the instantaneous screw and the acquired twist velocity.

The problem will be rendered more concise by the conception of the reduced wrench (§ 96). It is to be remembered, that as the body is only partially free, there are an infinite number of screws on which wrenches would make the body commence to twist about a given screw on the cylindroid. For, let θ be an impulsive screw situated anywhere, and let an impulsive wrench on θ cause the body to commence to move by twisting about some screw, α, on the cylindroid. Let λ, μ, ν, ρ be any four screws reciprocal to the cylindroid. Then any wrench on a screw belonging to the system defined by these five screws will make the body commence to move by twisting about α. Let ϵ be that one screw on the cylindroid which is reciprocal to θ, then ϵ is reciprocal to the whole system defined by λ, μ, ν, ρ, θ, and, conversely, each screw of this system will be reciprocal to ϵ. We thus see that any screw, wherever situated, provided only that it is reciprocal to ϵ, will be an impulsive screw corresponding to α as an instantaneous screw. Any one of this system may, with perfect generality, be chosen as the impulsive screw. Among them there is one which has a special feature. It is that screw, ϕ, on the cylindroid which is reciprocal to η; and hence we have the following theorem (§ 128):—

Given any screw, α, on the cylindroid, then there is in general another screw, ϕ, also on the cylindroid, such that an impulsive wrench administered on ϕ will make the body twist about α.

This correspondence of the two systems of screws must be of the one-to-one type; for, suppose that two impulsive screws on the cylindroid had the same instantaneous screw, it would then be possible for two impulsive wrenches, of properly chosen intensities on two different screws, to produce equal and opposite twist velocities on the common instantaneous screw. The body would then not move, and therefore the two impulsive wrenches must equilibrate. But this is impossible, if they are on two different screws.

137. Two Homographic Systems.

From what has been shown it might be expected that the points corresponding to the instantaneous screws and those corresponding to the impulsive screws should, on the representative circle, form two homographic systems. That this is so we shall now prove.

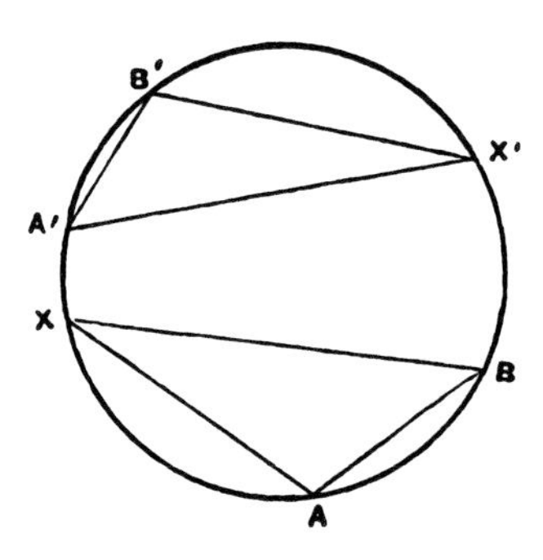

Fig. 20.

Let A, B (fig. 20) be a pair of impulsive screws, and let A', B' be respectively the corresponding pair of instantaneous screws, i.e. an impulsive wrench on A will make the body commence to twist about A', and similarly for B and B'. Let an impulsive wrench on A, of unit intensity, generate a twist velocity, $\dot{\alpha}$, about A', and let $\dot{\beta}$ be the similar quantity for B and B'.

Let X be any other screw on which an impulsive wrench is to be applied to the body supposed quiescent. The body will commence to twist about some other screw, X', with a certain twist velocity $\dot{\omega}$. We can determine $\dot{\omega}$ in the following manner:—The unit impulsive wrench on X can be replaced by two component wrenches on A and B, the intensities of these being

$$\frac{BX}{AB}, \quad \frac{AX}{AB},$$

respectively.

These impulsive wrenches will generate about A', B' twist velocities respectively equal to

$$\dot{\alpha}\frac{BX}{AB}, \quad \dot{\beta}\frac{AX}{AB}:$$

these components must, when compounded, produce the twist velocity $\dot{\omega}$ about X', and, accordingly, we have

$$\dot{\alpha}\frac{BX}{AB} = \dot{\omega}\frac{B'X'}{A'B'}; \quad \dot{\beta}\frac{AX}{AB} = \dot{\omega}\frac{A'X'}{A'B'}.$$

Retaining A, B, A', B', as before, let us now introduce a second pair of points, Y and Y', instead of X and X', and writing $\dot{\omega}'$ instead of $\dot{\omega}$, we have

$$\dot{\alpha}\frac{BY}{AB} = \dot{\omega}'\frac{B'Y'}{A'B'}; \quad \dot{\beta}\frac{AY}{AB} = \dot{\omega}'\frac{A'Y'}{A'B'};$$

whence, eliminating $\dot{\alpha}$, $\dot{\beta}$, $\dot{\omega}$, $\dot{\omega}'$, we have

$$\frac{BX}{AX} : \frac{BY}{AY} :: \frac{B'X'}{A'X'} : \frac{B'Y'}{A'Y'}.$$

As the length of a chord is proportional to the sine of the subtended angle, we see that the anharmonic ratio of the pencil, subtended by the four points A, B, X, Y at a point on the circumference, is equal to that subtended by their four correspondents, A', B', X', Y'. We thus learn the following important theorem:—

A system of points on the representative circle, regarded as impulsive screws, and the corresponding system of instantaneous screws, form two homographic systems.

138. The Homographic Axis.

Let A, B, C, D (fig. 21) represent four impulsive screws, and let A', B', C', D' be the four corresponding instantaneous screws. Then, by the well-

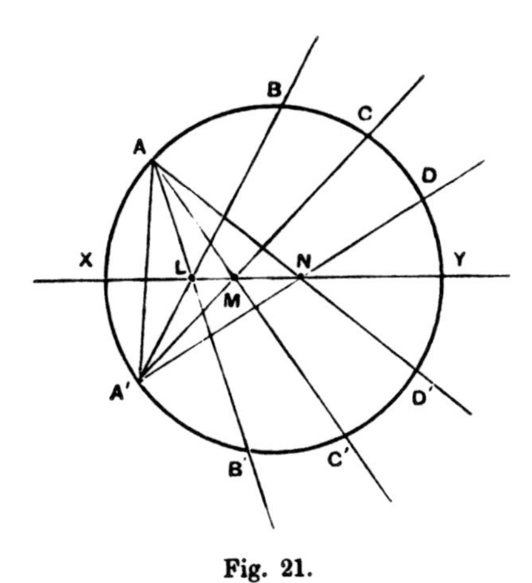

Fig. 21.

known homographic properties of the circle, the three points, L, M, N, will be collinear, and we have the following theorem:—

If A and B be any two impulsive screws, and if A' and B' be the corresponding instantaneous screws, then the chords AB' and BA' will always intersect upon the fixed right line XY.

This right line is called the homographic axis. It intersects the circle in two points, X and Y, which are the double points of the homographic systems. These points enjoy a special dynamical significance. They are the two Principal Screws of Inertia, and hence—

The homographic axis intersects the circle in two points, each of which possesses the property, that an impulsive wrench administered on that screw will make the body commence to move by twisting about the same screw.

The method by which we have been conducted to the Principal Screws of Inertia shows how there are in general two, and only two, of these screws on the cylindroid. The homographic axis is the Pascal line, for the Hexagon $AA'BB'CC'$, and thus we have a dynamical significance for Pascal's theorem.

139. Determination of the Homographic Axis.

The two principal screws of inertia must be reciprocal, and must also be conjugate screws of inertia (§ 84). The homographic axis must therefore comply with the conditions thus prescribed. We have already shown (§ 58) the condition that two screws be reciprocal, and (§ 135) the condition that two screws be conjugate screws of inertia, and, accordingly, we see—

1°. That the homographic axis must pass through O, the pole of the axis of pitch.

2°. That the homographic axis must pass through O', the pole of the axis of inertia.

The points O and O' having been already determined we have accordingly, as the simplest construction for the homographic axis, the chord joining O and O'.

140. Construction for Instantaneous Screws.

The points O and O' afford a simple construction for the instantaneous screw, corresponding to a given impulsive screw. The construction depends upon the following theorem (§ 81):—

If two conjugate screws of inertia be regarded as instantaneous screws, then the impulsive screw corresponding to either is reciprocal to the other.

Let A be an impulsive screw (fig. 22); if we join AO we obtain H, the screw reciprocal to A; and if we join HO' we obtain A', the conjugate screw of inertia to H. But, as A is the only screw reciprocal to H, it is necessary, by the theorem just given, that an impulsive wrench on A must make the body commence to move by twisting about A'.

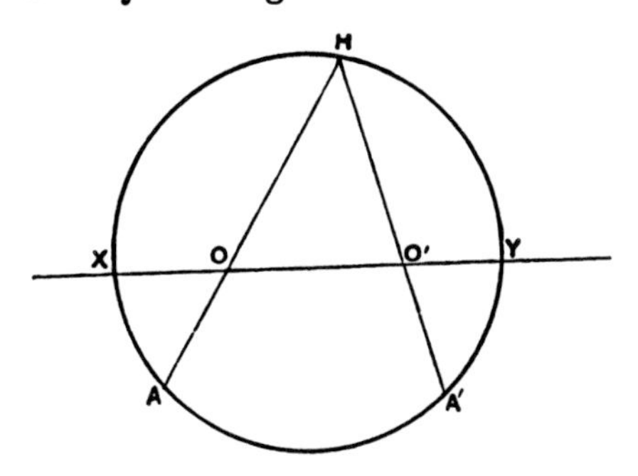

Fig. 22.

As O and O' are fixed, it follows from a well-known theorem, that as otherwise proved in § 137, A and A' form two homographic systems.

141. Twist Velocity acquired by an Impulse.

We can obtain a geometrical expression for the twist velocity acquired about A' by a unit impulsive wrench on A (Fig. 22).

It appears, from § 90 (see also § 147), that the twist velocity acquired on α by an impulsive wrench on η, is proportional to

$$\frac{\varpi_{\eta\alpha}}{u_\alpha^2}.$$

The numerator being the virtual coefficient is proportional to $AO \,.\, A'H$ (§ 63), and as u_α^2 is proportional to $A'O'.\,A'H$ (§ 134), we see that the required ratio varies as $AO \div A'O'$ which itself varies as

$$\frac{HO'}{HO};$$

hence we obtain the following theorem:—

The impulsive wrench on A, of intensity proportional to HO, generates a twist motion about A', with velocity proportional to HO'.

The geometrical representation of the effect of impulsive forces is thus completely determined both as regards the instantaneous screw, and the instantaneous twist velocity acquired.

142. Another Construction for the Twist Velocity.

A still more concise method of determining the instantaneous screw can be obtained if we discard the points O and O', and introduce a new fixed point, Ω, also on the homographic axis.

Let X, Y (Fig. 23) be the two principal screws of inertia. Let A be an impulsive screw, and A' the corresponding instantaneous screw. Draw

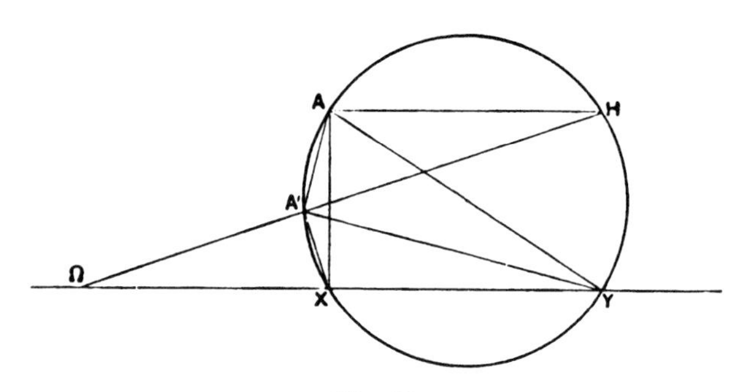

Fig. 23.

through A the line AH parallel to XY. Join HA', and produce it to meet the homographic axis at Ω. Let α be the twist velocity generated by an impulsive wrench of unit intensity at X, and let β be the corresponding quantity for Y.

It may be easily shown that the triangle $AA'X$ is similar to $YA'\Omega$, and that the triangle $AA'Y$ is similar to $XA'\Omega$; whence we obtain

$$\frac{A'X}{AX} = \frac{\Omega A'}{\Omega Y}; \quad \frac{A'Y}{AY} = \frac{\Omega A'}{\Omega X}.$$

The unit wrench on A can be decomposed into components on X and Y of respective intensities

$$\frac{AY}{XY}, \quad \frac{AX}{XY}.$$

These will generate twist velocities

$$\alpha \frac{AY}{XY}, \quad \beta \frac{AX}{XY}.$$

Let ω be the resulting twist velocity on A', then the components on X and Y must be equal to the quantities just written; whence

$$\omega \frac{A'Y}{XY} = \alpha \frac{AY}{XY},$$

$$\omega \frac{A'X}{XY} = \beta \frac{AX}{XY},$$

and we obtain

$$\alpha = \omega \frac{\Omega A'}{\Omega X}; \quad \beta = \omega \frac{\Omega A'}{\Omega Y};$$

or,

$$\alpha : \beta :: \Omega Y : \Omega X;$$

we thus see that Ω is a fixed point wherever A and A' may be.

It also follows that

$$\omega \Omega A'$$

is constant; whence we have the following theorem:—

Draw through the impulsive screw A a ray AH parallel to the homographic axis, then the ray from H to a fixed point Ω on the homographic axis will cut the circle in the instantaneous screw A', and the acquired twist velocity will be inversely proportional to $\Omega A'$.

If the twist velocity to be acquired by A' from a unit impulsive wrench on A be assigned, then $\Omega A'$ is determined: there will be two screws A', and two corresponding impulsive screws, either of which will solve the problem. The diameter through Ω indicates the two screws about which the body will acquire the greatest and the least velocities respectively with a given intensity for the impulsive wrench.

143. Twist Velocities on the Principal Screws.

The quantities α and β, which are the twist velocities acquired by unit impulsive wrenches on the principal screws, can be expressed geometrically as follows (Fig. 22):—

Let ω be the twist velocity acquired on A' by the wrench on A, then, by the last article,

$$\alpha AY = \omega A'Y,$$

$$\beta AX = \omega A'X;$$

whence

$$\alpha : \beta :: \frac{A'Y}{A'X} : \frac{AY}{AX}.$$

This ratio is the anharmonic ratio of the four points X, Y, A, A', that is, of X, Y, O, O'; whence, finally,

$$\alpha : \beta :: \frac{O'Y}{O'X} : \frac{OY}{OX}.$$

144. Another Investigation of the Twist Velocity acquired by an Impulse.

We have just seen that

$$\alpha AY = \omega A'Y,$$

$$\beta AX = \omega A'X;$$

whence

$$\alpha\beta AX \,.\, AY = \omega^2 A'X \,.\, A'Y.$$

Let fall perpendiculars AP, $A'P'$, HQ on the homographic axis (Fig. 24). Then, by the properties of the circle,

$$AX \,.\, AY : A'X \,.\, A'Y :: AP : A'P';$$

so that

$$\alpha\beta AP = \omega^2 A'P'.$$

By similar triangles,

$$\alpha\beta \frac{AO}{OH} \cdot HQ = \omega^2 \frac{O'A'}{O'H} \cdot HQ;$$

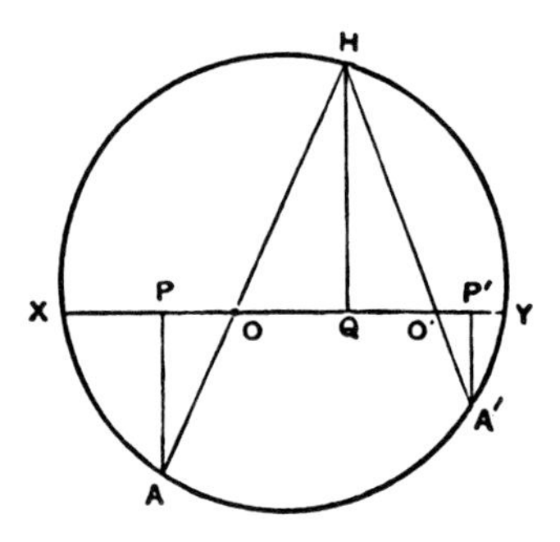

Fig. 24.

whence $$\omega^2 = \alpha\beta \frac{OA \cdot O'H}{O'A' \cdot OH} \propto \frac{O'H^2}{OH^2};$$

or, as before (§ 141), $$\omega \propto \frac{O'H}{OH}.$$

It will be noticed that, for this investigation, H may have been chosen *arbitrarily* on the circle. We thus see that, besides the two points O and O', there will be a system of pairs of points of which any one may be employed for finding the instantaneous screw, and for determining the instantaneous twist velocity.

If we choose any two points (Fig. 25), Ω and Ω', so that

$$(X\Omega\Omega'Y) = (XOO'Y);$$

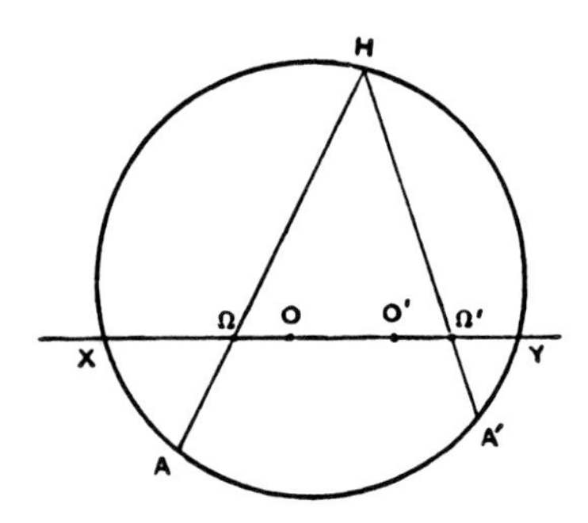

Fig. 25.

then A being given, $A\Omega$ determines H, and $H\Omega'$ determines A', while the twist velocity is proportional to $\Omega'H \div \Omega H$. We can suppose Ω at infinity, and thus obtain the construction used in § 142. A similar construction is obtained when Ω' is at infinity.

The two points, A and A', will divide the arc cut off by XY in a constant anharmonic ratio, for the pencil $H(X\Omega\Omega'Y)$ always preserves the same anharmonic ratio as H moves round the circle.

145. A Special Case.

If η be an impulsive screw, and if α be the corresponding instantaneous screw, it will *not* usually happen that when α is the impulsive screw η is the corresponding instantaneous screw. If, however, in even a single case, it be true that the impulsive screw and the instantaneous screw are interchangeable, then the relation will be universally true.

Let Ω and Ω' (Fig. 26) be a pair of points belonging to the system described in § 144. Then A being given, A' is found. If A' is similarly to

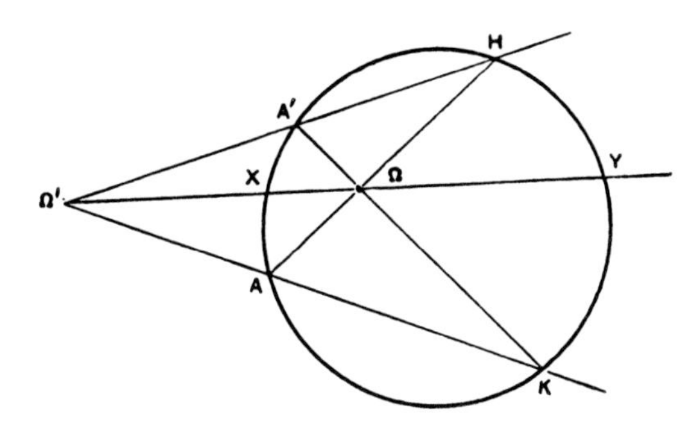

Fig. 26.

determine A, then the figure shows that Ω must lie on the polar of Ω', and, consequently, Ω and Ω' are conjugate points with respect to the circle; or, what comes to the same thing, they divide XY harmonically. The same must be true of each pair of points Ω and Ω', and therefore of O and O', and we have the following theorem:—

If the points O and O' be harmonic conjugates of the points where the homographic axis intersects the circle, then every pair of instantaneous and impulsive screws on the cylindroid are interchangeable.

We might, perhaps, speak of this condition of the system as one of *dynamical involution.* In this remarkable case an impulsive wrench of unit intensity applied to one of the principal screws of inertia will generate a velocity equal and opposite to that which would have been produced if the wrench had been applied to the other principal screw. The construction for the pairs of related screws becomes still more simplified by the theorem, that—

When the system is one of dynamical involution, the chord joining an impulsive screw with its instantaneous screw passes through the pole of the homographic axis.

We may take the opportunity of remarking, that dynamical involution is not confined to the system of the second order. It may be extended to a rigid body with any number of degrees of freedom, or even to any system of rigid bodies. Whenever it happens that the relation of impulsive screw and instantaneous screw is interchangeable in one case, it is interchangeable in every case.

For, let $\theta_1, \ldots \theta_n$ be the co-ordinates of an instantaneous screw, then (§ 97) the corresponding impulsive screw has for co-ordinates,

$$\frac{u_1^2}{p_1}\theta_1, \ldots \frac{u_n^2}{p_n}\theta_n;$$

and if this latter were regarded as an instantaneous screw, then its impulsive screw would be

$$\frac{u_1^4}{p_1^2}\theta_1, \ldots \frac{u_n^4}{p_n^2}\theta_n;$$

but as this is to be only

$$\theta_1, \ldots \theta_n,$$

we must have

$$\frac{u_1^4}{p_1^2} = \frac{u_2^4}{p_2^2} = \ldots \frac{u_n^4}{p_n^2},$$

which shows that if the theorem be true for one pair, it is true for all. The conditions, of course, are, that any one of the following systems of equations be satisfied:—

$$\pm\frac{u_1^2}{p_1} = \pm\frac{u_2^2}{p_2} = \ldots \pm\frac{u_n^2}{p_n}.$$

146. Another Construction for the Twist Velocity acquired by an Impulse.

Reverting to the general case, we find that the chord AA' (Fig. 27) is cut by the homographic axis at T, so that the square of the acquired twist velocity is proportional to the ratio of TA to TA'.

For, with the construction in § 142, draw HQ parallel to AT; then,

$$HQ : A'T :: H\Omega : A'\Omega,$$

$$\frac{AT}{A'T} = \frac{H\Omega}{A'\Omega} \propto \frac{1}{A'\Omega^2};$$

but we showed, in the article referred to, that $A'\Omega$ varies inversely as the acquired twist velocity, whence the theorem is proved.

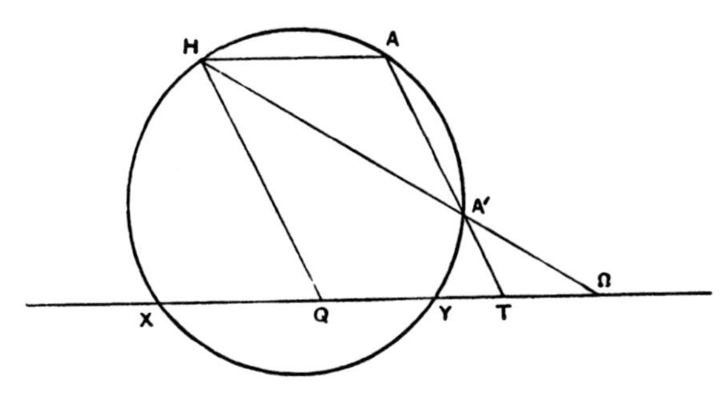

Fig. 27.

This is, in one respect, the simplest construction, for it only involves the chord AA' and the homographic axis.

The chord AA' must envelop a conic having double contact with the circle (Fig. 28), for this is a general property of the chord uniting two corresponding points, A and A', of two homographic systems. Let I be the

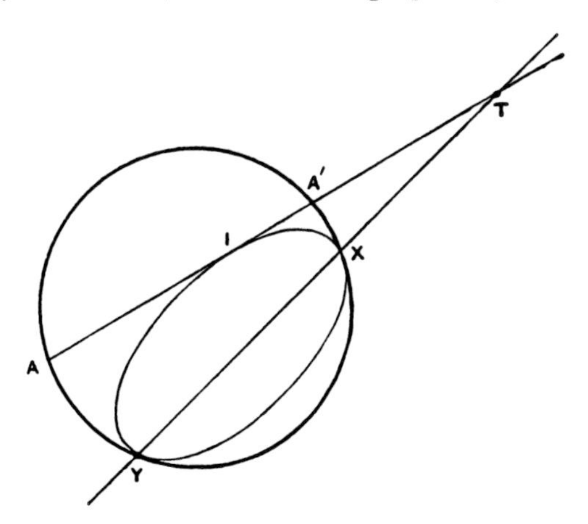

Fig. 28.

point of contact of the chord and conic (Fig. 28). Then AA' is divided harmonically in I and T; for, if XY be projected to infinity, the two conics become concentric circles, and the tangent to one meets it at the middle point of the chord in the other; the ratio is therefore harmonic, and must be so in every projection; whence,

$$\frac{AI}{A'I} = \frac{AT}{A'T};$$

but the last varies as the square of the twist velocity acquired, and hence we see that—

The chord joining any impulsive screw A to the corresponding instantaneous screw A' envelops a conic, and the point of contact, I, divides the chord into segments, so that the ratio of AI to $A'I$ is proportional to the square of the twist velocity acquired about A' by the unit impulsive wrench on A.

147. Constrained Motion.

We can now give another demonstration of the theorem in § 90, which is thus stated:—

If a body, *constrained* to twist about the screw α, be acted upon by an impulsive wrench on the screw η, then the twist velocity acquired varies as

$$\frac{\varpi_{\eta\alpha}}{u_\alpha^2}.$$

The numerator in this expression is the virtual coefficient of the two screws, and the denominator is the function of § 134, which is proportional to the kinetic energy of the body when twisting about α with the unit of twist velocity.

Let α and η be represented by A' and I respectively (Fig. 29), and let A

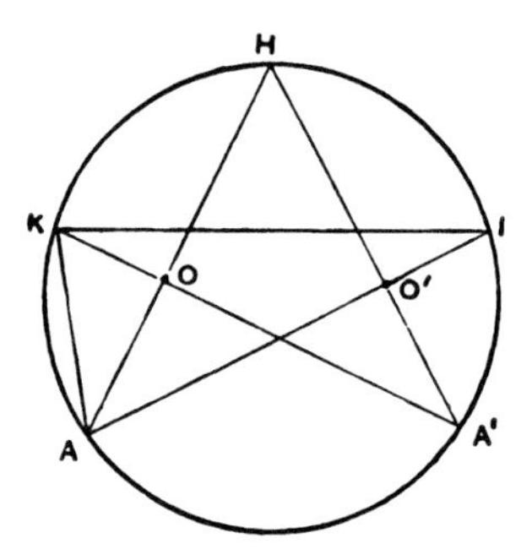

Fig. 29.

be the impulsive screw which would correspond to A' if the body had been free to twist about any screw whatever on the cylindroid defined by A and A'. Let K be reciprocal to A'.

The impulsive wrench on I is decomposed into components on K and A. The former is neutralized by the constraints; the latter has the intensity

$$\frac{KI}{KA};$$

whence the twist velocity ω, acquired by A', is (§ 141) proportional to

$$\frac{KI}{KA} \cdot \frac{HO'}{HO};$$

but, by geometry,

$$KA = \frac{A'H}{OA'} \cdot OA,$$

whence, § 63,

$$\omega \propto \frac{KI \cdot HO' \cdot OA'}{HO \cdot HA' \cdot OA},$$

$$\propto \frac{KI \cdot OA'}{O'A' \cdot HA'},$$

$$\propto \frac{\varpi_{a\eta}}{u_a^2},$$

which is the required result.

148. Energy acquired by an Impulse.

The kinetic energy acquired by a given impulse, using the same notation as before, is (§ 91) proportional to

$$\frac{\varpi^2_{\eta a}}{u_a^2}.$$

Let A be the impulsive screw, and A' the screw about which the body is *constrained* to twist. Draw the chord AOH (Fig. 30), then, as A' varies,

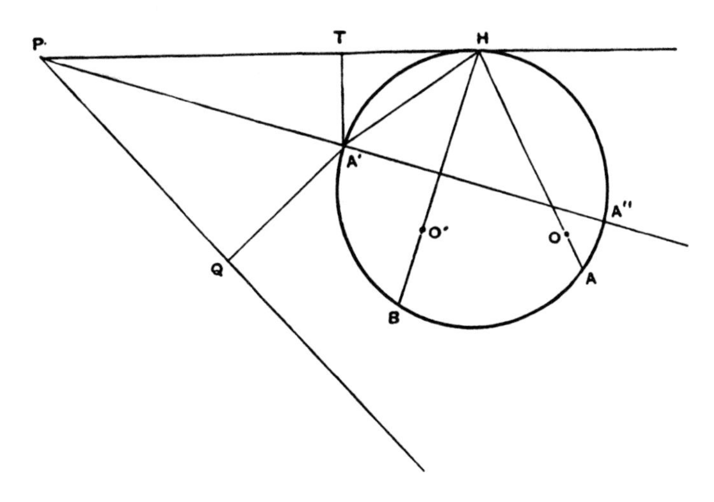

Fig. 30.

while A is fixed, the virtual coefficient of A and A' varies as $A'H$ (§ 63). The square of this is proportional to $A'T$, the length of the perpendicular from A' on the tangent at H. If PQ be the axis of inertia, the value of u_a^2 is proportional to the perpendicular $A'Q$, and, accordingly, the kinetic energy acquired is proportional to

$$\frac{A'T}{A'Q}.$$

Any ray through P, the intersection of the axis of inertia with the tangent at H, cuts the circle in two points, A' and A'', either of which will receive the same kinetic energy from the given impulse.

149. Euler's Theorem.

If the body be permitted to select the screw about which it will commence to twist, then, as already mentioned, § 94, Euler's theorem states that the body will commence to move with a greater kinetic energy than if it be restricted to some other screw. By drawing the tangent from P (not, however, shown in the figure) we obtain the point of contact B, where it is obvious that the ratio of the perpendiculars on PH and PQ is a maximum, and, consequently, the kinetic energy is greatest. It follows from Euler's theorem that B will be the instantaneous screw corresponding to A as the impulsive screw. The line BH is the polar of P, and, consequently, BH must contain O', the pole of the axis of inertia. We are thus again led to the construction (§ 140) for the instantaneous screw B; that is, draw AOH, and then $HO'B$.

150. To determine a Screw that will acquire a given Twist Velocity under a given Impulse.

The impulsive screw being given, and the intensity of the impulsive wrench being one unit, the acquired twist velocity (§ 147) will vary as (Fig. 30),

$$\frac{A'H}{A'Q}.$$

If, therefore, the twist velocity be given, this ratio is given. A' must then lie on a given ellipse, with H as the focus and the axis of inertia as the directrix. This ellipse will intersect the circle in *four* points, any one of which gives a screw which fulfils the condition proposed in the problem.

The relation between the intensity of the impulsive wrench and the twist velocity generated can be also investigated as follows:

Let P, Q, R, S be points on the circle (Fig. 31) corresponding to four impulsive screws, and let P', Q', R', S' be the four corresponding instantaneous screws deduced by the construction already given. Let p, q, r, s denote the intensities of the impulsive wrenches on P, Q, R, S, which will give the units of twist velocity on P', Q', R', S'. Supposing that impulsive wrenches on P, Q, R neutralize, then the corresponding twist velocities generated on P', Q', R' must neutralize also. In the former case, the intensities must be proportional to the sides of the triangle PQR; in the latter, the twist velocities must be proportional to the sides of the triangle $P'Q'R'$. Introducing another quantity d, we have

$$rP'Q' = dPQ,$$
$$qP'R' = dPR,$$
$$pQ'R' = dQR.$$

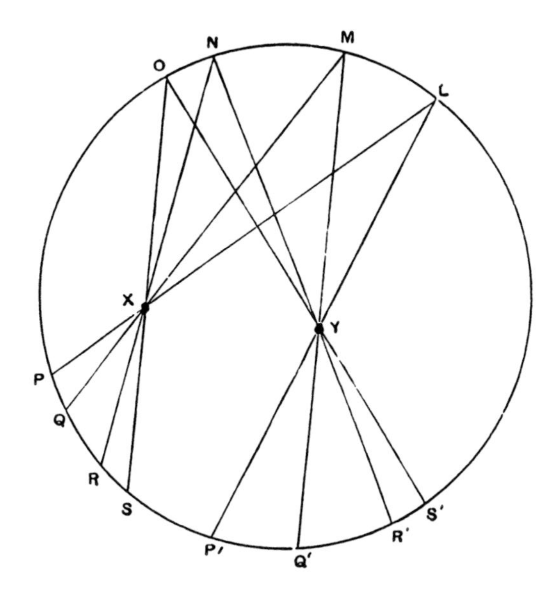

Fig. 31.

The three other groups of equations are similarly obtained

$$rQ'S' = aQS, \qquad qP'S' = cPS, \qquad rP'S' = bPS,$$
$$qR'S' = aRS, \qquad pQ'S' = cQS, \qquad pR'S' = bRS,$$
$$sR'Q' = aRQ, \qquad sQ'P' = cQP, \qquad sR'P' = bRP.$$

Whence we easily deduce

$$ap = bq = cr = ds = hpqrs,$$

where h is a new quantity. We hence obtain from the first equation

$$P'Q' = hPQpq.$$

As this is absolutely independent of R and S, it follows that h must be independent of the special points chosen, and that consequently for *any* two points on the circle P and Q, with their corresponding points P' and Q', we must have

$$pq \propto \frac{P'Q'}{PQ}.$$

In the limit we allow P and Q to coalesce, in which case, of course, P' and Q' coalesce, and p and q become coincident; but obviously we have then

$$PQ : ML :: PX : LX,$$

$$P'Q' : ML :: P'Y : LY;$$

whence
$$\frac{P'Q'}{PQ} = \frac{P'Y}{PX} = \frac{LX}{LY};$$

and as
$$P'Y \propto \frac{1}{LY} \text{ and } PX \propto \frac{1}{LX},$$

we have finally
$$p \propto \frac{LX}{LY}.$$

The result is, of course, the same as that of § 141. Being given the impulsive screw corresponding to P, we find P' by drawing PXL and LYP'; and then to produce a unit twist velocity on P', the intensity of the impulsive wrench on P must be proportional to $LX \div LY$. It is obvious that by a proper adjustment of the units of length, force and twist velocity, LX may be the intensity of the impulsive wrench, and LY the acquired twist velocity.

151. Principal Screws of the Potential.

Let us suppose that a body having two degrees of freedom is in a position of stable equilibrium under the influence of a conservative system of forces. If the body be displaced by a small twist, it will no longer be in a position of equilibrium, and a wrench has commenced to act upon it. This wrench can always, by suitable composition with the reactions of the constraints, be replaced by an equivalent wrench on a screw of the cylindroid (see § 96).

For every point H, corresponding to a displacement screw, we have a related point, H', corresponding to the screw about which the wrench is evoked. The relation is of the one-to-one type, and it will now be proved that the system of screws H is homographic with the corresponding system H'. The proof is obtained in the same manner as that already given in § 137, for impulsive and instantaneous screws.

Let E be a displacement screw about which a twist of unit magnitude evokes a wrench of intensity e on E'; let f be the similar quantity for another pair of screws, F and F'.

A twist of unit amplitude about H may be decomposed into components,

$$\frac{HF}{EF}, \frac{HE}{EF},$$

about E and F, respectively.

These will evoke wrenches on E' and F' of the intensities

$$e\frac{HF}{EF},\ f\frac{HE}{EF},$$

respectively. But this pair of wrenches are to compound into a wrench of intensity h on H', and consequently we have

$$h\frac{H'F'}{E'F'}=e\frac{HF}{EF},$$

$$h\frac{H'E'}{E'F'}=f\frac{HE}{EF};$$

whence $$\frac{HF}{HE}:\frac{H'F'}{H'E'}::f:e.$$

If we take another pair of points, K and K', we have

$$\frac{HF}{HE}:\frac{KF}{KE}::\frac{H'F'}{H'E'}:\frac{K'F'}{K'E'};$$

whence $$(HKFE)=(H'K'F'E').$$

Thus, the anharmonic ratio of any four points in one system is equal to that of their correspondents, and the two systems are homographic.

The homographic axis intersects the circle in two points, which are the principal screws of the potential, *i.e.* a twist about either evokes a wrench on the same screw. Of course this homographic axis is distinct from that in § 139. But this homographic axis, like the former one, passes through the pole of the axis of pitch because the principal screws of the potential are reciprocal.

152. Work done by a Twist.

Suppose that the body, when in equilibrium under the system of forces, receives a twist of small amplitude α' about any screw α, a quantity of work is expended, which we shall denote by

$$Fv_\alpha^2\alpha'^2.$$

In this, F is a constant, whose dimensions are a mass divided by the square of a time, and v_α is a linear magnitude specially appropriate to the screw α, and depending also upon the system of forces (§ 102). We may compare and contrast the three quantities, p_α, u_α, v_α: each is a linear magnitude specially correlated to the screw α. The first and simplest, p_α, is the pitch of the screw, and depends on the geometrical nature of the constraints; u_α involves also the mass of the body, and the distribution of the mass relatively to α; v_α, still more complicated, depends also on the system of forces.

153. Law of Distribution of v_α.

As we follow the screw α around the circle, it becomes of interest to study the corresponding variations of the linear magnitude v_α. We have already found a very concise representation of p_α and u_α by the axis of pitch and the axis of inertia, respectively. We shall now obtain a similar representation of v_α by the aid of the *axis of potential.*

It is shown (§ 102) that v_α^2 must be a quadratic function of the co-ordinates; we may therefore apply to this function the same reasoning as we applied to u_α^2 (§ 134). We learn that v_α^2 is at each point proportional to the perpendicular on a ray, which is the axis of potential.

Thus, if A (Fig. 32) be the screw, the value of v_α^2 is proportional to AP, the perpendicular on PT; if O'' be the pole of the axis of potential, then, as in § 59, we can also represent the value of v_α^2 by the product $AO''.AA'$.

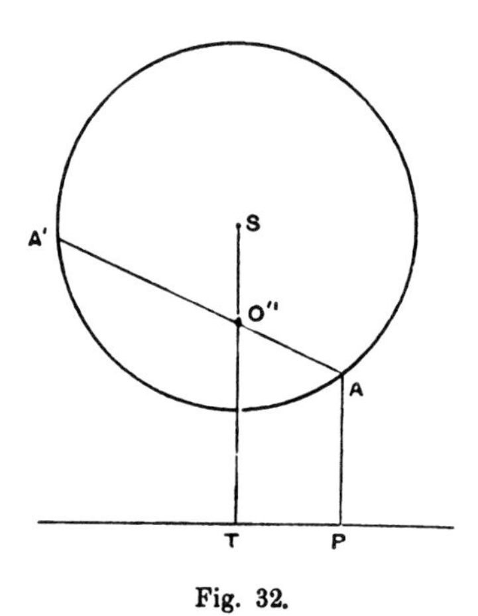

Fig. 32.

154. Conjugate Screws of Potential.

In general the energy expended by a small twist from a position of equilibrium can be represented by a quadratic function of the co-ordinates of the screw. If, moreover, the two screws of reference form what are called *conjugate screws of potential* (§ 100), then the energy is simply the sum of two square terms. The necessary and sufficient condition that the two screws shall be so related is, that their chord shall pass through O''.

Another property of two conjugate screws of potential is also analogous to that of two conjugate screws of inertia. If A and A' be two conjugate screws of potential, then the wrench evoked by a twist round A is reciprocal to A', and the wrench evoked by a twist around A' is reciprocal to A.

155. Determination of the Wrench evoked by a Twist.

The theorem just enunciated provides a simple means of discovering the wrench which would be evoked by a small twist which removes the body from a position of equilibrium.

Let A (Fig. 33) be the given screw; join AO'', and find H; then the required screw A' must be reciprocal to H, and is, accordingly, found by drawing the chord HA' through O.

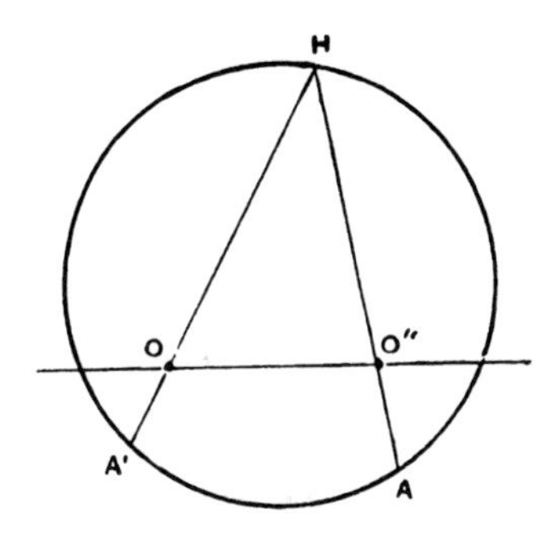

Fig. 33.

The axis OO'' is of course the homographic axis of § 151. We need not here repeat the demonstration of § 141, which will apply, *mutatis mutandis*, to the present problem. We see that the ratio of the intensity of the wrench to the amplitude of the twist is proportional to

$$\frac{HO}{HO''}.$$

The other constructions of a like character can also be applied to this case.

156. Harmonic Screws.

If after displacement the rigid body be released, and small oscillations result, the present geometrical method permits us to study the resulting movements.

It has been shown (§ 130) that there are two special screws on the surface, each of which possesses the property of being a *harmonic screw*. If a body be displaced from rest by a small twist about a harmonic screw, and if it also receive any small initial twist velocity about the same screw, then the body will continue for ever to perform harmonic twist oscillations about the same screw.

The two harmonic screws are X and Y, where the circle is intersected by the axis passing through the pole of the axis of inertia O', and the pole of the axis of potential O'' (Fig. 34).

For, suppose the body receives a small initial displacement about X, this will evoke a wrench on H, found by drawing $XO''Y$ and YOH (§ 155). But the

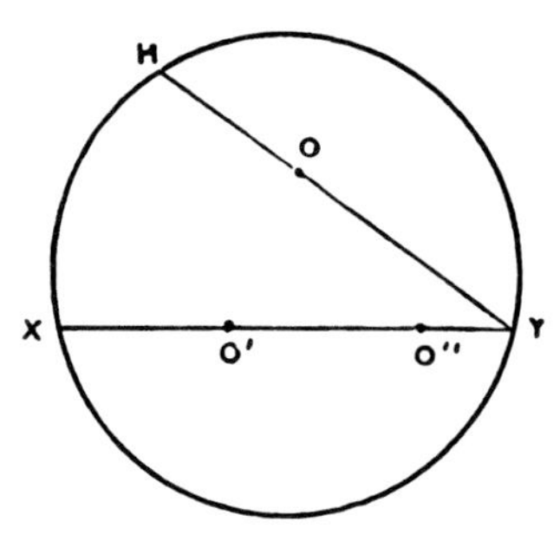

Fig. 34.

effect of a wrench on H will be to produce twist velocity about a screw found by drawing HOY and $YO'X$, *i.e.* X itself (§ 140). Hence the wrench evoked can only make the body still continue to twist about X, and harmonic vibration on X will be the result. Similar reasoning, of course, applies to Y.

157. Small Oscillations in general.

The initial displacement, and the initial twist velocity of the body, can always be decomposed into their respective components on X and Y. The resulting small oscillations can thus be produced by compounding simple harmonic twist oscillations about X and Y.

If it should happen that O' and O'' become coincident, then every screw would be a harmonic screw.

If O and O' coincided, then every screw would be a principal screw of inertia (§ 86).

If O and O'' coincided, then every screw would be a principal screw of potential.

158. Conclusion.

The object proposed in this Chapter has now been completed. It has been demonstrated that the representative circle affords a concise method of exhibiting many problems in the Dynamics of a Rigid System with two degrees of freedom, so long as the body remains near its initial position. The geometrical interest of the enquiry is found mainly to depend on the completely *general* nature of the constraints. If the constraints be specialized to those with which mechanical problems have made us familiar, it will frequently be found that the geometrical theory assumes some extreme and uninteresting type. For instance, a case often quoted as an illustration of two degrees of freedom, is that of a body free to rotate around an axis, and to slide along it. The representative circle has then an infinite radius, and the finite portion thereof is merely a ray perpendicular to the axis of pitch. The geometrical theory then retains merely a vestige of its interest.

CHAPTER XIII.

THE GEOMETRY OF THE CYLINDROID.

159. Another investigation of the Cylindroid.

The laws of the composition of twists and wrenches are of such fundamental importance in the present subject that the following method* of investigating the cylindroid seems worthy of attention. This method is not, however, presented as a substitute for that already given (Chap. II.) which is certainly both more simple and more direct.

Let α and β be any two screws, then if a body receives a twist about α, followed by another twist about β, the position arrived at could have been reached by a single twist about a third screw γ. If the amplitudes of the twists about α and β are given, then the position of γ, as well as the amplitude of the resultant twist thereon, are, of course, both determined. If, however, the amplitudes of the twists on α and β are made to vary while the screws α and β themselves remain fixed, then the position of γ, no less than the amplitude of the resultant twist, must both vary. It has however been shown in § 9 that the *position* and pitch of γ remain constant so long as the ratio of the amplitudes of the twists about α and β remains unchanged. As this ratio varies, the position of γ will vary, so that this position is a function of a single parameter; and, accordingly, γ must be restricted to be one of the generators of a certain ruled surface S, which includes α and β as extreme cases in which the ratio is zero and infinity respectively.

Let θ_1 be a screw which is reciprocal both to α and β, then θ_1 must also be reciprocal to every screw γ on S. Let θ_2, θ_3, θ_4 be three other screws also reciprocal to S. Since a screw is defined by five conditions, it is plain that a screw which fulfils the four conditions of being reciprocal to θ_1, θ_2, θ_3, θ_4 will have one disposable parameter, and must, therefore, be generally confined to a certain ruled surface. This surface must include S, inasmuch as all the screws on S are reciprocal to θ_1, θ_2, θ_3, θ_4; but further, it cannot include any

* *Proceedings of the Royal Irish Academy*, 2nd Ser., Vol. IV. p. 518 (1885).

screw ϵ not on S; for as ϵ and *any* screw γ on S are reciprocal to θ_1, θ_2, θ_3, θ_4, it will follow that any screw on the surface made from ϵ and γ, just as S is made from α and β, must also be reciprocal to θ_1, θ_2, θ_3, θ_4. As γ may be selected arbitrarily on S, we should thus find that the screws reciprocal to θ_1, θ_2, θ_3, θ_4 were not limited to one surface, but constituted a whole group of surfaces, which is contrary to what has been already shown. It is therefore the same thing to say that a screw lies on S, as to say that it is reciprocal to θ_1, θ_2, θ_3, θ_4 (§ 24).

Since the condition of reciprocity involves the pitches of the two screws in an expression containing only their sum, it follows that if all the pitches on θ_1, θ_2, θ_3, θ_4 be diminished by any constant m, and all those on S be increased by m, the reciprocity will be undisturbed. Hence, if the pitches of all the screws on S be increased by $+m$, the surface so modified will still retain the property, that twists about any three screws will neutralize each other if the amplitudes be properly chosen.

We can now show that there cannot be more than two screws of equal pitch on S; for suppose there were three screws of pitch m, apply the constant $-m$ to all, thus producing on S three screws of zero pitch. It must therefore follow that three *forces* on S can be made to neutralize; but this is obviously impossible, unless these forces intersect in a point and lie on a plane. In this case the whole surface degrades to a plane, and the case is a special one devoid of interest for our present purpose. It will, however, be seen that in general S does possess *two* screws of any given pitch. We can easily show that a wrench can always be decomposed into two forces in such a way that the line of action of one of these forces is arbitrary. Suppose that S only possessed one screw λ of pitch m. Reduce this pitch to zero; then any other wrench must be capable of decomposition into a force on λ (i.e. a wrench of pitch zero), and a force on some other line which must lie on S; therefore in its transformed character there must be a second screw of zero pitch on S, and, therefore, in its original form there must have been two screws of the given pitch m.

Intersecting screws are reciprocal if they are rectangular, or if their pitches be equal and opposite; hence it follows that a screw θ reciprocal to S must intersect S in certain points, the screws through which are either at right angles to θ or have an equal and opposite pitch thereto.

From this we can readily show that S must be of a higher degree than the second; for suppose it were a hyperboloid and that the screws lay on the generators of one species A, a screw θ which intersected two screws of equal pitch m must, when it receives the pitch $-m$, be reciprocal to the entire system A. We can take for θ one of the generators on the hyperboloid belonging to the species B; θ will then intersect every screw of the surface; it must also be reciprocal to all these; and, as there are only two screws of the given pitch, it will follow that θ must cut at right angles every generator of the species A. The same would have to be true for any other reciprocal screw ϕ similarly chosen; but it is obvious that two lines θ and ϕ cannot be found which will cut all the generators at right angles, unless, indeed, in the extreme case when all these are coplanar and parallel. In the general case it would require two common perpendiculars to two rays, which is, of course, impossible. We hence see that S cannot be a surface of the second degree.

We have thus demonstrated that S must be at least of the third degree—in other words, that a line which pierces the surface in two points will pierce it in at least one more. Let α and β be two screws on S of equal pitch m, and let θ be a screw of pitch $-m$ which intersects α and β. It follows that θ is reciprocal both to α and β, and therefore it must be reciprocal to every screw of S. Let θ cut S in a third point through which the screw γ is to be drawn, then θ and γ are reciprocal; but they cannot have equal and opposite pitches, because then the pitch of γ should be equal to that of α and β. We should thus have three screws on the surface of the same pitch, which is impossible. It is therefore necessary that θ shall always intersect γ at right angles. From this it will be easily seen that S must be of the third degree; for suppose that θ intersected S in a fourth point, through which a screw δ passed, then θ would have to be reciprocal to δ, because it is reciprocal to all the screws of S; and it would thus be necessary for θ to be at right angles to δ. Take then the four rays α, β, γ, δ, and draw across them the two common transversals θ and ϕ. We can show, in like manner, that ϕ is at right angles to γ and δ. We should thus have θ and ϕ as two common perpendiculars to the two rays γ and δ. This is impossible, unless γ and δ were in the same plane, and were parallel. If, however, γ and δ be so circumstanced, then twists about them can only produce a resultant twist also parallel to γ and δ, and in the same plane. The entire surface S would thus degenerate into a plane.

We are thus conducted to the result that S must be a ruled surface of the third degree, and we can ascertain its complete character. Since any transversal θ across α, β, and γ must be a reciprocal screw, if its pitch be equal and opposite to those of α and β, it will follow that each such transversal must be at right angles to γ. This will restrict the situation of γ, for unless it be specially placed with respect to α and β, the transversal θ will not always fulfil this condition. Imagine a plane perpendicular to γ, then this plane contains a line I at infinity, and the ray θ must intersect I as the necessary condition that it cuts γ at right angles. As θ changes its position, it traces out a quadric surface, and as I is one of the generators of that quadric, it must be a hyperbolic paraboloid. The three rays α, β, γ, belonging to the other system on the paraboloid must also be parallel to a plane, being that defined by the other generator I', in which the plane at infinity cuts the quadric.

Let PQ be a common perpendicular to α and γ, then since it intersects γ at right angles, it must also intersect I; and since PQ cuts the three generators of the paraboloid α, γ and I, it must be itself a generator, and therefore intersects β. But α, β, γ are all parallel to the same plane, and hence the common perpendicular to α and γ must be also perpendicular to β. We hence deduce the important result, that all the screws on the surface S must intersect the common perpendicular to α and β, and be at right angles thereto.

The geometrical construction of S is then as follows:—Draw two rays α and β, and also their common perpendicular λ. Draw any third ray θ, subject only to the condition that it shall intersect both α and β. Then the common perpendicular ρ to both θ and λ will be one of the required generators, and as θ varies this perpendicular will trace out the surface. It might at first appear that there should be a doubly infinite series of common perpendiculars ρ to λ and to θ. Were this so, of course S would not be a surface. The difficulty is removed by the consideration that *every* transversal across ρ, α, β is perpendicular to ρ. Each ρ thus corresponds to a singly infinite number of screws θ, and all the rays ρ form only a singly infinite series, i.e. a surface.

A simple geometrical relation can now be proved. Let the perpendicular distance between ρ and α be d_1, and the angle between ρ and α be A_1; let d_2 and A_2 be the similar quantities for ρ and β, then it will be obvious that

$$d_1 : d_2 :: \tan A_1 : \tan A_2;$$

or

$$d_1 + d_2 : d_1 - d_2 :: \sin(A_1 + A_2) : \sin(A_1 - A_2),$$

if z be the distance of ρ from the central point of the perpendicular h between α and β; and if ϵ be the angle between α and β, and θ be the angle made by ρ with a parallel to the bisector of the angle ϵ, then we have from the above

$$z : h :: \sin 2\phi : \sin 2\epsilon.$$

The equation of the surface S is now deduced for

$$\tan\phi = \frac{x}{y};$$

whence we obtain the equation of the cylindroid in the well-known form

$$z(x^2 + y^2) = \frac{2h}{\sin 2\epsilon} xy.$$

The law of the distribution of pitch upon the cylindroid can also be deduced from the same principles. If α and β are screws of zero pitch, then any reciprocal transversal θ will be also of zero pitch; and as ρ must be reciprocal to θ, it will follow that the pitch of ρ must be equal to the product of the shortest perpendicular distance between ρ and θ, and the tangent of the angle between the two lines. In short, the pitch of ρ must simply be equal to what is sometimes called the *moment* between ρ and θ.

We are also led to the following construction for the cylindroid.

Draw a plane pencil of rays and another ray L, situated anywhere. Then the common perpendiculars to L and the several rays of the pencil trace out the cylindroid.

I have already mentioned (p. 20) that the first model of the cylindroid was made by Plücker in illustration of his *Neue Geometrie des Raumes*. The model of the surface which is represented in the Frontispiece was made from my design by Sir Howard Grubb, the cost being defrayed by a grant from the Scientific Fund of the Royal Irish Academy. A hollow cylinder was mounted on a dividing engine and holes were drilled at the calculated points. Silver wires were then stretched across in the positions of the generators and a beautiful model is the result.

The equation to the tangent cone drawn from the point x', y', z' to the surface,

$$z(x^2 + y^2) - 2mxy = 0,$$

is of the fourth degree and is given by equating to zero the discriminant of the following function in ω,

$$\omega^3(xz' - zx') - \omega^2\{yz' - zy' + 2m(x - x')\} + \omega\{xz' - zx' + 2m(y - y')\} + zy' - yz'.$$

This cone has three cuspidal edges, and accordingly the model exhibits in every aspect a remarkable tricuspid arrangement.

I here give the details of the construction of the much simpler model of the cylindroid figured in Plate II.* A boxwood cylinder, $0^{m}\cdot15$ long and $0^{m}\cdot05$ in diameter, is chucked to the mandril of a lathe furnished with a dividing plate. A drill is mounted on the slide rest, and driven by overhead gear. The parameter $p_\alpha - p_\beta$ (in the present case $0^{m}\cdot066$) is divided into one hundred parts. By the screw, which moves the slide rest parallel to the bed of the lathe, the drill can be moved to any number z of these parts from its original position at the centre of the length of the cylinder. Four holes are to be drilled for each value of z. These consist of two pairs of diametrically opposite holes. The directions of the holes intersect the

* See *Transactions of Royal Irish Academy*, Vol. xxv. p. 216 (1871); and also *Phil. Mag.* Vol. XLII. p. 181 (1871); also *B. A. Report, Edinburgh*, 1871.

axis of the cylinder at right angles. The following table will enable the work to be executed with facility. l is the angle of § 13 :—

z	l	$90-l$	$180+l$	$270-l$
0·0	0	90	180	270
17·4	5	85	185	265
34·2	10	80	190	260
50·0	15	75	195	255
64·3	20	70	200	250
76·5	25	65	205	245
86·6	30	60	210	240
94·0	35	55	215	235
98·5	40	50	220	230
100·0	45	45	225	225

For example, when the slide has been moved 34·2 parts from the centre of the cylinder, the dividing plate is to be set successively to 10°, 80°, 190°, 260°, and a hole is to be drilled in at each of these positions. The slide rest is then to be moved on to 50 parts, and holes are to be drilled in at 15°, 75°, 195°, 255°. Steel wires, each about $0^{m}\cdot 3$ long, are to be forced into the holes thus made, and half the surface is formed. The remaining half can be similarly constructed: a length of $0^{m}\cdot 066 \cos 2l$ is to be coloured upon each wire to show the pitch. The sign of the pitch is indicated by using one colour for positive, and another colour for negative pitches.

Among the various other representations of the cylindroid I can now do no more than refer to an ingenious plan described by Goebel in his *Neueren Statik*, by which a model of this surface in card-board can be made with facility. There is also a model in the collection of the Cavendish Laboratory at Cambridge, and another belonging to the Mathematical Society of London, which, like that figured in Plate II., was made by myself. Sir Howard Grubb has also made a second model with the same dimensions as that figured in the frontispiece but mounted in a different manner. This exquisite exhibition of a ruled surface is the property of Mr G. L. Cathcart, Fellow of Trinity College, Dublin.

A suggestive construction for the cylindroid has been also given by Professor G. Minchin in his well-known book on Statics, and we have already mentioned (note to § 50) the construction given by Mr Lewis.

§ 160. Equation to plane section of Cylindroid.

Each generator of the cylindroid is the abode of a certain screw, and accordingly each point in a plane section will lie on one screw, and generally on only one. We may, accordingly, regard the several points of the cubic as in correspondence with the several screws on the cylindroid. It will often be convenient to speak of the points on the section as synonymous with the screws themselves which pass through those points.

We must first investigate the equation* to the cubic curve produced by cutting the cylindroid by a plane situated in any arbitrary position.

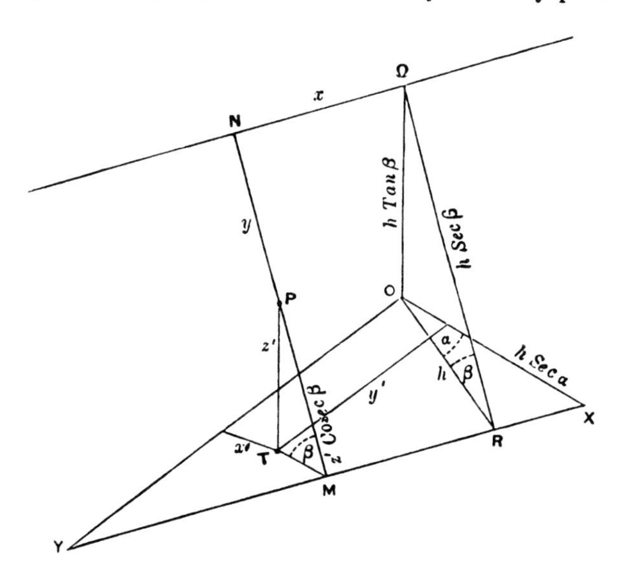

Fig. 35.

Let OX and OY (Fig. 35) be the two principal screws of the cylindroid of which $O\Omega$ is the nodal line. Let $XY\Omega$ be the arbitrary plane of section. The position of this plane is defined by the magnitudes h, α, β, whereof h is the length of the perpendicular from O on XY, α is the angle between OR and OX, and β is the angle $OR\Omega$, or the inclination of the plane of section to the principal plane of the cylindroid.

Draw through Ω the line ΩN parallel to XY; then we shall adopt ΩN as the new axis of x and ΩR as the new axis of y, so that if P be any point on the surface, we have $PN = y$ and $\Omega N = x$. The dotted letters, x', y', z' refer to the original axes of the cylindroid. Let fall PT perpendicular on the plane of OXY, and TM perpendicular to XY. Then we have $MN = R\Omega$, whence

$$y + z' \operatorname{cosec} \beta = h \sec \beta \quad \text{.........................(i)},$$

* *Transactions of the Royal Irish Academy*, Vol. XXIX. p. 1 (1887).

while, if θ be the angle XOT, and OT be denoted by r, we have

$$x = r \sin(\theta - \alpha),$$

or

$$x = y' \cos\alpha - x' \sin\alpha \quad \text{.......................(ii)};$$

but, obviously,

$$OR = r\cos(\theta - \alpha) + MT;$$

whence

$$h = x'\cos\alpha + y'\sin\alpha + z'\cot\beta \quad \text{..................(iii)}.$$

Solving the equations (i), (ii), (iii), we obtain

$$x' = -x\sin\alpha + y\cos\beta\cos\alpha,$$

$$y' = +x\cos\alpha + y\cos\beta\sin\alpha,$$

$$z' = \quad h\tan\beta - y\sin\beta.$$

It appears from these that

$$x'^2 + y'^2 = x^2 + y^2\cos^2\beta,$$

$$x'y' = xy\cos\beta\cos 2\alpha + (y^2\cos^2\beta - x^2)\sin\alpha\cos\alpha.$$

The equation of the cylindroid gives

$$z'(x'^2 + y'^2) = 2mx'y';$$

whence we deduce, as the required equation of the section,

$$(h\tan\beta - y\sin\beta)(x^2 + y^2\cos^2\beta)$$
$$= 2mxy\cos\beta\cos 2\alpha + 2m\sin\alpha\cos\alpha(y^2\cos^2\beta - x^2);$$

or, arranging the terms,

$$\sin\beta\cos^2\beta y^3 + \sin\beta y x^2 - (m\sin 2\alpha + h\tan\beta)x^2 + 2mxy\cos\beta\cos 2\alpha$$
$$+ (m\sin 2\alpha\cos^2\beta - h\sin\beta\cos\beta)y^2 = 0.$$

It is often convenient to use the expressions

$$x = h\tan(\theta - \alpha) - m\sin 2\theta\cot\beta\tan(\theta - \alpha),$$

$$y = h\sec\beta \qquad - m\sin 2\theta\operatorname{cosec}\beta,$$

from which, if θ be eliminated, the same equation for the cubic is obtained; or, still more concisely, we may write

$$x = y\cos\beta\tan(\theta - \alpha),$$

$$y = h\sec\beta - m\sin 2\theta\operatorname{cosec}\beta.$$

This cubic has one real asymptote, the equation of which is

$$y\sin\beta = m\sin 2\alpha + h\tan\beta,$$

and the asymptote cuts the curve in the finite point for which

$$x = -\tan 2\alpha(h + m\sin 2\alpha\cot\beta).$$

The value of θ at this point is $-\alpha$.

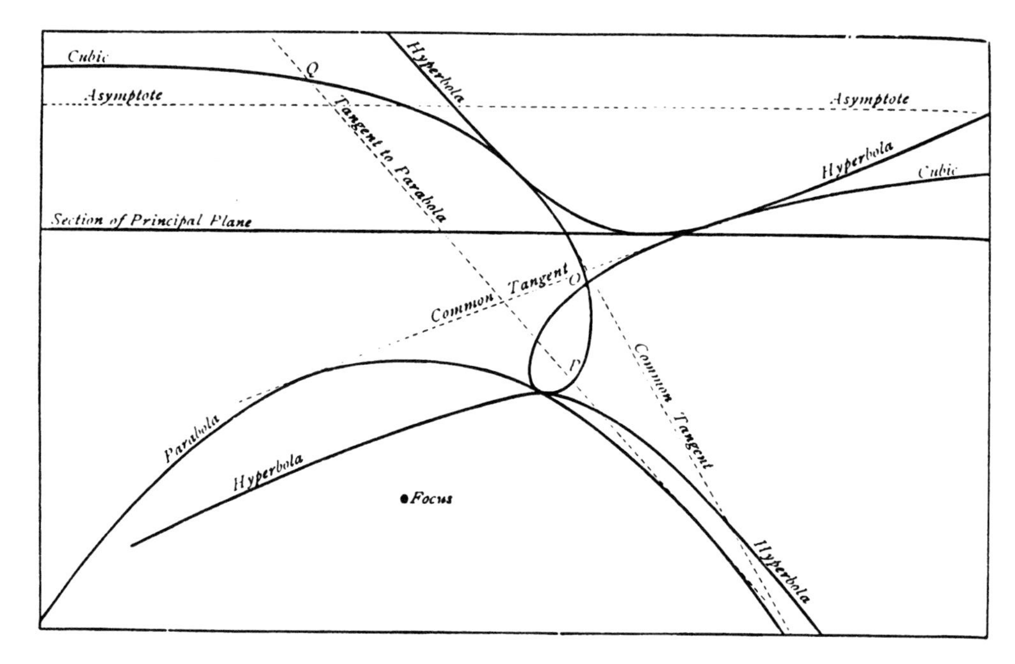

Fig. 36.

EXPLANATION OF FIG. 36.

General Section of the Cylindroid, showing—

(1) Cubic with the double point O.

(2) Asymptote of the cubic.

(3) The parabola, which is the envelope of the chords joining screws of equal pitch.

(4) Hyperbola having triple contact with the cubic, being envelope of reciprocal chords.

(5) Section of the principal plane. It is a tangent to the hyperbola.

(6) A tangent to the parabola, showing two screws, P and Q, of equal pitch.

(7) Common tangents to the parabola and the cubic, touching the latter at the two principal screws.

(8) Any tangent to the hyperbola intersects the cubic in three points, two of which belong to reciprocal screws (not shown).

Equations of Cubic.

$$x = \cdot 9y \tan(\theta - 25°),$$
$$y = 20 - 66 \sin 2\theta.$$

Equation of Parabola.

$$y = -31 - \left(5 + \frac{x}{15}\right)^2.$$

Equations of Hyperbola.

$$x = 1\cdot 6 \pm 21\cdot 6 \sec\phi \pm 47\cdot 5 \tan\phi,$$
$$y = -12\cdot 8 + 32\cdot 8 \sec\phi.$$

In Fig. 36 will be found a drawing of this curve. The following are the values of the constants adopted:

$$\alpha = 25^\circ; \quad \beta = 26^\circ; \quad h = 18; \quad m = 28{\cdot}9;$$

with which the equations become

$$x = {\cdot}9y \tan(\theta - 25^\circ),$$
$$y = 20 - 66 \sin 2\theta.$$

The curve was plotted down on "papier millimétrique," and has been copied in reduced size in the figure. The constants were selected after several trials, in order to give a curve that should be at once characteristic, and of manageable dimensions.

The distribution of pitch upon the screws of the cylindroid is of fundamental importance in the theory, so that we must express the pitches appropriate to the several points on the cubic.

Let p denote the pitch; then, from the known property of the cylindroid,

$$p = p_0 + m \cos 2\theta,$$

where p_0 is a constant. Transforming this result into the co-ordinates of the point on the cubic, we have

$$p = p_0 - m \frac{(x^2 - y^2 \cos^2\beta) \cos 2\alpha + 2xy \cos\beta \sin 2\alpha}{x^2 + y^2 \cos^2\beta}.$$

161. Chord joining Two Screws of Equal Pitch.

As the pitches of the two screws, defined by $+\theta$ and $-\theta$, are equal, the chord in question is found by drawing the line through the points x', y' and x'', y'', respectively, where

$$x' = y' \cos\beta \tan(\theta - \alpha),$$
$$y' = h \sec\beta - m \operatorname{cosec}\beta \sin 2\theta,$$
$$x'' = y'' \cos\beta \tan(-\theta - \alpha),$$
$$y'' = h \sec\beta + m \operatorname{cosec}\beta \sin 2\theta.$$

After a few reductions, the required equation is found to be

$$xm(\cos 2\theta + \cos 2\alpha) + y(h \sin\beta + m \cos\beta \sin 2\alpha) - h^2 \tan\beta + m^2 \cot\beta \sin^2 2\theta = 0.$$

If this chord passes through the origin, then

$$-h^2 \tan^2\beta + m^2 \sin^2 2\theta = 0;$$

or,

$$h \tan\beta \pm m \sin 2\theta = 0.$$

But this is obviously necessary; for from the geometry of the cylindroid it is plain that θ must then fulfil the required condition.

We can also determine the chord in a somewhat different manner, which has the advantage of giving certain other expressions that may be of service.

Let $U=0$ be the cubic curve.

Let $V=0$ be the equation of the two straight lines from the origin to the points of intersection with the two equal pitch screws $\pm\theta$.

Let $L=0$ be the chord joining the two intersections of U and V, distinct from the origin: this is, of course, the chord now sought for. Then we must have an identity of the type

$$cU = VX + LY;$$

where c is some constant. For the conditions $L=0$ and $V=0$ imply $U=0$, and L cuts U in three points, two of which lie on V, and the third point, called I, must lie on X. The line X is otherwise arbitrary, and we may, for convenience, take it to be the line ΩI from the origin to I. The product VX thus contains only terms of the third degree, and accordingly the terms of the second degree in U must be sought in LY.

Let $U=u_3+u_2$ where u_3 and u_2 are of the third and second degrees respectively, then cu_2 must be the quadratic part of the product LY. As L does not pass through the origin, it must have an absolute term, consequently Y must not contain either an absolute term or a term of the first degree. If, therefore, c be the absolute term in L, it is plain that Y must be simply u_2, and we have accordingly,

$$c(u_3+u_2) = VX + (L'+c)\,u_2,$$

where L' denotes the value of L without the absolute term: we have consequently the identity

$$cu_3 = VX + L'u_2.$$

In this equation we know u_2, u_3, V, and the other quantities have to be found.

If we substitute

$$x = -y\cos\beta\tan(\alpha\pm\theta),$$

we make V vanish, and representing L' by $\lambda x+\mu y$, we find

$$\lambda\cos\beta\tan(\alpha\pm\theta)-\mu = c\frac{\sin\beta}{h\tan\beta\pm m\sin 2\theta};$$

and after a few steps

$$\frac{\lambda}{c} = \frac{m\cos 2\alpha + m\cos 2\theta}{-h^2\tan\beta + m^2\cot\beta\sin^2 2\theta},$$

$$\frac{\mu}{c} = \frac{h\sin\beta + m\cos\beta\sin 2\alpha}{-h^2\tan\beta + m^2\cot\beta\sin^2 2\theta}.$$

We can also obtain X; for let it be $Px + Qy$, then severally identifying the coefficients of x^3 and y^3, we have,

$$P = m \sin 2\alpha + h \tan \beta,$$

$$Q = -m \cos \beta (\cos 2\alpha + \cos 2\theta);$$

finally, resuming the various results, we obtain the identity

$$cU = VX + LY;$$

wherein,

$$c = -h^2 \tan \beta + m^2 \cot \beta \sin^2 2\theta,$$

$$U = \sin \beta \cos^2 \beta y^3 + \sin \beta y x^2,$$
$$-(m \sin 2\alpha + h \tan \beta) x^2 + 2mxy \cos \beta \cos 2\alpha$$
$$+(m \sin 2\alpha \cos^2 \beta - h \sin \beta \cos \beta) y^2.$$

$$V = mx^2 (\cos 2\alpha + \cos 2\theta) + 2mxy \sin 2\alpha \cos \beta + my^2 \cos^2 \beta (\cos 2\theta - \cos 2\alpha),$$

$$X = x (m \sin 2\alpha + h \tan \beta) - my \cos \beta (\cos 2\alpha + \cos 2\theta),$$

$$L = mx (\cos 2\alpha + \cos 2\theta) + y (h \sin \beta + m \cos \beta \sin 2\alpha)$$
$$- h^2 \tan \beta + m^2 \cot \beta \sin^2 2\theta,$$

$$Y = x^2 (-m \sin 2\alpha - h \tan \beta) + 2mxy \cos \beta \cos 2\alpha$$
$$+ y^2 (m \sin 2\alpha \cos^2 \beta - h \sin \beta \cos \beta).$$

162. Parabola.

The screws reciprocal to a cylindroid intersect two screws of equal pitch on the surface. Any chord in the section which cuts the cubic in two points of equal pitch must thus be the residence of a screw reciprocal to the surface; accordingly the chord

$$mx (\cos 2\alpha + \cos 2\theta) + y (h \sin \beta + m \cos \beta \sin 2\alpha)$$
$$- h^2 \tan \beta + m^2 \cot \beta \sin^2 2\theta = 0,$$

when it receives a pitch equal to

$$-p_0 - m \cos 2\theta,$$

forms a screw reciprocal to the cylindroid.

It is easily shown that the envelope of this chord is a parabola; differentiating with respect to θ we have

$$x = 2m \cot \beta \cos 2\theta.$$

Eliminating θ we obtain

$$x^2 + 4mx \cot \beta \cos 2\alpha + 4y (h \cos \beta + m \cos \beta \cot \beta \sin 2\alpha) - 4h^2 + 4m^2 \cot^2 \beta = 0.$$

The vertex of the parabola is at the point

$$x = -2m \cot \beta \cos 2\alpha; \quad y = h \sec \beta - m \operatorname{cosec} \beta \sin 2\alpha.$$

The latus rectum is

$$4\cos\beta\,(h + m\cot\beta\sin 2\alpha).$$

The values of the two equal pitches (p) on the pair of screws are thus expressed in terms of the abscissa x of the point in which the chord touches the envelope by means of the equation

$$p = p_0 + \tfrac{1}{2}x\tan\beta.$$

From any point P, on the cubic, two tangents can be drawn to the parabola. Each of these tangents must intersect the cubic in a pair of screws of equal pitch. One tangent will contain the other screw whose pitch is equal to that of P. The second tangent passes through two screws of equal pitch in the two other points which, with P, make up the three intersections with the cubic. As the principal screws of the cylindroid are those of maximum and minimum pitch respectively, it follows that the tangents at these points will also touch the parabola. These common tangents are shown in the figure.

This parabola is drawn to scale in Fig. 36. The equation employed was

$$y = -31 - \left(5 + \frac{x}{15}\right)^2.$$

When the figure was complete, it was obvious that the parabola touched the cubic, and thus the following theorem was suggested:—

The parabola, which is the envelope of chords joining screws of equal pitch, touches the cubic in three points.

The demonstration is as follows:—To seek the intersections of the parabola with the cubic, we substitute, in the equation of the parabola, the values

$$x = h\tan(\theta - \alpha) - m\cot\beta\sin 2\theta\tan(\theta - \alpha),$$

$$y = h\sec\beta - m\,\mathrm{cosec}\,\beta\sin 2\theta.$$

This would, in general, give an equation of the sixth degree for $\tan\theta$. It will, however, be found in this case that the expression reduces to a perfect square. The six points in which the parabola meets the cubic must thus coalesce into three, of which two are imaginary. The values of θ for these three points are given by the equation

$$h\tan(\theta - \alpha) - m\cot\beta\,\{\sin 2\theta\tan(\theta - \alpha) + 2\cos 2\theta\} = 0.$$

We can also prove geometrically that the parabola touches the cubic at three points.

In general, a cone of screws reciprocal to the cylindroid can be drawn from any external point. If the point O happen to lie on the cylindroid, he cone breaks up into two planes. The nature of these planes is easily een. One of them, A, must be the plane perpendicular to the generator hrough O; the other, B, is the plane containing O, and the screw of equal ›itch to that of the screw through O. These planes intersect in a ray, L, ınd it must first be shown that L is a tangent to the cylindroid.

Any ray intersecting one screw on a cylindroid at right angles must cut he surface again in two screws of equal pitch; consequently L can only neet the surface in two distinct points, each of which has the pitch of the generator through O. It follows that L must intersect the surface at two coincident points O, i.e. that it is a tangent to the cylindroid at O.

Let any plane of section be drawn through O. This plane will, in general, intersect A and B in two distinct rays: these are the two screws reciprocal to the cylindroid, and they are accordingly the two tangents from O to the parabola we have been discussing. The only case in which these two rays could coalesce would occur when the plane of section was drawn through L; but the two tangents to a parabola from a point only coalesce when that point lies on the parabola. At a point where the parabola meets the cubic, L must needs be a tangent both to the parabola and to the cubic, which can only be the case if the two curves are touching. We have thus proved that the parabola must have triple contact with the cubic.

There are thus three points on the cubic which have the property that the tangent intersects the curve again in a point of equal pitch to that of the point of contact. We thus learn that all the screws of a four-system which lie in a plane touch a parabola having triple contact with the reciprocal cylindroid.

From any point P, on the cubic, two tangents can be drawn to the parabola. Each of these tangents must intersect the cubic in a pair of screws of equal pitch. One tangent will contain the two screws whose pitch is equal to that of P. The other tangent passes through two screws of equal pitch in the two other points, which, with P, make up the three intersections with the cubic.

As the principal screws of the cylindroid are those of maximum and minimum pitch, respectively, it follows that the tangents at these points will also touch the parabola. These common tangents are shown in Fig. 36.

From the equation of the cylindroid,

$$z(x^2 + y^2) = 2mxy,$$

it follows that the plane at infinity cuts the surface in three straight lines on the planes,

$$z = 0,$$

$$x \pm iy = 0.$$

The line at infinity on the plane of z is of course intersected by all the real generators of the cylindroid, inasmuch as they are parallel to z. The points at infinity on the planes $x \pm iy = 0$ are each the residence of an imaginary screw, also belonging to the surface. The pitches of both these screws are infinite.

We may deduce the two screws of infinite pitch on the surface in another way. The equations of a screw are

$$y = x \tan \theta,$$
$$z = m \sin 2\theta,$$

while the pitch is

$$p_0 + m \cos 2\theta.$$

If $\tan \theta$ be either $\pm i$, we find both z infinite and the pitch infinite. We thus see that through the infinitely distant point I, on the nodal line of the cylindroid, two screws belonging to the surface can be drawn, just as at any finite point. The peculiarity of the two screws through I is, that their pitches are equal, i.e. both infinite, and this is not the case with any other pair of intersecting screws.

It is now obvious why the envelope just considered turned out to be a parabola rather than any other conic section. Every plane section will have the line at infinity for a transversal cutting two screws of equal pitch; the envelope of such transversals must thus have the line at infinity for a tangent, i.e. must be a parabola.

163. Chord joining Two Points.

If θ' and θ'' be the angles by which two points on the cubic are defined, then the equation to the chord joining those points is

$$Ax + By + C = 0;$$

where
$$A = 2m \cos(\theta' - \alpha)\cos(\theta'' - \alpha)\cos(\theta' + \theta''),$$
$$B = h \sin\beta + m\cos\beta\cos(\theta' + \theta'')\sin(2\alpha - \theta' - \theta'')$$
$$- m\cos\beta\sin(\theta' + \theta'')\cos(\theta' - \theta''),$$
$$C = -\tan\beta\,(h - m\cot\beta\sin 2\theta')\,(h - m\cot\beta\sin 2\theta'').$$

If in these expressions we make $\theta' + \theta'' = 0$, we obtain the equation for the chord joining screws of equal pitch, as already obtained.

We shall find that, in particular sections, these expressions become considerably simplified. Suppose, for example, that the plane of section be a tangent plane to the cylindroid. The cubic then degenerates to a straight line and a conic. The condition for this will be obvious from the equation of the cubic. If the coefficient of x^2 become zero, the required decomposition takes place, for y is then a factor. The necessary and sufficient condition for the plane of section being a tangent, is therefore

$$m \sin 2\alpha + h \tan \beta = 0.$$

When this is the case, the three expressions of A, B, C may be divided by a common factor,

$$2m \cos(\theta' - \alpha) \cos(\theta'' - \alpha),$$

and we have

$$A = \cos(\theta' + \theta''),$$

$$B = -\cos\beta \sin(\theta' + \theta''),$$

$$C = -2m \cot\beta \sin(\alpha + \theta') \sin(\alpha + \theta'').$$

If the screws be of equal pitch, $\theta' + \theta'' = 0$, the coefficient of y disappears, and we see that all the chords are merely lines parallel to the axis of y, which is parallel to one of the axes of the ellipse.

The equation to the chord then becomes

$$(\cos 2\alpha + \cos 2\theta)\{x + m \cot\beta(\cos 2\alpha - \cos 2\theta)\} = 0.$$

For a given value of x there are two values of θ corresponding to the two chords that can be drawn through the point. One of these chords is parallel to y, and has a θ obtained from the equation

$$x + m \cot\beta(\cos 2\alpha - \cos 2\theta) = 0.$$

The other value of θ is $\frac{\pi}{2} - \alpha$, from the equation

$$\cos 2\alpha + \cos 2\theta = 0.$$

This is independent of x, as might have been foreseen from the fact that the two screws of equal pitch are in this case the line in the section and the other screw of equal pitch. The latter cuts the section in a certain point, and, of course, all chords through this point meet the curve in two screws of equal pitch.

164. Reciprocal Screws.

Another branch of the subject must now be considered. We shall first investigate the following general problem:—

From any point, P, a series of transversals is drawn across each pair of reciprocal screws on the cylindroid. It is required to determine the cone which is the locus of these transversals. We shall show that this is a cone of the second degree.

Let α, β, γ be the co-ordinates of the point. Then the plane through this point, and the generator of the cylindroid defined by the equations

$$y = x \tan \theta,$$

$$z = m \sin 2\theta,$$

is $$(y - x \tan \theta)(\gamma - m \sin 2\theta) = (\beta - \alpha \tan \theta)(z - m \sin 2\theta);$$

or, if we arrange in powers of $\tan \theta$, we obtain

$$A \tan^3 \theta + B \tan^2 \theta + C \tan \theta + D = 0,$$

in which

$$A = \alpha z - \gamma x; \quad D = \gamma y - \beta z,$$

$$B = \gamma y - \beta z + 2mx - 2m\alpha; \quad C = \alpha z - \gamma x + 2m\beta - 2my.$$

If the same transversal also crosses the generator defined by θ', then,

$$A \tan^3 \theta' + B \tan^2 \theta' + C \tan \theta' + D = 0.$$

When the two screws defined by θ and θ' are reciprocal,

$$\tan \theta \tan \theta' = H,$$

when H is a constant.

By eliminating and rejecting the factor

$$\tan \theta - \tan \theta'$$

we obtain

$$-A^2H^3 + ACH^2 - BDH + D^2 = 0.$$

And as this is of the second degree in x, y, z, the required theorem has been proved.

All these cones must pass through the centre of the cylindroid, inasmuch as the two principal screws of the cylindroid are reciprocal. If a constant be added to the pitches of all the screws on the cylindroid, then the pairs of reciprocals alter, inasmuch as H alters. The cone changes accordingly, and thus there would be through each point a family of cones, all of which, however, agree in having, as a generator, the ray from the vertex to the centre of the cylindroid. Thus, even when the cylindroid is given, we must further have the pitch of a stated screw given before the cone becomes definite. This state of things may be contrasted with that presented by the cone of reciprocal screws which may be drawn through a point. The latter depends only upon the cylindroid itself, and is not altered if all the pitches be modified by a constant increment.

The discriminant of the cubic

$$-A^2H^3 + ACH^2 - BDH + D^2 = 0$$

is

$$A^2D^2\left(A^2D^2 + \tfrac{4}{27}B^3D + \tfrac{4}{27}AC^3 - \tfrac{1}{27}B^2C^2 - \tfrac{2}{3}ABCD\right).$$

Omitting the factor A^2D^2, we have, for the envelope of the system of cones, the cone of the fourth order, found by equating the expression in the bracket to zero. It may be noted that the same cone is the envelope of the planes

$$AH^3 \pm BH^2 + CH \pm D = 0.$$

165. Application to the Plane Section.

We next study the chord joining a pair of reciprocal points on the cubic of § 160. Take any point in the plane of the section; then, as we have just seen, a cone of screws can be drawn through this point, each ray of which crosses two reciprocal screws. This cone is cut by the plane of section in two lines, and, accordingly, we see that through any point in the plane of section two chords can be drawn through a pair of reciprocal points. The actual situation of these chords is found by drawing a pair of tangents to a certain hyperbola. This will now be proved.

The values of θ and θ', which correspond to a pair of reciprocal points, fulfil the condition

$$\tan\theta \tan\theta' = H;$$

whence,

$$\cos(\theta - \theta') = \lambda \cos(\theta + \theta');$$

where, for brevity, we write λ instead of

$$\frac{1+H}{1-H}.$$

If, further, we make $\theta + \theta' = \psi$, we shall find, for the equation of the chord,

$$Px + Qy + Rz = 0;$$

in which,

$$P = \frac{m}{2}(\lambda + \cos 2\alpha) + \frac{m}{2}\sin 2\alpha \sin 2\psi + \frac{m}{2}(\lambda + \cos 2\alpha)\cos 2\psi,$$

$$Q = h \sin\beta + \frac{m}{2}\cos\beta \sin 2\alpha - \frac{m\cos\beta}{2}(\lambda + \cos 2\alpha)\sin 2\psi$$

$$+ \frac{m}{2}\cos\beta \sin 2\alpha \cos 2\psi,$$

$$R = -h^2\tan\beta - \frac{m^2}{2}(\lambda^2 - 1)\cot\beta + \lambda hm \sin 2\psi - \frac{m^2}{2}(\lambda^2 - 1)\cot\beta\cos 2\psi.$$

The envelope of this chord is found to be

$$
\begin{aligned}
0 = &+ x^2 \,|\, \tfrac{1}{4} m^2 \sin^2 2\alpha, \\
&+ y^2 \,|\, \tfrac{1}{4} m^2 \cos^2\beta \cos^2 2\alpha - hm \sin\beta \cos\beta \sin 2\alpha - h^2 \sin^2\beta \\
&\qquad\qquad + \tfrac{1}{2} m^2 \lambda \cos 2\alpha \cos^2\beta + \tfrac{1}{4} m^2 \lambda^2 \cos^2\beta, \\
&+ xy \,|\, -\frac{m^2}{2} \sin 2\alpha \cos 2\alpha \cos\beta - mh \cos 2\alpha \sin\beta \\
&\qquad\qquad - \tfrac{1}{2} m^2 \lambda \sin 2\alpha \cos\beta - mh\lambda \sin\beta, \\
&+ x \,|\, + mh^2 \cos 2\alpha \tan\beta + hm^2 \lambda \sin 2\alpha + mh^2 \lambda \tan\beta, \\
&+ y \,|\, - m^2 h \cos\beta + mh^2 \sin\beta \sin 2\alpha + 2h^3 \sin\beta \tan\beta - m^2 h\lambda \cos\beta \cos 2\alpha, \\
&+ \,|\, m^2 h^2 - h^4 \tan^2\beta.
\end{aligned}
$$

Using the data already assumed in § 160, and, with the addition now made of taking λ to be $-\frac{1}{3}$, the equation reduces to

$$122x^2 - 203y^2 - 161xy - 2417x - 5003y + 245436 = 0;$$

which, for convenience of calculation, I change into

$$x = 1{\cdot}6 + 21{\cdot}6 \sec\phi \pm 47{\cdot}5 \tan\phi,$$

$$y = -12{\cdot}8 + 32{\cdot}8 \sec\phi.$$

This hyperbola has been plotted down in Fig. 36. It obviously touches the cubic at three points. I had not anticipated this until the curves were carefully drawn; but, when the theorem was suggested in this manner, it was easy to provide the following demonstration:—

The cone of reciprocal chords drawn through any point P breaks up into a pair of planes when P lies on the cylindroid. (I use the expression, reciprocal *chord*, to signify the transversal drawn across a pair of reciprocal screws on the cylindroid. This is very different from a screw reciprocal to the cylindroid.) For, take the screw reciprocal to that which passes through P. Then the plane X, through P and this screw, is obviously one part of the locus. Draw through P any transversal across a pair of reciprocals on the cylindroid, then the plane Y, through the centre and this transversal, will be the other part of the locus. This pair of planes, X and Y, intersect in a ray which we shall call S.

A plane of section through P will, of course, usually cut the two planes in two rays, and these will be the two reciprocal chords through P. But suppose the plane of section happened to pass through S, then there will be only one reciprocal chord through P, and this will, of course, be S. Now, S must be a tangent to the cylindroid at P. Every chord through P, in the plane of Y, must cut the surface again in a pair of reciprocal points. To this S must be no exception, and as it lies in X, it intersects the screw reciprocal to that through P; therefore the third intersection of S with the surface must coalesce with P, or, in other words, S must be a tangent to the surface at P. We have thus shown that in the case where two reciprocal chords through a point on the cubic coalesce into one, that one must be the tangent to the cubic at P.

But the two reciprocal chords through a point will only coalesce when the point lies on the hyperbola, in which case the two chords unite into the tangent to the hyperbola. Consider, then, the case where the hyperbola meets the cubic at a point P, inasmuch as P lies on the hyperbola, the two chords coalesce into a tangent thereto, but because they do coalesce, this line must needs be also a tangent to the cubic; hence, whenever the hyperbola meets the cubic the two curves must have a common tangent. Altogether the curves meet in six points, which unite into three pairs, thus giving the required triple contact between the hyperbola and the cubic.

If a constant h be added to all the pitches of the screws on a cylindroid, then, as is well known, the screws so altered still represent a possible cylindroid (§ 18). The variations of h produce no alteration in the cubic section of the cylindroid; but, of course, the hyperbola just considered varies with each change of h. In every case, however, it has the triple contact, and there is also a fixed tangent which must touch every hyperbola. This is the chord joining the two principal screws on the cylindroid; for, as these are reciprocal, notwithstanding any augmentation to the pitches, their chord must always touch the hyperbola. The system of hyperbolæ, corresponding to the variations of h, is thus concisely represented; they must all touch this fixed line, and have triple contact with a fixed curve: that is, they must each fulfil four conditions, leaving one more disposable quantity for the complete definition of a conic. See Appendix, note 4.

We write the tangent to the hyperbola or the reciprocal chord in the form

$$L \cos 2\psi + M \sin 2\psi + N = 0.$$

If a pair of values can be found for x and y, which will simultaneously satisfy

$$L = 0, \quad M = 0, \quad N = 0,$$

then every chord of the type

$$L \cos 2\psi + M \sin 2\psi + N = 0$$

must pass through this point. The condition for this is, that the discriminant of the hyperbola is zero, and we find the discriminant to be

$$\tfrac{1}{4} m^2 h \sin \beta \, (\lambda^2 - 1) \, (h \tan \beta + m \sin 2\alpha).$$

There are two critical cases in which this expression vanishes. It does so if

$$h \tan \beta + m \sin 2\alpha = 0;$$

where the plane of section is tangential to the cylindroid.

But we also note that the discriminant will vanish if

$$h = 0,$$

i.e. if the plane of section passes through the centre of the cylindroid.

We might have foreseen this from the results of the last article; for the plane of Y is a central section, and the hyperbola has evidently degenerated, for all the reciprocal chords, instead of touching an hyperbola, merely pass through the common apex P. The case of the central section is therefore of special interest.

166. The Central Section of the Cylindroid.

By this we mean a section of the surface, special in no other sense, save that it passes through the centre of the surface. The equation to the central section is (§ 160)

$$y^3 \sin \beta \cos^2 \beta + yx^2 \sin \beta - mx^2 \sin 2\alpha + 2mxy \cos \beta \cos 2\alpha + my^2 \sin 2\alpha \cos^2 \beta = 0.$$

The chord joining points of equal pitch $\pm \theta$ is

$$x (\cos 2\theta + \cos 2\alpha) + y \cos \beta \sin 2\alpha + m \cot \beta \sin^2 2\theta = 0.$$

The apex P through which all reciprocal chords pass is

$$x = \frac{m (\lambda^2 - 1) \cot \beta (\lambda + \cos 2\alpha)}{1 + 2\lambda \cos 2\alpha + \lambda^2},$$

$$y = \frac{m (\lambda^2 - 1) \operatorname{cosec} \beta . \sin 2\alpha}{1 + 2\lambda \cos 2\alpha + \lambda^2};$$

and in general the co-ordinates of a point on the cubic are

$$x = y \cos \beta \tan (\theta - \alpha),$$

$$y = - m \operatorname{cosec} \beta \sin 2\theta.$$

One of these curves may be conveniently drawn to scale, from the equations

$$x = \cdot 9y \tan (\theta - 25),$$

$$y = - 66 \sin 2\theta.$$

The parabola, which is the envelope of equal pitch-chords, would in this case have as its equation

$$y = - 51 - \left(6 + \frac{x}{13}\right)^2.$$

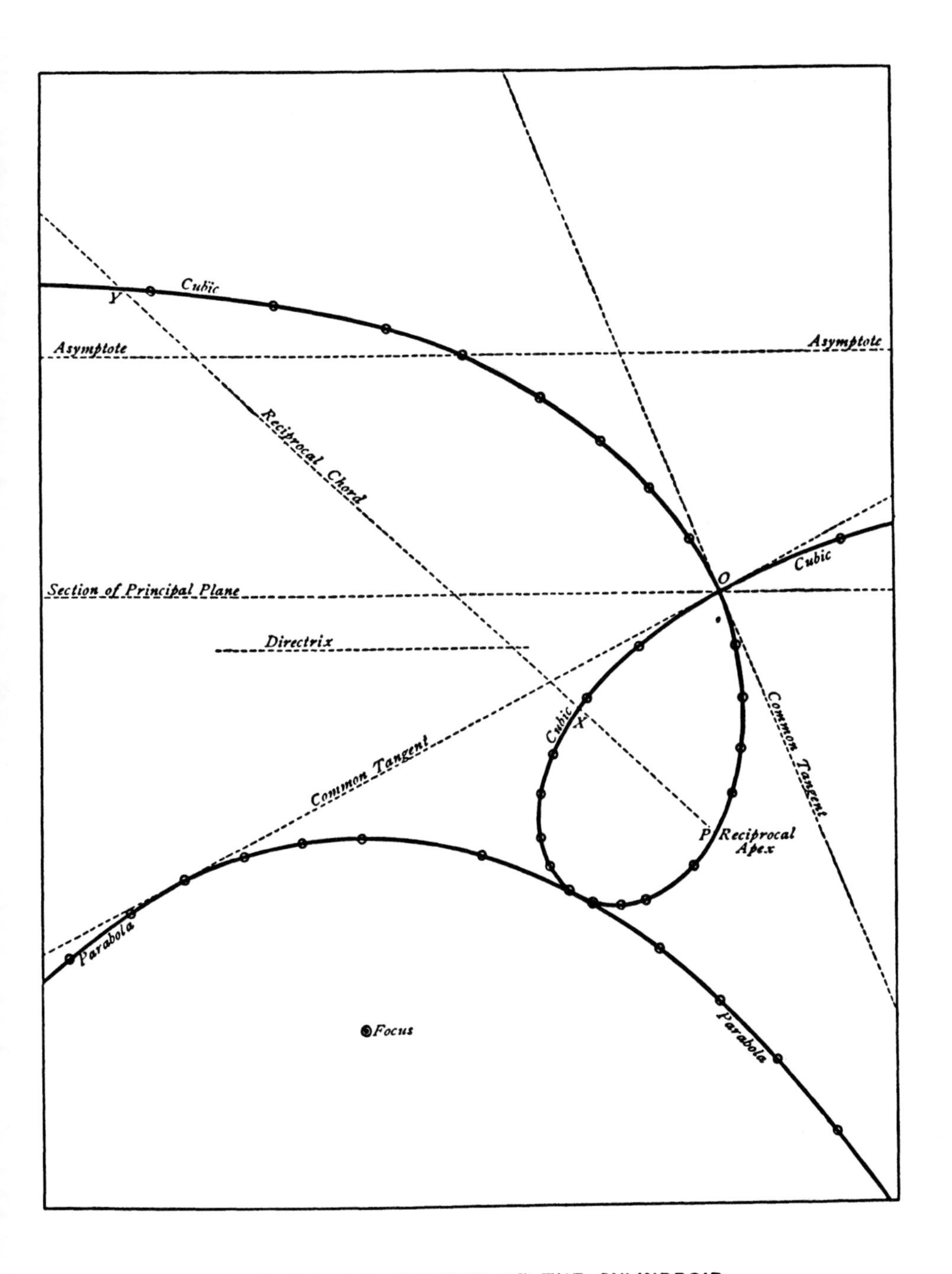

THE CENTRAL SECTION OF THE CYLINDROID.

The principal screws on the cylindroid both pass through the double point; the two tangents to the curve at this point must therefore both touch the parabola.

The leading feature of the central section is expressed by the important property possessed by the chords joining reciprocal screws. If we add any constant to the pitches, then we alter λ, and, accordingly, the point P, through which all reciprocal chords pass, moves along the curve.

The tangent to the cubic at P meets the cubic again in the point reciprocal to P. Two tangents, real or imaginary, can be drawn from P to the cubic touching it in the points T_1, T_2, respectively: as these must each correspond to a screw reciprocal to itself, it follows that T_1 and T_2 are the screws of zero pitch. We hence see that the two tangents from any point on the cubic touch the cubic in points of equal pitch.

Let α and β be two screws, and γ and δ another pair of screws, and let the two chords, $\alpha\beta$ and $\gamma\delta$, intersect again on the cubic. If d and θ be the perpendicular distance and angle between the first pair, and d' and θ' the corresponding quantities for the second pair, then there must be some quantity ω, which, if added to all the pitches on the cylindroid, will make α and β reciprocal, and also γ and δ reciprocal. We thus have

$$(p_\alpha + p_\beta + 2\omega)\cos\theta - d\sin\theta = 0,$$

$$(p_\gamma + p_\delta + 2\omega)\cos\theta' - d'\sin\theta' = 0;$$

whence,
$$p_\alpha + p_\beta - d\tan\theta = p_\gamma + p_\delta - d'\tan\theta';$$

in other words, for every pair of screws, α and β, whose chords belong to a pencil diverging from a common point on the surface, the expression

$$p_\alpha + p_\beta - d\tan\theta$$

is a constant. The value of this constant is double the pitch of the screw of either of the points of contact of the two tangents from P to the curve.

167. Section Parallel to the Nodal Line.

If the node on the cubic be at infinity, the form of equation to the cubic hitherto employed will be illusory. The nature of this section must therefore be studied in another way, as follows:—

Let the plane cut the two perpendicular screws in A and B. Let l be the perpendicular OC from O upon C, and let η be the inclination of this perpendicular to the axis of x. Then, taking OA as the new axis of x, in which case z will be the new y, we have

$$x' = l\tan(\eta - \theta),$$

$$y' = m\sin 2\theta.$$

Eliminating θ, and omitting the accents, we have, as the equation of the cubic,

$$yx^2 + mx^2 \sin 2\eta + l^2 y + 2lmx \cos 2\eta - ml^2 \sin 2\eta = 0.$$

The chord joining the two points θ and θ' has, as its equation,

$$mx \cos(\theta + \theta')[\cos(2\eta - \theta - \theta') + \cos(\theta - \theta')] + ly$$
$$- ml \sin(2\eta - \theta - \theta') \cos(\theta + \theta') - ml \cos(\theta - \theta') \sin(\theta + \theta') = 0.$$

If this be the chord joining the screws of equal pitch, then

$$\theta + \theta' = 0,$$

and the equation reduces to

$$mx(\cos 2\eta + \cos 2\theta) + ly - ml \sin 2\eta = 0.$$

We thus see that this chord, which in the general section envelops a parabola, now passes constantly through the fixed point

$$x = 0,$$
$$y = + m \sin 2\eta.$$

This result could have been foreseen; for, consider that screw on the cylindroid (and there must always be one) normal to the plane which it intersects at a point P, any ray in the plane through P is perpendicular to this screw, and, therefore, by a well-known property of the cylindroid, must intersect the curve again in two points of equal pitch. This point P is, of course, the point whose existence we have demonstrated above.

168. Relation between Two Conjugate Screws of Inertia.

We have found the relation between a pair of conjugate screws of inertia so important in the dynamical part of the theory, that it is worth while to investigate the properties of the chord joining two such points in the central section. It can readily be shown that this chord must envelop a conic. This conic and the point P on the cubic through which all reciprocal chords will pass, will enable the impulsive screw, corresponding to any instantaneous screw, to be immediately determined. For, draw through any point S that tangent to the conic which gives S as one of the two conjugate screws of inertia which must lie upon it; let S' be the other conjugate screw; then the chord PS' will cut the cubic again in the required impulsive screw. The two principal screws of inertia are found by drawing from P that tangent to the conic which has *not* P as one of the two conjugate screws of inertia. The two intersections of this tangent, with the cubic, are the required principal screws of inertia.

We can also determine the relation between the impulsive screw and the instantaneous screw with regard to any section whatever. We have here to consider two conics connected with the cubic, viz. the reciprocal conic, which is the envelope of reciprocal chords, and the inertia conic, which is the envelope of chords of conjugate screws of inertia. We must provide a means of discriminating the two tangents from a point P on the cubic to either conic; any ray, of course, cuts the cubic in three points, of which two possess the characteristic relation. If P be one of these two, we may call this tangent the '*odd* tangent.' The other tangent will have, as its significant points, the two remaining intersections; leaving out P, we can then proceed, as follows, to determine the impulsive screw corresponding to P as the instantaneous screw:—

Draw the odd tangent from P to the inertia conic, and from the conjugate point thus found draw the odd tangent to the reciprocal conic. The reciprocal point Q thus found is the impulsive screw corresponding to P as the instantaneous screw.

In general there are four common tangents to the two conics. Of these tangents there is only one possessing the property, that the same two of its three intersections with the cubic are the correlative points with respect to each of the conics. These two intersections are the principal screws of inertia.

To determine the small oscillations we find the *potential* conic, the tangents to which are chords joining two conjugate screws of the potential (§ 100). The two harmonic screws are then to be found on one of the two common tangents to the two conics. It can be shown that both the inertia conic and the potential conic will, like the reciprocal conic, have triple contact with the cubic.

CHAPTER XIV.

FREEDOM OF THE THIRD ORDER*.

169. Introduction.

The dynamics of a rigid body which has freedom of the third order, possesses a special claim to attention, for, included as a particular case, we have the celebrated problem of the rotation of a rigid body about a fixed point. In the theory of screws the screw system of the third order is characterised by the feature that the reciprocal screw system is also of the third order, and this is a fertile source of interesting theorems.

We shall first study the screw system of the third order, and its reciprocal. We shall then show how the instantaneous screw, corresponding to a given impulsive screw, can be determined for a rigid body whose movements are prescribed by any screw system of the third order. We shall also point out the three principal screws of inertia, of which the three principal axes are only special cases, and we shall determine the kinetic energy acquired by a given impulse. Finally, we shall determine the three harmonic screws, and we shall apply these principles to the discussion of the small oscillations of a rigid body about a fixed point under the influence of gravity.

A screw system of the first order consists of course of one screw. A screw system of the second order consists of all the screws on a certain ruled surface (the cylindroid). Ascending one step higher, we find that in a screw system of the third order the screws are so numerous that a finite number (three) can be drawn through every point in space. In the screw system of the fourth order a cone of screws can be drawn through every point, while to a screw system of the fifth order belongs a screw of suitable pitch on every straight line in space.

170. Screw System of the Third Order.

We shall now consider the collocation of the screws in space which constitute a screw system of the third order. A free rigid body can receive

* *Transactions of the Royal Irish Academy*, Vol. xxv. p. 191 (1871).

six independent displacements. Its position is, therefore, to be specified by six co-ordinates. If, however, the body be so constrained that its six co-ordinates must always satisfy three equations of condition, there are then only three really independent co-ordinates, and any position possible for a body so circumstanced may be attained by twists about three fixed screws, provided that twists about these screws are permitted by the constraints.

Let A be an initial position of a rigid body M. Let M be moved from A to a closely adjacent position, and let x be the screw by twisting about which this movement has been effected; similarly let y and z be the two screws, twists about which would have brought the body from A to two other independent positions. We thus have three screws, x, y, z, which completely specify the circumstances of the body so far as its capacity for movement is considered.

Since M can be twisted about each and all of x, y, z, it must be capable of twisting about a doubly infinite number of other screws. For suppose that by twists of amplitude x', y', z', the final position V is attained. This position could have been reached by twisting about some screw v, so as to come from A to V by a single twist. As the ratios of x' to y', and z', are arbitrary, and as a change in either of these ratios changes v, the number of v screws is doubly infinite.

All the screws of which v is a type form what we call *a screw system of the third order*. We may denote this screw system by the symbol S.

171. The Reciprocal Screw System.

A wrench which acts on a screw η will not be able to disturb the equilibrium of M, provided η be reciprocal to x, y, z. If η be reciprocal to three independent screws of the system S, it will be reciprocal to every screw of S. Since η has thus only three conditions to satisfy in order that it may be reciprocal to S, and since five quantities determine a screw, it follows that η may be any one of a doubly infinite number of screws which we may term the *reciprocal screw system* S'. Remembering the property of reciprocal screws (§ 20) we have the following theorem (§ 73).

A body only free to twist about all the screws of S cannot be disturbed by a wrench on any screw of S'; and, conversely, a body only free to twist about the screws of S' cannot be disturbed by a wrench on any screw of S.

The reaction of the constraints by which the freedom is prescribed constitutes a wrench on a screw of S'.

172. Distribution of the Screws.

To present a clear picture of all the movements which the body is competent to execute, it will be necessary to examine the mutual connexion of the doubly infinite number of screws which form the screw system. It will be most convenient in the first place to classify the screws in the system according to their pitches; the first theorem to be proved is as follows:—

All the screws of given pitch $+k$ in a three-system lie upon a hyperboloid of which they form one system of generators, while the other system of generators with the pitch $-k$ belong to the reciprocal screw system.

This is proved as follows:—Draw three screws, p, q, r, of pitch $+k$ belonging to S. Draw three screws, l, m, n, each of which intersects the three screws p, q, r, and attribute to each of l, m, n, a pitch $-k$. Since two intersecting screws of equal and opposite pitches are reciprocal, it follows that p, q, r, must all be reciprocal to l, m, n. Hence, since the former belong to S, the latter must belong to S'. Every other screw of pitch $+k$ intersecting l, m, n, must be reciprocal to S', and must therefore belong to S.

But the locus of a straight line which intersects three given straight lines is a hyperboloid of one sheet, and hence the required theorem has been proved.

173. The Pitch Quadric.

One member of the family of hyperboloids obtained by varying k presents exceptional interest. It is *the locus of the screws of zero pitch* belonging to the screw complex. As this quadric has an important property (§ 176) besides that of being the locus of the screws of zero pitch, it is desirable to denote it by the special phrase *pitch quadric*.

We shall now determine the equation of the pitch quadric. Let one of the principal axes of the pitch quadric be denoted by x, this will intersect the surface in two points through each of which a pair of generators can be drawn. One generator of each pair will belong to S, and the other to S'. Each pair of generators will be parallel to the asymptotes of the section of the pitch quadric by the plane containing the remaining principal axes y and z. Let μ, ν be the two generators belonging to S, then lines bisecting internally and externally the angle between two lines in the plane of y and z, parallel to μ, ν will be two of the principal axes of the pitch quadric. Draw the cylindroid ($\mu\nu$). The two screws of zero pitch on the cylindroid are equidistant from the centre of the cylindroid, and the two rectangular screws of the cylindroid bisect internally and externally the angle between the lines parallel to the screws of zero pitch. Hence it follows that the two rectangular screws of the cylindroid ($\mu\nu$) must be on the axes of y and z of the pitch quadric. We shall denote these screws by β and γ, and their pitches by p_β and p_γ. From the properties of the cylindroid it appears that a, the semiaxis of the pitch quadric, must be determined from the equations

$$a = (p_\beta - p_\gamma) \sin l \cos l,$$

$$p_\beta \cos^2 l + p_\gamma \sin^2 l = 0;$$

whence eliminating l, we deduce

$$a = \sqrt{-p_\beta p_\gamma},$$

with of course similar values of b and c. Substituting these values in the equation of the quadric

$$\frac{x^2}{a^2} + \frac{y^2}{b^2} + \frac{z^2}{c^2} = 1$$

we deduce the important result which may be thus stated:—

The three principal axes of the pitch quadric, when furnished with suitable pitches p_α, p_β, p_γ, constitute screws belonging to the screw system of the third order, and the equation of the pitch quadric has the form

$$p_\alpha x^2 + p_\beta y^2 + p_\gamma z^2 + p_\alpha p_\beta p_\gamma = 0.$$

We can also show conversely that *every* screw θ of zero pitch, which belongs to the screw system of the third order, must be one of the generators of the pitch quadric. For θ must be reciprocal to *all* the screws of zero pitch on the reciprocal system of generators of the pitch quadric; and since two screws of zero pitch cannot be reciprocal unless they intersect either at a finite or infinite distance, it follows that θ must intersect the pitch quadric in an infinite number of points, and must therefore be entirely contained thereon.

174. The Family of Quadrics.

It has been shown that all the screws of given pitch belonging to a system of the third order are the generators of a certain hyperboloid. There is of course a different hyperboloid for each pitch. We have now to show that all these hyperboloids are concentric.

Take any two screws whatever belonging to the system and draw the cylindroid which passes through those screws. This cylindroid contains two screws of every pitch. It must therefore have two generators in common with every hyperboloid of the family. But from the known symmetrical arrangement of the screws of equal pitch on a cylindroid, it follows that the centre of that surface must lie at the middle point of the shortest distance between each two screws of equal pitch. The centres of the hyperboloids for all possible pitches must therefore lie in the principal plane of any cylindroid of the system. Take any three cylindroids of the system.

The centres of all the hyperboloids coincide with the intersection of the three principal planes of the cylindroids. It will be convenient to call this point the *centre of the three-system.*

We hence see that whenever three screws of a three-system are given, the centre of the system is determined as the intersection of the principal planes of the three cylindroids defined by each pair of screws taken successively.

We may also show that not only are the family of hyperboloids concentric, but that they have also their three principal axes coincident in direction and situation with the principal axes of the pitch quadric.

Draw any principal axis z of the pitch quadric. Two screws of zero pitch belonging to the system will be intersected by z and we draw the cylindroid through these two screws. Let L_1 and L_2 be the two screws of equal pitch p on this cylindroid. Let O be the centre of the cylindroid, this same point being also the centre of the pitch quadric, and therefore as shown above of every p-pitch hyperboloid S_p. As the centre bisects every diameter, it follows that the plane OL_2 cuts the hyperboloid S_p again in a ray L_1' which is perpendicular to z and crosses L_1 at its intersection with z. The plane containing L_1 and L_1' is therefore a tangent to S_p at the point where the plane is cut by z. As z is perpendicular to this plane it follows that the diameter is perpendicular to its conjugate plane. Hence z is a principal axis of S_p, and the required theorem is proved.

Let now S denote a screw system of the third order, where α, β, γ are the three screws of the system on the principal axes of the pitch quadric. Diminish the pitches of all the screws of S by any magnitude k. Then the quadric

$$(p_\alpha - k)x^2 + (p_\beta - k)y^2 + (p_\gamma - k)z^2 + (p_\alpha - k)(p_\beta - k)(p_\gamma - k) = 0\ldots$$

must be the locus of screws of zero pitch in the altered system, and therefore of pitch $+k$ in the original system (§ 110).

Regarding k as a variable parameter, the equation just written represents *the family of quadrics* which constitute the screw system S and the reciprocal screw system S'. Thus all the generators of one system on each quadric, with pitch $+k$, constitute screws about which the body, with three degrees of freedom, can be twisted; while all the generators of the other system, with pitch $-k$, constitute screws, wrenches about which would be neutralized by the reaction of the constraints.

For the quadric to be a real surface it is plain that k must be greater than the least, and less than the greatest of the three quantities $p_\alpha, p_\beta, p_\gamma$.

Hence the pitches of all the real screws of the screw system S are intermediate between the greatest and least of the three quantities p_α, p_β, p_γ.

175. Construction of a three-system from three given Screws.

If a family of quadric surfaces have one pair of generators (which do not intersect) in common, then the centre of the surface will be limited to a certain locus. We may investigate this conveniently by generalizing the question into the search for the locus of the pole of a fixed plane with respect to the several quadrics.

Let A be the given plane, I be the ray which joins the two points in which the given pair of generators intersect A, X be the plane through I and the first generator, Y the plane through I and the second generator, B the plane through I which is the harmonic conjugate of A with respect to X and Y. Then B is the required locus.

For, draw any quadric through the two given generators, and let O be the pole of A with respect to that quadric.

Draw a transversal through O cutting the plane A in the point A_1 and the first and second generators in X_1 and Y_1 respectively. Since A_1 is on the polar of O it follows that $OX_1A_1Y_1$ is an harmonic section. But the transversal must be cut harmonically by the pencil of planes $I(BXAY)$ and hence O must lie in B, which proves the theorem.

In the particular case when A is the plane at infinity, then O is the centre of the quadric. A plane parallel to the two generators cuts the plane at infinity in the line I, and the planes X, Y and B must also contain I. Then A, B, X, Y are parallel planes. Any transversal across X and Y is cut harmonically by B and A, and as A is at infinity, the transversal must be bisected at B. It thus appears that when a family of quadrics have one pair of non-intersecting generators in common, then the plane which bisects at right angles the shortest distance between these generators is the locus of the centres of the quadrics.

If therefore three generators of a quadric are given, the three planes determined by each pair of the quadrics determine the centre by their intersection. The construction of the axes of the quadric may be effected geometrically in the following manner. Draw three transversals Q_1, Q_2, Q_3 across the three given generators R_1, R_2, R_3. Draw also two other transversals R_4, R_5 across Q_1, Q_2, Q_3. Construct the conic which passes through the five points in which R_1, R_2, R_3, R_4, R_5 intersect the plane at infinity. Find the common conjugate triangle to this conic and to the circle which is the intersection of every sphere with the plane at infinity. Then the three rays from the centre of the quadric to the vertices of this triangle are the three principal axes of the quadric.

We thus prove again that if α and β be any two screws of a three-system, the centre of the pitch-quadric must lie in the principal plane of the cylindroid through α and β. For the common perpendicular to any two screws of equal pitch on the cylindroid will be bisected by the principal plane and therefore any hyperboloid through these two screws of equal pitch must have its centre in that plane.

176. Screws through a Given Point.

We shall now show that three screws belonging to S, and also three screws belonging to S', can be drawn through any point x', y', z'. Substitute x', y', z', in the equation of § 175 and we find a cubic for k. This shows that three quadrics of the system can be drawn through each point of space. The three tangent planes at the point each contain two generators, one belonging to S, and the other to S'. It may be noticed that these three tangent planes intersect in a straight line.

From the form of the equation it appears that the sum of the pitches of three screws through a point is constant and equal to $p_\alpha + p_\beta + p_\gamma$.

Two intersecting screws can only be reciprocal if they be at right angles, or if the sum of their pitches be zero. It is hence easy to see that, if a sphere be described around any point as centre, the three screws belonging to S, which pass through the point, intersect the sphere in the vertices of a spherical triangle which is the polar of the triangle similarly formed by the lines belonging to S'.

We shall now show that *one* screw belonging to S can be found parallel to any given direction. All the generators of the quadric are parallel to the cone

$$(p_\alpha - k)\, x^2 + (p_\beta - k)\, y^2 + (p_\gamma - k)\, z^2 = 0,$$

and k can be determined so that this cone shall have one generator parallel to the given direction; the quadric can then be drawn, on which two generators will be found parallel to the given direction; one of these belongs to S, while the other belongs to S'.

It remains to be proved that *each screw of S has a pitch which is proportional to the inverse square of the parallel diameter of the pitch quadric**.

* This theorem is connected with the linear geometry of Plücker, who has shown (*Neue Geometrie des Raumes*, p. 130) that $k_1x^2 + k_2y^2 + k_3z^2 + k_1k_2k_3 = 0$, is the locus of lines common to three linear complexes of the first degree. The axes of the three complexes are directed along the co-ordinate axes, and the parameters of the complexes are k_1, k_2, k_3; the same author has also proved that the parameter of any complex belonging to the "dreigliedrige Gruppe" is proportional to the inverse square of the parallel diameter of the hyperboloid.

Let r be the intercept on a generator of the cone

$$(p_\alpha - k)\,x^2 + (p_\beta - k)\,y^2 + (p_\gamma - k)\,z^2 = 0\,;$$

by the pitch quadric

$$p_\alpha x^2 + p_\beta y^2 + p_\gamma z^2 + p_\alpha p_\beta p_\gamma = 0\,;$$

then

$$k = -\frac{p_\alpha p_\beta p_\gamma}{r^2}\,;$$

but k is the pitch of the screw of S, which is parallel to the line r.

Nine constants (§ 75) are required for the determination of a screw system of the third order. This is the same number as that required for the specification of a quadric surface. We hence infer, what is indeed otherwise manifest, viz., that when the pitch quadric is known the entire screw system of the third order is determined.

Another interesting property of the pitch quadric is thus enunciated. *Any three co-reciprocal screws of a given screw system of the third order are parallel to a triad of conjugate diameters of its pitch quadric.*

Take *any* three co-reciprocal screws of the system as screws of reference, and let p_1, p_2, p_3 be their pitches. If then the co-ordinates of any screw ρ belonging to the system be denoted by ρ_1, ρ_2, ρ_3, we shall have for the pitch of ρ (§ 95)

$$p_\rho = p_1\rho_1^2 + p_2\rho_2^2 + p_3\rho_3^2.$$

If a parallelopiped be constructed, of which the three lines parallel to the reciprocal screws, drawn through the centre of the pitch quadric, are conterminous edges, and of which the line parallel to ρ is the diagonal, and if x, y, z be the lengths of the edges, and r the length of the diagonal, then we have (§ 35)

$$\frac{x}{r} = \rho_1, \quad \frac{y}{r} = \rho_2, \quad \frac{z}{r} = \rho_3.$$

It follows that p_ρ must be proportional to the inverse square of the parallel diameter of the quadric surface

$$p_1x^2 + p_2y^2 + p_3z^2 = H.$$

But p_ρ must be proportional to the inverse square of the parallel diameter of the pitch quadric, and hence the equation last written must actually be the equation of the pitch quadric, when H is properly chosen. But the equation is obviously referred to three conjugate diameters, and hence three conjugate diameters of the pitch quadric are parallel to three co-reciprocal screws of the screw system.

We see from this that *the sum of the reciprocals of the pitches of three co-reciprocal screws is constant.* This theorem will be subsequently generalised.

177. Locus of the feet of perpendiculars on the generators*.

If p be the pitch of the screw of the three-system which makes angles α, β, γ with the three principal screws, it is then easy to show that the equation of the screw is

$$(p-a)\cos\alpha + z\cos\beta - y\cos\gamma = 0,$$

$$-z\cos\alpha + (p-b)\cos\beta + x\cos\gamma = 0,$$

$$+y\cos\alpha - x\cos\beta + (p-c)\cos\gamma = 0.$$

If perpendiculars be let fall from the origin on the several screws of the system, then if x, y, z be the foot of one of the perpendiculars

$$x\cos\alpha + y\cos\beta + z\cos\gamma = 0.$$

Eliminating $\cos\alpha$, $\cos\beta$, $\cos\gamma$ from this equation and the two last of those above, we have

$$\begin{vmatrix} x & y & z \\ -z & p-b & x \\ +y & -x & p-c \end{vmatrix} = 0,$$

or
$$(p-b)(p-c)x + x(x^2+y^2+z^2) + yz(b-c) = 0;$$

from this and the two similar equations we have, by elimination of p^2 and p and denoting $x^2+y^2+z^2$ by r^2,

$$\begin{vmatrix} x, & (b+c)x, & bcx + (b-c)yz + xr^2 \\ y, & (c+a)y, & cay + (c-a)zx + yr^2 \\ z, & (a+b)z, & abz + (a-b)xy + zr^2 \end{vmatrix} = 0;$$

multiplying the first column by r^2 and subtracting it from the last, we have

$$\begin{vmatrix} x, & (b+c)x, & bcx + (b-c)yz \\ y, & (c+a)y, & cay + (c-a)zx \\ z, & (a+b)z, & abz + (a-b)xy \end{vmatrix} = 0,$$

which may be written

$$(a-b)^2x^2y^2 + (b-c)^2y^2z^2 + (c-a)^2z^2x^2 = (a-b)(b-c)(c-a)xyz.$$

* This Article is due to Professor C. Joly, 'On the theory of linear vector functions,' *Transactions of the Royal Irish Academy*, Vol. xxx. pp. 601 and 617 (1895), where a profound discussion of Steiner's surface is given. See also by the same author 'Bishop Law's Mathematical Prize Examination,' *Dublin University Examination Papers*, 1898.

This equation denotes a form of Steiner's surface:

$$\sqrt{\alpha}+\sqrt{\beta}+\sqrt{\gamma}+\sqrt{\delta}=0,$$

where

$$\alpha=\frac{2x}{b-c}+\frac{2y}{c-a}+\frac{2z}{a-b}+1,$$

$$\beta=\frac{2x}{b-c}-\frac{2y}{c-a}-\frac{2z}{a-b}+1,$$

$$\gamma=-\frac{2x}{b-c}+\frac{2y}{c-a}-\frac{2z}{a-b}+1,$$

$$\delta=-\frac{2x}{b-c}-\frac{2y}{c-a}+\frac{2z}{a-b}+1.$$

From the form of its equation it appears that this surface has three double lines which meet in a point, viz. the three axes OX, OY, OZ. This being so any plane will cut the surface in a quartic curve with three double points, being those in which the plane cuts the axes. If the plane touch the surface, the point of contact is an additional double point on the section, that is, the section will be a quartic curve with four double points, i.e. a pair of conics. The projections of the origin on the generators of any cylindroid belonging to the system lie on a plane ellipse (§ 23). This ellipse must lie on the Steiner quartic. Hence the plane of the ellipse must cut the quartic in two conics and must be a tangent plane. See note on p. 182.

178. Screws of the Three-System parallel to a Plane.

Up to the present we have been analysing the screw system by classifying the screws into groups of constant pitch. Some interesting features will be presented by adopting a new method of classification. We shall now divide the general system into groups of screws which are parallel to the same plane.

We shall first prove that each of these groups is in general a cylindroid. For suppose a screw of infinite pitch normal to the plane, then all the screws of the group parallel to the plane are reciprocal to this screw of infinite pitch. But they are also reciprocal to any three screws of the original reciprocal system; they, therefore, form a screw system of the second order (§ 72)—that is, they constitute a cylindroid.

We shall prove this in another manner.

A quadric containing a line must touch every plane passing through the line. The number of screws of the system which can lie in a given plane is, therefore, equal to the number of the quadrics of the system which can be drawn to touch that plane.

The quadric surface whose equation is

$$(p_\alpha - k)\,x^2 + (p_\beta - k)\,y^2 + (p_\gamma - k)\,z^2 + (p_\alpha - k)\,(p_\beta - k)\,(p_\gamma - k) = 0,$$

touches the plane $Px + Qy + Rz + S = 0$, when the following condition is satisfied:

$$P^2(p_\beta - k)(p_\gamma - k) + Q^2(p_\alpha - k)(p_\gamma - k) + R^2(p_\alpha - k)(p_\beta - k) + S^2 = 0;$$

whence it follows that two values of k can be found, or that two quadrics can be made to touch the plane, and that, therefore, two screws of the system, and, of course, two reciprocal screws, lie in the plane.

From this it follows that all the screws of the system parallel to a plane must in general lie upon a cylindroid. For, take any two screws parallel to the plane, and draw a cylindroid through these screws. Now, this cylindroid will be cut by any plane parallel to the given plane in two screws, which must belong to the system; but this plane cannot contain any other screws; therefore, all the screws parallel to a given plane must lie upon the same cylindroid.

179. Determination of a Cylindroid.

We now propose to solve the following problem:—Given a plane, determine the cylindroid which contains all the screws, selected from a screw system of the third order, which are parallel to that plane.

Draw through O the centre of the pitch quadric a plane A parallel to the given plane. We shall first show that the centre of the cylindroid required lies in A (§ 174).

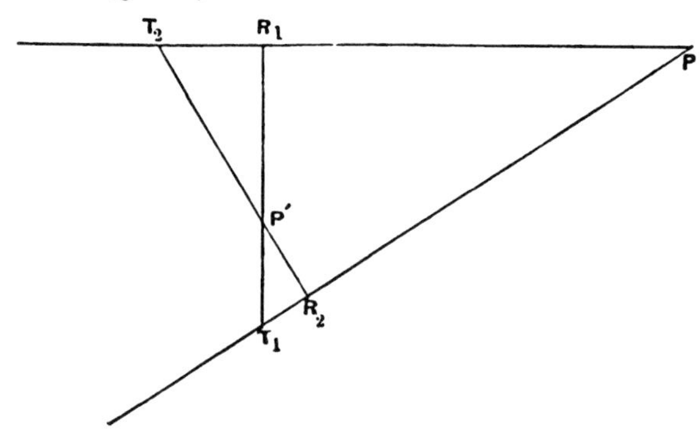

Fig. 37.

Let T_1, T_2 (Fig. 37) be two points in which the two quadrics of constant pitch touch the plane of the paper, which may be regarded as any plane parallel to A; then P is the intersection of the pair of screws belonging to the system PT_1, PT_2, which lie in that plane, and P' is the intersection of the pair of reciprocal screws $P'R_1$, $P'R_2$ belonging to the reciprocal system. Since $P'R_1$ is to be reciprocal to PT_2, it is essential that R_1 be a right angle; similarly R_2 is a right angle. The reciprocal cylindroid, whose axis passes through P', will be identical with the cylindroid belonging to the system whose axis passes through P; but the two will be differently posited. If the angle at P be a right angle, the points T_1 and T_2 are at infinity; therefore, the plane touches the quadrics at infinity; it must, therefore, touch the asymptotic cone, and must, therefore, pass through the centre of the pitch quadric O; but P is the centre of the cylindroid in this case, and, therefore, the centre of the cylindroid must lie in the plane A.

The position of the centre of the cylindroid in the plane A is to be found by the following construction:—Draw through the centre O a diameter of the pitch quadric conjugate to the plane A. Let this line intersect the pitch quadric in the points P_1, P_2, and let S, S' (Fig. 38) be the feet of the perpendiculars let fall from P_1, P_2 upon the plane A. Draw the asymptotes OL, OM to the section of the pitch quadric, made by the plane A. Through S and S' draw lines in the plane A, ST, ST', $S'T$, $S'T'$, parallel to the asymptotes, then T' and T are the centres of the two required cylindroids which belong to the two reciprocal screw systems.

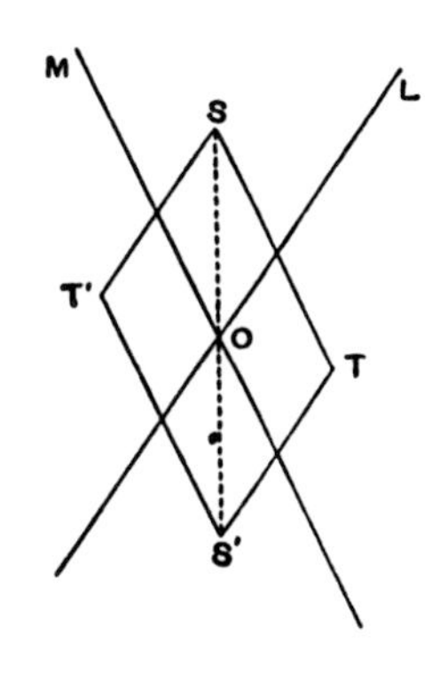

Fig. 38.

This construction is thus demonstrated:—

The tangent planes at P_1, P_2 each intersect the surface in lines parallel to OL, OM. Let us call these lines P_1L_1, P_1M_1 through the point P_1, and P_2L_2, P_2M_2 through the point P_2. Then P_1L_1, P_2M_2 are screws belonging to the system, and P_1M_1, P_2L_2 are reciprocal screws.

Since OL is a tangent to the pitch quadric, it must pass through the intersection of two rectilinear generators, which both lie in a plane which contains OL; but since OL touches the pitch quadric at infinity, the two generators in question must be parallel to OL, and therefore their projections on the plane of A must be $S'T$, ST'. Similarly for ST, $S'T'$; hence ST' and $S'T'$ are the projections of two screws belonging to the system, and therefore the centre of the cylindroid is at T'. In a similar way it is proved that the centre of the reciprocal cylindroid is at T.

Having thus determined the centre of the cylindroid, the remainder of the construction is easy. The pitches of two screws on the surface must be proportional to the inverse square of the parallel diameters of the section of the pitch quadric made by A. Therefore, the greatest and least pitches will be on screws parallel to the principal axes of the section. Hence, lines drawn through T' parallel to the external and internal bisectors of the angle between the asymptotes are the two rectangular screws of the cylindroid. Thus the problem of finding the cylindroid is completely solved.

It is easily seen that each cylindroid touches each of the quadrics in two points.

We may also note that a screw of the system perpendicular to the plane passes through T. Thus given any cylindroid of the system the position of the screw of the system parallel to the axis of the cylindroid is determined*.

180. Miscellaneous Remarks.

We are now in a position to determine the actual situation of a screw θ belonging to a screw system of the third order of which the direction is given. The construction is as follows:—Draw through O the centre of the pitch quadric a radius vector OR parallel to the given direction of θ, and cutting the pitch quadric in R. Draw a tangent plane to the pitch quadric in R. Then the plane A through OR, of which the intersection with the tangent plane is perpendicular to OR, is the plane which contains θ. For the section in which A cuts the pitch quadric has for a tangent at R a line perpendicular to OR; hence the line OR is a principal axis of the section, and hence (§ 179) one of the two screws of the system in the plane A must be parallel to OR. It remains to find the actual situation of θ in the plane A.

Since the direction of θ is known, its pitch is determinate, because it is inversely proportional to the square of OR. Hence the quadric can be constructed, which is the locus of all the screws which have the same pitch as θ. This quadric must be intersected by the plane A in two parallel

* In a letter (10 April 1899) Professor C. Joly writes as follows:—Any plane through the origin contains one pair of screws A and B belonging to the system intersecting at right angles and another pair A' and B' belonging to the reciprocal system. The group A, B, A', B' form a rectangle of which the origin O is the centre. The feet of the perpendiculars from O on A and on B and the point of intersection of A' and B' will lie on the Steiner's quartic

$$(b-c)^2 y^2z^2 + (c-a)^2 z^2x^2 + (a-b)^2 x^2y^2 = +(b-c)(c-a)(a-b)\,xyz.$$

The point of intersection of A and B and the feet of the perpendiculars on A' and B' will lie on the new Steiner's quartic

$$(b-c)^2 y^2z^2 + (c-a)^2 z^2x^2 + (a-b)^2 x^2y^2 = -(b-c)(c-a)(a-b)\,xyz.$$

The locus of the feet of the perpendiculars on the screws of a three-system from any arbitrary origin whatever is still a Steiner's quartic, but its three double lines are no longer mutually rectangular. They are coincident with the three screws of the reciprocal three-system which passed through the origin. This quartic is likewise the locus of the intersection of the pairs of screws of the reciprocal system which are coplanar with the origin. There is a second Steiner's quartic whose double lines coincide with the three screws of the given system which pass through the origin and which is the locus of intersection of those pairs of screws of the given system which lie in planes through the origin. It is also the locus of the feet of perpendiculars on the screws of the reciprocal system.

lines. One of these lines is the required residence of the screw θ, while the other line, with a pitch equal in magnitude to that of θ, but opposite in sign, belonging, as it does, to one of the other system of generators, is a screw reciprocal to the system.

The family of quadric surfaces of constant pitch have the same planes of circular section, and therefore every plane through the centre cuts the quadrics in a system of conics having the same directions of axes.

The cylindroid which contains all the screws of the screw system parallel to one of the planes of circular section must be composed of screws of equal pitch. A cylindroid in this case reduces to a plane pencil of rays passing through a point. We thus have two points situated upon a principal axis of the pitch quadric, through each of which a plane pencil of screws can be drawn, which belong to the screw system. All the screws passing through either of these points have equal pitch. The pitches of the two pencils are equal in magnitude, but opposite in sign. The magnitude is that of the pitch of the screw situated on the principal axis of the pitch quadric*.

181. Virtual Coefficients.

Let ρ be a screw of the screw system which makes angles whose cosines are f, g, h, with the three screws of reference α, β, γ upon the axes of the pitch quadric. Then, reference being made to any six co-reciprocals, we have for the co-ordinates of ρ,

$$\rho_1 = f\alpha_1 + g\beta_1 + h\gamma_1,$$
$$\&c., \&c.,$$
$$\rho_6 = f\alpha_6 + g\beta_6 + h\gamma_6.$$

Let η be any given screw. The virtual coefficient of ρ and η is

$$f\varpi_{\alpha\eta} + g\varpi_{\beta\eta} + h\varpi_{\gamma\eta}.$$

Draw from the centre of the pitch quadric a radius vector r parallel to ρ, and equal to the virtual coefficient just written; then the locus of the extremity of r is the sphere

$$x^2 + y^2 + z^2 = x\varpi_{\alpha\eta} + y\varpi_{\beta\eta} + z\varpi_{\gamma\eta}.$$

The tangent plane to the sphere obtained by equating the right-hand side of this equation to zero is the principal plane of that cylindroid which contains all the screws of the screw system which are reciprocal to η.

* If a, b, c be the three semiaxes of the pitch quadric, and $\pm d$ the distances from the centre, on a, of the two points in question, it appears from § 179 that $a^2d^2 = (a^2 - b^2)(a^2 - c^2)$, which shows that d is the fourth proportional to the primary semiaxis of the surface, and to those of its focal ellipse and hyperbola.

182. Four Screws of the Screw System.

Take *any* four screws α, β, γ, δ of the screw system of the third order. Then we shall prove that the cylindroid (α, β) must have a screw in common with the cylindroid (γ, δ). For twists of appropriate amplitudes about α, β, γ, δ must neutralise, and hence the twists about α, β must be counteracted by those about γ, δ; but this cannot be the case unless there is some screw common to the cylindroids (α, β) and (γ, δ).

This theorem provides a convenient test as to whether four screws belong to a screw system of the third order.

183. Geometrical notes.

The following theorem may be noted:

Any ray η which crosses at right angles two screws α, β of a three-system is the seat of a screw reciprocal to the system.

For, draw the cylindroid α, β, then of course η, whatever be its pitch, is reciprocal to all the screws on this cylindroid. Through any point P on η there are two screws of the system which lie on the cylindroid, and there must be a third screw γ of the system through P, which, certainly, does not lie on the cylindroid. If, therefore, we give η a pitch $-p_\gamma$, it must be reciprocal to the three-system.

In general, one screw of a three-system can be found which intersects at right angles any screw whatever η.

For η must, of course, cut each of the quadrics containing the screws of equal pitch in two points. Take, for example, the quadric with screws of pitch p. There are, therefore, two screws, α and β of pitch p belonging to the system, which intersect η. The cylindroid α, β must belong to the system, and from the known property of the cylindroid the ray η, which crosses the two equal pitch screws (§ 22), must cross at right angles some third screw γ on this cylindroid; but this belongs to the three-system, and therefore the theorem has been proved.

184. Cartesian Equation of the Three-System.

If we are given the co-ordinates of any three screws of a three-system with reference to six canonical co-reciprocals, we can calculate in the following manner the equation to the family of pitch quadrics of which the three-system is constituted.

Let the three given screws be α, β, γ, with co-ordinates respectively

$\alpha_1, \ldots \alpha_6$; $\beta_1, \ldots \beta_6$; $\gamma_1, \ldots \gamma_6$. Then if λ, μ, ν be three variable parameters, the co-ordinates of the other screws of the three-system will be

$$\lambda\alpha_1 + \mu\beta_1 + \nu\gamma_1, \quad \lambda\alpha_2 + \mu\beta_2 + \nu\gamma_2, \ldots \lambda\alpha_6 + \mu\beta_6 + \nu\gamma_6.$$

We shall denote the pitch of this screw by p, and from § 43 we have for the equations of this screw with reference to the associated Cartesian axes:

$$\begin{aligned} 0 = &+ (\lambda\alpha_5 + \mu\beta_5 + \nu\gamma_5 + \lambda\alpha_6 + \mu\beta_6 + \nu\gamma_6)\, y \\ &- (\lambda\alpha_3 + \mu\beta_3 + \nu\gamma_3 + \lambda\alpha_4 + \mu\beta_4 + \nu\gamma_4)\, z \\ &- (\lambda\alpha_1 + \mu\beta_1 + \nu\gamma_1 - \lambda\alpha_2 - \mu\beta_2 - \nu\gamma_2)\, a \\ &+ (\lambda\alpha_1 + \mu\beta_1 + \nu\gamma_1 + \lambda\alpha_2 + \mu\beta_2 + \nu\gamma_2)\, p, \end{aligned}$$

with two similar equations.

From these we eliminate λ, μ, ν and the determinant thus arising admits of an important reduction.

To effect this we multiply it by the determinant

$$\begin{vmatrix} \alpha_1 + \alpha_2, & \alpha_3 + \alpha_4, & \alpha_5 + \alpha_6 \\ \beta_1 + \beta_2, & \beta_3 + \beta_4, & \beta_5 + \beta_6 \\ \gamma_1 + \gamma_2, & \gamma_3 + \gamma_4, & \gamma_5 + \gamma_6 \end{vmatrix}$$

For brevity we introduce the following notation:

$$\begin{aligned} P = &\, x\,[(\beta_5 + \beta_6)(\gamma_3 + \gamma_4) - (\beta_3 + \beta_4)(\gamma_5 + \gamma_6)] \\ &+ y\,[(\beta_1 + \beta_2)(\gamma_5 + \gamma_6) - (\beta_5 + \beta_6)(\gamma_1 + \gamma_2)] \\ &+ z\,[(\beta_3 + \beta_4)(\gamma_1 + \gamma_2) - (\beta_1 + \beta_2)(\gamma_3 + \gamma_4)], \end{aligned}$$

with similar values for Q and R by cyclical interchange.

We also make

$$L_{\alpha\beta} = a\,(\alpha_1 + \alpha_2)(\beta_1 - \beta_2) + b\,(\alpha_3 + \alpha_4)(\beta_3 - \beta_4) + c\,(\alpha_5 + \alpha_6)(\beta_5 - \beta_6),$$
$$L_{\beta\alpha} = a\,(\beta_1 + \beta_2)(\alpha_1 - \alpha_2) + b\,(\beta_3 + \beta_4)(\alpha_3 - \alpha_4) + c\,(\beta_5 + \beta_6)(\alpha_5 - \alpha_6),$$

with similar values for $L_{\alpha\gamma}$, $L_{\gamma\alpha}$, $L_{\beta\gamma}$, $L_{\gamma\beta}$ by cyclical interchange.

The equation to the family of pitch quadrics is then easily seen to be

$$0 = \begin{vmatrix} p_\alpha - p &, -R + L_{\alpha\beta} - p\cos(\alpha\beta), & +Q + L_{\alpha\gamma} - p\cos(\alpha\gamma) \\ +R + L_{\beta\alpha} - p\cos(\alpha\beta), & p_\beta - p &, -P + L_{\beta\gamma} - p\cos(\beta\gamma) \\ -Q + L_{\gamma\alpha} - p\cos(\alpha\gamma), & +P + L_{\gamma\beta} - p\cos(\beta\gamma), & p_\gamma - p \end{vmatrix}$$

If the three given screws α, β, γ had been co-reciprocal, then as

$$L_{\alpha\beta} + L_{\beta\alpha} = 2\varpi_{\alpha\beta} = 0,$$

it follows that $L_{\alpha\beta}$ and $L_{\beta\alpha}$ only differ in sign, so that if

$$P' = P + L_{\gamma\beta}; \; Q' = Q + L_{\alpha\gamma}; \; R' = R + L_{\beta\alpha},$$

the equation becomes

$$0 = \begin{vmatrix} p_\alpha - p & , & -R' - p\cos(\alpha\beta), & +Q' - p\cos(\alpha\gamma) \\ R' - p\cos(\alpha\beta), & & p_\beta - p \quad , & -P' - p\cos(\beta\gamma) \\ -Q' - p\cos(\alpha\gamma), & & +P' - p\cos(\beta\gamma), & p_\gamma - p \end{vmatrix}.$$

By expanding this as a cubic for p we see that the coefficient of p^2 divided by that of p^3 with its sign changed is

$$\frac{p_\alpha \sin^2(\beta\gamma) + p_\beta \sin^2(\gamma\alpha) + p_\gamma \sin^2(\alpha\beta)}{4\sin\frac{1}{2}[(\beta\gamma)+(\gamma\alpha)+(\alpha\beta)]\sin\frac{1}{2}[(\beta\gamma)+(\gamma\alpha)-(\alpha\beta)]\sin\frac{1}{2}[(\beta\gamma)-(\gamma\alpha)+(\alpha\beta)]\sin\frac{1}{2}[-(\beta\gamma)+(\gamma\alpha)+(\alpha\beta)]}.$$

This is accordingly the constant sum of the three pitches of the screws of the system which can be drawn through any point.

185. Equilibrium of Four Forces applied to a Rigid Body.

If the body be free, the four forces must be four wrenches on screws of zero pitch which are members of a screw system of the third order. The forces must therefore be generators of a hyperboloid, all belonging to the same system (§ 132).

Three of the forces, P, Q, R, being given in position, S must then be a generator of the hyperboloid determined by P, Q, R. This proof of a well-known theorem (due to Möbius) is given to show the facility with which such results flow from the Theory of Screws.

Suppose, however, that the body have only freedom of the fifth order, we shall find that somewhat more latitude exists with reference to the choice of S. Let X be the screw reciprocal to the screw system by which the freedom is defined. Then for equilibrium it will only be necessary that S belong to the system of the fourth order defined by the four screws

$$P, \ Q, \ R, \ X.$$

A cone of screws can be drawn through every point in space belonging to this system, and on that cone one screw of zero pitch can always be found (§ 123). Hence one line can be drawn through every point in space along which S might act.

If the body have freedom of the fourth order, the latitude in the choice of S is still greater. Let X_1, X_2 be two screws reciprocal to the system, then S is only restrained by the condition that it belong to the screw system of the fifth order defined by the screws

$$P, \ Q, \ R, \ X_1, \ X_2.$$

Any line in space when it receives the proper pitch is a screw of this system. Through any point in space a plane can be drawn such that every line in the plane passing through the point with zero pitch is a screw of the system (§ 110).

Finally, if the body has only freedom of the third order, the four equilibrating forces P, Q, R, S may be situated anywhere.

The positions of the forces being given, their magnitudes are determined; for draw three screws X_1, X_2, X_3 reciprocal to the system, and find (§ 28) the intensities of the seven equilibrating wrenches on

$$P,\ Q,\ R,\ S,\ X_1,\ X_2,\ X_3.$$

The last three are neutralised by the reactions of the constraints, and the four former must therefore equilibrate.

Given any four screws in space, it is possible for four wrenches of proper intensities on these screws to hold a body having freedom of the third order in equilibrium. For, take the four given screws, and three reciprocal screws. Wrenches of proper intensities on these seven screws will equilibrate; but those on the reciprocal screws are destroyed by the reactions, and, therefore, the four wrenches on the four screws equilibrate. It is manifest that this theorem may be generalised into the following:—If a body have freedom of the kth order, then properly selected wrenches about any $k+1$ screws (not reciprocal to the screw system) will hold the body in equilibrium.

That a rigid body with freedom of the third order may be in equilibrium under the action of gravity, we have the necessary and sufficient condition, which is thus stated:—

The vertical through the centre of inertia must be one of the reciprocal system of generators on the pitch quadric.

We see that the centre of inertia must, therefore, lie upon a screw of zero pitch which belongs to the screw system; whence we have the following theorem:—The restraints which are necessary for the equilibrium of a body which has freedom of the third order under the action of gravity, would permit rotation of the body round one definite line through the centre of inertia.

186. The Ellipsoid of Inertia.

The momental ellipsoid, which is of such significance in the theory of the rotation of a rigid body about a fixed point, is presented in the Theory of Screws as a particular case of another ellipsoid, called the ellipsoid of inertia, which is of great importance in connexion with the general screw system of the third order.

If we take three conjugate screws of inertia from the screw system as screws of reference, then we have seen (§ 97) that, if θ_1, θ_2, θ_3, be the co-ordinates of a screw θ, we have

$$u_\theta^2 = u_1^2\theta_1^2 + u_2^2\theta_2^2 + u_3^2\theta_3^2,$$

where u_1, u_2, u_3 are the values of u_θ with reference to the three conjugate screws of inertia.

Draw from any point lines parallel to θ, and to the three conjugate screws of inertia. If then a parallelopiped be constructed of which the diagonal is the line parallel to θ, and of which the three lines parallel to the conjugate screws are conterminous edges, and if r be the length of the diagonal, and x, y, z the lengths of the edges, then we have

$$\frac{x}{r} = \theta_1, \quad \frac{y}{r} = \theta_2, \quad \frac{z}{r} = \theta_3.$$

We see, therefore, that the parameter u appropriate to any screw θ is inversely proportional to the parallel diameter of the ellipsoid

$$u_1^2x^2 + u_2^2y^2 + u_3^2z^2 = H,$$

where H is a certain constant.

Hence we have the following theorem:—*The kinetic energy of a rigid body, when twisting with a given twist velocity about any screw of a system of the third order, is proportional to the inverse square of the parallel diameter of a certain ellipsoid, which may be called the ellipsoid of inertia; and a set of three conjugate diameters of the ellipsoid are parallel to a set of three conjugate screws of inertia which belong to the screw system.*

We might also enunciate the property in the following manner:—*Any diameter of the ellipsoid of inertia is proportional to the twist velocity with which the body should twist about the parallel screw of the screw system, so that its kinetic energy shall be constant.*

187. The Principal Screws of Inertia.

It will simplify matters to consider that the ellipsoid of inertia is concentric with the pitch quadric. It will then be possible to find a triad of common conjugate diameters to the two ellipsoids. We can then determine three screws of the system parallel to these diameters (§ 180), and these three screws will be co-reciprocal, and also conjugate screws of inertia. They will, therefore, (§ 87), form what we have termed the principal screws of inertia. When the screw system reduces to a pencil of screws of zero pitch passing through a point, then the principal screws of inertia reduce to the well-known principal axes.

188. Lemma.

If from a screw system of the nth order we select n screws $A_1, \ldots, A_n$, which are conjugate screws of inertia (§ 87), and if S_1 be *any* screw which is reciprocal to $A_2, \ldots, A_n$, then an impulsive wrench on S_1 will cause the body, when only free to twist about the screws of the system, to commence to twist about A_1. Let R_1 be the screw which, if the body were perfectly free, would be the impulsive screw corresponding to A_1 as the instantaneous screw. R_1 must be reciprocal to $A_2, \ldots, A_n$ (§ 81). Take also $6-n$ screws of the reciprocal system $B_1, \ldots, B_{6-n}$. Then the $8-n$ screws $R_1, S_1, B_1, \ldots, B_{6-n}$ must be reciprocal to the $n-1$ screws $A_2, \ldots, A_n$, and therefore the $8-n$ screws must belong to a screw system of the $(7-n)$th order. Hence an impulsive wrench upon the screw S_1 can be resolved into components on $R_1, B_1, \ldots, B_{6-n}$. Of these all but the first are neutralised by the reactions of the constraints, and by hypothesis the effect of an impulsive wrench upon R_1 is to make the body commence to twist about A_1, and therefore an impulsive wrench on S_1 would make the body twist about A_1.

189. Relation between the Impulsive Screw and the Instantaneous Screw.

A quiescent rigid body which possesses freedom of the third order is acted upon by an impulsive wrench about a given screw η. It is required to determine the instantaneous screw θ, about which the body will commence to twist.

The screws which belong to the system, and are at the same time reciprocal to η, must all lie upon a cylindroid, as they each fulfil the condition of being reciprocal to four screws. All the screws on the cylindroid are parallel to a certain plane drawn through the centre of the pitch quadric, which may be termed the *reciprocal plane* with respect to the screw η. The reciprocal plane having been found, the diameter conjugate to this plane in the ellipsoid of inertia is parallel to the required screw θ.

For let μ and ν denote two screws of the system parallel to a pair of conjugate diameters of the ellipsoid of inertia in the reciprocal plane. Then θ, μ, ν are a triad of conjugate screws of inertia; but η is reciprocal to μ and ν, and, therefore, by the lemma of the last article, an impulsive wrench upon η will make the body commence to twist about θ.

190. Kinetic Energy acquired by an Impulse.

We shall now consider the following problem:—A quiescent rigid body of mass M receives an impulsive wrench of intensity η''' on a screw η. We have now to determine the locus of a screw θ belonging to a screw system of the third order, such that, if the body be constrained to twist about θ, it shall acquire a given kinetic energy E, in consequence of the impulsive wrench.

We have from § 91 the equation

$$E = \frac{1}{M} \frac{\eta'''^2}{u_\theta^2} \varpi^2_{\eta\theta}.$$

We can assign a geometrical interpretation to this equation, which will lead to some interesting results.

Through the centre O of the pitch quadric the plane A reciprocal to η is to be drawn. A sphere (§ 181) is to be described touching the plane A at the origin O, the diameter of the sphere being so chosen that the intercept OP made by the sphere on a radius vector parallel to any screw θ is equal to $\varpi_{\eta\theta}$ (§ 181). The quantity u_θ is inversely proportional to the radius vector OQ of the ellipsoid of inertia, which is parallel to θ (§ 186). Hence for all the screws of the screw system which acquire a given kinetic energy in consequence of a given impulse, we must have the product $OP \cdot OQ$ constant.

From a well-known property of the sphere, it follows that all the points Q must lie upon a plane A', parallel to A. This plane cuts the ellipsoid of inertia in an ellipse, and all the screws required must be parallel to the generators of the cone of the second degree, formed by joining the points of this ellipse to the origin, O.

Since we have already shown how, when the direction of a screw belonging to a screw system of the third order is given, the actual situation of that screw is determined (§ 180), we are now enabled to ascertain all the screws θ on which the body acted upon by a given impulse would acquire a given kinetic energy.

The distance between the planes A and A' is proportional to $OP \cdot OQ$, and therefore to the square root of E. Hence, when the impulse is given, the kinetic energy acquired on a screw determined by this construction is greatest when A and A' are as remote as possible. For this to happen, it is obvious that A' will just touch the ellipsoid of inertia. The group of screws will, therefore, degenerate to the single screw parallel to the diameter of the ellipsoid of inertia conjugate to A. But we have seen (§ 130) that the screw so determined is the screw which the body will naturally select if permitted to make a choice from all the screws of the system of the third order. We thus see again what Euler's theorem (§ 94) would have also told us, viz., that when a quiescent rigid body which has freedom of the third order is set in motion by the action of a given impulsive wrench, the kinetic energy which the body acquires is greater than it would have been had the body been restricted to any other screw of the system than that one which it naturally chooses.

191. Reaction of the Constraints.

An impulsive wrench on a screw η acts upon a body with freedom of the third order, and the body commences to move by twisting upon a screw θ. It is required to find the screw λ, a wrench on which constitutes the initial reaction of the constraints. Let ϕ denote the impulsive screw which, if the body were free, would correspond to θ as the instantaneous screw. Then λ must lie upon the cylindroid (ϕ, η), and may be determined by choosing on (ϕ, η) a screw reciprocal to any screw of the given screw system.

192. Impulsive Screw is Indeterminate.

Being given the instantaneous screw θ in a system of the third order, the corresponding impulsive screw η is indeterminate, because the impulsive wrench may be compounded with any reactions of the constraints. In fact η may be any screw selected from a screw system of the fourth order, which is thus found. Draw the diametral plane conjugate to a line parallel to θ in the ellipsoid of inertia, and construct the cylindroid which consists of screws belonging to the screw system parallel to this diametral plane. Then *any* screw which is reciprocal to this cylindroid will be an impulsive screw corresponding to θ as an instantaneous screw.

Thus we see that through any point in space a whole cone of screws can be drawn, an impulsive wrench on any one of which would make the body commence to twist about the same screw.

One impulsive couple can always be found which would make the body commence to twist about any given screw of the screw system. For a couple in a plane perpendicular to the nodal line of a cylindroid may be regarded as a wrench upon a screw reciprocal to the cylindroid; and hence a couple in a diametral plane of the ellipsoid of inertia, conjugate to the diameter parallel to the screw θ, will make the body commence to twist about the screw θ.

It is somewhat remarkable that a force directed along the nodal line of the cylindroid must make the body commence to twist about precisely the same screw as the couple in a plane perpendicular to the nodal line.

If a cylindroid be drawn through two of the principal screws of inertia, then an impulsive wrench on any screw of this cylindroid will make the body commence to twist about a screw on the same cylindroid. For the impulsive wrench may be resolved into wrenches on the two principal screws. Each of these will produce a twisting motion about the same screw, which will, of course, compound into a twisting motion about a screw on the same cylindroid.

193. Quadric of the Potential.

A body which has freedom of the third order is in equilibrium under the influence of a conservative system of forces. The body receives a twist of small amplitude θ' about a screw θ of the screw system. It is required to determine a geometrical representation for the quantity of work which has been done in effecting the displacement. We have seen that to each screw θ corresponds a certain linear parameter v_θ (§ 102), and that the work done is represented by

$$Fv_\theta^2\theta'^2.$$

We have also seen that the quantity v_θ^2 may be represented by

$$v_1^2\theta_1^2 + v_2^2\theta_2^2 + v_3^2\theta_3^2;$$

where $\theta_1, \theta_2, \theta_3$ are the co-ordinates of the screw θ referred to three conjugate screws of the potential, and v_1, v_2, v_3, denote the values of v_θ for each of the three screws of reference (§ 102).

Drawing through the centre of the pitch quadric three axes parallel to the three screws of reference, we can then construct the quadric of which the equation is

$$v_1^2x^2 + v_2^2y^2 + v_3^2z^2 = H,$$

which proves the following theorem :—

The work done in giving the body a twist of given amplitude from a position of equilibrium about any screw of a system of the third order, is proportional to the inverse square of the parallel diameter of a certain quadric which we may call the *quadric of the potential*, and three conjugate diameters of this quadric are parallel to three conjugate screws of the potential in the screw system.

194. The Principal Screws of the Potential.

The three common conjugate diameters of the pitch hyperboloid, and the quadric of the potential, are parallel to three screws of the system which we call the *principal screws of the potential*. If the body be displaced by a twist about a principal screw of the potential from a position of stable equilibrium, then the reduced wrench which is evoked is upon the same screw.

The three principal screws of the potential must not be confounded with the three screws of the system which are parallel to the principal axes of the ellipsoid of the potential. The latter are the screws on which a twist of given amplitude requires a maximum or minimum consumption of energy, and they are rectangular, which, of course, is not in general the case with the principal screws of the potential.

195. Wrench evoked by Displacement.

By the aid of the quadric of the potential we shall be able to solve the problem of the determination of the screw on which a wrench is evoked by a twist about a screw θ from a position of stable equilibrium. The construction which will now be given will enable us to determine the screw of the system on which the reduced wrench acts.

Draw through the centre of the pitch quadric a line parallel to θ. Construct the diametral plane A of the quadric of the potential conjugate to this line, and let λ, μ be any two screws of the system parallel to a pair of conjugate diameters of the quadric of the potential which lie in the plane A. Then the required screw ϕ is parallel to that diameter of the pitch quadric which is conjugate to the plane A.

For ϕ will then be reciprocal to both λ and μ; and as λ, μ, θ are conjugate screws of the potential, it follows that a twist about θ must evoke a reduced wrench on ϕ.

196. Harmonic Screws.

When a rigid body has freedom of the third order, it must have (§ 106) three harmonic screws, or screws which are conjugate screws of inertia, as well as conjugate screws of the potential. We are now enabled to construct these screws with facility, for they must be those screws of the screw system which are parallel to the triad of conjugate diameters common to the ellipsoid of inertia, and the quadric of the potential.

We have thus a complete geometrical conception of the small oscillations of a rigid body which has freedom of the third order. If the body be once set twisting about one of the harmonic screws, it will continue to twist thereon for ever, and in general its motion will be compounded of twisting motions upon the three harmonic screws.

If the displacement of the body from its position of equilibrium has been effected by a small twist about a screw on the cylindroid which contains two of the harmonic screws, then the twist can be decomposed into components on the harmonic screws, and the instantaneous screw about which the body is twisting at any epoch will oscillate backwards and forwards upon the cylindroid, from which it will never depart.

If the periods of the twist oscillations on two of the harmonic screws coincided, then every screw on the cylindroid which contains those harmonic screws would also be a harmonic screw.

If the periods of the three harmonic screws were equal, then every screw of the system would be a harmonic screw.

197. Oscillations of a Rigid Body about a Fixed Point*.

We shall conclude the present Chapter by applying the principles which it contains to the development of a geometrical solution of the following important problem:—

A rigid body, free to rotate in every direction around a fixed point, is in stable equilibrium under the influence of gravity. The body is slightly disturbed: it is required to determine its small oscillations.

Since three co-ordinates are required to specify the position of a body when rotating about a point, it follows that the body has freedom of the third order. The screw system, however, assumes a very extreme type, for the pitch quadric has become illusory, and the screw system reduces to a pencil of screws of zero pitch radiating in all directions from the fixed point.

The quantity u_θ appropriate to a screw θ reduces to the radius of gyration when the pitch of the screw is zero; hence the ellipsoid of inertia reduces in the present case to the well-known momental ellipsoid.

The quadric of the potential (§ 193) assumes a remarkable form in the present case. The work done in giving the body a small twist is proportional to the vertical distance through which the centre of inertia is elevated. In the position of equilibrium the centre of inertia is vertically beneath the point of suspension, it is therefore obvious from symmetry that the ellipsoid of the potential must be a surface of revolution about a vertical axis. It is further evident that the vertical radius vector of the cylinder must be infinite, because no work is done in rotating the body around a vertical axis.

Let O be the centre of suspension, and I the centre of inertia, and let OP be a radius vector of the quadric of the potential. Let fall IQ perpendicular on OP, and PT perpendicular upon OI. It is extremely easy to show that the vertical height through which I is raised is proportional to $IQ^2 \times OP^2$; whence the area of the triangle OPI is constant, and therefore the locus of P must be a right circular cylinder of which OI is the axis.

We have now to find the triad of conjugate diameters common to the momental ellipsoid, and the circular cylinder just described. A group of three conjugate diameters of the cylinder must consist of the vertical axis, and any two other lines through the origin, which are conjugate diameters of the ellipse in which their plane cuts the cylinder. It follows that the triad required will consist of the vertical axis, and of the pair of conjugate

* *Trans. Roy. Irish Acad.*, Vol. XXIV. Science, p. 593 (1870).

diameters common to the two ellipses in which the plane conjugate to the vertical axis in the momental ellipsoid cuts the momental ellipsoid and the cylinder. These three lines are the three harmonic axes.

As to that vertical axis which appears to be one of the harmonic axes, the time of vibration about it would be infinite. The three harmonic screws which are usually found in the small oscillations of a body with freedom of the third order are therefore reduced in the present case to two, and we have the following theorem:—

A rigid body which is free to rotate about a fixed point is at rest under the action of gravity. If a plane S be drawn through the point of suspension O, conjugate to the vertical diameter OI of the momental ellipsoid, then the common conjugate diameters of the two ellipses in which S cuts the momental ellipsoid, and a circular cylinder whose axis is OI, are the two harmonic axes. If the body be displaced by a small rotation about one of these axes, the body will continue for ever to oscillate to and fro upon this axis, just as if the body had been actually constrained to move about this axis.

To complete the solution for any initial circumstances of the rigid body, a few additional remarks are necessary.

Assuming the body in any given position of equilibrium, it is first to be displaced by a small rotation about an axis OX. Draw the plane containing OI and OX, and let it cut the plane S in the line OY. The small rotation around OX may be produced by a small rotation about OI, followed by a small rotation about OY. The effect of the small rotation about OI is merely to alter the azimuth of the position, but not to disturb the equilibrium. Had we chosen this altered position as that position of equilibrium from which we started, the initial displacement would be communicated by a rotation around OY. We may, therefore, without any sacrifice of generality, assume that the axis about which the initial displacement is imparted lies in the plane S. We shall now suppose the body to receive a small angular velocity about any other axis. This axis must be in the plane S, if small oscillations are to exist at all, for the initial angular velocity, if not capable of being resolved into components about the two harmonic axes, will have a component around the vertical axis OI. An initial rotation about OI would give the body a continuous rotation around the vertical axis, which is not admissible when small oscillations only are considered.

If, therefore, the body performs small oscillations only, we *may* regard the initial *axis of displacement* as lying in the plane S, while we *must* have the initial *instantaneous axis* in that plane. The initial displacement may be resolved into two displacements, one on each of the harmonic axes, and

13—2

the initial angular velocity may also be resolved into two angular velocities on the two harmonic axes. The entire motion will, therefore, be found by compounding the vibrations about the two harmonic axes. Also the instantaneous axis will at every instant be found in the plane of the harmonic axes, and will oscillate to and fro in their plane.

Since conjugate diameters of an ellipse are always projected into conjugate diameters of the projected ellipse, it follows that the harmonic axes must project into two conjugate diameters of a circle on any horizontal plane. Hence we see that two vertical planes, each containing one of the harmonic axes, are at right angles to each other.

We have thus obtained a complete solution of the problem of the small oscillations of a body about a fixed point under the influence of gravity.

CHAPTER XV.

THE PLANE REPRESENTATION OF FREEDOM OF THE THIRD ORDER*.

198. A Fundamental Consideration.

Let x, y, z denote the Cartesian co-ordinates of a point in the body referred to axes fixed in space. When the body moves into an adjacent position these co-ordinates become, respectively, $x + \delta x$, $y + \delta y$, $z + \delta z$, and we have, by a well-known consequence of the rigidity of the body,

$$\delta x = a + gz - hy,$$

$$\delta y = b + hx - fz,$$

$$\delta z = c + fy - gx,$$

where a, b, c, f, g, h may be regarded as expressing the six generalized co-ordinates of the twist which the body has received.

If the body has only three degrees of freedom, its position must be capable of specification by three independent co-ordinates, which we shall call θ_1, θ_2, θ_3. The six quantities, a, b, c, f, g, h, must each be a function of θ_1, θ_2, θ_3, so that when the latter are given the former are determined. As all the movements are infinitely small, it is evident that these equations must in general be linear, and of the type

$$a = A_1\theta_1 + A_2\theta_2 + A_3\theta_3,$$

in which A_1, A_2, A_3 are constants depending on the character of the constraints. We should similarly have

$$b = B_1\theta_1 + B_2\theta_2 + B_3\theta_3,$$

and so on for all the others.

It is a well-known theorem that the new position of the body defined by θ_1, θ_2, θ_3 may be obtained by a twist about a screw of which the axis is defined by the equations

$$\frac{a + gz - hy}{f} = \frac{b + hx - fz}{g} = \frac{c + fy - gx}{h}.$$

* *Trans. Roy. Irish Acad.*, Vol. XXIX. p. 247 (1888).

The angle through which the body has been rotated is

$$(f^2 + g^2 + h^2)^{\frac{1}{2}},$$

and the distance of translation is

$$\frac{af + bg + ch}{(f^2 + g^2 + h^2)^{\frac{1}{2}}},$$

while the pitch of the screw is

$$\frac{af + bg + ch}{f^2 + g^2 + h^2}.$$

Every distinct set of three quantities, θ_1, θ_2, θ_3, will correspond to a definite position of the rigid body, and to a group of such sets there will be a corresponding group of positions. Let ρ denote a variable parameter, and let us consider the variations of the set,

$$\rho\theta_1, \quad \rho\theta_2, \quad \rho\theta_3,$$

according as ρ varies. To each value of ρ a corresponding position of the rigid body is appropriate, and we thus have the change of ρ associated with a definite progress of the body through a series of positions. We can give geometrical precision to a description of this movement. The equations to the axis of the screw, as well as the expression of its pitch, only involve the *ratios* of a, b, c, f, g, h. We have also seen that these quantities are each linear and homogeneous functions of θ_1, θ_2, θ_3. If, therefore, we substitute for θ_1, θ_2, θ_3 the more general values

$$\rho\theta_1, \quad \rho\theta_2, \quad \rho\theta_3,$$

the screw would remain unaltered, both in position and in pitch, though the angle of rotation and the distance of translation will each contain ρ as a factor.

Thus we demonstrate that the several positions denoted by the set $\rho\theta_1$, $\rho\theta_2$, $\rho\theta_3$ are all occupied in succession as we twist the body continuously around one particular screw.

199. The Plane Representation.

All possible positions of the body correspond to the triply infinite triad

$$\theta_1, \quad \theta_2, \quad \theta_3.$$

If, for the moment, we regard these three quantities as the co-ordinates of a point in space, then every point of space will be correlated to a position of the rigid body. We shall now sort out the triply infinite multitude of positions into a doubly infinite number of sets each containing a singly infinite number of positions.

If we fix our glance upon the *screws* about which the body is free to twist, the principle of classification will be obvious. Take an arbitrary triad

$$\theta_1, \quad \theta_2, \quad \theta_3,$$

and then form the infinite group of triads

$$\rho\theta_1, \quad \rho\theta_2, \quad \rho\theta_3$$

for every value of ρ from zero up to any finite magnitude: all these triads will correspond to the positions attainable by twisting about a single screw. We may therefore regard

$$\theta_1, \quad \theta_2, \quad \theta_3$$

as the co-ordinates of a screw, it being understood that only the ratios of these quantities are significant.

We are already familiar with a set of three quantities of this nature in the well-known trilinear co-ordinates of a point in a plane. We thus see that the several screws about which a body with three degrees of freedom can be twisted correspond, severally, with the points of a plane. Each of the points in a plane corresponds to a perfectly distinct screw, about which it is possible for a body with three degrees of freedom to be twisted. Accordingly we have, as the result of the foregoing discussion, the statement that—

To each screw of a three-system corresponds one point in the plane.

To develope this correspondence is the object of the present Chapter.

200. The Cylindroid.

A twist of amplitude θ' on the screw θ has for components on the three screws of reference

$$\theta'\theta_1, \quad \theta'\theta_2, \quad \theta'\theta_3;$$

a twist of amplitude ϕ' on some other screw ϕ has the components

$$\phi'\phi_1, \quad \phi'\phi_2, \quad \phi'\phi_3.$$

When these two twists are compounded they will unite into a single twist upon a screw of which the co-ordinates are proportional to

$$\theta'\theta_1 + \phi'\phi_1, \quad \theta'\theta_2 + \phi'\phi_2, \quad \theta'\theta_3 + \phi'\phi_3.$$

If the ratio of ϕ' to θ' be λ, we see that the twists about θ and ϕ unite into a twist about the screw whose co-ordinates are proportional to

$$\theta_1 + \lambda\phi_1, \quad \theta_2 + \lambda\phi_2, \quad \theta_3 + \lambda\phi_3.$$

By the principles of trilinear co-ordinates this point lies on the straight line joining the points θ and ϕ. As the ratio λ varies, the corresponding screw moves over the cylindroid and the corresponding point moves over the straight line. Hence we obtain the following important result:—

The several screws on a cylindroid correspond to the points on a straight line.

In general two cylindroids have no screw in common. If, however, the two cylindroids be each composed of screws taken from the same three-system, then they will have one screw in common. This is demonstrated by the fact that the two straight lines corresponding to these cylindroids necessarily intersect in a point which corresponds to the screw common to the two surfaces.

Three twist velocities about three screws will neutralize and produce rest, provided that the three corresponding points lie in a straight line, and that the amount of each twist velocity is proportional to the sine of the angle between the two non-corresponding screws.

Three wrenches will equilibrate when the three points corresponding to the screws are collinear, and when the intensity of each wrench is proportional to the sine of the angle between the two non-corresponding screws.

201. The Screws of the Three-system.

In any three-system there are three principal screws at right angles to each other, and intersecting in a point (§ 173). It is natural to choose these as the screws of reference, and also as the axes for Cartesian co-ordinates. The pitches of these screws are p_1, p_2, p_3, and we shall, as usual, denote the screw co-ordinates by θ_1, θ_2, θ_3. The displacement denoted by this triad of co-ordinates is obtained by rotating the body through angles θ_1, θ_2, θ_3 around three axes, and then by translating it through distances $p_1\theta_1$, $p_2\theta_2$, $p_3\theta_3$ parallel to these axes. As these quantities are all small, we have, for the displacements produced in a point x, y, z,

$$\delta x = p_1\theta_1 + z\theta_2 - y\theta_3,$$

$$\delta y = p_2\theta_2 + x\theta_3 - z\theta_1,$$

$$\delta z = p_3\theta_3 + y\theta_1 - x\theta_2;$$

these displacements correspond to a twist about a screw of which the axis has the equations

$$\frac{p_1\theta_1 + z\theta_2 - y\theta_3}{\theta_1} = \frac{p_2\theta_2 + x\theta_3 - z\theta_1}{\theta_2} = \frac{p_3\theta_3 + y\theta_1 - x\theta_2}{\theta_3},$$

while the pitch p is thus given:

$$p = \frac{p_1\theta_1^2 + p_2\theta_2^2 + p_3\theta_3^2}{\theta_1^2 + \theta_2^2 + \theta_3^2}.$$

We have now to investigate the locus of the screws of given pitch, *and as p is presumed to be a determinate quantity,* we have

$$(p_1 - p)\,\theta_1 + z\theta_2 - y\theta_3 = 0,$$

$$-z\theta_1 + (p_2 - p)\,\theta_2 + x\theta_3 = 0,$$

$$+y\theta_1 - x\theta_2 + (p_3 - p)\,\theta_3 = 0,$$

whence, by eliminating θ_1, θ_2, θ_3 we obtain, as the locus of the screws of pitch p, the quadric otherwise found in the previous chapter

$$(p_1 - p)\,x^2 + (p_2 - p)\,y^2 + (p_3 - p)\,z^2 + (p_1 - p)(p_2 - p)(p_3 - p) = 0.$$

According as p varies, this family of quadrics will exhibit all the screws of the three-system which possess a *definite pitch.*

202. Imaginary Screws.

To complete the inventory of the screws it is, however, necessary to add those of *indefinite pitch,* i.e. those whose co-ordinates satisfy both the equations

$$p_1\theta_1^2 + p_2\theta_2^2 + p_3\theta_3^2 = 0,$$

$$\theta_1^2 + \theta_2^2 + \theta_3^2 = 0.$$

There are four triads of co-ordinates which satisfy these conditions, and, remembering that only the *ratios* are concerned, the values of θ_1, θ_2, θ_3 may be written thus:

$$+(p_2 - p_3)^{\frac{1}{2}}, \quad +(p_3 - p_1)^{\frac{1}{2}}, \quad +(p_1 - p_2)^{\frac{1}{2}},$$

$$-(p_2 - p_3)^{\frac{1}{2}}, \quad +(p_3 - p_1)^{\frac{1}{2}}, \quad +(p_1 - p_2)^{\frac{1}{2}},$$

$$+(p_2 - p_3)^{\frac{1}{2}}, \quad -(p_3 - p_1)^{\frac{1}{2}}, \quad +(p_1 - p_2)^{\frac{1}{2}},$$

$$+(p_2 - p_3)^{\frac{1}{2}}, \quad +(p_3 - p_1)^{\frac{1}{2}}, \quad -(p_1 - p_2)^{\frac{1}{2}}.$$

The equations of the axis written without p are

$$x\,(\theta_2^2 + \theta_3^2) - y\theta_1\theta_2 - z\theta_1\theta_3 + (p_2 - p_3)\,\theta_2\theta_3 = 0,$$

$$y\,(\theta_3^2 + \theta_1^2) - z\theta_2\theta_3 - x\theta_2\theta_1 + (p_3 - p_1)\,\theta_3\theta_1 = 0,$$

$$z\,(\theta_1^2 + \theta_2^2) - x\theta_3\theta_1 - y\theta_3\theta_2 + (p_1 - p_2)\,\theta_1\theta_2 = 0,$$

of which two are independent.

If we substitute the values of θ_1, θ_2, θ_3 for the first indeterminate screw, the three equations just written reduce to

$$x\,(p_2 - p_3)^{\frac{1}{2}} + y\,(p_3 - p_1)^{\frac{1}{2}} + z\,(p_1 - p_2)^{\frac{1}{2}} - (p_2 - p_3)^{\frac{1}{2}}(p_3 - p_1)^{\frac{1}{2}}(p_1 - p_2)^{\frac{1}{2}} = 0.$$

If we make

$$\alpha = (p_2 - p_3)^{\frac{1}{2}}; \quad \beta = (p_3 - p_1)^{\frac{1}{2}}; \quad \gamma = (p_1 - p_2)^{\frac{1}{2}},$$

the equations of the four planes are expressed in the form

$$+\alpha x + \beta y + \gamma z - \alpha\beta\gamma = 0,$$

$$-\alpha x + \beta y + \gamma z - \alpha\beta\gamma = 0,$$

$$+\alpha x - \beta y + \gamma z - \alpha\beta\gamma = 0,$$

$$+\alpha x + \beta y - \gamma z - \alpha\beta\gamma = 0.$$

It is remarkable that the three equations of the axis for each of these screws here coalesce to a single one. The screw of indeterminate pitch is thus limited, not to a line, but to a plane. The same may be said of each of the other three screws of indeterminate pitch; they also are each limited to a plane found by giving variety of signs to the radicals in the equations just written. We have thus discovered that the complete locus of the screws of a three-system consists, not only of the family of quadrics, which contain the screws of real or imaginary, but definite pitch, but that it also contains a tetrahedron of four imaginary planes, each plane being the locus of one of the four screws of indefinite pitch.

203. Relation of the Four Planes to the Quadrics.

The planes have an interesting geometrical connexion with the family of quadrics, which we shall now develop. The first theorem to be proved is, that each of the quadrics touches each of the planes. This is geometrically obvious, inasmuch as each quadric contains all the screws of the system which have a given pitch p; but each of the planes contains a system of screws of every pitch, among which there must be one of pitch p. There will thus be a ray in the plane, which is also a generator of the hyperboloid—but this, of course, requires that the plane be a tangent to the hyperboloid.

It is easy to verify this by direct calculation.

Write the quadric,

$$(p_1 - p)x^2 + (p_2 - p)y^2 + (p_3 - p)z^2 + (p_1 - p)(p_2 - p)(p_3 - p) = 0.$$

The tangent plane to this, at the point x', y', z', is

$$(p_1 - p)xx' + (p_2 - p)yy' + (p_3 - p)zz' + (p_1 - p)(p_2 - p)(p_3 - p) = 0.$$

If we identify this with the equation

$$\alpha x + \beta y + \gamma z - \alpha\beta\gamma = 0,$$

we shall obtain

$$x' = -\frac{(p_2 - p)(p_3 - p)}{(p_3 - p_1)^{\frac{1}{2}}(p_1 - p_2)^{\frac{1}{2}}},$$

$$y' = -\frac{(p_3 - p)(p_1 - p)}{(p_1 - p_2)^{\frac{1}{2}}(p_2 - p_3)^{\frac{1}{2}}},$$

$$z' = -\frac{(p_1 - p)(p_2 - p)}{(p_2 - p_3)^{\frac{1}{2}}(p_3 - p_1)^{\frac{1}{2}}},$$

and as these values satisfy the equation of the quadric, the theorem has been proved.

The family of quadric surfaces are therefore inscribed in a common tetrahedron, and they have four common points, as well as four common tangent planes. For, write the two cones

$$x^2 + y^2 + z^2 = 0,$$

$$p_1x^2 + p_2y^2 + p_3z^2 = 0.$$

These cones have four generators in common, and the four points in which these generators cut the plane at infinity will lie on every surface of the type

$$(p_1 - p)x^2 + (p_2 - p)y^2 + (p_3 - p)z^2 + (p_1 - p)(p_2 - p)(p_3 - p) = 0.$$

We now see the distribution of the screws in the imaginary planes. In each one of these planes there are a system of parallel lines; each line of this system passes through the same point at infinity, which is, of course, one of the four points just referred to. Every line of the parallel set, when it receives appropriate pitch, belongs to the three-system.

It thus appears that the ambiguity in the pitches of the screws in the planes is only apparent. The system of screw co-ordinates which usually defines a screw with absolute definiteness, loses that definiteness for the screws in these planes. Each plane contains a whole pencil of screws, radiating from a point at infinity, but the co-ordinates can only represent these screws collectively, for the three co-ordinates then represent, not a single screw, but a whole pencil of screws. As the pitches vary on every screw of the pencil, the co-ordinates can only meet this difficulty by representing the pitch as indeterminate.

The proof that only a single screw of each pitch is found in the pencil is easily given. If there were two, then the same hyperboloid would have two generators in this plane of equal pitch; but this is impossible, because, from the known properties of the three-system, only one of these generators belongs to the three-system, and the other to the reciprocal system.

204. The Pitch Conics.

The discussion in § 203 will prepare us for the plane representation of the screws of given pitch p, for we have

$$p(\theta_1^2+\theta_2^2+\theta_3^2)-p_1\theta_1^2-p_2\theta_2^2-p_3\theta_3^2=0.$$

This, of course, represents a conic section, and, accordingly, we have the following theorem:—

The locus of points corresponding to screws of given pitch is a conic section.

A special case is that where the pitch is zero, in which case the locus is given by

$$p_1\theta_1^2+p_2\theta_2^2+p_3\theta_3^2=0.$$

This we shall often refer to as the *conic of zero-pitch.*

Another important case is that where p is infinite, in which case the equation is

$$\theta_1^2+\theta_2^2+\theta_3^2=0.$$

The conic of zero-pitch and the conic of infinite pitch intersect in four points, and through these four points all the other conics must pass. The points, of course, correspond to the screws of indeterminate pitch: we may call them P_1, P_2, P_3, P_4.

Any conic through these four critical points will be a conic of equal-pitch screws.

As a straight line cuts a conic in two points, we see the well-known theorem, that every cylindroid will contain two screws of each pitch.

The two principal screws on a cylindroid are those of maximum and minimum pitch; they will be found by drawing through P_1, P_2, P_3, P_4, the two conics touching the straight line corresponding to the cylindroid. The two points of contact are the screws required.

If α and β are the two principal screws on a cylindroid, then any pair of harmonic conjugates to α and β represent a pair of screws of equal pitch.

For if $S+kS'=0$ be a system of conics, then it is well known that the pairs of points in which a fixed ray is cut by this system form a system in involution. The double points of this involution are the points of contact of the two conics of the system which touch the line.

205. The Angle between Two Screws.

From the equations of the screw given in § 201, we see that the direction cosines are proportional to θ_1, θ_2, θ_3; for if we take the point infinitely distant we find that the equations reduce to

$$\frac{x}{\theta_1} = \frac{y}{\theta_2} = \frac{z}{\theta_3}.$$

Accordingly, the line drawn parallel to the screw through the origin has its direction cosines proportional to θ_1, θ_2, θ_3, and hence the actual direction cosines are

$$\frac{\theta_1}{\sqrt{\theta_1^2 + \theta_2^2 + \theta_3^2}}, \quad \frac{\theta_2}{\sqrt{\theta_1^2 + \theta_2^2 + \theta_3^2}}, \quad \frac{\theta_3}{\sqrt{\theta_1^2 + \theta_2^2 + \theta_3^2}}.$$

The cosine of the angle between two screws, θ and ϕ, will therefore be

$$\frac{\theta_1\phi_1 + \theta_2\phi_2 + \theta_3\phi_3}{\sqrt{\theta_1^2 + \theta_2^2 + \theta_3^2}\sqrt{\phi_1^2 + \phi_2^2 + \phi_3^2}}.$$

By the aid of the conic of infinite pitch we can give to this a geometrical interpretation.

The co-ordinates of a screw on the straight line joining θ and ϕ will be

$$\theta_1 + \lambda\phi_1, \quad \theta_2 + \lambda\phi_2, \quad \theta_3 + \lambda\phi_3.$$

If we substitute this in the equation to the conic of infinite pitch we obtain

$$\theta_1^2 + \theta_2^2 + \theta_3^2 + 2\lambda(\theta_1\phi_1 + \theta_2\phi_2 + \theta_3\phi_3) + \lambda^2(\phi_1^2 + \phi_2^2 + \phi_3^2) = 0.$$

Writing this in the form

$$a\lambda^2 + 2b\lambda + c = 0,$$

of which λ_1 and λ_2 are the roots, we have, as the four values of λ, corresponding, respectively, to the points θ and ϕ, and to the points in which their chord cuts the conic of infinite pitch,

$$\lambda_1, \quad \lambda_2, \quad 0, \quad \infty.$$

The anharmonic ratio is

$$\frac{\lambda_1}{\lambda_2},$$

or

$$\frac{b - \sqrt{b^2 - ac}}{b + \sqrt{b^2 - ac}}.$$

If ω be the angle between the two screws, θ and ϕ, then

$$\cos\omega = \frac{b}{\sqrt{ac}},$$

and the anharmonic ratio reduces to

$$e^{-2i\omega},$$

whence we deduce the following theorem:—

The angle between two screws is equal to $\frac{1}{2}i$ times the logarithm of the anharmonic ratio in which their corresponding chord is divided by the infinite pitch conic.

The reader will be here reminded of the geometry of non-Euclidian space, in which a magnitude, which in Chapter XXVI. is called the Intervene, analogous to the distance between two points, is equal to $\frac{1}{2}i$ times the logarithm of the anharmonic ratio in which their chord is divided by the absolute. We have only to call the conic of infinite pitch the absolute, and the angle between two screws is the intervene between their corresponding points.

206. Screws at Right Angles.

If two screws, θ and ϕ, be at right angles, then

$$\theta_1\phi_1 + \theta_2\phi_2 + \theta_3\phi_3 = 0.$$

In other words, θ and ϕ are conjugate points of the conic of infinite pitch,

$$\theta_1^2 + \theta_2^2 + \theta_3^2 = 0.$$

All the screws at right angles to a given screw lie on the polar of the point with regard to the conic of infinite pitch. Hence we see that all the screws perpendicular to a given screw lie on a cylindroid. This is otherwise obvious, for a screw can always be found with an axis parallel to a given direction. If, therefore, a cylindroid of the system be taken, a screw of the system parallel to the nodal axis of that cylindroid can also be found, and thus we have the cylindroid and the screw, which stand in the relation of the pole and the polar to the conic of infinite pitch.

A point on the conic of infinite pitch must represent a screw at right angles to itself. Every straight line cuts the conic of infinite pitch in two points, and thus every cylindroid has two screws of infinite pitch, and each of these screws is at right angles to itself.

In general, the direction cosines of the nodal axis of a cylindroid are proportional to the co-ordinates of the pole of the line corresponding to the cylindroid with respect to the conic of infinite pitch.

207. Reciprocal Screws.

If θ_1, θ_2, θ_3 be the co-ordinates of a screw, and ϕ_1, ϕ_2, ϕ_3 those of another screw, then it is known, § 37, that the condition for these two screws to be reciprocal is

$$p_1\theta_1\phi_1 + p_2\theta_2\phi_2 + p_3\theta_3\phi_3 = 0.$$

We are thus led to the following theorem, which is of fundamental importance in the present investigation :—

A pair of reciprocal screws are conjugate points with respect to the zero-pitch conic.

From this theorem we can at once draw the following conclusions:—

All the screws of the system reciprocal to a given screw θ lie upon a cylindroid.

For the locus of points conjugate to θ is, of course, the polar of θ with respect to the zero-pitch conic, and this polar will correspond to a cylindroid.

On any cylindroid one screw can always be found reciprocal to a given screw θ. For this will be the intersection of the polar of θ with the line corresponding to the given cylindroid.

A triad of co-reciprocal screws will correspond to a self-conjugate triangle of the conic of zero-pitch.

208. The Principal Screws of the System.

Draw the conic of zero-pitch A, and the conic of infinite pitch B, which intersect in the four screws of indeterminate pitch, P_1, P_2, P_3, P_4 (see fig. 39). Draw the diagonals of the complete quadrilateral, and let them intersect in

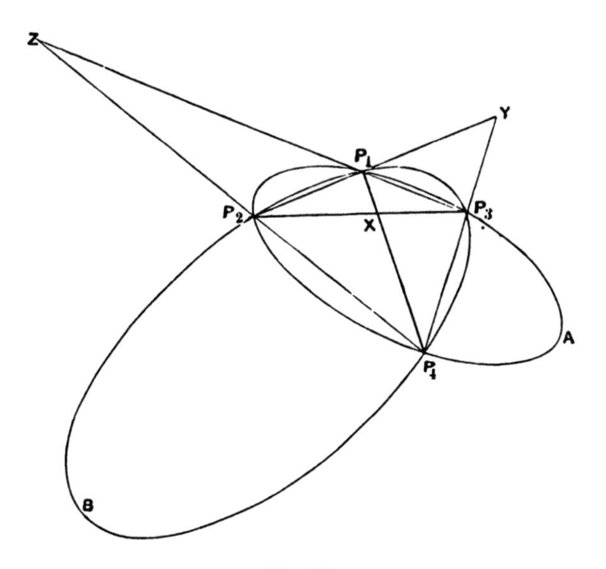

Fig. 39.

the points X, Y, Z. These three points are significant. Take any pair of them, X and Y; then, by the known properties of conics, X and Y are conjugate points with respect to both of the conics A and B. The screws X

and Y must therefore be mutually perpendicular and reciprocal. If p_θ and p_ϕ be the pitches of these screws; if ω be the angle between them, and d their perpendicular distance, then the virtual coefficient is one-half of

$$(p_\theta + p_\phi)\cos\omega - d_{\theta\phi}\sin\omega;$$

as they are reciprocal, this is zero, and as ω is a right angle, we must have $d = 0$; in other words, the screws corresponding to X and Y must intersect at right angles. The same may be proved of either of the pairs X and Z or Y and Z. The points X, Y, Z must therefore correspond to three screws of the system mutually perpendicular, and intersecting at a point. But in the whole system there is only a single triad of screws possessing these properties. They are the axes in the equations of § 201, and are known as the principal screws of the system. The three points X, Y, Z being the vertices of a self-conjugate triangle with respect to both the conics A and B, and hence to the whole system, we have the following theorem :—

The vertices of the conjugate triangle common to the system of pitch conics correspond to the three principal screws of the three-system.

209. Expression for the Pitch.

Each conic drawn through the four points of indeterminate pitch, P_1, P_2, P_3, P_4, is the locus of screws with a given pitch belonging to the system. We are thus led to connect the constancy of the pitch at each point of this conic with another feature of constancy, viz. that of the anharmonic ratio subtended by a variable point of the conic with the four fixed points. The connexion between the pitch and the anharmonic ratio will now be demonstrated.

Let θ_1, θ_2, θ_3 be the co-ordinates of any point on the conic, and let α, β, γ be the co-ordinates of one of the four points, say P_1; then if ϕ_1, ϕ_2, ϕ_3 be the current co-ordinates, the equation of the line joining θ to P_1 is

$$\begin{vmatrix} \phi_1, & \phi_2, & \phi_3 \\ \theta_1, & \theta_2, & \theta_3 \\ \alpha, & \beta, & \gamma \end{vmatrix} = 0.$$

As we are dealing with the anharmonic ratio of a pencil, we may take any section for the calculation of the ratios, and, accordingly, make

$$\phi_3 = 0;$$

and we have for the co-ordinates of the point in which the line joining θ and P_1 intersects $\phi_3 = 0$, the conditions,

$$\frac{\phi_1}{\phi_2} = -\frac{\theta_3\alpha - \theta_1\gamma}{\theta_2\gamma - \theta_3\beta}.$$

By changing the sign of α, and then changing the signs of β and of γ, we shall obtain the four points in which the pencil $\theta(P_1, P_2, P_3, P_4)$ cuts the axis $\phi_3 = 0$. If four values of $\frac{\phi_1}{\phi_2}$ be represented by k, l, m, n, the required anharmonic ratio is, of course,

$$\frac{(n-l)(m-k)}{(n-m)(l-k)},$$

and after a few reductions we find that this becomes

$$\theta(P_1 P_2 P_3 P_4) = \frac{\beta^2(\gamma^2\theta_1^2 - \alpha^2\theta_3^2)}{\alpha^2(\gamma^2\theta_2^2 - \beta^2\theta_3^2)}.$$

But we have

$$p = \frac{p_1\theta_1^2 + p_2\theta_2^2 + p_3\theta_3^2}{\theta_1^2 + \theta_2^2 + \theta_3^2}$$

Eliminating θ_1^2, we find that θ_2^2 and θ_3^2 disappear also when we make

$$\alpha^2 = p_2 - p_3, \quad \beta^2 = p_3 - p_1, \quad \gamma^2 = p_1 - p_2,$$

and we obtain the following result:

$$\theta(P_1 P_2 P_3 P_4) = \frac{p_3 - p_1}{p_3 - p_2} \times \frac{p_2 - p}{p_1 - p},$$

which gives the following theorem:—

Measure off distances p_1, p_2, p_3, p, *from an arbitrary point on a straight line, then the anharmonic ratio of the four points thus obtained is equal to the anharmonic ratio subtended by any point of the p-pitch conic at the four points of indeterminate pitch.*

It is possible without any sacrifice of generality to make the zero-pitch conic a circle. For take three angles A, B, C whose sum is 180° and such that the equations

$$\frac{\sin 2A}{p_1} = \frac{\sin 2B}{p_2} = \frac{\sin 2C}{p_3},$$

are satisfied where p_1, p_2, p_3 are the three principal pitches of the three-system. If the fundamental triangle has A, B, C for its angles then the equation

$$\alpha_1^2 \sin 2A + \alpha_2^2 \sin 2B + \alpha_3^2 \sin 2C = 0,$$

is the equation of the zero-pitch conic. It is however a well-known theorem in conics that this equation represents a circle with its centre at the orthocentre, that is, the intersection of the perpendiculars from the vertices of the triangle on its opposite sides.

We thus have as the system of pitch conics

$$\alpha_1^2 \sin 2A + \alpha_2^2 \sin 2B + \alpha_3^2 \sin 2C - p(\alpha_1^2 + \alpha_2^2 + \alpha_3^2) = 0.$$

The centre of a conic of the system has for co-ordinates

$$\frac{\sin A}{\sin 2A - p},\quad \frac{\sin B}{\sin 2B - p},\quad \frac{\sin C}{\sin 2C - p}.$$

The locus of the centres of the system of conics is easily seen to pass through the vertices of the triangle of reference. It must also pass through the orthocentre, and hence by a well-known property it must be an equilateral hyperbola. It can also be easily shown that this hyperbola must pass through the "symmedian" point of the triangle, *i.e.* through the centre of gravity of three particles at the vertices of the triangle when the mass of each particle is proportional to the square of the sine of the corresponding angle.

In Fig. 40 a system of pitch conics has been shown drawn to scale. The sides of the fundamental triangle are represented by the numbers 117, 189, 244 respectively. The equation to the system of conics, expressed in Cartesian co-ordinates for convenience of calculation, is

$$0 = x^2\left(\cdot 7864115 - \frac{\cdot 6852041}{p}\right) + y^2\left(2\cdot 2135882 - \frac{\cdot 6852041}{p}\right)$$

$$+ \cdot 1649571xy - 9\cdot 46700x - 430\cdot 5179y$$

$$+ 25199\cdot 54 + \frac{5726\cdot 031}{p} = 0.$$

Among critical conics of the system we may mention:

1. The two parabolas for which the pitches are respectively

$$\cdot 8766 \text{ and } \cdot 3089.$$

2. The three cases in which the conic breaks up into a pair of straight lines for the pitches $\sin 2A$, $\sin 2B$, $\sin 2C$, respectively. Of these, the first alone is a real pair corresponding to the pitch $\cdot 8256034$. The equations of these lines are

$$x = 7\cdot 84y - 978,$$

$$x = -4\cdot 06y + 760.$$

For convenience in laying down the curves the current co-ordinates on each conic are expressed by means of an auxiliary angle; thus, for example, in computing points on the hyperbola with the pitch $\cdot 748984$ I used the equations

$$x = 66 + 26\cdot 1 \sec\theta + 132\tan\theta,$$

$$y = 161\cdot 5 + 40\cdot 7\sec\theta.$$

The ellipse with pitch $\cdot 9$ was constructed from the equations

$$x = -367 - 168\cos\theta \pm 351\sin\theta,$$

$$y = +169 + 51\cos\theta.$$

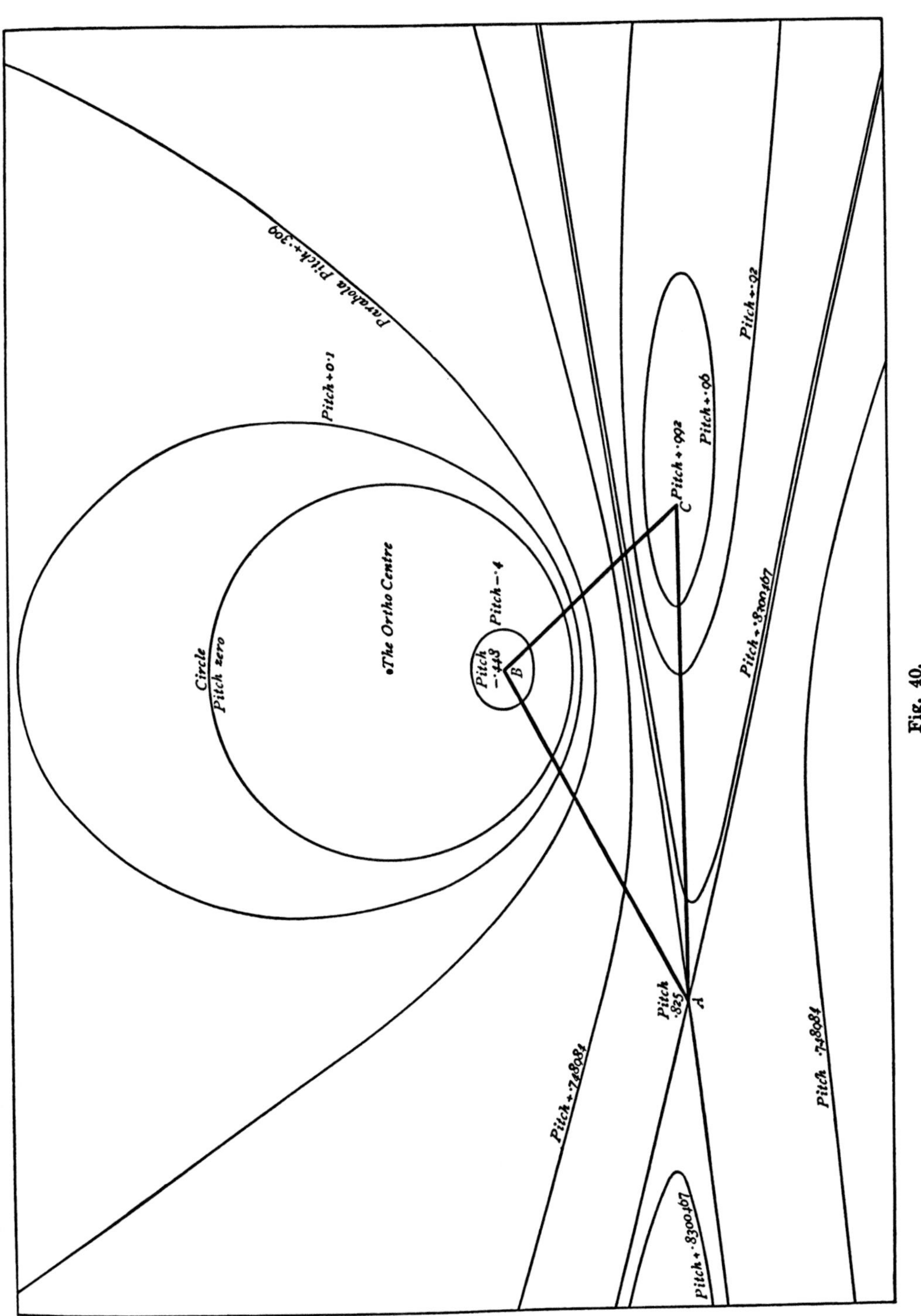

Parabola Pitch +·309
Pitch +0·1
Circle
Pitch zero
•The Ortho Centre
Pitch −·4
Pitch −·443
B
C
Pitch +·992
Pitch +·96
Pitch +·92
Pitch +·8300467
Pitch ·825
A
Pitch +·748084
Pitch ·748084
Pitch +·8300467

Fig. 40.

The following are the pitches of the several points and curves represented

Vertices of the Triangle $+\cdot992$, $+\cdot825$, $-\cdot448$,

Ellipses $+\cdot96$, $+\cdot92$, $+\cdot1$, $-\cdot4$,

Parabola $+\cdot309$,

Hyperbolas $+\cdot748984$, $+\cdot8300467$.

The locus of the centres of the pitch conics is

$$\alpha_1\alpha_2 \sin(A-B)\sin 2C + \alpha_2\alpha_3 \sin(B-C)\sin 2A + \alpha_3\alpha_1 \sin(C-A)\sin 2B = 0.$$

210. Intersecting Screws in a Three-System.

If two screws of a three-system intersect, then their two corresponding points must fulfil some special condition which we propose to investigate.

Let α be one screw supposed fixed, then we shall investigate the locus of the point θ which expresses a screw which intersects α. We can at once foresee a certain character of this locus. A ray through α can only cut it in one other point, for if it cut in two points, we should have three co-cylindroidal screws intersecting, which is not generally possible. The locus is, as we shall find, a cubic and the necessary condition is secured by the fact that α is a double point on the cubic, so that a ray through α has only one more point of intersection with the curve. We can indeed prove that this curve must be a cubic from the fact that any screw meets a cylindroid in three points. Draw then a ray (§ 200) corresponding to a cylindroid of the three-system. There must be in general three points of the locus on this ray. Therefore the locus must be a cubic.

As α and θ intersect we have since $d_{\theta\alpha} = 0$

$$2\varpi_{\theta\alpha} = (p_\alpha + p_\theta)\cos(\alpha\theta).$$

By substituting the values of the different quantities in terms of the co-ordinates we have the following homogeneous equation of the cubic:

$$\begin{aligned} 0 = {} & 2(\theta_1^2 + \theta_2^2 + \theta_3^2)(p_1\alpha_1\theta_1 + p_2\alpha_2\theta_2 + p_3\alpha_3\theta_3)(\alpha_1^2 + \alpha_2^2 + \alpha_3^2) \\ & - (p_1\theta_1^2 + p_2\theta_2^2 + p_3\theta_3^2)(\alpha_1\theta_1 + \alpha_2\theta_2 + \alpha_3\theta_3)(\alpha_1^2 + \alpha_2^2 + \alpha_3^2) \\ & - (\theta_1^2 + \theta_2^2 + \theta_3^2)(p_1\alpha_1^2 + p_2\alpha_2^2 + p_3\alpha_3^2)(\alpha_1\theta_1 + \alpha_2\theta_2 + \alpha_3\theta_3). \end{aligned}$$

We first note that this cubic must pass through the four points of intersection of

$$0 = \theta_1^2 + \theta_2^2 + \theta_3^2,$$

$$0 = p_1\theta_1^2 + p_2\theta_2^2 + p_3\theta_3^2.$$

But this might have been expected, because as we have shown (§ 203) each of these four points corresponds, not to a single screw, but to a plane of screws.

In a certain sense therefore α must intersect each of these four screws, and accordingly the cubic has to pass through the four points.

To prove that α is a double point we write for brevity

$$P=\theta_1^2+\theta_2^2+\theta_3^2, \qquad R=p_1\alpha_1\theta_1+p_2\alpha_2\theta_2+p_3\alpha_3\theta_3,$$

$$Q=p_1\theta_1^2+p_2\theta_2^2+p_3\theta_3^2, \qquad S=\alpha_1\theta_1+\alpha_2\theta_2+\alpha_3\theta_3,$$

$$L=p_1\alpha_1^2+p_2\alpha_2^2+p_3\alpha_3^2, \qquad H=\alpha_1^2+\alpha_2^2+\alpha_3^2,$$

and the equation is

$$P(2HR-LS)-SQH=0.$$

Differentiating with respect to θ_1, θ_2, θ_3 respectively and equating the results to zero we have

$$0=2\theta_1[2RH-LS-aHS]+\alpha_1[2aPH-LP-HQ],$$

$$0=2\theta_2[2RH-LS-bHS]+\alpha_2[2bPH-LP-HQ],$$

$$0=2\theta_3[2RH-LS-cHS]+\alpha_3[2cPH-LP-HQ].$$

These are satisfied by $\theta_1=\alpha_1$, $\theta_2=\alpha_2$, $\theta_3=\alpha_3$ which proves that α is a double point.

The cubic equation is satisfied by the conditions

$$0=\alpha_1\theta_1+\alpha_2\theta_2+\alpha_3\theta_3,$$

$$0=p_1\alpha_1\theta_1+p_2\alpha_2\theta_2+p_3\alpha_3\theta_3.$$

This might have been expected because these equations mean that α and θ are both reciprocal and rectangular, in which case they must intersect. Thus we obtain the following result:

If α_1, α_2, α_3 are the co-ordinates of a screw α in the plane representation, then the co-ordinates of the screw which, together with α, constitute the principal screws of a cylindroid of the system are respectively

$$\frac{p_3-p_2}{\alpha_1}, \quad \frac{p_1-p_3}{\alpha_2}, \quad \frac{p_2-p_1}{\alpha_3}.$$

The following theorem may also be noted. Among the screws of a three-system which intersect one screw of that system there will generally be two screws of any given pitch.

For the cubic which indicates by its points the screws that intersect α will cut any pitch conic in general in six points. Four of these are of course the four imaginary points referred to already. The two remaining intersections indicate the two screws of the pitch appropriate to the conic which intersect α.

The cubic

$$P(2HR-LS)-SQH=0,$$

of course passes through the two points which lie at the intersections of $2HR - LS = 0$ and $Q = 0$. Hence we deduce the following result.

The cylindroid through the two screws of zero pitch which intersect α corresponds to the ray whose equation is

$$2(\alpha_1^2+\alpha_2^2+\alpha_3^2)(p_1\alpha_1\theta_1+p_2\alpha_2\theta_2+p_3\alpha_3\theta_3)-(p_1\alpha_1^2+p_2\alpha_2^2+p_3\alpha_3^2)(\alpha_1\theta_1+\alpha_2\theta_2+\alpha_3\theta_3)=0,$$

α will of course intersect a third screw on this cylindroid and be perpendicular thereto.

The cubic passes through the two points defined by

$$0 = \alpha_1\theta_1 + \alpha_2\theta_2 + \alpha_3\theta_3,$$

$$0 = \theta_1^2 + \theta_2^2 + \theta_3^2,$$

but these points are the points of contact of the pair of tangents drawn from the point α to the conic of infinite pitch.

These two points in addition to the four points on the same conic defined by its intersection with the conic of zero pitch, make six known points on the cubic. If however we are given six points on a cubic and also a double point on the curve, then it is determinate. Thus we obtain the following geometrical construction.

In the plane representation of a three-system we draw the conic of zero pitch C and the conic of infinite pitch U. To find the locus of the screws which intersect a given screw α, it is necessary to draw the two tangents from α to U and through the two points of contact, and also the four points common to C and U then draw that single cubic which has a double point at α. See Appendix, note 5.

211. Application to Dynamics.

By the aid of the plane representation we are enabled to solve certain problems in the dynamics of a rigid body which has freedom of the third order.

Let an impulsive wrench act upon a quiescent rigid body; it is required to determine the instantaneous screw about which the body will commence to twist.

It has been shown (§ 96) that the impulsive wrench, *wherever situated*, can generally be adequately represented by the reduced impulsive wrench on a screw of the three-system. The problem is therefore reduced to the determination of the point corresponding to the instantaneous screw, when that corresponding to the impulsive screw is known.

We have first to draw the conic of which the equation is

$$u_1^2\theta_1^2 + u_2^2\theta_2^2 + u_3^2\theta_3^2 = 0.$$

This conic is of course imaginary, being in fact the locus of screws about which, if the body were twisting with the unit of twist velocity, the kinetic energy would nevertheless be zero. If two points θ, ϕ are conjugate with respect to this conic, then

$$u_1^2\theta_1\phi_1 + u_2^2\theta_2\phi_2 + u_3^2\theta_3\phi_3 = 0.$$

The screws corresponding to θ and ϕ are then what we have called *conjugate screws of inertia.*

This conic is referred to a self-conjugate triangle, the vertices of which are three conjugate screws of inertia. There is one triangle self-conjugate both to the conic of zero pitch, and to the conic of inertia just considered. The vertices of this triangle are of especial interest. Each pair of them correspond to a pair of screws which are reciprocal, as well as being conjugate screws of inertia. They are therefore what we have designated as the *principal screws of inertia* (§ 87). They degenerate into the principal axes of the body when the freedom degenerates into the special case of rotation around a fixed point.

When referred to this self-conjugate triangle, the relation between the impulsive point and the corresponding instantaneous point can be expressed with great simplicity. Thus the impulsive point ϕ, whose co-ordinates are

$$\theta_1 u_1^2 \div p_1; \quad \theta_2 u_2^2 \div p_2; \quad \theta_3 u_3^2 \div p_3,$$

corresponds to the instantaneous point whose co-ordinates are θ_1, θ_2, θ_3. The geometrical construction is sufficiently obvious when derived from the theorem thus stated.

If ϕ denote an impulsive screw, and θ the corresponding instantaneous screw, then the polar of ϕ with regard to the conic of zero pitch is the same straight line as the polar of θ with regard to the conic of inertia.

If H be the virtual coefficient of two screws θ and η, then

$$H^2(\theta_1^2 + \theta_2^2 + \theta_3^2) = (p_1\theta_1\eta_1 + p_2\theta_2\eta_2 + p_3\theta_3\eta_3)^2.$$

It follows that the locus of the points which have a given virtual coefficient with a given point is a conic touching the conic of infinite pitch at two points. If ψ be the screw whose polar with regard to the conic of infinite pitch is identical with the polar of η with regard to the conic of zero pitch, then all the screws θ which have a given virtual coefficient with η are equally inclined to ψ. It hence follows that all the screws of a three-system which have a given virtual coefficient with a given screw are parallel to the generators of a right circular cone. All the screws reciprocal to η form a cylindroid, and ψ is the one screw of the system which is parallel to the nodal line of the cylindroid. The virtual coefficient of ψ and η is greater than that of η with any other screw.

If θ be a screw about which, when a body is twisting with a given twist velocity it has a given kinetic energy, then we must have

$$u_1^2\theta_1^2 + u_2^2\theta_2^2 + u_3^2\theta_3^2 - E(\theta_1^2 + \theta_2^2 + \theta_3^2) = 0,$$

where E is a constant proportional to the energy. It follows that the locus of θ must be a conic passing through the four points of intersection of the two conics

$$u_1^2\theta_1^2 + u_2^2\theta_2^2 + u_3^2\theta_3^2 = 0,$$

$$\theta_1^2 + \theta_2^2 + \theta_3^2 = 0.$$

The four points in which these two conics intersect correspond to the screws about which the body can twist with indefinite kinetic energy. These four points A, B, C, D being known, the kinetic energy appropriate to every point P can be readily ascertained. It is only necessary to measure the anharmonic ratio subtended by P, at A, B, C, D, and to set off on a straight line distances u_1^2, u_2^2, u_3^2, h^2, so that the anharmonic ratio of the four points shall be equal to that subtended by P. This will determine h^2, which is proportional to the kinetic energy due to the unit twist velocity about the screw corresponding to P.

A quiescent rigid body of mass M receives an impulsive wrench of given intensity on a given screw η; we investigate the locus of the screw θ belonging to the three-system, such that if the body be constrained to twist about θ, it shall acquire a given kinetic energy.

It follows at once (§ 91) that we must have

$$E(u_1^2\theta_1^2 + u_2^2\theta_2^2 + u_3^2\theta_3^2) = (p_1\theta_1\eta_1 + p_2\theta_2\eta_2 + p_3\theta_3\eta_3)^2,$$

where E is proportional to the kinetic energy. The required locus is therefore a conic having double contact with the conic of inertia.

It is easy to prove from this that E will be a maximum if

$$u_1^2\theta_1 : p_1\eta_1 = u_2^2\theta_2 : p_2\eta_2 = u_3^2\theta_3 : p_3\eta_3;$$

whence again we have Euler's well-known theorem that if the body be allowed to select the screw about which it will twist, the kinetic energy acquired will be larger than when the body is constrained to a screw other than that which it naturally chooses (§ 94).

A somewhat curious result arises when we seek the interpretation of a tangent to the conic of infinite pitch. This tangent must, like any other straight line, correspond to a cylindroid; and since it is the polar of the point of contact, it follows that every screw on the cylindroid must be at right angles to the direction corresponding to the point of contact. The co-ordinates of the point of contact must therefore be proportional to the direction cosines of the nodal line of the cylindroid.

If the body be in equilibrium under the action of a conservative system of forces, then there is a conic (analogous to the conic of inertia) which denotes the locus of screws about which the body can be displaced to a neighbouring position, so that even as far as the second order of small quantities no energy is consumed. The vertices of the triangle self-conjugate both to this conic and the conic of inertia correspond to the harmonic screws about which, if the body be once displaced, it will continue to oscillate.

CHAPTER XVI.

FREEDOM OF THE FOURTH ORDER.

212. Screw System of the Fourth Order.

The most general type of a screw system of the fourth order is exhibited by the set of screws which are reciprocal to an arbitrary cylindroid (§ 75). To obtain certain properties of this screw system it is, therefore, only necessary to re-state a few results already obtained.

All the screws which belong to a screw system of the fourth order and which can be drawn through a given point are generators of a certain cone of the second degree (§ 23).

All the screws of the same pitch which belong to a screw system of the fourth order must intersect two fixed lines, viz. those two screws which, lying on the reciprocal cylindroid, have pitches equal in magnitude but opposite in sign to the given pitch (§ 22).

One screw of given pitch and belonging to a given screw system of the fourth order can be drawn through each point in space (§ 123).

As we have already seen that two screws belonging to a screw system of the *third* order can be found in any plane (§ 178), so we might expect to find that a singly infinite number of screws belonging to a screw system of the *fourth* order can be found in any plane. We shall now prove that *all these screws envelope a parabola.* A theorem equivalent to this has been already proved in a different manner in § 162.

Take any point P in the plane, then the screws through P reciprocal to the cylindroid form a cone of the second order, which is cut by the plane in two lines. Thus two screws belonging to a given screw system of the fourth order can be drawn in a given plane through a given point. But it can be easily shown that only *one* screw of the system parallel to a given line can be found in the plane. Therefore from the point at infinity only a single finite tangent to the curve can be drawn. Therefore the other tangent from the point must be the line at infinity itself, and as the line at infinity touches the conic, the envelope must be a parabola.

In general there is *one* line in each screw system of the fourth order, which forms a screw belonging to the screw system, whatever be the pitch assigned to it. The line in question is the nodal line of the cylindroid reciprocal to the four-system. The kinematical statement is as follows:—

When a rigid body has freedom of the fourth order, there is in general one straight line, about which the body can be rotated, and parallel to which it can be translated.

A body which has freedom of the fourth order may be illustrated by the particular case where one point P of the body is forbidden to depart from a given curve. The position of the body will then be specified by four quantities, which may be, for example, the arc of the curve from a fixed origin up to P, and three rotations about three axes intersecting in P. The reciprocal cylindroid will in this case assume an extreme form; it has degenerated to a plane, and in fact consists of screws of zero pitch on all the normals to the curve at P.

It is required to determine the locus of screws parallel to a given straight line L, and belonging to a screw system of the fourth order. The problem is easily solved from the principle that each screw of the screw system must intersect at right angles a screw of the reciprocal cylindroid (§ 22). Take, therefore, that one screw θ on the cylindroid which is perpendicular to L. Then a plane through θ parallel to L is the required locus.

213. Equilibrium with freedom of the Fourth Order.

When a rigid body has freedom of the fourth order, it is both necessary and sufficient for equilibrium, that the forces shall constitute a wrench upon a screw of the cylindroid reciprocal to the given screw system. Thus, if a single force can act on the body without disturbing equilibrium, then this force must lie on one of the two screws of zero pitch on the cylindroid. If there were no real screws of zero pitch on the cylindroid—that is, if the pitch conic were an ellipse, then it would be impossible for equilibrium to subsist under the operation of a single force. It is, however, worthy of remark, that if one force could act without disturbing the equilibrium, then in general another force (on the other screw of zero pitch) could also act without disturbing equilibrium.

A couple which is in a plane perpendicular to the nodal line can be neutralized by the reaction of the constraints, and is, therefore, consistent with equilibrium. In no other case, however, can a body which has freedom of the fourth order be in equilibrium under the influence of a couple.

We can also investigate the conditions under which five forces applied to a free rigid body can neutralize each other. The five forces must, as the body is free, belong to a screw system of the fourth order. Draw the cylindroid reciprocal to the system. The five forces must, therefore, intersect both the screws of zero pitch on the cylindroid. We thus prove the well-known theorem that if five forces equilibrate two straight lines can be drawn which intersect each of the five forces. Four of the forces will determine the two transversals, and therefore the fifth force may enjoy any liberty consistent with the requirement that it also intersects the same two lines.

If $A_1, \dots A_5$ be the five forces, the ratio of any pair, let us say for example, $A_1 : A_2$ is thus determined.

Let P, Q be the two screws of zero pitch upon the cylindroid, *i.e.* the two common transversals of $A_1, \dots A_5$.

Choose any two screws X and Y reciprocal to both A_1 and A_2, but not reciprocal to A_3, A_4 or A_5.

Choose any screw Z reciprocal to A_3, A_4, A_5, but not reciprocal to A_1 or A_2.

Construct (§ 25) the single screw I reciprocal to the five screws

$$X,\ Y,\ P,\ Q,\ Z.$$

The four screws X, Y, P, Q are reciprocal to the cylindroid A_1, A_2; therefore I, which is reciprocal to X, Y, P, Q, must lie upon the cylindroid (A_1, A_2) (§ 24).

Since P, Q, Z are all reciprocal to A_3, A_4, A_5, it follows that I being reciprocal to P, Q, Z must belong to the screw system A_3, A_4, A_5. Hence I is found on the cylindroid (A_1, A_2), and it must also belong to the system (A_3, A_4, A_5). If, therefore, forces along $A_1, \dots A_5$ are to equilibrate, the forces along A_1, A_2 must compound into a wrench on I.

But I being known by construction the angles A_2I and A_1I are known, and consequently the ratio of the sines of these angles, *i.e.* the relative intensities of the forces on A_1 and A_2 are determined (§ 14).

If a free rigid body is acted upon by five forces, the preceding considerations will show in what manner the body could be moved so that it shall not do work against nor receive energy from any one of the forces.

Let $A_1, \dots A_5$ be the five forces. Draw two transversals L, M intersecting $A_1, \dots A_4$. Construct the cylindroid of which L, M are the screws of zero pitch; find, upon this cylindroid, the screw X reciprocal to A_5. Then the only movement which the body can receive, so as to fulfil the prescribed conditions, is a twist about the screw X. For X is then reciprocal to $A_1, \ldots A_5$, and therefore a body twisted about X will do no work against forces directed along $A_1, \ldots A_5$.

From the theory of reciprocal screws it follows that a body rotated around any of the lines $A_1, \ldots A_5$ will not do work against nor receive energy from a wrench on X.

In the particular case, where $A_1, \ldots A_5$ have a common transversal, then X is that transversal, and its pitch is zero. In this case it is sufficiently obvious that forces on $A_1, \ldots A_5$ cannot disturb the equilibrium of a body only free to rotate about X.

214. Screws of Stationary Pitch.

We begin by investigating the screws in an n-system of which the pitch is *stationary* in the sense employed in the Theory of Maximum and Minimum. We take the case of $n = 4$.

The co-ordinates $\theta_1, \ldots \theta_6$ of the screws of a four-system have to satisfy the two linear equations defining the system. We may write these equations in the form

$$A_1\theta_1 + \ldots + A_6\theta_6 = 0,$$

$$B_1\theta_1 + \ldots + B_6\theta_6 = 0.$$

The screws of reference being co-reciprocal, we have for the pitch p_θ the equation

$$\Sigma p_1\theta_1^2 - Rp_\theta = 0,$$

where R is the homogeneous function of the second degree in the co-ordinates which is replaced by unity (§ 35) in the formulae after differentiation.

If the pitch be stationary, then by the ordinary rules of the differential calculus (§ 38),

$$\left(2p_1\theta_1 - \frac{dR}{d\theta_1}p_\theta\right)\delta\theta_1 + \ldots + \left(2p_6\theta_6 - \frac{dR}{d\theta_6}p_\theta\right)\delta\theta_6 = 0.$$

As however θ belongs to the four-system, the variations of its co-ordinates must satisfy the two conditions

$$A_1\delta\theta_1 + \ldots + A_6\delta\theta_6 = 0,$$

$$B_1\delta\theta_1 + \ldots + B_6\delta\theta_6 = 0.$$

Following the usual process we multiply the first of these equations by some indeterminate multiplier λ, the second by another quantity μ, and then add the products to the former equation. We can then equate the coefficients of $\delta\theta_1, \ldots \delta\theta_6$ severally to zero, thus obtaining

$$2p_1\theta_1 - \frac{dR}{d\theta_1} p_\theta + \lambda A_1 + \mu B_1 = 0,$$

$$\cdots\cdots\cdots\cdots\cdots\cdots\cdots\cdots$$

$$2p_6\theta_6 - \frac{dR}{d\theta_6} p_\theta + \lambda A_6 + \mu B_6 = 0.$$

Choose next from the four-system any screw whatever of which the coordinates are $\phi_1, \ldots \phi_6$. Multiply the first of the above six equations by ϕ_1, the second by ϕ_2, &c. and add the six products. The coefficients of λ and μ vanish, and we obtain

$$2\varpi_{\theta\phi} - \left(\phi_1 \frac{dR}{d\theta_1} + \ldots + \phi_6 \frac{dR}{d\theta_6}\right) p_\theta = 0.$$

The coefficient of p_θ is however merely double the cosine of the angle between θ and ϕ. This is obvious by employing canonical co-reciprocals in which

$$R = (\theta_1 + \theta_2)^2 + (\theta_3 + \theta_4)^2 + (\theta_5 + \theta_6)^2,$$

whence

$$\phi_1 \frac{dR}{d\theta_1} + \ldots + \phi_6 \frac{dR}{d\theta_6}$$

$$= 2(\phi_1 + \phi_2)(\theta_1 + \theta_2) + 2(\phi_3 + \phi_4)(\theta_3 + \theta_4) + (\phi_5 + \phi_6)(\theta_5 + \theta_6) = 2\cos(\theta\phi).$$

We thus obtain the following theorem, which must obviously be true for other values of n besides four.

If ϕ be any screw of an n-system and if θ be a screw of stationary pitch in the same system then $\varpi_{\theta\phi} = \cos(\theta\phi)\, p_\theta$.

Suppose that there were two screws of stationary pitch θ and ϕ in an n-system. Then

$$\varpi_{\theta\phi} = \cos(\theta\phi)\, p_\theta,$$

$$\varpi_{\theta\phi} = \cos(\theta\phi)\, p_\phi.$$

If p_θ and p_ϕ are different these equations require that

$$\varpi_{\theta\phi} = 0; \quad \cos(\theta\phi) = 0;$$

i.e. the screws are both reciprocal and rectangular and must therefore intersect.

We have thus shown that if there are two stationary screws of different pitches in any n-system, then these screws must intersect at right angles.

In general we learn that if any screw ϕ of an n-system has a pitch equal to that of a screw θ of stationary pitch in the same system, then θ and ϕ must intersect. For the general condition

$$\varpi_{\theta\phi} = \cos(\theta\phi)\, p_\theta$$

is of course

$$(p_\theta + p_\phi)\cos(\theta\phi) - \sin(\theta\phi)\, d_{\theta\phi} = 2p_\theta \cos(\theta\phi),$$

or

$$(p_\phi - p_\theta)\cos(\theta\phi) - \sin(\theta\phi)\, d_{\theta\phi} = 0.$$

If then $p_\phi = p_\theta$ we must have $\sin(\theta\phi)\, d_{\theta\phi} = 0$, which requires that θ and ϕ must intersect at either a finite or an infinite distance.

In the case where ϕ is at right angles to θ it follows from the formula $\varpi_{\theta\phi} = \cos(\theta\phi)\, p_\theta$ that $\varpi_{\theta\phi} = 0$, or that θ and ϕ are reciprocal. But two screws which are at right angles and also reciprocal must intersect, and hence we have the following theorem.

If θ be a screw of stationary pitch in an n-system, then any other screw belonging to the n-system and at right angles to θ must intersect θ.

If ϕ belongs to an n-system its co-ordinates must, on that account, satisfy $6 - n$ linear equations. If it be further assumed that ϕ has to be perpendicular to θ, then the co-ordinates of ϕ have to satisfy yet one more equation, *i.e.*

$$\phi_1 \frac{dR}{d\theta_1} + \dots \phi_6 \frac{dR}{d\theta_6} = 0.$$

In this case ϕ is subjected to $7 - n$ linear equations. It follows (§ 76) that ϕ will have as its locus a certain $(n-1)$-system, whence we have the following general theorem.

If θ is a screw of stationary pitch in an n-system P then among the $(n-1)$-systems included in P there is one Q such that every screw of Q intersects θ at right angles.

These theorems can also be proved by geometrical considerations. If a screw θ have stationary pitch in an n-system it follows *a fortiori* that θ must have stationary pitch on any cylindroid through θ and belonging wholly to the n-system. This means that θ must be one of the two principal screws on such a cylindroid. Choose any other screw ϕ of the system and draw the cylindroid (θ, ϕ) then θ is a principal screw, and if θ and the other principal screw on the cylindroid be two of the co-reciprocal screws of reference, then the co-ordinate of ϕ with respect to θ is $\cos(\theta\phi)$ (§ 40). But that co-ordinate must also have the general form $\varpi_{\phi 1} \div p_1$, whence at once we obtain

$$\varpi_{\theta\phi} = \cos(\theta\phi)\, p_\theta.$$

Let θ be a screw of stationary pitch in a three-system, and let ϕ and ψ be any two other screws in that system. Then θ is one of the principal screws on the cylindroid $(\theta\phi)$; let σ be the other principal screw on that cylindroid. In like manner let ρ be the other principal screw of the cylindroid ($\theta\psi$). Then ρ and σ determine the cylindroid ($\rho\sigma$) which belongs to the system, θ must lie on the common perpendicular to ρ and σ, and hence the screws of the cylindroid ($\rho\sigma$) each intersect θ at right angles.

If θ is a screw of stationary pitch in a four-system, it can be shown that three screws ρ, σ, τ not on the same cylindroid can be found in the same system, and such that they intersect θ at right angles. In this case ρ, σ, τ will determine a three-system, every screw of which intersects θ at right angles.

215. Application to the Two-System.

The principles of the last article afford a simple proof of many fundamental propositions in the theory. We take as the first illustration the well-known fact (§ 76) that if the co-ordinates of a screw satisfy four linear equations then the locus of that screw is a cylindroid.

From the general theorem we see by the case of $n = 2$ that in any two-system a screw of stationary pitch will be intersected at right angles by another screw of the two-system.

These two screws may be conveniently taken as the first and third of the canonical co-reciprocal system lying on the axes of x and y. Hence we have as the co-ordinates of a screw of the system θ_1, 0, θ_3, 0, 0, 0.

The investigation has thus assumed a very simple form inasmuch as the four linear equations express that of the six co-ordinates of a screw of the system four are actually zero.

Let λ, μ, ν be the direction angles of the screw θ with respect to the associated Cartesian axes then (§ 44),

$$\theta_1 = \frac{(p_\theta + a)\cos\lambda - d_{\theta 1}\sin\lambda}{a}; \quad 0 = \frac{(p_\theta - a)\cos\lambda - d_{\theta 1}\sin\lambda}{-a};$$

$$\theta_3 = \frac{(p_\theta + b)\cos\mu - d_{\theta 2}\sin\mu}{b}; \quad 0 = \frac{(p_\theta - b)\cos\mu - d_{\theta 2}\sin\mu}{-b};$$

$$0 = \frac{(p_\theta + c)\cos\nu - d_{\theta 3}\sin\nu}{c}; \quad 0 = \frac{(p_\theta - c)\cos\nu - d_{\theta 3}\sin\nu}{-c}.$$

The two last of these equations give

$$\cos\nu = 0; \quad d_{\theta 3} = 0.$$

Hence we learn that θ must intersect the axis of z at right angles. θ is thus parallel to the plane of xy at a distance $d_{\theta 1} = d_{\theta 2} = z$, and accordingly we have the equations

$$(p_\theta - a)\cos\lambda - z\sin\lambda = 0,$$

$$(p_\theta - b)\cos\mu - z\sin\mu = 0,$$

whence eliminating z and observing that $\lambda - \mu = 90°$ we obtain,

$$p_\theta = a \cos^2 \lambda + b \sin^2 \lambda,$$

and eliminating p_θ,

$$(b - a) \sin \lambda \cos \lambda = z.$$

If we desire the equation of the surface we have

$$y = x \tan \lambda,$$

and hence finally

$$(b - a)\, xy = z\,(x^2 + y^2).$$

Thus again we arrive at the well-known equation of the cylindroid.

We can also prove in the following manner the fundamental theorem that among the screws belonging to any two-system there are two which intersect at right angles (§ 13).

Let θ be any screw of the two-system, and accordingly the six co-ordinates of θ must satisfy four linear equations which may be written

$$A_1\theta_1 + \ldots + A_6\theta_6 = 0,$$

$$B_1\theta_1 + \ldots + B_6\theta_6 = 0,$$

$$C_1\theta_1 + \ldots + C_6\theta_6 = 0,$$

$$D_1\theta_1 + \ldots + D_6\theta_6 = 0.$$

If ϕ be a screw which intersects θ at right angles, then we must also have

$$\theta_1 \frac{dR}{d\phi_1} + \ldots + \theta_6 \frac{dR}{d\phi_6} = 0,$$

$$p_1\theta_1\phi_1 + \ldots + p_6\theta_6\phi_6 = 0,$$

inasmuch as these screws are reciprocal as well as rectangular.

From these six equations $\theta_1, \ldots \theta_6$ can be eliminated, and we have the resulting equation in the co-ordinates of ϕ,

$$\begin{vmatrix} \frac{dR}{d\phi_1}, & \frac{dR}{d\phi_2}, & \frac{dR}{d\phi_3}, & \frac{dR}{d\phi_4}, & \frac{dR}{d\phi_5}, & \frac{dR}{d\phi_6} \\ p_1\phi_1, & p_2\phi_2, & p_3\phi_3, & p_4\phi_4, & p_5\phi_5, & p_6\phi_6 \\ A_1, & A_2, & A_3, & A_4, & A_5, & A_6 \\ B_1, & B_2, & B_3, & B_4, & B_5, & B_6 \\ C_1, & C_2, & C_3, & C_4, & C_5, & C_6 \\ D_1, & D_2, & D_3, & D_4, & D_5, & D_6 \end{vmatrix} = 0.$$

This equation involves the co-ordinates of ϕ in the second degree. If this equation stood alone it would merely imply that ϕ belonged to the quadratic five-system (§ 223) which included all the screws that intersected at right angles any one of the screws of the given cylindroid. If we further assume that ϕ is to be a screw on the given cylindroid, then we have

$$A_1\phi_1 \ldots + A_6\phi_6 = 0,$$

$$B_1\phi_1 \ldots + B_6\phi_6 = 0,$$

$$C_1\phi_1 \ldots + C_6\phi_6 = 0,$$

$$D_1\phi_1 \ldots + D_6\phi_6 = 0.$$

From these five equations two sets of values of ϕ can be found. Thus among the system of screws which satisfy four linear equations there must be two screws which intersect at right angles. These are of course the two principal screws of the cylindroid.

216. Application to the Three-System.

The equations of the three-system can be also deduced from the principle employed in § 214 which enunciated for this purpose is as follows.

If θ be a screw of stationary pitch in a three-system P then there is a cylindroid belonging to P such that every screw of the cylindroid intersects θ at right angles.

It is obvious that this condition could only be complied with if θ lies on the axis of the cylindroid, and as the cylindroid has two intersecting screws at right angles we have thus a proof that in any three-system there must be one set of three screws which intersect rectangularly. Let their pitches be a, b, c, then on the first we may put a screw of pitch $-a$, on the second a screw of pitch $-b$, and on the third a screw of pitch $-c$. Thus we arrive at a set of canonical co-reciprocals specially convenient for the particular three-system.

We have therefore learned that whatever be the three linear equations defining the three-system it is always possible without loss of generality to employ a set of canonical co-reciprocals such that the 1st, 3rd and 5th screws shall belong to the system.

These three screws will define the system. Any other screw of the system can be produced by twists about these three screws. Hence we see that for every screw of the system we must have

$$\theta_2 = 0; \quad \theta_4 = 0; \quad \theta_6 = 0.$$

If λ, μ, ν be the direction angles of θ we have therefore (§ 44)

$$\theta_2 = \frac{(p_\theta - a)\cos\lambda - d_{\theta 1}\sin\lambda}{a} = 0,$$

$$\theta_4 = \frac{(p_\theta - b)\cos\mu - d_{\theta 3}\sin\mu}{b} = 0,$$

$$\theta_6 = \frac{(p_\theta - c)\cos\nu - d_{\theta 5}\sin\nu}{c} = 0.$$

The direction cosines of the common perpendicular to θ and 1 are

$$0, \quad \frac{\cos\nu}{\sin\lambda}, \quad -\frac{\cos\mu}{\sin\lambda},$$

whence the cosine of the angle between this perpendicular and the radius vector to a point x, y, z on θ is

$$\frac{y}{r}\cdot\frac{\cos\nu}{\sin\lambda} - \frac{z}{r}\frac{\cos\mu}{\sin\lambda} = \frac{d_{\theta 1}}{r},$$

or
$$d_{\theta 1}\sin\lambda = y\cos\nu - z\cos\mu.$$

We have thus the three conditions

$$(p_\theta - a)\cos\lambda \quad + z\cos\mu \quad - y\cos\nu = 0 \ldots\ldots\ldots\ldots \text{(i)},$$

$$-z\cos\lambda + (p_\theta - b)\cos\mu \quad + x\cos\nu = 0 \ldots\ldots\ldots\ldots \text{(ii)},$$

$$+y\cos\lambda \quad - x\cos\mu + (p_\theta - c)\cos\nu = 0 \ldots\ldots\ldots\ldots \text{(iii)},$$

whence eliminating $\cos\lambda$, $\cos\mu$, $\cos\nu$ we obtain

$$(p_\theta - a)(p_\theta - b)(p_\theta - c) + (p_\theta - a)x^2 + (p_\theta - b)y^2 + (p_\theta - c)z^2 = 0.$$

Thus we deduce the equation otherwise obtained in § 174, for the family of pitch-hyperboloids on which are arranged according to their pitches the several screws of the three-system.

217. Principal pitches of the Reciprocal Cylindroid.

From a system of the fourth order a system of canonical co-reciprocals can in general be selected which possesses exceptional facilities for the investigation of the properties of the screws which form that four-system.

Let OA and OB be the axes of the two principal screws of the reciprocal cylindroid. Let a and b be the pitches of these two principal screws and let c be any third linear magnitude. Let OC be the axis of the cylindroid.

Then the canonical co-reciprocal system now under consideration consists of

Two screws on OA with pitches $+a$ and $-a$.

Two screws on OB with pitches $+b$ and $-b$.

Two screws on OC with pitches $+c$ and $-c$.

Of these the four screws with pitches $-a$, $-b$, $+c$, $-c$ respectively are each reciprocal to the cylindroid. Each of these four screws must thus belong to the four-system. Further these four screws are co-reciprocal.

If $\theta_1 \ldots \theta_6$ be the six co-ordinates of a screw in the four-system referred to these canonical co-reciprocals, then we have

$$\theta_1 = 0, \quad \theta_3 = 0.$$

For $\theta_1 = \frac{\varpi_{1\theta}}{a}$, but as the first screw of reference belongs to the reciprocal cylindroid we must have $\varpi_{1\theta} = 0$. In like manner $\varpi_{3\theta} = 0$, and therefore $\theta_1 = 0$ and $\theta_3 = 0$ are the two linear equations which specify this particular four-system.

The pitch of any screw on the four-system expressed in terms of its co-ordinates is

$$\frac{-a\theta_2^2 - b\theta_4^2 + c(\theta_5^2 - \theta_6^2)}{\theta_2^2 + \theta_4^2 + (\theta_5 + \theta_6)^2},$$

of which the four stationary values are $-a$, $-b$, $+\infty$, $-\infty$.

We may remark that if the four co-ordinates here employed be taken as a system of quadriplanar co-ordinates of a point we have a representation of the four-system by the points in space. Each point corresponds to one screw of the system. The screws of given pitch p_θ are found on the quadric surfaces

$$U + p_\theta V = 0,$$

where $U = 0$ is the quadric whose points correspond to the screws of zero pitch and where $V = 0$ is an imaginary cone whose points correspond to the screws of infinite pitch. Conjugate points with respect to $U = 0$ will correspond to reciprocal screws. A plane will correspond to a three-system and a straight line to a two-system.

The general theorem proved in § 214 states that when θ is a screw of stationary pitch in an n-system to which any other screw ϕ belongs, then

$$\varpi_{\theta\phi} = p_\theta \cos \theta\phi.$$

Let us now take a four-system referred to any four co-reciprocals and choose for ϕ in the above formula each one of the four co-reciprocals in succession, we then have

$$\left.\begin{aligned}
p_1\theta_1 &= p_\theta\{\theta_1 + \theta_2\cos(12) + \theta_3\cos(13) + \theta_4\cos(14)\}\\
p_2\theta_2 &= p_\theta\{\theta_1\cos(12) + \theta_2 + \theta_3\cos(23) + \theta_4\cos(24)\}\\
p_3\theta_3 &= p_\theta\{\theta_1\cos(13) + \theta_2\cos(23) + \theta_3 + \theta_4\cos(34)\}\\
p_4\theta_4 &= p_\theta\{\theta_1\cos(14) + \theta_2\cos(24) + \theta_3\cos(34) + \theta_4\}
\end{aligned}\right\}.$$

Eliminating θ_1, θ_2, θ_3, θ_4 we deduce a biquadratic for p_θ. But we have already seen that two of the roots of this must be infinite, whence this equation reduces to a quadratic, and its roots are as we have seen equal but opposite in sign to the pitches of the principal screws of the reciprocal cylindroid.

After a few reductions and replacing p_θ by $-\rho$ we obtain the following equation

$$\rho^2 (p_1 p_2 \sin^2(34) + p_1 p_3 \sin^2(24) + \ldots)$$
$$+ \rho (p_1 p_2 p_3 + p_1 p_2 p_4 + p_2 p_3 p_4 + p_1 p_3 p_4)$$
$$+ p_1 p_2 p_3 p_4 = 0.$$

We thus deduce from any four co-reciprocal screws the quadratic equation which gives the pitches of the two principal screws of the cylindroid to which the given four-system is reciprocal.

218. Equations to the screw in a four-system.

The screws of the four-system are defined by the equations

$$(p_\theta + a) \cos \lambda + z \cos \mu - y \cos \nu = 0,$$
$$- z \cos \lambda + (p_\theta + b) \cos \mu + x \cos \nu = 0,$$

where p_θ is the pitch where $\cos \lambda$, $\cos \mu$, $\cos \nu$ are the direction cosines and where x, y, z is a point on the screw. By these equations the properties of the various screws of the system can be easily investigated.

If p_θ be eliminated we obtain

$$x \cos \lambda \cos \nu + y \cos \mu \cos \nu - z (\cos^2 \lambda + \cos^2 \mu) - (a - b) \cos \lambda \cos \mu,$$

whence we obtain for the equation to the cone of screws which belongs to the four-system, and has its vertex at x_0, y_0, z_0

$$x_0 (x - x_0)(z - z_0) + y_0 (y - y_0)(z - z_0)$$
$$- z_0 \{(x - x_0)^2 + (y - y_0)^2\} - (a - b)(x - x_0)(y - y_0) = 0.$$

This is of course the cone which has been referred to in § 123.

219. Impulsive Screws and Instantaneous Screws.

A body which is free to twist about all the screws of a screw system of the fourth order receives an impulsive wrench on the screw η, the impulsive intensity being η'''. It is required to calculate the co-ordinates of the screw θ about which the body will commence to twist, and also the initial reactions of the constraints.

Let λ and μ be any two screws on the reciprocal cylindroid, then the impulsive reaction of the constraints may be considered to consist of impulsive wrenches on λ, μ of respective intensities λ''', μ'''. If we adopt the six absolute principal screws of inertia as screws of reference, (§ 79) then the body will commence to move as if it were free, but had been acted upon by a wrench of which the co-ordinates are proportional to $p_1\theta_1, \ldots, p_6\theta_6$. It follows that the given impulsive wrench, when compounded with the reactions of the constraints, must constitute the wrench of which the co-ordinates have been just written; whence if h be a certain quantity which is the same for each co-ordinate, we have the six equations

$$hp_1\theta_1 = \eta'''\eta_1 + \lambda'''\lambda_1 + \mu'''\mu_1,$$
$$\cdots\cdots\cdots\cdots\cdots\cdots$$
$$hp_6\theta_6 = \eta'''\eta_6 + \lambda'''\lambda_6 + \mu'''\mu_6.$$

Multiply the first of these equations by λ_1, the second by λ_2, &c.: adding the six equations thus obtained, and observing that θ is reciprocal to λ, and that consequently

$$\Sigma p_1\theta_1\lambda_1 = 0,$$

we obtain

$$\eta'''\Sigma\eta_1\lambda_1 + \lambda'''\Sigma\lambda_1^2 + \mu'''\Sigma\lambda_1\mu_1 = 0,$$

and similarly multiplying the original equations by $\mu_1, \ldots, \mu_6$ and adding, we obtain

$$\eta'''\Sigma\eta_1\mu_1 + \lambda'''\Sigma\mu_1\lambda_1 + \mu'''\Sigma\mu_1^2 = 0.$$

From these two equations the unknown quantities λ''', μ''' can be found, and thus the initial reaction of the constraints is known. Substituting the values of λ''', μ''' in the six original equations, the co-ordinates of the required screw θ are determined.

220. Principal Screws of Inertia in the Four-System.

We have already given in Chapter VII. the general methods of determining the principal screws of inertia in an n-system. The following is a different process which though of general application is in this chapter set down for the case of the four-system.

Choose four co-reciprocal screws α, β, γ, δ of the four-system and let their co-ordinates be as usual $\alpha_1, \ldots, \alpha_6$; $\beta_1, \ldots, \beta_6$; $\gamma_1, \ldots, \gamma_6$; $\delta_1, \ldots, \delta_6$; referred to the six absolute principal screws of inertia (§ 79).

Let an impulsive wrench on one of the principal screws of inertia θ in the four-system be decomposed into components on α, β, γ, δ, and let the impulsive intensities be α''', β''', γ''', δ'''.

Let λ, μ be any two screws on the reciprocal cylindroid. Then the body will move as if it had been free and had received impulsive wrenches on the absolute principal screws of inertia, the impulsive intensities being

$$\alpha'''\alpha_1 + \beta'''\beta_1 + \gamma'''\gamma_1 + \delta'''\delta_1 + \lambda'''\lambda_1 + \mu'''\mu_1$$
$$\cdots\cdots\cdots\cdots\cdots\cdots\cdots\cdots\cdots\cdots\cdots\cdots$$
$$\alpha'''\alpha_6 + \beta'''\beta_6 + \gamma'''\gamma_6 + \delta'''\delta_6 + \lambda'''\lambda_6 + \mu'''\mu_6.$$

The co-ordinates of θ are proportional to

$$\alpha'''\alpha_1 + \beta'''\beta_1 + \gamma'''\gamma_1 + \delta'''\delta_1$$
$$\cdots\cdots\cdots\cdots\cdots\cdots$$
$$\alpha'''\alpha_6 + \beta'''\beta_6 + \gamma'''\gamma_6 + \delta'''\delta_6.$$

As θ is to be a principal screw of inertia it follows that the expressions last written multiplied severally by $p_1, \ldots, p_6$ must be proportional to the intensities of the impulsive wrenches received by the body: whence we have the following equations in which h is a quantity which is the same for each of the co-ordinates.

$$hp_1(\alpha'''\alpha_1 + \beta'''\beta_1 + \gamma'''\gamma_1 + \delta'''\delta_1) = \alpha'''\alpha_1 + \beta'''\beta_1 + \gamma'''\gamma_1 + \delta'''\delta_1 + \lambda'''\lambda_1 + \mu'''\mu_1,$$
$$\cdots\cdots\cdots\cdots\cdots\cdots\cdots\cdots\cdots\cdots\cdots\cdots$$
$$hp_6(\alpha'''\alpha_6 + \beta'''\beta_6 + \gamma'''\gamma_6 + \delta'''\delta_6) = \alpha'''\alpha_6 + \beta'''\beta_6 + \gamma'''\gamma_6 + \delta'''\delta_6 + \lambda'''\lambda_6 + \mu'''\mu_6.$$

We are now to multiply these equations by $\alpha_1, \ldots, \alpha_6$ respectively, and add. If we repeat the process using $\beta_1, \ldots, \beta_6$; $\gamma_1, \ldots, \gamma_6$; $\delta_1, \ldots, \delta_6$; $\lambda_1, \ldots, \lambda_6$; $\mu_1, \ldots, \mu_6$ and if we remember that α is reciprocal to β, γ, δ because the system is co-reciprocal and that α is reciprocal to λ and μ because λ and μ belong to the reciprocal system, then observing that like conditions hold for β, γ, and δ, we have the equations

$$'(\Sigma\alpha_1^2 - hp_\alpha) + \beta'''\Sigma\alpha_1\beta_1 + \gamma'''\Sigma\alpha_1\gamma_1 + \delta'''\Sigma\alpha_1\delta_1 + \lambda'''\Sigma\alpha_1\lambda_1 + \mu'''\Sigma\alpha_1\mu_1 = 0,$$
$$'\Sigma\alpha_1\beta_1 + \beta'''(\Sigma\beta_1^2 - hp_\beta) + \gamma'''\Sigma\beta_1\gamma_1 + \delta'''\Sigma\beta_1\delta_1 + \lambda'''\Sigma\beta_1\lambda_1 + \mu'''\Sigma\beta_1\mu_1 = 0,$$
$$'\Sigma\alpha_1\gamma_1 + \beta'''\Sigma\gamma_1\beta_1 + \gamma'''(\Sigma\gamma_1^2 - hp_\gamma) + \delta'''\Sigma\gamma_1\delta_1 + \lambda'''\Sigma\gamma_1\lambda_1 + \mu'''\Sigma\gamma_1\mu_1 = 0,$$
$$'\Sigma\alpha_1\delta_1 + \beta'''\Sigma\delta_1\beta_1 + \gamma'''\Sigma\delta_1\gamma_1 + \delta'''(\Sigma\delta_1^2 - hp_\delta) + \lambda'''\Sigma\delta_1\lambda_1 + \mu'''\Sigma\delta_1\mu_1 = 0,$$
$$'\Sigma\alpha_1\lambda_1 + \beta'''\Sigma\lambda_1\beta_1 + \gamma'''\Sigma\lambda_1\gamma_1 + \delta'''\Sigma\lambda_1\delta_1 + \lambda'''\Sigma\lambda_1^2 + \mu'''\Sigma\lambda_1\mu_1 = 0,$$
$$'\Sigma\alpha_1\mu_1 + \beta'''\Sigma\mu_1\beta_1 + \gamma'''\Sigma\mu_1\gamma_1 + \delta'''\Sigma\mu_1\delta_1 + \lambda'''\Sigma\mu_1\lambda_1 + \mu'''\Sigma\mu_1^2 = 0.$$

From these equations α''', β''', γ''', δ''', λ''', μ''' can be eliminated and the result is to give a biquadratic for h. Thus we have the four roots for the equation. Each of these roots will give a corresponding set of values for α''', β''', γ''', δ''', λ''', μ'''; thus we obtain

$$\alpha'''\alpha_1 + \beta'''\beta_1 + \gamma'''\gamma_1 + \delta'''\delta_1,$$
$$\cdots\cdots\cdots\cdots\cdots\cdots$$
$$\alpha'''\alpha_6 + \beta'''\beta_6 + \gamma'''\gamma_6 + \delta'''\delta_6,$$

which are proportional to the co-ordinates of the corresponding principal screw of inertia.

The values of λ''' and μ''' determine the impulsive reaction of the constraints.

221. Application of Euler's Theorem.

It may be of interest to show how the co-ordinates of the instantaneous screw corresponding to those of a given impulsive screw can be deduced from Euler's theorem (§ 94). If a body receive an impulsive wrench on a screw η while the body is constrained to twist about a screw θ, then we have seen in § 91 that the kinetic energy acquired is proportional to

$$\frac{\varpi^2_{\eta\theta}}{u^2_\theta}.$$

If θ_1, θ_2, θ_3, θ_4 be the co-ordinates of θ referred to the four principal screws of inertia belonging to the screw system of the fourth order, then (§§ 95, 97)

$$\varpi_{\eta\theta}^2 = (p_1\eta_1\theta_1 + p_2\eta_2\theta_2 + p_3\eta_3\theta_3 + p_4\eta_4\theta_4)^2,$$

$$u_\theta^2 = u_1^2\theta_1^2 + u_2^2\theta_2^2 + u_3^2\theta_3^2 + u_4^2\theta_4^2.$$

Hence we have to determine the four independent variables θ_1, θ_2, θ_3, θ_4, so that

$$\frac{(p_1\eta_1\theta_1 + p_2\eta_2\theta_2 + p_3\eta_3\theta_3 + p_4\eta_4\theta_4)^2}{u_1^2\theta_1^2 + u_2^2\theta_2^2 + u_3^2\theta_3^2 + u_4^2\theta_4^2},$$

shall be stationary. This is easily seen to be the case when θ_1, θ_2, θ_3, θ_4 are respectively proportional to

$$\frac{p_1}{u_1^2}\eta_1, \quad \frac{p_2}{u_2^2}\eta_2, \quad \frac{p_3}{u_3^2}\eta_3, \quad \frac{p_4}{u_4^2}\eta_4.$$

These are accordingly, as we already know (§ 97), the co-ordinates of the screw about which the body will commence to twist after it has received an impulsive wrench on η.

This method might of course be applied to any order of freedom.

222. General Remarks.

It has been shown in § 80 how the co-ordinates of the instantaneous screw corresponding to a given impulsive screw can be determined when the rigid body is perfectly free. It will be observed that the connexion between the two screws depends only upon the three principal axes through the centre of inertia, and the radii of gyration about these axes. We may express this result more compactly by the well-known conception of the momental ellipsoid. The centre of the momental ellipsoid is at the centre of inertia of the rigid body, the directions of the principal axes of the ellipsoid are the same as the principal axes of inertia, and the lengths of the axes of the ellipsoid are inversely proportional to the corresponding radii of gyration. When, therefore, the impulsive screw is given, the momental ellipsoid alone must be capable of determining the corresponding instantaneous screw.

A family of rigid bodies may be conceived which have a common momental ellipsoid; every rigid body which fulfils nine conditions will belong to this family. If an impulsive wrench applied to a member of this family cause it to twist about a screw θ, then the same impulsive wrench applied to any other member of the same family will cause it likewise to twist about θ. If we added the further condition that the masses of all the members of the family were equal, then it would be found that the twist velocity, and the kinetic energy acquired in consequence of a given impulse, would be the same to whatever member of the family the impulse were applied (§§ 90, 91).

223. Quadratic n-systems.

We have always understood by a screw system of the nth order or briefly an n-system, the collection of screws whose co-ordinates satisfy a certain system of $6-n$ linear homogeneous equations. We have now to introduce the conception of a screw system of the nth order and second degree or briefly a *quadratic n-system* ($n < 6$). By this expression we are to understand a collection of screws such that their co-ordinates satisfy $6-n$ homogeneous equations; of these equations $5-n$, that is to say, all but one are linear; the remaining equation involves the co-ordinates in the second degree.

Let $\theta_1, \ldots, \theta_6$ be the co-ordinates of a screw belonging to a quadratic n-system. We may suppose without any loss of generality that the $5-n$ linear equations have been transformed into

$$\theta_{n+2} = 0\,; \quad \theta_{n+3} = 0\,; \ \ldots\ \theta_6 = 0.$$

The remaining equation of the second degree is accordingly obtained by equating to zero a homogeneous quadratic function of

$$\theta_1 \ldots \theta_{n+1}.$$

We express this equation which characterizes the quadratic n-system as

$$U_\theta = 0.$$

All the screws whose co-ordinates satisfy the $5-n$ linear equations must themselves form a screw system of the $6-(5-n) = (n+1)$th system. This screw system may be regarded as an *enclosing system* from which the screws are to be selected which further satisfy the equation of the second degree $U_\theta = 0$. The enclosing system comprises the screws which can be formed by giving all possible values to the co-ordinates $\theta_1, \ldots, \theta_{n+1}$.

Of course there may be as many different screw systems of the nth order and second degree comprised within the same enclosing system as there can be different quadratic forms obtained by annexing coefficients to the several squares and products of $n+1$ co-ordinates. If $n = 5$, the enclosing system would consist of every screw in space.

224. Properties of a Quadratic Two-system.

The quadratic two-system is constituted of screws whose coordinates satisfy three linear equations and one quadratic equation, and these screws lie generally on a surface of the sixth degree (§ 225). If we take the plane representation of the three-system given in Chapter XV., then any conic in the plane corresponds to a quadratic two-system and all the points in the plane correspond to the enclosing three-system. Since any straight line in the plane corresponds to a cylindroid in the enclosing system and the straight line will, in general, cut a conic in the plane in two points, we have the following theorem.

A quadratic two-system has two screws in common with any cylindroid belonging to the enclosing three-system.

A pencil of four rays in the plane will correspond to four cylindroids with a common screw, which we may term a pencil of cylindroids. Any fifth transversal cylindroid belonging also to the same three-system will be intersected by a pencil of four cylindroids in four screws, which have the same anharmonic ratio whatever be the cylindroid of the three-system which is regarded as the transversal. We thus infer from the well-known anharmonic property of conics the following theorem relative to the screws of a quadratic two-system.

If four screws α, β, γ, δ be taken on a quadratic two-system, and also any fifth screw η belonging to the same system, then the pencil of cylindroids $(\eta\alpha)$, $(\eta\beta)$, $(\eta\gamma)$, $(\eta\delta)$ will have the same anharmonic ratio whatever be the screw η. (See Appendix, note 6.)

The plane illustration will also suggest the instructive theory of Polar screws which will presently be stated more generally. Let $U=0$ be the conic representing the quadratic two-system and let $V=0$ be the conic representing the screws of zero pitch belonging to the enclosing three-system. Let P be a point in the plane corresponding to an arbitrary screw θ of the three-system. Draw the polar of P with respect to $U=0$ and let Q be the pole of this straight line with respect to $V=0$, then Q will correspond to some screw ϕ of the enclosing three-system. From any given screw θ, then by the help of the quadratic two-system a corresponding screw ϕ is determined. We may term ϕ the *polar screw* of θ with respect to $U=0$. Three screws of the enclosing system will coincide with their polars. These will be the vertices of the triangle which is self-conjugate with respect both to U and to V.

A possible difficulty may be here anticipated. The equation $V=0$ is itself of course equivalent to a certain quadratic two-system and therefore should correspond to a surface of the sixth degree. We know however (§ 173) that the locus of the screws of zero pitch in a three-system is an hyperboloid, so that in this case the expectation that the surface would rise to the sixth degree seems not to be justified. It is however shown in § 202 that this hyperboloid is really not more than a part of the locus. There are also four imaginary planes which with the hyperboloid complete the locus, and the combination thus rises to the sixth degree.

225. The Quadratic Systems of Higher Orders.

If we had taken $n = 3$, then of course the quadratic three-system would mean the collection of screws whose four co-ordinates satisfied an equation which in form resembles that of a quadric surface in quadriplanar co-ordinates. A definite number of screws belonging to the quadratic three-system can in general be drawn through every point in space.

We shall first prove that the number of those screws is six. Let $\theta_1, \ldots, \theta_6$ be the co-ordinates of any screw θ referred to a canonical co-reciprocal system. Then if x', y', z' be a point on θ, we have (§ 43)

$$(\theta_5 + \theta_6) y' - (\theta_3 + \theta_4) z' = a (\theta_1 - \theta_2) - p_\theta (\theta_1 + \theta_2),$$

$$(\theta_1 + \theta_2) z' - (\theta_5 + \theta_6) x' = b (\theta_3 - \theta_4) - p_\theta (\theta_3 + \theta_4),$$

$$(\theta_3 + \theta_4) x' - (\theta_1 + \theta_2) y' = c (\theta_5 - \theta_6) - p_\theta (\theta_5 + \theta_6).$$

If we express that θ belongs to the enclosing four-system we shall have two linear equations to be also satisfied by the co-ordinates of θ. These equations may be written without loss of generality in the form

$$\theta_2 = 0; \quad \theta_4 = 0.$$

We have finally the equation $U_\theta = 0$ characteristic of the quadratic three-system. From these equations the co-ordinates are to be eliminated. But the eliminant of k equations in $(k-1)$ independent variables is a homogeneous function of the coefficients of each equation whose order is, in general, equal to the product of the degrees of all the remaining equations*. In the present case, the coefficient of each of the first three equations must be of the second degree in the eliminant and hence, the resulting equation for p_θ is of the sixth degree, so that we have the following theorem.

Of the screws which belong to a quadratic three-system, six can be drawn through any point.

As the enclosing system in this case is of the fourth order, the screws of the enclosing system drawn through any point must lie on a cone of the second degree (§ 218). Hence it follows that the six screws just referred to must all lie on the surface of a cone of the second degree.

We may verify the theorem just proved by the consideration that if the function U_θ could be decomposed into two linear factors, each of those factors

* Salmon, *Modern Higher Algebra*, p. 76, 4th Edition (1885).

equated to zero would correspond to a three-system selected from the enclosing four-system. We know (§ 176) that three screws of a three-system can be drawn through each point. We have, consequently, three screws through the point for each of the two factors of U_θ, *i.e.* six screws in all.

The equation of the 6th degree in p_θ contains also the co-ordinates x', y', z' in the sixth degree. Taking these as the current co-ordinates we may regard this equation as expressing the family of surfaces which, taken together, contain all the screws of the quadratic three-system. The screws of this system which have the same pitch p_θ are thus seen to be ranged on the generators of a ruled surface of the sixth degree. All these screws belong of course to the enclosing four-system, and as they have the same pitches, they must all intersect the same pair of screws on the reciprocal cylindroid (§ 212). It follows that each of these pitch surfaces of the sixth degree must have inscribed upon it a pair of generators of the reciprocal cylindroid.

Ascending one step higher in the order of the enclosing system we see that the quadratic four-system is composed of those screws whose co-ordinates satisfy one linear homogeneous equation $L=0$, and one homogeneous equation of the second degree $U=0$. We may study these screws as follows.

Let the direction cosines of a screw θ be $\cos\lambda$, $\cos\mu$, $\cos\nu$. If the reference be made, as usual, to a set of canonical co-reciprocals we have

$$\cos\lambda = \theta_1 + \theta_2;\ \cos\mu = \theta_3 + \theta_4;\ \cos\nu = \theta_5 + \theta_6.$$

We therefore have for a point x', y', z' on θ the equations (§ 218)

$$2a\theta_1 = (a + p_\theta)\cos\lambda - z'\cos\mu + y'\cos\nu,$$

$$2a\theta_2 = (a - p_\theta)\cos\lambda + z'\cos\mu - y'\cos\nu,$$

with similar expressions for θ_3, θ_4, θ_5, θ_6.

Substituting these expressions in $L=0$ and $U=0$ and eliminating p_θ, we obtain an homogeneous equation of the fourth degree in $\cos\lambda$, $\cos\mu$, $\cos\nu$. If we substitute for these quantities $x-x'$, $y-y'$, $z-z'$, we obtain the equation of the cone of screws which can be drawn through x', y', z'; this cone is accordingly of the fourth degree. We verify this conclusion by noticing that if $U=0$ were the product of two linear functions, this cone would decompose into two cones of the second degree, as should clearly be the case (§ 218).

It remains to consider the Quadratic Five-system. In this case the enclosing system includes every screw in space, and the six co-ordinates of the screw θ are subjected to no other relation than that implied by the quadratic relation

$$U_\theta = 0.$$

As before we may substitute for $\theta_1, \ldots, \theta_6$ from the equations

$$2a\theta_1 = (a + p_\theta)\cos\lambda - z'\cos\mu + y'\cos\nu,$$

$$2a\theta_2 = (a - p_\theta)\cos\lambda + z'\cos\mu - y'\cos\nu,$$

with similar expressions for $2b\theta_1$, $2b\theta_2$, &c.

Introducing these values into

$$U_\theta = 0,$$

we obtain a result which may be written in the form

$$Ap_\theta^2 + 2Bp_\theta + C = 0,$$

where A, B, C contain $\cos\lambda$, $\cos\mu$, $\cos\nu$ in the second degree, and where x', y', z' enter linearly into B and in the second degree into C.

Hence we see that on any straight line in space there will be in general two screws belonging to any quadratic five-system. For the straight line being given x', y', z' are given, and so are $\cos\lambda$, $\cos\mu$, $\cos\nu$. The equation just written gives two values for a pitch which will comply with the necessary conditions.

If we consider p_θ and also x', y', z' as given, and if we substitute for $\cos\lambda$, $\cos\mu$, $\cos\nu$ the expressions $x - x'$, $y - y'$, $z - z'$ respectively, we obtain the equation of a cone of the second degree. Thus we learn that for each given pitch any point in space may be the vertex of a cone of the second degree such that the generators of the cone when they have received the given pitch are screws belonging to a given quadratic five-system.

If the equation

$$AC - B^2 = 0$$

be satisfied, then the straight lines which satisfy this condition will be singular, inasmuch as each contains but a single screw belonging to the quadratic five-system. As $\cos\lambda$, $\cos\mu$, $\cos\nu$ enter to the fourth degree into this equation it appears that each point in space is the vertex of a cone of the fourth degree, the generators of which when proper pitches are assigned to them will be singular screws of the quadratic five-system.

If we regard $\cos\lambda$, $\cos\mu$, $\cos\nu$ as given quantities in the equation

$$AC - B^2 = 0,$$

then this will represent a quadric surface inasmuch as x', y', z' enter to the second degree. This quadric is the locus of those singular screws of the quadratic five-system which are parallel to a given direction. Hence the equation must represent a cylinder.

If $B=0$ the two roots of the equation in p_θ will be equal, but with opposite signs; as $\cos\lambda$, $\cos\mu$, $\cos\nu$ enter to the second degree in B it follows that through any point in space as vertex a cone of the second degree can be drawn such that each generator of this cone when the proper pitch is assigned to it will equally belong to the quadratic five-system, whether that pitch be positive or negative.

If $B=0$ and $C=0$, then both values of p_θ must be zero. Regarding x', y', z' as fixed, each of these equations will correspond to a cone with vertex at x', y', z'; these cones will have four common generators, and hence we see that through any point in space four straight lines can in general be drawn such that with the pitch zero but not with any other pitch, these screws will be members of a given quadratic five-system.

226. Polar Screws.

The general discussion of the quadratic screw-systems is a subject of interest both geometrical and physical. We shall here be content with a few propositions which are of fundamental importance.

Let as before

$$U_\theta = 0$$

be the homogeneous relation between the co-ordinates $\theta_1, \ldots, \theta_{n+1}$ of the screws which constitute a quadratic n-system.

Let η and ζ denote any two screws other than θ and chosen from the enclosing n-system, from which the screws of the quadratic n-system are selected by the aid of the condition $U_\theta = 0$. If then we adopt the fertile method of investigation introduced by Joachimsthal, we shall substitute in $mU_\theta = 0$ for $\theta_1, \ldots, \theta_{n+1}$ the respective values

$$l\eta_1 + m\zeta_1,\ l\eta_2 + m\zeta_2 \ldots\ldots l\eta_{n+1} + m\zeta_{n+1}.$$

The result will be

$$l^2 U_\eta + lm U_{\eta\zeta} + m^2 U_\zeta = 0,$$

where

$$U_{\eta\zeta} = \zeta_1 \frac{dU_\eta}{d\eta_1} + \ldots + \zeta_{n+1} \frac{dU_\eta}{d\eta_{n+1}}.$$

Solving this quadratic equation for $l \div m$ we obtain two values of this ratio and hence (§ 119) we deduce the following theorem.

Any cylindroid of a given $(n+1)$-system will possess generally two screws belonging to every quadratic n-system which the given $(n+1)$-system encloses.

If the two screws η and ζ had been so selected that they satisfied the condition

$$U_{\eta\zeta} = \zeta_1 \frac{dU_\eta}{d\eta_1} + \ldots + \zeta_{n+1} \frac{dU_\eta}{d\eta_{n+1}} = 0,$$

then the two roots of the quadratic are equal but with opposite signs, and hence (§ 119) we have the following theorem.

If the condition $U_{\eta\zeta}=0$ is satisfied by the co-ordinates of two screws η and ζ which belong to the enclosing $(n+1)$-system, then these two screws η, ζ and the two screws which, lying on the cylindroid (η, ζ), also belong to the quadratic n-system $U_\theta=0$, will be parallel to the four rays of an harmonic pencil.

We are now to develop the conception of polar screws alluded to in § 224, and this may be most conveniently done by generalizing from a well-known principle in geometry.

Let O be a point and S a quadric surface. Let any straight line through O cut the quadric in the two points X_1 and X_2. Take on this straight line a point P so that the section OXP_1X_2 is harmonic; then for the different straight lines through O the locus of P is a plane. This plane is of course the well-known polar of P. We have an analogous conception in the present theory which appears as follows.

Take any screw η in the enclosing $(n+1)$-system. Draw a pencil of n cylindroids through η, all the screws of each cylindroid lying in the enclosing $(n+1)$-system. Each of these cylindroids will have on it two screws which belong to the quadratic n-system $U_\theta=0$. On each of these cylindroids a screw ζ can be taken which is the harmonic conjugate with respect to η with reference to the two screws of the quadratic n-system which are found on the cylindroid. We thus have n screws of the ζ type, and these n screws will define an n-system which is of course included within the enclosing $(n+1)$-system.

The equation of this n-system is obviously

$$U_{\eta\zeta}=\zeta_1\frac{dU_\eta}{d\eta_1} \qquad\qquad + \zeta_{n+1}\frac{dU_\eta}{d\eta_{n+1}}=0.$$

This equation is analogous to the polar of a point with regard to a quadric surface. We have here within a given enclosing $(n+1)$-system a certain n-system which is the polar of a screw η with respect to a certain quadratic n-system.

The conception of reciprocal screws enables us to take a further important step which has no counterpart in the ordinary theory of poles and polars. The linear equation for the co-ordinates of ζ, namely

$$U_{\eta\zeta}=0,$$

is merely the analytical expression of the fact that ζ is reciprocal to the screw of the enclosing $(n+1)$-system whose co-ordinates are proportional to

$$\frac{1}{p_1}\frac{dU_\eta}{d\eta_1}, \ldots \frac{1}{p_{n+1}}\frac{dU_\eta}{d\eta_{n+1}}.$$

This we shall term the *polar screw* of η with respect to the quadratic n-system. It is supposed, of course, that the screws of reference are co-reciprocals.

If α and β be two screws of an enclosing $(n+1)$-system, and if η and ζ be their respective polar screws with reference to a quadratic n-system, then when α is reciprocal to ζ we shall have β reciprocal to η. For we have, where h is a common factor,

$$h\eta_1 = \frac{1}{p_1}\frac{dU_\alpha}{d\alpha_1}, \ldots h\eta_{n+1} = \frac{1}{p_{n+1}}\frac{dU_\alpha}{d\alpha_{n+1}},$$

whence

$$h(p_1\eta_1\beta_1 + \ldots + p_{n+1}\eta_{n+1}\beta_{n+1}) = U_{\alpha\beta}.$$

If therefore β and η are reciprocal the left-hand member of this equation is zero and so must the right-hand member be zero. But the symmetry shows that ζ and α are in this case also reciprocal. We may in such a case regard α and β as two *conjugate screws* of the quadratic n-system.

As a first illustration of the relation between a screw and its polar, we shall take for $U_\alpha = 0$, the form

$$p_1\alpha_1^2 + p_2\alpha_2^2 + \ldots + p_6\alpha_6^2 - p\,(\alpha_1^2 + \alpha_2^2 + \ldots + \alpha_6^2 + 2\alpha_1\alpha_2 \cos(12) \ldots) = 0.$$

This means of course that $U_\alpha = 0$ denotes *every* screw which has the pitch p.

Take any screw α and draw a cylindroid through α. The two screws of pitch p on this cylindroid belong to U and a fourth screw θ may be taken on this cylindroid so that α, θ, and the two screws of pitch p form an harmonic pencil.

By drawing another cylindroid through α another screw of the θ-system can be similarly constructed. If these five cylindroids be drawn through α we can construct five different screws of the θ-system. To these one screw will be reciprocal, and this is the polar of α. We have thus the means of constructing the polar of α.

Seeing however that $U_\alpha = 0$ includes nothing more or less than all the p-pitch screws in the universe and that in the construction just given for the polar of α there has been no reference to the screws of reference, symmetry requires that the polar of α must be a screw which though different from α must be symmetrically placed with reference thereto. The only method of securing this is for the polar of α with respect to this particular function to lie on the same straight line as α.

Hence we deduce that the screw with co-ordinates

$$\alpha_1, \; \alpha_2, \; \ldots \; \alpha_6,$$

and the screw with co-ordinates proportional to

$$\frac{1}{p_1}\frac{dU}{d\alpha_1}, \quad \frac{1}{p_2}\frac{dU}{d\alpha_2}, \ldots \frac{1}{p_6}\frac{dU}{d\alpha_6},$$

in which U is the expression

$$p_1\alpha_1^2 + p_2\alpha_2^2 \ldots + p_6\alpha_6^2 + \lambda\,(\alpha_1^2 + \alpha_2^2 \ldots + 2\alpha_1\alpha_2\cos(12)\ldots)$$

must be collinear, and this is true for all values of λ.

We hence see that the co-ordinates of a screw collinear with α must be proportional to

$$\alpha_1 + \frac{\lambda}{2p_1}\frac{dR}{d\alpha_1}, \quad \alpha_2 + \frac{\lambda}{2p_2}\frac{dR}{d\alpha_1} \ldots,$$

where

$$R = \alpha_1^2 + \alpha_2^2 + \ldots + 2\alpha_1\alpha_2\cos(12) + \ldots$$

Thus we obtain the results of § 47 in a different manner.

227. Dynamical application of Polar Screws.

We have seen (§ 97) that the kinetic energy of a body twisting about a screw θ with a twist velocity $\frac{d\theta'}{dt}$ and belonging to a n-system is

$$M\left(\frac{d\theta'}{dt}\right)^2 (u_1^2\theta_1^2 + \ldots + u_n^2\theta_n^2),$$

the screws of reference being the principal screws of inertia.

If we make $u_1^2\theta_1^2 + \ldots + u_n^2\theta_n^2 = 0$, then θ must belong to a quadratic n-system. This system is, of course, imaginary, for the kinetic energy of the body when twisting about any screw which belongs to it is zero*.

The polar η of the screw θ, with respect to this quadratic n-system, has co-ordinates proportional to

$$\frac{u_1^2}{p_1}\theta_1, \ldots \frac{u_n^2}{p_n}\theta_n.$$

Comparing this with § 97, we deduce the following important theorem:

A quiescent rigid body is free to twist about all the screws of an enclosing $(n+1)$-system A. If the body receive an impulsive wrench on a screw η

* In a letter to the writer, Professor Klein pointed out many years ago the importance of the above screw system. He was led to it by expressing the condition that the impulsive screw should be reciprocal to the corresponding instantaneous screw.

belonging to A, then the body will commence to twist about the screw θ, of which η is the polar with respect to the quadratic n-system composed of the imaginary screws about which the body would twist with zero kinetic energy.

If a rigid body which has freedom of the nth order be displaced from a position of stable equilibrium under the action of a system of forces by a twist of given amplitude about a screw θ, of which the co-ordinates referred to the n principal screws of the potential are $\theta_1, \ldots \theta_n$, then the potential energy of the new position may, as we have seen (§ 103) be expressed by

$$v_1^2\theta_1^2 + \ldots + v_n^2\theta_n^2.$$

If this expression be equated to zero, it denotes a quadratic n-system, which is of course imaginary. We may term it the potential quadratic n-system.

The potential quadratic n-system possesses a physical importance in every respect analogous to that of the kinetic quadratic n-system: by reference to (§ 102) the following theorem can be deduced.

If a rigid body be displaced from a position of stable equilibrium by a twist about a screw θ, then a wrench acts upon the body in its new position on a screw which is the polar of θ with respect to the potential quadratic n-system.

The constructions by which the harmonic screws were determined in the case of the second and the third orders have no analogies in the fourth order. We shall, therefore, here state a general algebraical method by which they can be determined.

Let $U = 0$ be the kinetic quadratic n-system, and $V = 0$ the potential quadratic n-system, then it follows from a well-known algebraical theorem that one set of screws of reference can in general be found which will reduce both U and V to the sum of n squares. These screws of reference are the harmonic screws.

We may here also make the remark, that any quadratic n-system can generally be transformed in one way to the sum of n square terms *with co-reciprocal screws of reference;* for if U_θ and p_θ be transformed so that each consists of the sum of n square terms, then the form for the expression of p_θ (§ 38) shows that the screws are co-reciprocal.

228. On the degrees of certain surfaces.

We have already had occasion (§ 210) to demonstrate that the general condition that two screws shall intersect involves the co-ordinates of each of the screws in the third degree. We can express this condition as a determinant by employing a canonical system of co-reciprocals. For if two screws θ and ϕ intersect, then there must be some point x, y, z which shall satisfy the six equations (§ 43):

$$(\alpha_5 + \alpha_6) y - (\alpha_3 + \alpha_4) z = a(\alpha_1 - \alpha_2) - p_\alpha (\alpha_1 + \alpha_2),$$
$$(\alpha_1 + \alpha_2) z - (\alpha_5 + \alpha_6) x = b(\alpha_3 - \alpha_4) - p_\alpha (\alpha_3 + \alpha_4),$$
$$(\alpha_3 + \alpha_4) x - (\alpha_1 + \alpha_2) y = c(\alpha_5 - \alpha_6) - p_\alpha (\alpha_5 + \alpha_6),$$
$$(\theta_5 + \theta_6) y - (\theta_3 + \theta_4) z = a(\theta_1 - \theta_2) - p_\theta (\theta_1 + \theta_2),$$
$$(\theta_1 + \theta_2) z - (\theta_5 + \theta_6) x = b(\theta_3 - \theta_4) - p_\theta (\theta_3 + \theta_4),$$
$$(\theta_3 + \theta_4) x - (\theta_1 + \theta_2) y = c(\theta_5 - \theta_6) - p_\theta (\theta_5 + \theta_6).$$

From these equations we eliminate the five quantities $x, y, z, p_\theta, p_\alpha$ and the required condition that θ and α shall intersect, is given by the equation

$$\begin{vmatrix} 0, & (\alpha_5 + \alpha_6), & -(\alpha_3 + \alpha_4), & (\alpha_1 + \alpha_2), & 0, & a(\alpha_1 - \alpha_2) \\ -(\alpha_5 + \alpha_6), & 0, & (\alpha_1 + \alpha_2), & (\alpha_3 + \alpha_4), & 0, & b(\alpha_3 - \alpha_4) \\ (\alpha_3 + \alpha_4), & -(\alpha_1 + \alpha_2), & 0, & (\alpha_5 + \alpha_6), & 0, & c(\alpha_5 - \alpha_6) \\ 0, & (\theta_5 + \theta_6), & -(\theta_3 + \theta_4), & 0, & (\theta_1 + \theta_2), & a(\theta_1 - \theta_2) \\ -(\theta_5 + \theta_6), & 0, & (\theta_1 + \theta_2), & 0, & (\theta_3 + \theta_4), & b(\theta_3 - \theta_4) \\ (\theta_3 + \theta_4), & -(\theta_1 + \theta_2), & 0, & 0, & (\theta_5 + \theta_6), & c(\theta_5 - \theta_6) \end{vmatrix} = 0.$$

Four homogeneous equations between the co-ordinates of θ indicate that the corresponding screw lies on a certain ruled surface. Let us suppose that the degrees of these equations are l, m, n, r respectively, then the degree of the ruled surface must not exceed $3lmnr$.

For express the condition that θ shall also intersect some given screw α, we then obtain a fifth homogeneous equation containing the co-ordinates of θ in the third degree. The determination of the ratios of the six co-ordinates $\theta_1, \ldots \theta_6$ is thus effected by five equations of the several degrees $l, m, n, r, 3$. For each ratio we obtain a system of values equal in number to the product of the degrees of the equations, *i.e.* to $3lmnr$. This is accordingly a major limit to the number of points in which in general α pierces the surface, that is to say, it is a major limit to the degree of the surface. Of course we might affirm that it *was* the degree of the surface save for the possibility that through one or more of the points in which α met the surface more than a single generator might pass.

As an example, we may take the simple case of the cylindroid, in which l, m, n, r being each unity the locus is of the third degree. The screws of a three-system which satisfy an equation of the nth degree must have as their locus a surface of degree not exceeding $3n$. The most important application of this is when $n=2$, in which case the screws form a quadratic two-system. The degree of this surface cannot exceed six, on the other hand, if the quadratic condition which we may write

$$A\theta_1^2 + B\theta_2^2 + C\theta_3^2 + 2F\theta_2\theta_3 + 2G\theta_1\theta_3 + 2H\theta_1\theta_2 = 0,$$

should break up into two linear factors each of these linear factors will correspond to a cylindroid, *i.e.* a surface of the third degree. Hence the degree of the surface must in general be neither less than six nor greater than six, and hence we learn that the surface which is the locus of the screws of a quadratic two-system is of the sixth degree.

A particular case of special importance arises when the pitches of all the screws on the surface are to be the same. The statement of this condition is of course one equation of the second degree in the co-ordinates of the screw. In the case of canonical co-reciprocals, this equation would be

$$p_1\theta_1^2 + \dots + p_6\theta_6^2 = p_\theta\{(\theta_1+\theta_2)^2 + (\theta_3+\theta_4)^2 + (\theta_5+\theta_6)^2\}.$$

But the condition that θ and α shall intersect will now submit to modification. We sacrifice no generality by making α of zero pitch, so that if θ has a given pitch p_θ, the condition that α and θ shall intersect is no longer of the third degree. It is the linear equation

$$2\varpi_{\alpha\theta} = p_\theta \cos(\alpha\theta).$$

If therefore the co-ordinates of θ satisfy three homogeneous equations of degrees l, m, n respectively, in addition to the equation of the second degree expressing that the pitch is a given quantity, then the locus is a surface of degree not exceeding $2lmn$.

As the simplest illustration of this result we observe that if l, m, n be each unity, the locus in question is the locus of the screws of given pitch in a three-system. This locus cannot therefore be above the second degree, and we know, of course (Chapter XIV.) that the locus is a quadric.

If l and m were each unity and if $n=2$ we should then have the locus of screws of given pitch belonging to a four-system and whose co-ordinates satisfied a certain equation of the second degree. This locus is a surface of the fourth degree. In the special case where the given pitch is zero, the surface so defined is known in the theory of the linear complex. It is there presented as the locus of lines belonging to the complex and whose co-ordinates further satisfy both a linear equation and a quadratic equation.

Mr A. Panton has kindly pointed out to me that in this particular case the surface has two double lines which are the screws of zero pitch on the cylindroid reciprocal to the enclosing four-system. It has also eight singular tangent planes (four through each double line) touching all along a generator. The cylindroid is a very special case of this surface in which one of the double lines is at infinity. If the system be reciprocal to the cylindroid $z(x^2+y^2)+kxy=0$, then the cylindroid to which the surface reduces is $z(x^2+y^2)-kxy=0$. The singular tangent planes are represented by the two tangent planes at the limits of the cylindroid.

CHAPTER XVII.

FREEDOM OF THE FIFTH ORDER.

229. **Screw Reciprocal to Five Screws.**

There is no more important theorem in the Theory of Screws than that which asserts the existence of one screw reciprocal to five given screws. At the commencement, therefore, of the chapter of which this theorem is the foundation, it may be well to give a demonstration founded on elementary principles.

Let one of the five given screws be typified by

$$\frac{x - x_k}{\alpha_k} = \frac{y - y_k}{\beta_k} = \frac{z - z_k}{\gamma_k} \text{ (pitch} = \rho_k),$$

while the desired screw is defined by

$$\frac{x - x'}{\alpha} = \frac{y - y'}{\beta} = \frac{z - z'}{\gamma} \text{ (pitch} = \rho).$$

The condition of reciprocity (§ 20) produces five equations of the following type:—

$$\begin{aligned}&\alpha\,[(\rho + \rho_k)\,\alpha_k + \gamma_k y_k - \beta_k z_k] + \beta\,[(\rho + \rho_k)\,\beta_k + \alpha_k z_k - \gamma_k x_k]\\&+ \gamma\,[(\rho + \rho_k)\,\gamma_k + \beta_k x_k - \alpha_k y_k] + \alpha_k\,(\gamma y' - \beta z') + \beta_k\,(\alpha z' - \gamma x')\\&+ \gamma_k\,(\beta x' - \alpha y') = 0.\end{aligned}$$

From these five equations the relative values of the six quantities

$$\alpha,\ \beta,\ \gamma,\ \gamma y' - \beta z',\ \alpha z' - \gamma x',\ \beta x' - \alpha y'$$

can be determined by linear solution. Introducing these values into the identity

$$\alpha\,(\gamma y' - \beta z') + \beta\,(\alpha z' - \gamma x') + \gamma\,(\beta x' - \alpha y') = 0,$$

gives the equation which determines ρ.

To express this equation concisely we introduce two classes of subsidiary magnitudes. We write one magnitude of each class as a determinant.

$$\begin{vmatrix} \rho_1\beta_1 + z_1\alpha_1 - x_1\gamma_1, & \rho_1\gamma_1 + x_1\beta_1 - y_1\alpha_1, & \alpha_1, & \beta_1, & \gamma_1 \\ \rho_2\beta_1 + z_2\alpha_2 - x_2\gamma_2, & \rho_2\gamma_2 + x_2\beta_2 - y_2\alpha_2, & \alpha_2, & \beta_2, & \gamma_2 \\ \rho_3\beta_3 + z_3\alpha_3 - x_3\gamma_3, & \rho_3\gamma_3 + x_3\beta_3 - y_3\alpha_3, & \alpha_3, & \beta_3, & \gamma_3 \\ \rho_1\beta_4 + z_4\alpha_4 - x_4\gamma_4, & \rho_4\gamma_4 + x_4\beta_4 - y_4\alpha_4, & \alpha_4, & \beta_4, & \gamma_4 \\ \rho_5\beta_5 + z_5\alpha_5 - x_5\gamma_5, & \rho_5\gamma_5 + x_5\beta_5 - y_5\alpha_5, & \alpha_5, & \beta_5, & \gamma_5 \end{vmatrix} = P.$$

By cyclical interchange the two analogous functions Q and R are defined.

$$\begin{vmatrix} \rho_1\alpha_1 + y_1\gamma_1 - z_1\beta_1, & \rho_1\beta_1 + z_1\alpha_1 - x_1\gamma_1, & \rho_1\gamma_1 + x_1\beta_1 - y_1\alpha_1, & \beta_1, & \gamma_1 \\ \rho_2\alpha_2 + y_2\gamma_2 - z_2\beta_2, & \rho_2\beta_2 + z_2\alpha_2 - x_2\gamma_2, & \rho_2\gamma_2 + x_2\beta_2 - y_2\alpha_2, & \beta_2, & \gamma_2 \\ \rho_3\alpha_3 + y_3\gamma_3 - z_3\beta_3, & \rho_3\beta_3 + z_3\alpha_3 - x_3\gamma_3, & \rho_3\gamma_3 + x_3\beta_3 - y_3\alpha_3, & \beta_3, & \gamma_3 \\ \rho_4\alpha_4 + y_4\gamma_4 - z_4\beta_4, & \rho_4\beta_4 + z_4\alpha_4 - x_4\gamma_4, & \rho_4\gamma_4 + x_4\beta_4 - y_4\alpha_4, & \beta_4, & \gamma_4 \\ \rho_5\alpha_5 + y_5\gamma_5 - z_5\beta_5, & \rho_5\beta_5 + z_5\alpha_5 - x_5\gamma_5, & \rho_5\gamma_5 + x_5\beta_5 - y_5\alpha_5, & \beta_5, & \gamma_5 \end{vmatrix} = L.$$

By cyclical interchange the two analogous functions M and N are defined.

The equation for ρ reduces to

$$(P^2 + Q^2 + R^2)\,\rho + PL + QM + RN = 0.$$

The reduction of this equation to the first degree is an independent proof of the principle, that one screw, and only one, can be determined which is reciprocal to five given screws; ρ being known, α, β, γ can be found, and also two linear equations between x', y', z', whence the reciprocal screw is completely determined.

For the study of the screws representing a five-system we may take the first screw of a set of canonical coreciprocals to be the screw reciprocal to the system. Then the co-ordinates of a screw in the system are

$$0,\ \theta_2,\ \theta_3, \ldots \theta_6,$$

while if λ, μ, ν be the direction cosines of θ and x, y, z a point thereon, and p_θ the pitch we have (§ 43)

$$(p_\theta + a)\cos\lambda - z\cos\mu + y\cos\nu = 0.$$

We can obtain at once the relation between the direction and the pitch of the screw belonging to the system and passing through a fixed point. If $p_\theta = 0$ and z and y be given, then the equation shows that the screw is limited to a plane (§ 110).

230. Six Screws Reciprocal to One Screw.

When six screws, A_1, ... A_6 are reciprocal to a single screw T, a certain relation must subsist between the six screws. This relation may be expressed by equating the determinant of § 39 to zero. The determinant (which may perhaps be called the *sexiant*) may be otherwise expressed as follows:—

The equations of the screw A_k are

$$\frac{x - x_k}{\alpha_k} = \frac{y - y_k}{\beta_k} = \frac{z - z_k}{\gamma_k} \text{ (pitch } \rho_k).$$

We shall presently show that we are justified in assuming for T the equations

$$\frac{x}{\alpha} = \frac{y}{\beta} = \frac{z}{\gamma} \text{ (pitch} = \rho).$$

The condition that A_k and T be reciprocal is

$$(\rho + \rho_k)(\alpha\alpha_k + \beta\beta_k + \gamma\gamma_k) + x_k(\gamma\beta_k - \beta\gamma_k) + y_k(\alpha\gamma_k - \gamma\alpha_k)$$
$$+ z_k(\beta\alpha_k - \alpha\beta_k) = 0.$$

Writing the six equations of this type, found by giving k the values 1 to 6, and eliminating the six quantities

$$\rho\alpha,\ \rho\beta,\ \rho\gamma,\ \alpha,\ \beta,\ \gamma,$$

we obtain the result:—

$$\begin{vmatrix}
\alpha_1\rho_1 + \gamma_1 y_1 - \beta_1 z_1, & \beta_1\rho_1 + \alpha_1 z_1 - \gamma_1 x_1, & \gamma_1\rho_1 + \beta_1 x_1 - \alpha_1 y_1, & \alpha_1, & \beta_1, & \gamma_1 \\
\alpha_2\rho_2 + \gamma_2 y_2 - \beta_2 z_2, & \beta_2\rho_2 + \alpha_2 z_2 - \gamma_2 x_2, & \gamma_2\rho_2 + \beta_2 x_2 - \alpha_2 y_2, & \alpha_2, & \beta_2, & \gamma_2 \\
\alpha_3\rho_3 + \gamma_3 y_3 - \beta_3 z_3, & \beta_3\rho_3 + \alpha_3 z_3 - \gamma_3 x_3, & \gamma_3\rho_3 + \beta_3 x_3 - \alpha_3 y_3, & \alpha_3, & \beta_3, & \gamma_3 \\
\alpha_4\rho_4 + \gamma_4 y_4 - \beta_4 z_4, & \beta_4\rho_4 + \alpha_4 z_4 - \gamma_4 x_4, & \gamma_4\rho_4 + \beta_4 x_4 - \alpha_4 y_4, & \alpha_4, & \beta_4, & \gamma_4 \\
\alpha_5\rho_5 + \gamma_5 y_5 - \beta_5 z_5, & \beta_5\rho_5 + \alpha_4 z_5 - \gamma_5 x_5, & \gamma_5\rho_5 + \beta_5 x_5 - \alpha_5 y_5, & \alpha_5, & \beta_5, & \gamma_5 \\
\alpha_6\rho_6 + \gamma_6 y_6 - \beta_6 z_6, & \beta_6\rho_6 + \alpha_6 z_6 - \gamma_6 x_6, & \gamma_6\rho_6 + \beta_6 x_6 - \alpha_6 y_6, & \alpha_6, & \beta_6, & \gamma_6
\end{vmatrix} = 0.$$

By transformation to *any parallel axes* the value of this determinant is *unaltered*. The evanescence of the determinant is therefore a necessary condition *whenever* the six screws are reciprocal to a single screw. Hence we sacrificed no generality in the assumption that T passed through the origin.

Since the sexiant is linear in x_1, y_1, z_1, it appears that all parallel screws of given pitch reciprocal to one screw lie in a plane. Since the sexiant is linear in α_1, β_1, γ_1, we have another proof of Möbius' theorem (§ 110).

The property possessed by six screws when their sexiant vanishes may be enunciated in different ways, which are precisely equivalent.

(*a*) The six screws are all reciprocal to one screw.

(*b*) The six screws are members of a screw-system of the fifth order and first degree.

(*c*) Wrenches of appropriate intensities on the six screws equilibrate, when applied to a free rigid body.

(*d*) Properly selected twist velocities about the six screws neutralize, when applied to a rigid body.

(*e*) A body might receive six small twists about the six screws, so that after the last twist the body would occupy the same position which it had before the first.

If seven wrenches equilibrate (or twists neutralize), then the intensity of each wrench (or the amplitude of each twist) is proportional to the sexiant of the six non-corresponding screws.

For a rigid body which has freedom of the fifth order to be in equilibrium, the necessary and sufficient condition is that the forces which act upon the body constitute a wrench upon that one screw to which the freedom is reciprocal. We thus see that it is not possible for a body which has freedom of the fifth order to be in equilibrium under the action of gravity unless the screw reciprocal to the freedom have zero pitch, and coincide in position with the vertical through the centre of inertia.

Sylvester has shown* that when six lines, P, Q, R, S, T, U, are so situated that forces acting along them equilibrate when applied to a *free* rigid body, a certain determinant vanishes, and he speaks of the six lines so related as being *in involution*†.

Using the ideas and language of the Theory of Screws, this determinant is the sexiant of the six screws, the pitches of course being zero.

If x_m, y_m, z_m, be a point on one of the lines, the direction cosines of the same line being $\alpha_m, \beta_m, \gamma_m$, the condition is

$$\begin{vmatrix} \alpha_1, & \beta_1, & \gamma_1, & y_1\gamma_1 - z_1\beta_1, & z_1\alpha_1 - x_1\gamma_1, & x_1\beta_1 - y_1\alpha_1 \\ \alpha_2, & \beta_2, & \gamma_2, & y_2\gamma_2 - z_2\beta_2, & z_2\alpha_2 - x_2\gamma_2, & x_2\beta_2 - y_2\alpha_2 \\ \alpha_3, & \beta_3, & \gamma_3, & y_3\gamma_3 - z_3\beta_3, & z_3\alpha_3 - x_3\gamma_3, & x_3\beta_3 - y_3\alpha_3 \\ \alpha_4, & \beta_4, & \gamma_4, & y_4\gamma_4 - z_4\beta_4, & z_4\alpha_4 - x_4\gamma_4, & x_4\beta_4 - y_4\alpha_4 \\ \alpha_5, & \beta_5, & \gamma_5, & y_5\gamma_5 - z_5\beta_5, & z_5\alpha_5 - x_5\gamma_5, & x_5\beta_5 - y_5\alpha_5 \\ \alpha_6, & \beta_6, & \gamma_6, & y_6\gamma_6 - z_6\beta_6, & z_6\alpha_6 - x_6\gamma_6, & x_6\beta_6 - y_6\alpha_6 \end{vmatrix} = 0.$$

* *Comptes Rendus*, tome 52, p. 816. See also p. 741.

† In our language a system of lines thus related consists of the *screws of equal pitch belonging to a five-system*. In the language of Plücker (*Neue Geometrie des Raumes*) a system of lines in involution forms a *linear complex*. It may save the reader some trouble to observe here that the word *involution* has been employed in a more generalised sense by Battaglini, and in quite a different sense by Klein.

It must always be possible to find a single screw X which is reciprocal to the six screws P, Q, R, S, T, U. Suppose the rigid body were only free to twist about X, then the six forces would not only collectively be in equilibrium, but severally would be unable to stir the body only free to twist about X.

In *general* a body able to twist about six screws (of any pitch) would have perfect freedom; but the body capable of rotating about each of the six lines, P, Q, R, S, T, U, which are in involution, is not necessarily perfectly free (Möbius).

If a rigid body were perfectly free, then a wrench about any screw could move the body; if the body be only free to rotate about the six lines in involution, then a wrench about every screw (except X) can move it.

The *conjugate axes* discussed by Sylvester are presented in the Theory of Screws as follows:—Draw *any* cylindroid which contains the reciprocal screw X, then the two screws of zero pitch on this cylindroid are a pair of conjugate axes. For a force on any transversal intersecting this pair of screws is reciprocal to the cylindroid, and is therefore in involution with the original system.

Draw any two cylindroids, each containing the reciprocal screw, then all the screws of the cylindroids form a screw system of the third order. Therefore the two pairs of conjugate axes, being four screws of zero pitch, must lie upon the same quadric. This theorem, due to Sylvester, is proved by him in a different manner.

The cylindroid also presents in a clear manner the solution of the problem of finding two *rotations* which shall bring a body from one position to any other given position. Find the twist which would effect the desired change. Draw *any* cylindroid through the corresponding screw, then the two screws of zero pitch on the cylindroid are a pair of axes that fulfil the required conditions. If one of these axes were given the cylindroid would be defined and the other axis would be determinate.

231. Four Screws of a Five-system on every Quadric.

On any single sheeted hyperboloid four screws of any given pitch p can in general be determined which belong to any given system of the fifth order. A pair of these screws lie on each kind of generator.

Let X be the screw reciprocal to the system. Take any three generators A, B, C of one system on the hyperboloid, and regarding them as screws of pitch $-p$ draw the cylindroid XA and take on this A' the second screw of pitch $-p$. Then the two screws of pitch p which can be drawn as transversals across A, B, C, A' are coincident with two generators of the hyperboloid while they are also reciprocal to the cylindroid because they cross two screws thereon with pitches equal in magnitude but opposite in sign. They are therefore reciprocal to X. In like manner it can be shown that two of the other system of generators possess the same property.

On every cylindroid there is as we know (§ 26) one screw of a given five-system. This important proposition may be otherwise proved as follows. Let θ be the co-ordinates of a screw on the cylindroid, then these co-ordinates must satisfy four linear equations. There must be a fifth equation in the six quantities $\theta_1, \ldots \theta_6$ inasmuch as θ is to lie on the given five-system. Thus from these five equations one set of values of $\theta_1, \ldots \theta_6$ can be determined.

On a quadratic two-system (§ 224) there will always be two screws belonging to any given five-system. For the quadratic two-system is the surface whose screws satisfy four homogeneous equations of which three are linear and one is quadratic. If another linear equation be added two screws on the surface can, in general, be found which will satisfy that equation.

232. Impulsive Screws and Instantaneous Screws.

We can determine the instantaneous screw corresponding to a given impulsive screw in the case of freedom of the fifth order by geometrical considerations. Let λ, as before, represent the screw reciprocal to the freedom, and let ρ be the instantaneous screw which would correspond to λ as an impulsive screw, if the body were perfectly free; let η be the screw on which the body receives an impulsive wrench, and let ξ be the screw about which the body would commence to twist in consequence of this impulse if it had been perfectly free.

The body when limited to the screw system of the fifth order will commence to move as if it had been free, but had been acted upon by a certain unknown wrench on λ, together with the given wrench on η. The movement which the body actually acquires is a twisting motion about a screw θ which must lie on the cylindroid (ξ, ρ). We therefore determine θ to be that one screw on the known cylindroid (ξ, ρ) which is reciprocal to the given screw λ. The twist velocity of the initial twisting motion about θ, as well as the intensity of the impulsive wrench on the screw λ produced by the reaction of the constraints, are also determined by the same construction. For by § 17 the relative twist velocities about θ, ξ, and ρ are known; but since the impulsive intensity η''' is known, the twist velocity about ξ is known (§ 90); and therefore, the twist velocity about θ is known; finally, from the twist velocity about ρ, the impulsive intensity λ''' is determined.

233. Analytical Method.

A quiescent rigid body which has freedom of the fifth order receives an impulsive wrench on a screw η: it is required to determine the instantaneous screw θ, about which the body will commence to twist.

Let λ be the screw reciprocal to the freedom, and let the co-ordinates be referred to the absolute principal screws of inertia. The given wrench compounded with a certain wrench on λ must constitute the wrench which, if the body were free, would make it twist about θ, whence we deduce the six equations (h being an unknown quantity)

$$hp_1\theta_1 = \eta'''\eta_1 + \lambda'''\lambda_1$$

$$\cdots\cdots\cdots\cdots\cdots\cdots$$

$$hp_6\theta_6 = \eta'''\eta_6 + \lambda'''\lambda_6.$$

Multiplying the first of these equations by λ_1, the second by λ_2, &c., adding the six equations thus produced, and remembering that θ and λ are reciprocal, we deduce

$$\eta'''\Sigma\eta_1\lambda_1 + \lambda'''\Sigma\lambda_1^2 = 0.$$

This equation determines λ''' the impulsive intensity of the reaction of the constraints. The co-ordinates of the required screw θ are, therefore, proportional to the six quantities

$$\frac{\eta_1\Sigma\lambda_1^2 - \lambda_1\Sigma\eta_1\lambda_1}{p_1}, \ldots \frac{\eta_6\Sigma\lambda_1^2 - \lambda_6\Sigma\eta_1\lambda_1}{p_6}.$$

234. Principal Screws of Inertia.

We can now determine the co-ordinates of the five principal screws of inertia; for if ξ be a principal screw of inertia, then in general

$$hp_1\xi_1 = \xi'''\xi_1 + \lambda'''\lambda_1,$$

whence

$$\xi_1 = \frac{\lambda'''\lambda_1}{hp_1 - \xi'''}$$

with similar values for $\xi_2, \ldots \xi_6$. Substituting these values in the equation

$$p_1\lambda_1\xi_1 + p_2\lambda_2\xi_2 + p_3\lambda_3\xi_3 + p_4\lambda_4\xi_4 + p_5\lambda_5\xi_5 + p_6\lambda_6\xi_6 = 0,$$

and making $\frac{\xi'''}{h} = x$, we have for x the equation

$$\frac{p_1\lambda_1^2}{p_1 - x} + \frac{p_2\lambda_2^2}{p_2 - x} + \frac{p_3\lambda_3^2}{p_3 - x} + \frac{p_4\lambda_4^2}{p_4 - x} + \frac{p_5\lambda_5^2}{p_5 - x} + \frac{p_6\lambda_6^2}{p_6 - x} = 0.$$

This equation is of the fifth degree, corresponding to the five principal screws of inertia. If x' denote one of the roots of the equation, then the corresponding principal screw of inertia has co-ordinates proportional to

$$\frac{\lambda_1}{p_1 - x'}, \frac{\lambda_2}{p_2 - x'}, \frac{\lambda_3}{p_3 - x'}, \frac{\lambda_4}{p_4 - x'}, \frac{\lambda_5}{p_5 - x'}, \frac{\lambda_6}{p_6 - x'}.$$

We can easily verify as in § 84 that these five screws are co-reciprocal and are also conjugate screws of inertia.

It is assumed in the deduction of this quintic that all the quantities $\lambda_1 \ldots \lambda_6$ are different from zero. If one of the quantities, suppose λ_1, had been zero this means that the first absolute principal Screw of Inertia would belong to the n-system expressing the freedom.

Let us suppose that $\lambda_1 = 0$ then the equations are

$$hp_1\xi_1 = \xi'''\xi_1$$
$$hp_2\xi_2 = \xi'''\xi_2 + \lambda'''\lambda_2$$
$$\cdots\cdots\cdots\cdots\cdots\cdots$$
$$hp_6\xi_6 = \xi'''\xi_6 + \lambda'''\lambda_6.$$

Of course one solution of this system will be $\lambda''' = 0$, $\xi_2 = 0 \ldots \xi_6 = 0$. This means that the first *absolute* principal Screw of Inertia is also one of the principal Screws of Inertia in the n-system, as should obviously be the case. For the others $\xi_1 = 0$ and we have an equation of the fourth degree in $x = \xi''' \div h$

$$\frac{p_2\lambda_2^2}{p_2 - x} + \ldots + \frac{p_6\lambda_6^2}{p_6 - x} = 0.$$

In the general case we can show that there are no imaginary roots in the quintic, for since the screws

$$\frac{\lambda_1}{p_1 - x'}, \quad \frac{\lambda_2}{p_2 - x'}, \quad \frac{\lambda_6}{p_6 - x'},$$

and

$$\frac{\lambda_1}{p_1 - x''}, \quad \frac{\lambda_2}{p_2 - x''}, \quad \frac{\lambda_6}{p_6 - x''}$$

are conjugate screws of inertia, we must have (§ 81)

$$\Sigma \frac{p_1^2\lambda_1^2}{(p_1 - x')(p_1 - x'')} = 0.$$

If $x' = \alpha + i\beta$; $y' = \alpha - i\beta$ then this equation reduces to

$$\Sigma \frac{p_1^2\lambda_1^2}{(p_1 - \alpha)^2 + \beta^2} = 0,$$

but as these are each positive terms their sum cannot be zero. This is a particular case of § 86. (See Appendix, note 2.)

235. The limits of the roots.

We can now show the limits between which the five roots, just proved to be real, must actually lie in the equation

$$\frac{p_1\lambda_1^2}{p_1 - x} + \ldots + \frac{p_6\lambda_6^2}{p_6 - x} = 0;$$

substitute
$$p_1 = \frac{1}{q_1},\quad p_2 = \frac{1}{q_2},\ \ldots\ p_6 = \frac{1}{q_6},\quad x = \frac{1}{y};$$

and suppose $q_1, q_2, q_3, q_4, q_5, q_6$ to be in descending order of magnitude.

Thus
$$\frac{\lambda_1^2}{y-q_1} + \frac{\lambda_2^2}{y-q_2} + \ldots + \frac{\lambda_6^2}{y-q_6} = 0.$$

That is
$$\lambda_1^2 (y-q_2)(y-q_3)(y-q_4)(y-q_5)(y-q_6) + \ldots$$
$$+ \lambda_6^2 (y-q_1)(y-q_2)(y-q_3)(y-q_4)(y-q_5) = 0.$$

.In the left-hand member of this equation substitute the values

$$q_1, q_2, q_3, q_4, q_5, q_6$$

successively for y; five of the six terms vanish in each case, and the values of the remaining term (and therefore of the whole member) are alternately positive and negative.

The five values of y must therefore lie in the intervals between the six quantities $q_1, q_2, \ldots q_6$, the roots are accordingly proved to be real and distinct (unless one of the quantities $\lambda_1, \lambda_2, \lambda_3, \lambda_4, \lambda_5, \lambda_6 = 0$ and a further condition hold, or unless some of the quantities $q_1, \ldots q_6$ be equal).

The values of $p_1, \ldots p_6$ are $\pm a$, $\pm b$, $\pm c$; and we suppose a, b, c, positive and $a > b > c$.

The values of y lie in the successive intervals between

$$\frac{1}{c},\ \frac{1}{b},\ \frac{1}{a},\ -\frac{1}{a},\ -\frac{1}{b},\ -\frac{1}{c};$$

and consequently of the roots of the equation in x.

Two are positive and lie between a and b, and between b and c respectively.

Two are negative and lie between $-a$ and $-b$, and between $-b$ and $-c$ respectively.

The last is either positive and $> a$ or negative and $< -a$.

236. The Pectenoid.

A surface of some interest in connection with the freedom of the fifth order may be investigated as follows.

Let a be the pitch of the one screw ω, to which the five system is reciprocal.

Take any point O on ω and draw through O any two right lines OY, and OZ which are at right angles and which lie in the plane perpendicular to ω.

Then if θ be a screw of the five-system with direction cosines $\cos\lambda$, $\cos\mu$, $\cos\nu$, and if x, y, z be a point on the screw θ and p_θ its pitch we must have (§ 216)

$$(p_\theta + a)\cos\lambda + z\cos\mu - y\cos\nu = 0.$$

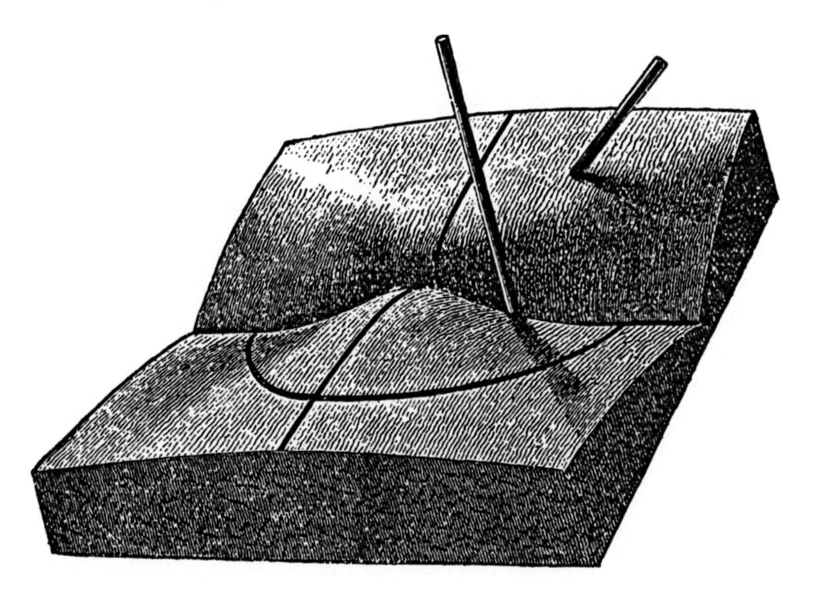

Fig. 41.

Everything that we wish to specify about the five-system may be conveniently inferred from this equation.

For example, let it be desired to find the locus of the screws of a five-system which can be drawn through a given point x', y', z' and have the given pitch p_θ.

We have $$(p_\theta + a)\cos\lambda + z'\cos\mu - y'\cos\nu = 0.$$

If x, y, z be a point on θ we may substitute $x' - x$, $y' - y$, $z' - z$ for $\cos\lambda$, $\cos\mu$, $\cos\nu$, and we obtain

$$(p_\theta + a)(x' - x) + z'(y' - y) - y'(z' - z) = 0,$$

whence we see that the locus is a plane, as has been already proved otherwise. (When the pitch is zero, this is Möbius' theorem, § 110.)

If we change the origin to some other point P which may with complete generality be that point whose co-ordinates are o, h, o and call X, Y, Z the co-ordinates with these new axes, the equation becomes

$$(p_\theta + a)\cos\lambda + Z\cos\mu - (Y + h)\cos\nu = 0.$$

Let the radius vector of length $p_\theta + a = R$ be marked off along each screw θ drawn through P, then the equation becomes

$$RX = hZ,$$

or squaring $$(X^2 + Y^2 + Z^2) X^2 = h^2 Z^2.$$

This represents a surface of the fourth degree. A model of the surface has been constructed*. It is represented in Fig. 41, and from its resemblance to the valves of a scallop shell the name *pectenoid* is suggested.

The geometrical nature of a pectenoid is thus expressed. Given a screw α of pitch p_α and a point O situated anywhere. If a screw θ drawn through O be reciprocal to α then the extremity of a radius vector from O along θ equal to $p_\alpha + p_\theta$ will trace out a pectenoid.

All pectenoids are similar surfaces, they merely differ in size in accordance with the variations of the quantity h. The perpendicular from O upon α is a nodal line, and this is the only straight line on the surface. The pectenoid though unclosed is entirely contained between the pair of parallel planes $Z = +h$, $Z = -h$. Sections parallel to the plane of Z are hyperbolas. Any plane through the nodal line cuts the pectenoid in a circle.

A straight wire at right angles to the nodal line marked on the model indicates the screw reciprocal to the five-system. A second wire starts from the origin and projects from the surface. It is introduced to show concisely what the pectenoid expresses. If this wire be the axis of a screw θ whose pitch when added to the pitch of the screw α, is equal to the intercept from the origin to the surface then the two screws are reciprocal. The interpretation of the nodal line is found in the obvious truth that when two screws intersect at right angles they are reciprocal whatever be the sum of their pitches. One of the circular sections made by a plane drawn through the nodal line is also indicated in the model. The physical interpretation is found in the theorem already mentioned, that all screws of the same pitch drawn through the same point and reciprocal to a given screw will lie in a plane.

With the help of the pectenoid we can give another proof of the theorem that all the screws of a four-system which can be drawn through a point lie on a cone of the second degree (§ 123).

Let O be the point and let α and β be two screws on the cylindroid reciprocal to the system. Let θ be a screw through O belonging to the four-system and therefore reciprocal to α and to β.

Then for the pectenoid relating to O and α, we have

$$(p_\theta + p_\alpha) M - kN = 0,$$

where $M = 0$, $N = 0$ represent planes passing through O.

* *Transactions of the Royal Irish Academy*, Vol. XXV. Plate XII. (1871).

In like manner for the pectenoid relating to O and β, we have

$$(p_\theta + p_\beta)M' - k'N' = 0,$$

where $M' = 0$, $N' = 0$ also represent planes passing through O, whence eliminating p_θ we have

$$(p_\alpha - p_\beta)MM' - kNM' + k'N'M = 0.$$

The equation of the pectenoid can also be deduced directly as follows.

Let five screws of the five-system be given. Take a point O on one of these screws (α) and through O draw four screws β, γ, δ, ϵ which belong to the four-system defined by the remaining four screws of the original five. Let there be any three rectangular axes drawn through O. Let α_1, α_2, α_3 be the direction cosines of α and let β_1, β_2, β_3 be the direction cosines of β, and similarly for γ, δ, ϵ. Let θ be some other screw of the five-system which passes through O and let θ_1, θ_2, θ_3 be its direction cosines, then if twists of amplitudes α', β', γ', δ', ϵ', θ' neutralize we must have

$$\alpha'\alpha_1 + \beta'\beta_1 + \gamma'\gamma_1 + \delta'\delta_1 + \epsilon'\epsilon_1 + \theta'\theta_1 = 0,$$
$$\alpha'\alpha_2 + \beta'\beta_2 + \gamma'\gamma_2 + \delta'\delta_2 + \epsilon'\epsilon_2 + \theta'\theta_2 = 0,$$
$$\alpha'\alpha_3 + \beta'\beta_3 + \gamma'\gamma_3 + \delta'\delta_3 + \epsilon'\epsilon_3 + \theta'\theta_3 = 0,$$

because the rotations neutralize, and also

$$\alpha' p_\alpha \alpha_1 + \beta' p_\beta \beta_1 + \gamma' p_\gamma \gamma_1 + \delta' p_\delta \delta_1 + \epsilon' p_\epsilon \epsilon_1 + \theta' p_\theta \theta_1 = 0,$$
$$\alpha' p_\alpha \alpha_2 + \beta' p_\beta \beta_2 + \gamma' p_\gamma \gamma_2 + \delta' p_\delta \delta_2 + \epsilon' p_\epsilon \epsilon_2 + \theta' p_\theta \theta_2 = 0,$$
$$\alpha' p_\alpha \alpha_3 + \beta' p_\beta \beta_3 + \gamma' p_\gamma \gamma_3 + \delta' p_\delta \delta_3 + \epsilon' p_\epsilon \epsilon_3 + \theta' p_\theta \theta_3 = 0,$$

whence by elimination of α', β', ... θ', we have

$$\begin{vmatrix} \alpha_1 & \beta_1 & \gamma_1 & \delta_1 & \epsilon_1 & \theta_1 \\ \alpha_2 & \beta_2 & \gamma_2 & \delta_2 & \epsilon_2 & \theta_2 \\ \alpha_3 & \beta_3 & \gamma_3 & \delta_3 & \epsilon_3 & \theta_3 \\ p_\alpha \alpha_1 & p_\beta \beta_1 & p_\gamma \gamma_1 & p_\delta \delta_1 & p_\epsilon \epsilon_1 & p_\theta \theta_1 \\ p_\alpha \alpha_2 & p_\beta \beta_2 & p_\gamma \gamma_2 & p_\delta \delta_2 & p_\epsilon \epsilon_2 & p_\theta \theta_2 \\ p_\alpha \alpha_3 & p_\beta \beta_3 & p_\gamma \gamma_3 & p_\delta \delta_3 & p_\epsilon \epsilon_3 & p_\theta \theta_3 \end{vmatrix} = 0.$$

This equation has the form

$$p_\theta = \frac{L\theta_1 + M\theta_2 + N\theta_3}{L'\theta_1 + M'\theta_2 + N'\theta_3}.$$

Let $p_\theta = \rho - h$ and $\theta_1 = x \div \rho$, $\theta_2 = y \div \rho$, $\theta_3 = z \div \rho$ then by reduction and transformation of axes we obtain

$$\rho z = ay,$$

where y and z are planes at right angles and a is constant. This is the equation of the pectenoid.

CHAPTER XVIII.

FREEDOM OF THE SIXTH ORDER.

237. Introduction.

When a rigid body has freedom of the sixth order, it is perfectly free. The screw system of the sixth order includes every screw in space. The statement that there is no reciprocal screw to such a system is merely a different way of asserting the obvious proposition that when a body is perfectly free it cannot remain in equilibrium, if the forces which act upon it have a resultant.

238. Impulsive Screws.

Let A_1, A_2, ... denote a series of instantaneous screws which correspond respectively to the impulsive screws R_1, R_2, ... the body being perfectly free. Corresponding to each pair A_1, R_1 is a certain specific parameter. This parameter may be conveniently defined to be the twist velocity produced about A_1 by an impulsive wrench on R_1, of which the intensity is one unit. If six pairs, A_1, R_1; A_2, R_2, ... be known, and also the corresponding specific parameters, then the impulsive wrench on any other screw R can be resolved into six impulsive wrenches on $R_1, \ldots R_6$, these will produce six known twist velocities on $A_1, \ldots A_6$, which being compounded determine the screw A, the twist velocity about A, and therefore the specific parameter of R and A. We thus see that it is only necessary to be given six corresponding pairs, and their specific parameters, in order to determine completely the effect of any other impulsive wrench.

If seven pairs of corresponding instantaneous and impulsive screws be given, then the relation between every other pair is absolutely determined. It appears from § 28 that appropriate twist velocities about $A_1, \ldots A_7$ can neutralise. When this is the case, the corresponding impulsive wrenches on $R_1, \ldots R_7$, must equilibrate, and therefore the relative values of the intensities are known. It follows that the specific parameter of each pair A_1, R_1 is proportional to the quotient obtained by dividing the sexiant of

$A_2, \ldots A_6$, by the sexiant of $R_2, \ldots R_6$. With the exception of a common factor, the specific parameter of every pair of screws is therefore known, when seven corresponding screws are known. It will be shown in Chap. XXI. that *three* corresponding pairs are really sufficient.

When seven instantaneous screws are known, and the corresponding seven impulsive screws, we are therefore enabled by geometrical construction alone to deduce the instantaneous screw corresponding to any eighth impulsive screw and *vice versâ*.

A precisely similar method of proof will give us the following theorem:—

If a rigid body be in position of stable equilibrium under the influence of a system of forces which have a potential, and if the twists about seven given screws evoke wrenches about seven other given screws, then, without further information about the forces, we shall be able to determine the screw on which a wrench is evoked by a twist about any eighth screw.

We may present the results of the present section in another form. We must conceive two corresponding systems of screws, of which the correspondence is completely established, when, to any seven screws regarded as belonging to one system, the seven corresponding screws in the other system are known. To every screw in space viewed as belonging to one system will correspond another screw viewed as belonging to the other system. Six screws can be found, each of which coincides with its correspondent. To a screw system of the nth order and mth degree in one system will correspond a screw system of the nth order and mth degree in the other system.

We add here a few examples to illustrate the use which may be made of screw co-ordinates.

239. Theorem.

When an impulsive force acts upon a free quiescent rigid body, the directions of the force and of the instantaneous screw are parallel to a pair of conjugate diameters in the momental ellipsoid.

Let $\eta_1, \ldots \eta_6$ be the co-ordinates of the force referred to the absolute principal screws of inertia, then (§ 35)

$$(\eta_1 + \eta_2)^2 + (\eta_3 + \eta_4)^2 + (\eta_5 + \eta_6)^2 = 1,$$

and from (§ 41) it follows that the direction cosines of η with respect to the principal axes through the centre of inertia are

$$(\eta_1 + \eta_2), \quad (\eta_3 + \eta_4), \quad (\eta_5 + \eta_6).$$

If a, b, c be the radii of gyration, then the instantaneous screw corresponding to η has for co-ordinates

$$+\frac{\eta_1}{a}, \quad -\frac{\eta_2}{a}, \quad +\frac{\eta_3}{b}, \quad -\frac{\eta_4}{b}, \quad +\frac{\eta_5}{c}, \quad -\frac{\eta_6}{c}.$$

The condition that η and its instantaneous screw shall be parallel to a pair of conjugate diameters of the momental ellipsoid is

$$a^2(\eta_1+\eta_2)\frac{\eta_1-\eta_2}{a}+b^2(\eta_3+\eta_4)\frac{\eta_3-\eta_4}{b}+c^2(\eta_5-\eta_6)\frac{\eta_5-\eta_6}{c}=0;$$

or

$$\Sigma p_1\eta_1^2 = p_\eta = 0.$$

But if the impulsive wrench on η be a force, then the pitch of η is zero, whence the theorem is proved.

240. Theorem.

When an impulsive wrench acting on a free rigid body produces an instantaneous rotation, the axis of the rotation must be perpendicular to the impulsive screw.

Let $\eta_1, \ldots \eta_6$ be the axis of the rotation, then

$$\Sigma p_1\eta_1^2 = 0,$$

or

$$a(\eta_1-\eta_2)(\eta_1+\eta_2)+b(\eta_3-\eta_4)(\eta_3+\eta_4)+c(\eta_5-\eta_6)(\eta_5+\eta_6)=0,$$

whence the screw of which the co-ordinates are $+a\eta_1$, $-a\eta_2$, $+b\eta_3, \ldots$ is perpendicular to η, and the theorem is proved.

From this theorem, and the last, we infer that, when an impulsive *force* acting on a rigid body produces an instantaneous *rotation*, the direction of the force, and the axis of the rotation, are parallel to the principal axes of a section of the momental ellipsoid.

241. Principal Axis.

If η be a principal axis of a rigid body, it is required to prove that

$$\Sigma p_1^3\eta_1^2 = 0,$$

reference being made to the absolute principal screws of inertia.

For in this case a force along a line θ intersecting η, compounded with a couple in a plane perpendicular to η, must constitute an impulsive wrench to which η corresponds as an instantaneous screw, whence we deduce (§ 120), h and k being the same for each coordinate,

$$\theta_1 = \frac{h}{p_1}\frac{dR}{d\eta_1} + kp_1\eta_1,$$

$$\ldots\ldots\ldots\ldots\ldots\ldots\ldots$$

$$\theta_6 = \frac{h}{p_6}\frac{dR}{d\eta_6} + kp_6\eta_6.$$

Expressing the condition that $p_\theta = 0$, we have

$$k^2 \Sigma p_1{}^3 \eta_1{}^2 + 2hk\Sigma p_1 \eta_1 \frac{dR}{d\eta_1} + h^2 \Sigma \frac{1}{p_1}\left(\frac{dR}{d\eta_1}\right)^2 = 0;$$

but we have already seen (§ 131) that the two last terms of this equation are zero, whence the required theorem is demonstrated.

The formula we have just proved may be written in the form

$$\Sigma p_1 \,.\, p_1\eta_1 \,.\, p_1\eta_1 = 0.$$

This shows that if the body were free, then an impulsive *force* suitably placed would make the body commence to rotate about η. Whence we have the following theorem*:—

A rigid body previously in unconstrained equilibrium in free space is supposed to be set in motion by a single impulsive force; if the initial axis of twist velocity be a principal axis of the body, the initial motion is a pure rotation, and conversely.

It may also be asked at what point of the body one of the three principal axes coincides with η? This point is the intersection of θ and η. To determine the co-ordinates of θ it is only necessary to find the relation between h and k, and this is obtained by expressing the condition that θ is reciprocal to η, whence we deduce

$$2h + ku_\eta{}^2 = 0.$$

Thus θ is known, and the required point is determined. If the body be fixed at this point, and then receive the impulsive couple perpendicular to η, the instantaneous reaction of the point will be directed along θ.

242. Harmonic Screws.

We shall conclude by stating for the sixth order the results which are included as particular cases of the general theorems in Chapter IX.

If a perfectly free rigid body be in equilibrium under the influence of a conservative system of forces, then six screws can generally be found such that each pair are conjugate screws of inertia, as well as conjugate screws of the potential, and these six screws are called harmonic screws. If the body be displaced from its position of equilibrium by a twist of small amplitude about a harmonic screw, and if the body further receive a small initial twisting motion about the same screw, then the body will continue for ever to perform small twist oscillations about that screw. And, more generally, whatever be the initial circumstances, the movement of the body is compounded of twist oscillations about the six harmonic screws.

* Townsend, *Educational Times Reprint*, Vol. XXI. p. 107.

CHAPTER XIX.

HOMOGRAPHIC SCREW SYSTEMS*.

243. Introduction.

Several of the most important parts of the *Theory of Screws* can be embraced in a more general theory. I propose in the present chapter to sketch this general theory. It will be found to have points of connexion with the modern higher geometry; in particular the theory of Homographic Screws is specially connected with the general theory of correspondence. I believe it will be of some interest to show how these abstract geometrical theories may be illustrated by dynamics.

244. On Plane Homographic Systems.

It may be convenient first to recite the leading principle of the purely geometrical theory of homography. We have already had to mention a special case in the Introduction.

Let α be any point in a plane, and let β be a corresponding point. Let us further suppose that the correspondence is of the one-to-one type, so that when one α is given then one β is known, when one β is given then it is the correspondent of a single α. The relation is not generally interchangeable. Only in very special circumstances will it be true that β, regarded as in the first system, will correspond to α in the second system.

The general relation between the points α and β can be expressed by the following equations, where α_1, α_2, α_3 are the ordinary trilinear co-ordinates of α, and β_1, β_2, β_3, the co-ordinates of β,

$$\beta_1 = (11)\,\alpha_1 + (12)\,\alpha_2 + (13)\,\alpha_3,$$
$$\beta_2 = (21)\,\alpha_1 + (22)\,\alpha_2 + (23)\,\alpha_3,$$
$$\beta_3 = (31)\,\alpha_1 + (32)\,\alpha_2 + (33)\,\alpha_3.$$

In these expressions (11), (12), &c., are the constants defining the particular character of the homographic system.

* *Proc. Roy. Irish Acad.* Ser. II. Vol. III. p. 435 (1881).

There are in general three points, which coincide with their correspondents. These are found by putting

$$\beta_1 = \rho\alpha_1; \quad \beta_2 = \rho\alpha_2; \quad \beta_3 = \rho\alpha_3.$$

Introducing these values, and eliminating α_1, α_2, α_3, we obtain the following equation for ρ:—

$$0 = \begin{vmatrix} (11) - \rho, & (12), & (13) \\ (21), & (22) - \rho, & (23) \\ (31), & (32), & (33) - \rho \end{vmatrix}$$

If we choose these three points of the vertices of the triangle of reference, the equations relating y with x assume the simple form,

$$\beta_1 = f_1\alpha_1; \quad \beta_2 = f_2\alpha_2; \quad \beta_3 = f_3\alpha_3,$$

where f_1, f_2, f_3 are three new constants.

245. Homographic Screw Systems.

Given one screw α, it is easy to conceive that another screw β corresponding thereto shall be also determined. We may, for example, suppose that the co-ordinates of β (§ 34) shall be given functions of those of α. We might imagine a geometrical construction by the aid of fixed lines or curves by which, when an α is given, the corresponding β shall be forthwith known: again, we may imagine a connexion involving dynamical conceptions such as that, when α is the seat of an impulsive wrench, β is the instantaneous screw about which the body begins to twist.

As α moves about, so the corresponding screw β will change its position and thus two *corresponding screw systems* are generated. Regarding the connexion between the two systems from the analytical point of view, the co-ordinates of α and β will be connected by certain equations. If it be invariably true that a single screw β corresponds to a single screw α, and that conversely a single screw α corresponds to a single screw β; then *the two systems of screws are said to be homographic.*

A screw α in the first system has one corresponding screw β in the second system; so also to β in the first system corresponds one screw α' in the second system. It will generally be impossible for α and α' to coincide, but cases may arise in which they do coincide, and these will be discussed further on.

246. Relations among the Co-ordinates.

From the fundamental property of two homographic screw systems the co-ordinates of β must be expressed by six equations of the type—

$$\beta_1 = f_1(\alpha_1, \ldots \alpha_6)$$

$$\ldots\ldots\ldots\ldots\ldots\ldots\ldots$$

$$\beta_6 = f_6(\alpha_1, \ldots \alpha_6).$$

If these six equations be solved for $\alpha_1, \ldots \alpha_6$ we must have

$$\alpha_1 = F_1(\beta_1, \ldots \beta_6)$$
$$\ldots\ldots\ldots\ldots\ldots\ldots$$
$$\alpha_6 = F_6(\beta_1, \ldots \beta_6).$$

As a single α is to correspond to a single β, and *vice versâ*, these equations must be linear: whence we have the following important result:—

In two homographic screw systems the co-ordinates of a screw in one system are linear functions with constant coefficients of the co-ordinates of the corresponding screw in the other system.

If we denote the constant coefficients by the notation (11), (22), &c., then we have the following system of equations:—

$$\beta_1 = (11)\,\alpha_1 + (12)\,\alpha_2 + (13)\,\alpha_3 + (14)\,\alpha_4 + (15)\,\alpha_5 + (16)\,\alpha_6,$$
$$\beta_2 = (21)\,\alpha_1 + (22)\,\alpha_2 + (23)\,\alpha_3 + (24)\,\alpha_4 + (25)\,\alpha_5 + (26)\,\alpha_6,$$
$$\ldots\ldots\ldots\ldots\ldots\ldots\ldots\ldots\ldots\ldots\ldots\ldots\ldots\ldots\ldots\ldots\ldots\ldots$$
$$\beta_6 = (61)\,\alpha_1 + (62)\,\alpha_2 + (63)\,\alpha_3 + (64)\,\alpha_4 + (65)\,\alpha_5 + (66)\,\alpha_6.$$

247. The Double Screws.

It is now easy to show that there are in general six screws which coincide with their corresponding screws; for if $\beta_1 = \rho\alpha_1$, $\beta_2 = \rho\alpha_2$, &c., we obtain an equation of the sixth degree for the determination of ρ. We therefore have the following result:—

In two homographic screw systems six screws can in general be found, each of which regarded as a screw in either system coincides with its correspondent in the other system.

248. The Seven Pairs.

In two homographic rows of points we have the anharmonic ratio of any four points equal to that of their correspondents. In the case of two homographic screw systems we have a set of eight screws in one of the systems specially related to the corresponding eight screws in the other system.

We first remark that, given seven pairs of corresponding screws in the two systems, then the screw corresponding to any other given screw is determined. For from the six equations just written by substitution of known values of $\alpha_1, \ldots \alpha_6$ and $\beta_1, \ldots \beta_6$, we can deduce six equations between (11), (12), &c. As, however, the co-ordinates are homogeneous and their ratios are alone involved, we can use only the ratios of the equations so that each pair of screws gives five relations between the 36 quantities (11), (12), &c. The seven pairs thus give 35 relations which suffice to determine linearly the ratios of the coefficients. The screw β corresponding to any other screw α is completely determined; we have therefore proved that

When seven corresponding pairs of screws are given, the two homographic screw systems are completely determined.

A perfectly general way of conceiving two homographic screw systems may be thus stated:—Decompose a wrench of given intensity on a screw α into wrenches on six arbitrary screws. Multiply the intensity of each of the six component wrenches by an arbitrary constant; construct the wrench on the screw β which is the resultant of the six components thus modified; then as α moves into every position in space, and has every fluctuation in pitch, so will β trace out the homographic screw system.

It is easily seen that in this statement we might have spoken of twist velocities instead of wrenches.

249. Homographic n-systems.

The seven pairs of screws of which the two systems are defined cannot be always chosen arbitrarily. If, for example, three of the screws were co-cylindroidal, then the three corresponding screws must be co-cylindroidal, and can only be chosen arbitrarily subject to this imperative restriction. More generally we shall now prove that if any $n+1$ screws belong to an n-system (§ 69), then the $n+1$ corresponding screws will also belong to an n-system. If $n+1$ screws belong to an n-system it will always be possible to determine the intensities of certain wrenches on the $n+1$ screws which when compounded together will equilibrate. The conditions that this shall be possible are easily expressed. Take, for example, $n=3$, and suppose that the four screws α, β, γ, δ are such that suitable wrenches on them, or twist velocities about them, neutralize. It is then obvious (§ 76) that each of the determinants must vanish which is formed by taking four columns from the expression

$$\left|\begin{array}{cccccc} \alpha_1, & \alpha_2, & \alpha_3, & \alpha_4, & \alpha_5, & \alpha_6 \\ \beta_1, & \beta_2, & \beta_3, & \beta_4, & \beta_5, & \beta_6 \\ \gamma_1, & \gamma_2, & \gamma_3, & \gamma_4, & \gamma_5, & \gamma_6 \\ \delta_1, & \delta_2, & \delta_3, & \delta_4, & \delta_5, & \delta_6 \end{array}\right|.$$

It is, however, easy to see that these determinants will equally vanish for the corresponding screws in the homographic system; for if we take as screws of reference the six common screws of the two systems, then we have at once for the co-ordinates of the screw corresponding to α

$$(11)\,\alpha_1, \quad (22)\,\alpha_2, \quad (33)\,\alpha_3, \quad (44)\,\alpha_4, \quad (55)\,\alpha_5, \quad (66)\,\alpha_6.$$

When these substitutions are made in the determinants it is plain that they still vanish; we hence have the important result that

The screws corresponding homographically to the screws of an n-system form another n-system.

Thus to the screws on a cylindroid will correspond the screws on a cylindroid. It is, however, important to notice that two reciprocal screws have not in general two reciprocal screws for their correspondents. We thus see that while two reciprocal screw systems of the nth and $(6-n)$th orders respectively have as correspondents systems of the same orders, yet that their connexion as reciprocals is divorced by the homographic transformation.

Reciprocity is not, therefore, an invariantive attribute of screws or screw systems. There are, however, certain functions of eight screws analogous to anharmonic ratios which are invariants. These functions are of considerable interest, and they are not without physical significance.

250. Analogy to Anharmonic Ratio.

We have already (§ 230) discussed the important function of six screws which is called the *Sexiant.* This function is most concisely written as the determinant $(\alpha_1\beta_2\gamma_3\delta_4\epsilon_5\zeta_6)$ where α, β, γ, δ, ϵ, ζ, are the screws. In Sylvester's language we may speak of the six screws as being in *involution* when their sexiant vanishes. Under these circumstances six wrenches on the six screws can equilibrate; the six screws all belong to a 5-system, and they possess one common reciprocal. In the case of eight screws we may use a very concise notation; thus $\overline{12}$ will denote the sexiant of the six screws obtained by *leaving out* screws 1 and 2. It will now be easy to show that functions of the following form are invariants, *i.e.* the same in both systems:—

$$\frac{\overline{12}\,.\,\overline{34}}{\overline{13}\,.\,\overline{24}}.$$

It is in the first place obvious that as the co-ordinates of each screw enter to the same degree in the numerator and the denominator, no embarrassment can arise from the arbitrary common factor with which the six co-ordinates of each screw may be affected. In the second place it is plain that if we replace each of the co-ordinates by those of the corresponding screw, the function will still remain unaltered, as all the factors (11), (22), &c., will divide out. We thus see that the function just written will be absolutely unaltered when each screw is changed into its corresponding screw.

By the aid of these invariant functions it is easy, when seven pairs of screws are given, to construct the screw corresponding to any given eighth screw. We may solve this problem in various ways. One of the simplest will be to write the five invariants

$$\frac{\overline{12}.\overline{38}}{\overline{13}.\overline{28}},\ \frac{\overline{13}.\overline{48}}{\overline{14}.\overline{38}},\ \frac{\overline{14}.\overline{58}}{\overline{15}.\overline{48}},\ \frac{\overline{15}.\overline{68}}{\overline{16}.\overline{58}},\ \frac{\overline{16}.\overline{78}}{\overline{17}.\overline{68}}.$$

These can be computed from the given eight screws of one system; hence we have five linear equations to determine the ratios of the coefficients of the required eighth screw of the other system.

It would seem that of all the invariants of eight screws, five alone can be independent. These five invariants are attributes of the eight-screw system, in the same way that the anharmonic ratio is an attribute of four collinear points.

251. A Physical Correspondence.

The invariants are also easily illustrated by considerations of a mechanical nature. To a wrench on one screw corresponds a twist on the corresponding screw, and the ratio of the intensities of the wrench and twist is to be independent of those intensities. We may take a particular case to illustrate the argument:—Suppose a free rigid body to be at rest. If that body be acted upon by an impulsive system of forces, those forces will constitute a wrench on a certain screw α. In consequence of these forces the body will commence to move, and its instantaneous motion cannot be different from a twist velocity about some other screw β. To one screw α will correspond one screw β, and (since the body is perfectly free) to one screw β will correspond one screw α. It follows, from the definition of homography, that as α moves over every screw in space, β will trace out an homographic system.... From the laws of motion it will follow, that if F be the intensity of the impulsive wrench, and if V be the twist velocity which that wrench evokes, then $F \div V$ will be independent of F and V, though, of course, it is not independent of the actual position of α and β.

252. Impulsive and Instantaneous Systems.

It is known (§ 230) that when seven wrenches equilibrate (or when seven twist velocities neutralize), the intensity of the wrench (or the twist velocity) on any one screw must be proportional to the sexiant of the six non-corresponding screws.

Let $F_{18}, F_{28}, \ldots F_{78}$ be the intensities of seven impulsive wrenches on the screws 1, 2, ... 7, which equilibrate, then we must have

$$\frac{F_{18}}{\overline{18}} = \frac{F_{28}}{\overline{28}} = \ldots = \frac{F_{78}}{\overline{78}}.$$

Similarly, by omitting the first screw, we can have seven impulsive wrenches which equilibrate, where

$$\frac{F_{12}}{\overline{12}} = \frac{F_{13}}{\overline{13}} = \frac{F_{14}}{\overline{14}} = \dots = \frac{F_{18}}{\overline{18}};$$

hence we have

$$\frac{\overline{12} \,.\, \overline{38}}{\overline{13} \,.\, \overline{28}} = \frac{F_{12} \,.\, F_{38}}{F_{13} \,.\, F_{28}}.$$

Let the instantaneous twist velocity corresponding to F_{18} be denoted by V_{18}, then, as when seven wrenches equilibrate, the seven corresponding twist velocities must also equilibrate, we must have in the corresponding system,

$$\frac{\overline{12} \,.\, \overline{38}}{\overline{13} \,.\, \overline{28}} = \frac{V_{12} V_{38}}{V_{13} V_{28}}.$$

But we must have the twist velocity proportional to the impulsive intensity; hence, from the second pair of screws we have

$$F_{28} : V_{28} :: F_{12} : V_{12},$$

and from the third pair,

$$F_{38} : V_{38} :: F_{13} : V_{13};$$

hence we deduce

$$\frac{V_{12} \,.\, V_{38}}{V_{13} \,.\, V_{28}} = \frac{F_{12} \,.\, F_{38}}{F_{13} \,.\, F_{28}},$$

and, consequently, the function of the eight impulsive screws

$$\frac{\overline{12} \,.\, \overline{38}}{\overline{13} \,.\, \overline{28}},$$

must be identical with the same function of the instantaneous screws.

It should, however, be remarked, that the impulsive and instantaneous screws do not exhibit the most general type of two homographic systems. A more special type of homography, and one of very great interest, characterizes the two sets of screws referred to.

253. Special type of Homography.

If the general linear transformation, which changes each screw α into its correspondent θ, be specialized by the restriction that the co-ordinates of θ are given by the equations

$$\theta_1 = \frac{1}{p_1} \frac{dU}{d\alpha_1},$$

$$\theta_2 = \frac{1}{p_2} \frac{dU}{d\alpha_2},$$

$$\dots\dots\dots\dots\dots$$

$$\theta_6 = \frac{1}{p_6} \frac{dU}{d\alpha_6},$$

where U is any homogeneous function of the second order in $\alpha_1, \ldots \alpha_6$, and where $p_1, \ldots p_6$ are the pitches of the screws of reference, then the two systems are related by the special type of homography to which I have referred.

The fundamental property of the two special homographic systems is thus stated:—

Let α and β be any two screws, and let θ and ϕ be their correspondents, then, when α is reciprocal to ϕ, β will be reciprocal to θ.

We may, without loss of generality, assume that the screws of reference are co-reciprocal, and in this case the condition that β and θ shall be co-reciprocal is

$$p_1\beta_1\theta_1 + p_2\beta_2\theta_2 - \ldots + p_6\beta_6\theta_6 = 0;$$

but by substituting for $\theta_1, \ldots \theta_6$, this condition reduces to

$$\beta_1 \frac{dU}{d\alpha} + \ldots + \beta_6 \frac{dU}{d\alpha} = 0.$$

Similarly, the condition that α and ϕ shall be reciprocal is

$$\alpha_1 \frac{dU}{d\beta_1} + \ldots + \alpha_6 \frac{dU}{d\beta_6} = 0.$$

It is obvious that as U is a homogeneous function of the second degree, these two conditions are identical, and the required property has been proved.

254. Reduction to a Canonical form.

It is easily shown that by suitable choice of the screws of reference the function U may, in general, be reduced to the sum of six square terms. We now proceed to show that this reduction is generally possible, while still retaining six co-reciprocals for the screws of reference.

The pitch p_α of the screw α is given by the equation (§ 38),

$$p_\alpha = p_1\alpha_1^2 + \ldots + p_6\alpha_6^2:$$

the six screws of reference being co-reciprocals, the function p_α must retain the same form after the transformation of the axes. The discriminant of the function

$$U + \lambda p_\alpha$$

equated to zero will give six values of λ; these values of λ will determine the coefficients of U in the required form. I do not, however, enter further into the discussion of this question, which belongs to the general theory of linear transformations.

The transformation having been effected, an important result is immediately deduced. Let the transformed function be denoted by

$$(11)\,\alpha_1^2 + \ldots + (66)\,\alpha_6^2,$$

then we have

$$\beta_1 = \frac{1}{p_1}(11)\,\alpha_1,$$

$$\ldots\ldots\ldots\ldots\ldots\ldots$$

$$\beta_6 = \frac{1}{p_6}(66)\,\alpha_6\,;$$

whence it appears that the six screws of reference are the common screws of the two systems. We thus find that in this special case of homography

The six common screws of the two systems are co-reciprocal.

The correspondence between impulsive screws and instantaneous screws is a particular case of the type here referred to. The six common screws of the two systems are therefore what we have called the *principal screws of inertia*, and they are co-reciprocal.

255. Correspondence of a Screw and a system.

We shall sometimes have cases in which each screw of a system corresponds not to a single screw but to a system of screws. For the sake of illustration, suppose the case of a quiescent rigid body with two degrees of freedom and let this receive an impulsive wrench on some screw situated anywhere in space. The movement which the body can accept is limited. It can, indeed, only twist about one of the singly infinite number of screws, which constitute a cylindroid. To any screw in space will correspond one screw on the cylindroid. But will it be correct to say, that to one screw on the cylindroid corresponds one screw in space? The fact is, that there are a quadruply infinite number of screws, an impulsive wrench on any one of which will make the body choose the same screw on the cylindroid for its instantaneous movement. The relation of this quadruply infinite group is clearly exhibited in the present theory. It is shown in § 128 that, given a screw α on the cylindroid, there is, in general one, but only one screw θ on the cylindroid, an impulsive wrench on which will make the body commence to twist about α. It is further shown that any screw whatever which fulfils the single condition of being reciprocal to a single specified screw on the cylindroid possesses the same property. The screws corresponding to α thus form a five-system. The correspondence at present before us may therefore be enunciated in the following general manner.

To one screw in space corresponds one screw on the cylindroid, and to one screw on the cylindroid corresponds a five-system in space.

256. Correspondence of m and n systems.

We may look at the matter in a more general manner. Consider an m-system (A) of screws, and an n-system (B) ($m > n$). (If we make $m = 6$ and $n = 2$, this system includes the system we have been just discussing.) To one screw in A will correspond one screw in B, but to one screw in B will correspond, not a single screw in A, but an $(m + 1 - n)$-system of screws.

If $m = n$, we find that one screw of one system corresponds to one screw of the other system. Thus, if $m = n = 2$, we have a pair of cylindroids, and one screw on one cylindroid corresponds to one screw on the other. If $m = 3$, and $n = 2$, we see that to each screw on the cylindroid will correspond a whole cylindroid of screws belonging to the three-system. For example, if a body have freedom of the second order and a screw be indicated on the cylindroid which defines the freedom, then a whole cylindroid full of screws can always be chosen from any three-system, an impulsive wrench on any one of which will make the body commence to twist about the indicated screw.

257. Screws common to the two systems.

The property of the screws common to the two homographic systems will of course require some modification when we are only considering an m-system and an n-system. Let us take the case of a three-system on the one hand, and a six-system, or all the screws in space, on the other hand. To each screw α of the three-system A must correspond, a four-system, B, so that a cone of the screws of this four-system can be drawn through every point in space. It is interesting to note that one screw β can be found, which, besides belonging to B, belongs also to A. Take any two screws reciprocal to B, and any three screws reciprocal to A, then the single screw β, which is reciprocal to the five screws thus found, belongs to both A and B. We thus see that to each screw α of A, one corresponding screw in the same system can be determined. The result just arrived at can be similarly shown generally, and thus we find that when every screw in space corresponds to a screw of an n-system, then each screw of the n-system will correspond to a $(7 - n)$-system, and among the screws of this system one can always be found which lies on the original n-system.

As a mechanical illustration of this result we may refer to the theorem (§ 96), that if a rigid body has freedom of the nth order, then, no matter what be the system of forces which act upon it, we may in general combine the resultant wrench with certain reactions of the constraints, so as to produce a wrench *on a screw of the n-system* which defines the freedom of the body, and this wrench will be dynamically equivalent to the given system of forces.

258. Corresponding Screws defined by Equations.

It is easy to state the matter analytically, and for convenience we shall take a three-system, though it will be obvious that the process is quite general.

Of the six screws of reference, let three screws be chosen on the three-system, then the co-ordinates of any screw on that system will be α_1, α_2, α_3, the other three co-ordinates being equal to zero. The co-ordinates of the corresponding screw β must be indeterminate, for any screw of a four-system will correspond to β. This provision is secured by β_4, β_5, β_6 remaining quite arbitrary, while we have for β_1, β_2, β_3 the definite values,

$$\beta_1 = (11)\,\alpha_1 + (12)\,\alpha_2 + (13)\,\alpha_3,$$

$$\beta_2 = (21)\,\alpha_1 + (22)\,\alpha_2 + (23)\,\alpha_3,$$

$$\beta_3 = (31)\,\alpha_1 + (32)\,\alpha_2 + (33)\,\alpha_3.$$

If we take β_4, β_5, β_6 all zero, then the values of β_1, β_2, β_3, just written, give the co-ordinates of the special screw belonging to the three-system, which is among those which correspond to α.

As α moves over the three-system, so will the other screw of that system which corresponds thereto. There will, however, be three cases in which the two screws coincide; these are found at once by making

$$\beta_1 = \rho\alpha_1; \quad \beta_2 = \rho\alpha_2; \quad \beta_3 = \rho\alpha_3,$$

whence we obtain a cubic for ρ.

It is thus seen that generally n screws can be found on an n-system, so that each screw shall coincide with its correspondent. As a dynamical illustration we may give the important theorem, that when a rigid body has n degrees of freedom, then n screws can always be found about any one of which the body will commence to twist when it receives an impulsive wrench on the same screw. These screws are of course the principal screws of inertia (§ 84).

259. Generalization of Anharmonic Ratio.

We have already seen the anharmonic equality between four screws on a cylindroid, and the four corresponding screws; we have also shown a *quasi* anharmonic equality between any eight screws in space and their correspondents. More generally, any $n+2$ screws of an n-system are connected with their $n+2$ correspondents, by relations which are analogous to anharmonic properties. The invariants are not generally so simple as in the eight-screw case, but we may state them, at all events, for the case of $n = 3$.

Five screws belonging to a three-system, and their five correspondents are so related, that when nine are given, the tenth is immediately determined; for this two data are required, that being the number required to specify a screw already known to belong to a given three-system.

We may, as before, denote by $\overline{12}$ the condition that the screws 3, 4, 5 shall be co-cylindroidal. This, indeed, requires no less than four distinct conditions, yet, as pointed out (§ 76), functions can be found whose evanescence will supply all that is necessary. Nor need this cause any surprise, when it is remembered that the evanescence of the sine of an angle between two lines contains the two conditions necessary that the direction cosines are identical. The function

$$\frac{\overline{12}\,.\,\overline{34}}{\overline{13}\,.\,\overline{24}}$$

can then be shown to be an invariant which retains its value unaltered when we pass from one set of five screws in a three-system to the corresponding set in the other system.

CHAPTER XX.

EMANANTS AND PITCH INVARIANTS.

260. The Dyname.

If we wish to speak of a magnitude which may be a twist or a wrench or a twist velocity it is convenient to employ the word *Dyname* used by Plücker* and by other writers. The Dyname α is completely expressed by its components $\alpha_1, \ldots \alpha_6$ on the six screws of reference. These six quantities are quite independent. They may be considered as the co-ordinates of the Dyname.

Let α' be the intensity of the Dyname on α; then α' is a factor in each of $\alpha_1, \ldots \alpha_6$, and if the Dyname be replaced by another on the same screw α, but of intensity $x\alpha'$, the co-ordinates of this new Dyname will be $x\alpha_1, \ldots x\alpha_6$.

Let β be a second Dyname on another screw quite arbitrary as to its position and as to its intensity β'. Let the co-ordinates of β, referred to the same screws of reference, be $\beta_1, \ldots \beta_6$. If we suppose a Dyname of intensity $y\beta'$ on the screw β, then its co-ordinates will be $y\beta_1, \ldots y\beta_6$. Let us now compound together the two Dynames of intensities $x\alpha'$ and $y\beta'$ on the screws α and β. They will, according to the laws for the composition of twists and wrenches (§ 14), form a single Dyname on a third screw lying on the same cylindroid as α and β. The position of the resultant screw is such that it divides the angle between α and β into parts whose sines have the ratio of y to x. The intensity of the resultant Dyname is also determined (as in the parallelogram of force) to be the diagonal where x and y are the sides, and the angle between them is the angle between α and β. It is important to notice that in the determination of this resultant the screws to which the co-ordinates are referred bear no part; the position of the resultant Dyname on the cylindroid as well as its intensity each depend solely upon the two original Dynames, and on the numerical magnitudes x and y.

* Plücker, 'Fundamental views regarding Mechanics,' *Phil. Trans.* 1866, Vol. CLVI. pp. 361—380.

We have now to form the co-ordinates of the resulting Dyname, or its components when decomposed along the six screws of reference. The first Dyname has a component of intensity $x\alpha_1$ on the first screw; and as the second Dyname has a component $y\beta_1$, it follows that the sum of these two must be the component of the resultant. Thus we have for the co-ordinates of the resultant Dyname the expressions

$$x\alpha_1 + y\beta_1, \ldots x\alpha_6 + y\beta_6.$$

261. Emanants.

Let us suppose that without in any particular altering either of the Dynames α and β we make a complete change of the six screws of reference. Let the co-ordinates of α with regard to these new screws be $\lambda_1, \ldots \lambda_6$, and those of β be $\mu_1, \ldots \mu_6$. Precisely the same argument as has just been used will show that the composition of the Dynames $x\alpha'$ and $y\beta'$ will produce a Dyname whose co-ordinates are $x\lambda_1 + y\mu_1, \ldots x\lambda_6 + y\mu_6$. We thus see that the Dyname defined by the co-ordinates $x\alpha_1 + y\beta_1, \ldots x\alpha_6 + y\beta_6$, referred to the first group of reference screws is absolutely the same Dyname as that defined by the co-ordinates $x\lambda_1 + y\mu_1, \ldots x\lambda_6 + y\mu_6$ referred to the second group of reference screws, and that this must remain true for every value of x and y.

In general, let $\theta_1, \ldots \theta_6$ denote the co-ordinates of a Dyname in the first system, and $\phi_1, \ldots \phi_6$ denote those of the same Dyname in the second system. Let $f(\theta_1, \ldots \theta_6)$ denote any homogeneous function of the first Dyname, and let $F(\phi_1, \ldots \phi_6)$ be the same function transformed to the other screws of reference. Then we have

$$f(\theta_1, \ldots \theta_6) = F(\phi_1, \ldots \phi_6)$$

as an identical equation which must be satisfied whenever the Dyname defined by $\theta_1, \ldots \theta_6$ is the same as that defined by $\phi_1, \ldots \phi_6$. We must therefore have

$$f(x\alpha_1 + y\beta_1, \ldots x\alpha_6 + y\beta_6) = F(x\lambda_1 + y\mu_1, \ldots x\lambda_6 + y\mu_6).$$

These expressions being homogeneous, they may each be developed in ascending powers of $\frac{y}{x}$. But as the identity must subsist for every value of this ratio, we must have the coefficients of the various powers equal on both sides. The expression of this identity gives us a series of equations which are all included in the form *

$$\left(\beta_1 \frac{d}{d\alpha_1} + \ldots + \beta_6 \frac{d}{d\alpha_6}\right)^n f = \left(\mu_1 \frac{d}{d\lambda_1} + \ldots + \mu_6 \frac{d}{d\lambda_6}\right)^n F.$$

* See *Proceedings Roy. Irish Acad.*, Ser. II. Vol. III.; *Science*, p. 661 (1882).

The functions thus arising are well known as "emanants" in the theory of modern algebra. The cases which we shall consider are those of $n = 1$ and $n = 2$. In the former case the emanant may be written

$$\beta_1 \frac{df}{d\alpha_1} + \ldots + \beta_6 \frac{df}{d\alpha_6}.$$

262. Angle between Two Screws.

It will of course be understood that f is perfectly arbitrary, but results of interest may be most reasonably anticipated when f has been chosen with special relevancy to the Dyname itself, as distinguished from the influence due merely to the screws of reference. We shall first take for f the square of the intensity of the Dyname, the expression for which is found (§ 35) to be

$$R = \alpha_1^2 + \ldots + \alpha_6^2 + 2\alpha_1\alpha_2(12) + \ldots,$$

where (12) denotes the cosine of the angle between the first and second screws of reference, which are here taken to be perfectly arbitrary. The second group of reference screws we shall take in a special form. They are to be a canonical co-reciprocal system, so that

$$R = (\lambda_1 + \lambda_2)^2 + (\lambda_3 + \lambda_4)^2 + (\lambda_5 + \lambda_6)^2.$$

Introducing these values, we have, as the first emanant,

$$\Sigma\alpha_1\beta_1 + \Sigma(\alpha_1\beta_2 + \alpha_2\beta_1)(12)$$
$$= (\mu_1 + \mu_2)(\lambda_1 + \lambda_2) + (\mu_3 + \mu_4)(\lambda_3 + \lambda_4) + (\mu_5 + \mu_6)(\lambda_5 + \lambda_6);$$

but in the latter form the expression obviously denotes the cosine of the angle between α and β where the intensities are both unity; hence, *whatever be the screws of reference*, we must have for the cosine of the angle between the two screws the result

$$\Sigma\alpha_1\beta_1 + \Sigma(\alpha_1\beta_2 + \alpha_2\beta_1)(12).$$

263. Screws at Right Angles.

In general we have the following formula for the cosine of the angle between two Dynames multiplied into the product of their intensities:—

$$\tfrac{1}{2}\theta_1 \frac{dR}{d\alpha_1} + \tfrac{1}{2}\theta_2 \frac{dR}{d\alpha_2} + \ldots + \tfrac{1}{2}\theta_6 \frac{dR}{d\alpha_6}.$$

This expression, equated to zero, gives the condition that the two Dynames be rectangular.

If three screws, α, β, γ, are all parallel to the same plane, and if θ be a screw normal to that plane, then we must have

$$\theta_1 \frac{dR}{d\alpha_1} \dots + \theta_6 \frac{dR}{d\alpha_6} = 0,$$

$$\theta_1 \frac{dR}{d\beta_1} \dots + \theta_6 \frac{dR}{d\beta_6} = 0,$$

$$\theta_1 \frac{dR}{d\gamma_1} \dots + \theta_6 \frac{dR}{d\gamma_6} = 0.$$

264. Conditions that Three Screws shall be parallel to a Plane.

Since a screw of a three-system can be drawn parallel to any direction, it will be possible to make any three of the quantities $\theta_1, \dots \theta_6$ equal to zero. Hence, we have as the condition that the three screws, α, β, γ shall be all parallel to a plane the evanescence of all the determinants of the type

$$\begin{vmatrix} \frac{dR}{d\alpha_1}, & \frac{dR}{d\alpha_2}, & \frac{dR}{d\alpha_3} \\ \frac{dR}{d\beta_1}, & \frac{dR}{d\beta_2}, & \frac{dR}{d\beta_3} \\ \frac{dR}{d\gamma_1}, & \frac{dR}{d\gamma_2}, & \frac{dR}{d\gamma_3} \end{vmatrix}.$$

265. Screws on the same Axis.

The locus of the screws θ perpendicular to α is represented by the equation

$$\theta_1 \frac{dR}{d\alpha_1} \dots + \theta_6 \frac{dR}{d\alpha_6} = 0.$$

If we assume that the screws of reference are co-reciprocal, then the equation just written can only denote all the screws reciprocal to the one screw whose co-ordinates are

$$\frac{1}{p_1} \frac{dR}{d\alpha_1}, \dots \frac{1}{p_6} \frac{dR}{d\alpha_6}.$$

It is manifest that all the screws perpendicular to a given line cannot be reciprocal to a single screw unless the pitch of that screw be infinite, otherwise the condition

$$(p_\alpha + p_\theta) \cos \phi - d \sin \phi = 0$$

could not be fulfilled. We therefore see that the co-ordinates just written can only denote those of a screw of infinite pitch parallel to α.

If x be a variable parameter, then the co-ordinates

$$\alpha_1 + \frac{x}{4p_1}\frac{dR}{d\alpha_1}, \ldots \alpha_6 + \frac{x}{4p_6}\frac{dR}{d\alpha_6}$$

must denote a screw of variable pitch x on the same screw as α. We are thus conducted to a more general form of the results previously obtained (§ 47).

These expressions may be written

$$\alpha_1 + \frac{x}{2p_1}\cos\alpha_1,\ \alpha_2 + \frac{x}{2p_2}\cos\alpha_2, \ldots$$

where α_1, α_2, ... are the angles which α makes with the screws of reference.

266. A general Expression for the Virtual Coefficient.

We may also consider that function of the co-ordinates of a Dyname which, being always proportional to the pitch, becomes exactly equal to the pitch when the intensity is equal to unity. More generally, we may define the function to be equal to the pitch multiplied into the square of the intensity, and it is easy to assign a physical meaning to this function. It is half the work done in a twist against a wrench, on the same screw, where the amplitude of the twist is equal to the intensity of the wrench. Referred to *any* co-ordinates, we denote this function by V expressed in terms of $\lambda_1, \ldots \lambda_6$. If we express the same function by reference to six co-reciprocal axes with co-ordinates $\alpha_1, \ldots \alpha_6$, we have the result

$$p_1\alpha_1^2 + \ldots p_6\alpha_6^2 = V.$$

Forming now the first emanant, we have

$$2p_1\alpha_1\beta_1 + \ldots + 2p_6\alpha_6\beta_6 = \mu_1\frac{dV}{d\lambda_1} \ldots + \mu_6\frac{dV}{d\lambda_6};$$

but the expression on the left-hand side denotes the product of the two intensities into double the virtual coefficient of the two screws; hence the right-hand member must denote the same. If, therefore, *after the differentiations* we make the intensities equal to unity, we have for the virtual coefficient between two screws λ and μ referred to *any* screws of reference whatever one-half the expression

$$\mu_1\frac{dV}{d\lambda_1} \ldots + \mu_6\frac{dV}{d\lambda_6}.$$

Suppose, for instance, that λ is reciprocal to the first screw of reference, then

$$\frac{dV}{d\lambda_1} = 0.$$

This can be verified in the following manner. We have

$$V = p\lambda'^2,$$

$$\frac{dV}{d\lambda_1} = \lambda'^2 \frac{dp}{d\lambda_1} + 2\lambda' p \frac{d\lambda'}{d\lambda_1};$$

and, therefore, if λ be reciprocal to the first screw of reference, the formula to be proved is

$$\lambda'^2 \frac{dp}{d\lambda_1} + 2\lambda' p \frac{d\lambda'}{d\lambda_1} = 0.$$

A few words will be necessary on the geometrical signification of the differentiation involved. Suppose a Dyname λ be referred to six co-ordinate screws of absolute generality, and let us suppose that one of these co-ordinates, for instance λ_1, be permitted to vary, the corresponding situation of λ also changes, and considering each one of the co-ordinates in succession, we thus have six routes established along which λ will travel in correspondence with the growth of the appropriate co-ordinate. Each route is, of course, a ruled surface; but the conception of a surface is not alone adequate to express the route. We must also associate a linear magnitude with each generator of the surface, which is to denote the pitch of the corresponding screw. Taking λ and another screw on one of the routes, we can draw a cylindroid through these two screws. It will now be proved that this cylindroid is itself the locus in which λ moves, when the co-ordinate correlated thereto changes its value. Let θ be the screw arising from an increase in the co-ordinate λ_1; a wrench on θ of intensity θ'' has components of intensities $\theta_1'', \dots \theta_6''$. A wrench on λ has components $\lambda_1'', \dots \lambda_6''$. But from the nature of the case,

$$\frac{\theta_2''}{\lambda_2''} = \frac{\theta_3''}{\lambda_3''} \dots = \frac{\theta_6''}{\lambda_6''}.$$

If therefore θ'' be suitably chosen, we can make each of these ratios -1, so that when θ'' and λ'' are each resolved along the six screws of reference, all the components except θ_1'', $-\lambda_1''$ shall neutralize. But this can only be possible if the first reference screw lie on the cylindroid containing θ and λ. Hence we deduce the result that each of the six cylindroids must pass through the corresponding screw of reference; and thus we have a complete view of the route travelled by a screw in correspondence with the variation of one of its co-ordinates.

Let the six screws of reference be 1, 2, 3, 4, 5, 6. Form the cylindroid (λ, 1), and find that one screw η on this cylindroid which has with 2, 3, 4, 5, 6, a common reciprocal (§ 26). From a point O draw a pencil of four rays parallel to four screws on the cylindroid. Let OA be parallel to one of the principal screws; $O\lambda$ be parallel to λ, $O\eta$ to η, and Oh to the first screw of reference.

Let the angle AOh be denoted by A, the angle $AO\eta$ by B, and the angle $AO\lambda$ by ϕ. To find the component λ_1 we must decompose λ', a twist on λ, into two components, one on η, the other on the first screw of reference. The component on η can be resolved along the other five screws of reference, since the six form one system with a common reciprocal. If we denote by η' the component on η, we then have

$$\frac{\lambda'}{\sin(B-A)} = \frac{\lambda_1}{\sin(\phi - B)} = \frac{\eta'}{\sin(\phi - A)};$$

and if a and b be the pitches of the two principal screws on the cylindroid, we have for the pitch of λ the equation

$$p = a\cos^2\phi + b\sin^2\phi;$$

also $\dfrac{dp}{d\lambda_1} = \dfrac{dp}{d\phi}\dfrac{d\phi}{d\lambda_1}$, because the effect of a change in λ_1 is to move the screw along this cylindroid.

We have
$$\lambda_1 = \eta' \frac{\sin(\phi - B)}{\sin(\phi - A)},$$

and as the other co-ordinates are to be left unchanged, it is necessary that η' be constant, so that

$$\frac{d\lambda_1}{d\phi} = \eta' \frac{\sin(B-A)}{\sin^2(\phi - A)},$$

and hence
$$\frac{dp}{d\lambda_1} = (b-a)\sin 2\phi \frac{\sin^2(\phi - A)}{\eta' \sin(B-A)}.$$

Also
$$\frac{d\lambda'}{d\lambda_1} = \frac{d\lambda'}{d\phi} \cdot \frac{d\phi}{d\lambda_1} = -\cos(\phi - A).$$

Hence, substituting in the equation

$$\lambda' \frac{dp}{d\lambda_1} + 2p\frac{d\lambda'}{d\lambda_1} = 0,$$

we deduce
$$a = b\tan\phi\tan A:$$

but this is the condition that λ and the first screw of reference shall be reciprocal (§ 40).

267. Analogy to Orthogonal Transformation.

The emanants of the second degree are represented by the equation

$$\left(\beta_1 \frac{d}{d\alpha_1} + \ldots + \beta_6 \frac{d}{d\alpha_6}\right)^2 f = \left(\mu_1 \frac{d}{d\lambda_1} + \ldots + \mu_6 \frac{d}{d\lambda_6}\right)^2 F,$$

when F is the function into which f becomes transformed when the co-ordinates are changed from one set of screws of reference to another. If we take for f either of the functions already considered, these equations reduce to an identity; but retaining f in its general form, we can deduce some results of very considerable interest. The discussion which now follows was suggested by the reasoning employed by Professor W. S. Burnside* in the theory of orthogonal transformations.

Let us suppose that we transform the function f from one set of co-reciprocal screws of reference to another system. Let $p_1, \ldots p_6$ be the pitches of the first set, and $q_1, \ldots q_6$ be those of the second set. Then we must have

$$p_1\beta_1^2 + \ldots + p_6\beta_6^2 = q_1\mu_1^2 + \ldots + q_6\mu_6^2,$$

for each merely denotes the pitch of the Dyname multiplied into the square of its intensity. Multiply this equation by any arbitrary factor x and add it to the preceding, and we have

$$\left(\beta_1 \frac{d}{d\alpha_1} + \ldots + \beta_6 \frac{d}{d\alpha_6}\right)^2 f + x\,(p_1\beta_1^2 + \ldots + p_6\beta_6^2)$$

$$= \left(\mu_1 \frac{d}{d\lambda_1} + \ldots + \mu_6 \frac{d}{d\lambda_6}\right)^2 f + x\,(q_1\mu_1^2 + \ldots + q_6\mu_6^2).$$

Regarding $\beta_1, \ldots \beta_6$ as variables, the first member of this equation equated to zero would denote a certain screw system of the second degree. If that system were "central" it would possess a certain screw to which the polars of all other screws would be reciprocal, and its discriminant would vanish; but the screw β being absolutely the same as μ, it is plain that the discriminant of the second side must in such case also vanish. We thus see that the ratios of the coefficients of the various powers of x in the following well-known form of determinant must remain unchanged when one co-reciprocal set of screws is exchanged for another. In writing the determinant we put 12 for $\dfrac{d^2f}{d\alpha_1 \,.\, d\alpha_2}$, &c. .

$$\begin{vmatrix} 11 + xp_1, & 12, & 13, & 14, & 15, & 16 \\ 21, & 22 + xp_2, & 23, & 24, & 25, & 26 \\ 31, & 32, & 33 + xp_3, & 34, & 35, & 36 \\ 41, & 42, & 43, & 44 + xp_4, & 45, & 46 \\ 51, & 52, & 53, & 54, & 55 + xp_5, & 56 \\ 61, & 62, & 63, & 64, & 65, & 66 + xp_6 \end{vmatrix} = 0.$$

Take for instance the coefficient of x^5 divided by that of x^6, which is easily seen to be

$$\frac{1}{p_1} \cdot \frac{d^2f}{d\alpha_1^2} + \ldots + \frac{1}{p_6} \cdot \frac{d^2f}{d\alpha_6^2};$$

* Williamson, *Differential Calculus*, p. 412.

and we learn that this expression will remain absolutely unaltered provided that we only change from one set of co-reciprocals to another. In this f is perfectly arbitrary.

268. Property of the Pitches of Six Co-reciprocals.

We may here introduce an important property of the pitches of a set of co-reciprocal screws selected from a screw system.

There is one screw on a cylindroid of which the pitch is a maximum, and another screw of which the pitch is a minimum. These screws are parallel to the principal axes of the pitch conic (§ 18). Belonging to a screw system of the third order we have, in like manner, three screws of maximum or minimum pitch, which lie along the three principal axes of the pitch quadric (§ 173). The general question, therefore, arises, as to whether it is always possible to select from a screw system of the nth order a certain number of screws of maximum or minimum pitch.

Let $\theta_1, \ldots \theta_6$ be the six co-ordinates of a screw referred to n co-reciprocal screws belonging to the given screw system. Then the function p_θ, or

$$p_1\theta_1^2 + \ldots + p_6\theta_6^2,$$

is to be a maximum, while, at the same time, the co-ordinates satisfy the condition (§ 35)

$$\Sigma\theta_1^2 + 2\Sigma\theta_1\theta_2\cos(12) = 1,$$

which for brevity we denote as heretofore by

$$R = 1.$$

Applying the ordinary rules for maxima and minima, we deduce the six equations

$$2p_1\theta_1 - p_\theta\frac{dR}{d\theta_1} = 0,$$

$$\cdots\cdots\cdots\cdots\cdots\cdots$$

$$2p_6\theta_6 - p_\theta\frac{dR}{d\theta_6} = 0.$$

From these six equations $\theta_1, \ldots \theta_6$ can be eliminated, and we obtain the determinantal equation which, by writing $x = 1 \div p_\theta$, becomes

$$0 = \begin{vmatrix} 1 - xp_1, & \cos(21), & \cos(31), & \cos(41), & \cos(51), & \cos(61) \\ \cos(12), & 1 - xp_2, & \cos(32), & \cos(42), & \cos(52), & \cos(62) \\ \cos(13), & \cos(23), & 1 - xp_3, & \cos(43), & \cos(53), & \cos(63) \\ \cos(14), & \cos(24), & \cos(34), & 1 - xp_4, & \cos(54), & \cos(64) \\ \cos(15), & \cos(25), & \cos(35), & \cos(45), & 1 - xp_5, & \cos(65) \\ \cos(16), & \cos(26), & \cos(36), & \cos(46), & \cos(56), & 1 - xp_6 \end{vmatrix}.$$

It is easily seen that this equation must reduce to the form

$$x^6 = 0.$$

In fact, seeing it expresses the solution of the problem of finding a screw of maximum pitch, and that the choice may be made from a system of the sixth order, that is to say, from all conceivable screws in the universe it is obvious that the equation could assume no other form.

What we now propose to study is the manner in which the necessary evanescence of the several coefficients is provided for. After the equation has been expanded we shall suppose that each term is divided by the coefficient of x^6 that is, by $p_1p_2p_3p_4p_5p_6$.

From any point draw a pencil of rays parallel to the six screws. On four of these rays, 1, 2, 3, 4, we can assign four forces which equilibrate at the point. Let these magnitudes be X_1, X_2, X_3, X_4. We can express the necessary relations by resolving these four forces along each of the four directions successively. Hence

$$\begin{aligned}
&X_1 + X_2\cos(12) + X_3\cos(13) + X_4\cos(14) = 0.\\
&X_1\cos(21) + X_2 + X_3\cos(23) + X_4\cos(24) = 0.\\
&X_1\cos(31) + X_2\cos(32) + X_3 + X_4\cos(34) = 0.\\
&X_1\cos(41) + X_2\cos(42) + X_3\cos(43) + X_4 = 0.
\end{aligned}$$

Eliminating the four forces we have

$$\begin{vmatrix} 1, & \cos(12), & \cos(13), & \cos(14) \\ \cos(21), & 1, & \cos(23), & \cos(24) \\ \cos(31), & \cos(32), & 1, & \cos(34) \\ \cos(41), & \cos(42), & \cos(43), & 1 \end{vmatrix} = 0.$$

Thus we learn that every determinant of this type vanishes identically.

Had we taken five or six forces at the point it would, of course, have been possible in an infinite number of ways to have adjusted five or six forces to equilibrate. Hence it follows that the determinants analogous to that just written, but with five and six rows of elements respectively, are all zero.

These theorems simplify our expansion of the original harmonic determinant. In fact, it is plain that the coefficients of x^2, of x, and of the absolute term vanish identically. The terms which remain are as follows:—

$$x^6 + Ax^5 + Bx^4 + Cx^3 = 0.$$

where

$$A = \Sigma \frac{1}{p_1},$$

$$B = \Sigma \frac{\sin^2(1,\ 2)}{p_1 p_2},$$

$$C = \Sigma \frac{E_{123}}{p_1 p_2 p_3},$$

in which

$$E_{123} = \begin{vmatrix} 1, & \cos(12), & \cos(13) \\ \cos(12), & 1, & \cos(23) \\ \cos(13), & \cos(23), & 1 \end{vmatrix}.$$

If by S (123) we denote the scalar of the product of three unit vectors along 1, 2, 3, then it is easy to show that

$$S^2(123) = E_{123}.$$

We thus obtain the following three relations between the pitches and the angular directions of the six screws of a co-reciprocal system*,

$$\Sigma \frac{1}{p_1} = 0,$$

$$\Sigma \frac{\sin^2(1,\ 2)}{p_1 p_2} = 0,$$

$$\Sigma \frac{S^2(1,\ 2,\ 3)}{p_1 p_2 p_3} = 0.$$

The first of these formulæ gives the remarkable result that, *the sum of the reciprocals of the pitches of the six screws of a co-reciprocal system is equal to zero.*

The following elegant proof of the first formula was communicated to me by my friend Professor Everett. Divide the six co-reciprocals into any two groups A and B of three each, then it appears from § 174 that the pitch quadric of each of these groups is identical. The three screws of A are parallel to a triad of conjugate diameters of the pitch quadric, and the sum of the reciprocals of the pitches is proportional to the sum of the squares of the conjugate diameters (§ 176). The three screws of B are parallel to another triad of conjugate diameters of the pitch quadric, and the sum of the reciprocals of the pitches, *with their signs changed*, is proportional to the sum of the squares of the conjugate diameters. Remembering that the sum of the squares of the two sets of conjugate diameters is equal, the required theorem is at once evident.

* *Proceedings of the Royal Irish Academy*, Series III. Vol. I. p. 375 (1890). A set of six screws are in general determined by 30 parameters. If those screws be reciprocal 15 conditions must be fulfilled. The above are three of the conditions, see also § 271.

269. Property of the Pitches of n Co-reciprocals.

The theorem just proved can be extended to show that *the sum of the reciprocals of the pitches of n co-reciprocal screws, selected from a screw system of the nth order, is a constant for each screw system.*

Let A be the given screw system, and B the reciprocal screw system. Take $6-n$ co-reciprocal screws on B, and *any* n co-reciprocal screws on A. The sum of the reciprocals of the pitches of these six screws must be always zero; but the screws on B may be constant, while those on A are changed, whence the sum of the reciprocals of the pitches of the n co-reciprocal screws on A must be constant.

Thus, as we have already seen from geometrical considerations, that the sum of the reciprocals of the pitches of co-reciprocals is constant for the screw system of the second and third order (§§ 40, 176), so now we see that the same must be likewise true for the fourth, fifth, and sixth orders.

The actual value of this constant for any given screw system is evidently a characteristic feature of that screw system.

270. Theorem as to Signs.

If in one set of co-reciprocal screws of an n-system there be k screws with negative pitch and $n-k$ screws with positive pitch, then in every set of co-reciprocal screws of the same system there will also be k screws with negative pitch and $n-k$ screws with positive pitch.

To prove this we may take the case of a five-system, and suppose that of five co-reciprocals A_1, A_2, A_3, A_4, A_5 the pitches of three are positive, say m_1^2, m_2^2, m_3^2, while the pitches of the two others are negative, say $-m_4^2$, $-m_5^2$.

Let S be any screw of the system, then if $\theta_1, \ldots \theta_5$ be its co-ordinates with respect to the five co-reciprocals just considered, we have for the pitch of θ the expression

$$m_1^2\theta_1^2 + m_2^2\theta_2^2 + m_3^2\theta_3^2 - m_4^2\theta_4^2 - m_5^2\theta_5^2.$$

Let us now take another set of five co-reciprocals B_1, B_2, B_3, B_4, B_5 belonging to the same system, then the pitches of three of these screws must be $+$ and the pitches of two must be $-$. For suppose this was not so, but that the five pitches were, let us say $n_1^2, n_2^2, n_3^2, n_4^2, -n_5^2$. Let the co-ordinates of S with respect to these new screws of reference be $\phi_1, \phi_2, \ldots \phi_5$, then the pitch will be

$$n_1^2\phi_1^2 + n_2^2\phi_2^2 + n_3^2\phi_3^2 + n_4^2\phi_4^2 - n_5^2\phi_5^2.$$

Equating these two values of the pitch we ought to have for every screw S

$$m_1^2\theta_1^2 + m_2^2\theta_2^2 + m_3^2\theta_3^2 + n_5^2\phi_5^2$$
$$= n_1^2\phi_1^2 + n_2^2\phi_2^2 + n_3^2\phi_3^2 + n_4^2\phi_4^2 + m_4^2\theta_4^2 + m_5^2\theta_5^2.$$

But it can easily be seen that this equation is impossible.

Let H be the screw to which all the screws of the five-system are reciprocal, and let us choose for S the screw reciprocal to A_1, A_2, A_3, B_5, H. The fact that S is reciprocal to H is of course implied in the assumption that S belongs to the five-system, while the fact that S is reciprocal to each of the screws A_1, A_2, A_3, B_5 gives us

$$\theta_1 = 0, \quad \theta_2 = 0, \quad \theta_3 = 0, \quad \phi_5 = 0.$$

Hence we would have the equation

$$n_1^2\phi_1^2 + n_2^2\phi_2^2 + n_3^2\phi_3^2 + n_4^2\phi_4^2 + m_4^2\theta_4^2 + m_5^2\theta_5^2 = 0,$$

which would require that all the co-ordinates were zero, which is impossible.

In like manner any other supposition inconsistent with the theorem of this article would be shown to lead to an absurdity. The theorem is therefore proved.

We can hence easily deduce the important theorem that three of the screws in a complete co-reciprocal system of six must have positive pitch and three must have negative pitch*.

For in the canonical system of co-reciprocals the pitches are $+a$, $-a$, $+b$, $-b$, $+c$, $-c$, i.e. three are positive and three are negative, and as in this case the n-system being the six-system includes every screw in space we see that of any six co-reciprocals three of the pitches must be positive and three must be negative.

271. Identical Formulæ in a Co-reciprocal System.

Let any screw α be inclined at angles $\alpha 1, \alpha 2, \dots \alpha 6$ to the respective six screws of a co-reciprocal system.

Then we have for the co-ordinate α_n

$$\alpha_n = \frac{(p_\alpha + p_n)\cos \alpha 1 - d_{\alpha 1} \sin \alpha 1}{2p_n}.$$

* This interesting theorem was communicated to me by Klein, who had proved it as a property of the parameters of "six fundamental complexes in involution" (*Math. Ann. Band.* i. p. 204).

If we substitute these values for $\alpha_1, \ldots \alpha_n$ in the expression

$$p_\alpha = \Sigma p_1 \alpha_1^2$$

we obtain the equation

$$0 = p_\alpha^2 \left[\frac{\cos^2 \alpha 1}{p_1} + \frac{\cos^2 \alpha 2}{p_2} + \ldots + \frac{\cos^2 \alpha 6}{p_6}\right]$$

$$+ 2p_\alpha \left[\frac{\cos \alpha 1\,(p_1 \cos \alpha 1 - d_{\alpha 1} \sin \alpha 1)}{p_1} + \ldots + \frac{\cos \alpha 6\,(p_6 \cos \alpha 6 - d_{\alpha 6} \sin \alpha 6)}{p_6} - 2\right]$$

$$+ \frac{(p_1 \cos \alpha 1 - d_{\alpha 1} \sin \alpha 1)^2}{p_1} + \ldots + \frac{(p_6 \cos \alpha 6 - d_{\alpha 6} \sin \alpha 6)^2}{p_6}.$$

As $\alpha 1$, &c., $d_{\alpha 1}$, &c., p_1, &c. are independent of p_α we must have the three co-efficients of this quadratic in p_α severally equal to zero.

272. Three Pitches Positive and Three Negative.

The equation

$$\frac{\cos^2 \alpha 1}{p_1} + \frac{\cos^2 \alpha 2}{p_2} + \ldots + \frac{\cos^2 \alpha 6}{p_6} = 0$$

also shows that three pitches of a set of six co-reciprocals must be positive and three must be negative. For, suppose that the pitches of four of the co-reciprocals had the same sign, and let α be a screw perpendicular to the two remaining co-reciprocals, then the identity just written would reduce to the sum of four positive terms equal to zero.

From this formula and also

$$\frac{1}{p_1} + \frac{1}{p_2} + \ldots + \frac{1}{p_6} = 0$$

we have

$$\frac{\sin^2 \alpha 1}{p_1} + \frac{\sin^2 \alpha 2}{p_2} + \ldots + \frac{\sin^2 \alpha 6}{p_6} = 0.$$

273. Linear Pitch Invariant Functions.

We propose to investigate the linear functions of the six co-ordinates of a screw which possess the property that they remain unaltered notwithstanding an alteration in the pitch of the screw which the co-ordinates denote. It will first be convenient to demonstrate a general theorem which introduces a property of the six screws of a co-reciprocal system.

The virtual coefficient of two screws is, as we know, represented by half the expression

$$(p_\alpha + p_\beta) \cos \theta - d \sin \theta,$$

where p_α and p_β are the pitches, θ is the angle between the two screws. and d the shortest perpendicular distance. The pitches only enter into this expression by their sum; and, consequently, if p_α be changed into $p_\alpha + x$, and p_β be changed into $p_\beta - x$, the virtual coefficient will remain unaltered whatever x may be.

We have found, however (§ 37), that the virtual coefficient admits of representation in the form

$$p_1\alpha_1\beta_1 + \ldots + p_6\alpha_6\beta_6.$$

To augment the pitch of α by x, we substitute for $\alpha_1, \alpha_2, \ldots$ the several values (§ 265),

$$\alpha_1 + \frac{x}{2p_1}\cos a_1, \quad \alpha_2 + \frac{x}{2p_2}\cos a_2, \ldots$$

where $a_1, a_2, \ldots$ are the angles made by the screw α with the screws of reference. Similarly, to diminish the pitch of β by x, we substitute for $\beta_1, \beta_2, \ldots$ the several values

$$\beta_1 - \frac{x}{2p_1}\cos b_1, \quad \beta_2 - \frac{x}{2p_2}\cos b_2, \text{ \&c.}$$

With this change the virtual coefficient, as above expressed, becomes

$$\Sigma p_1\left(\alpha_1 + \frac{x}{2p_1}\cos a_1\right)\left(\beta_1 - \frac{x}{2p_1}\cos b_1\right),$$

or,

$$\Sigma p\alpha_1\beta_1 + \frac{x}{2}(\beta_1\cos a_1 + \beta_2\cos a_2 + \ldots - \alpha_1\cos b_1 - \alpha_2\cos b_2 - \ldots)$$
$$- \frac{x^2}{4}\left(\frac{\cos a_1\cos b_1}{p_1} + \frac{\cos a_2\cos b_2}{p_2} + \ldots + \frac{\cos a_6\cos b_6}{p_6}\right).$$

We have already shown that such a change must be void of effect upon the virtual coefficient for all values of x. It therefore follows that the coefficients of both x and x^2 in the expressions just written must be zero. Hence we obtain the two following properties:

$$0 = (\beta_1\cos a_1 + \ldots + \beta_6\cos a_6) - (\alpha_1\cos b_1 + \ldots + \alpha_6\cos b_6),$$

$$0 = \frac{\cos a_1\cos b_1}{p_1} + \ldots + \frac{\cos a_6\cos b_6}{p_6},$$

The second of the two formulæ is the important one for our present purpose. It will be noted that though the two screws, α and β, are completely arbitrary, yet the six direction cosines of α with regard to the screws of reference, and the six direction cosines of β with regard to the same screws of reference, must be connected by this relation. Of course the equation in this form is only true when the six screws of reference are *co-reciprocal*. In the more general case the equivalent identity would be of a much more complicated type.

274. A Pitch Invariant.

Let $h_1, \dots h_6$ be the direction angles of any ray whatever with regard to six co-reciprocal screws of reference, then the function

$$U = \alpha_1 \cos h_1 + \dots + \alpha_6 \cos h_6$$

is a pitch invariant.

For, if we augment the pitch of α by x, we have to write for $\alpha_1, \dots \alpha_6$ the expressions

$$\alpha_1 + \frac{x}{2p_1} \cos a_1 \dots \alpha_6 + \frac{x}{2p_6} \cos a_6,$$

and then U becomes

$$\alpha_1 \cos h_1 + \dots + \alpha_6 \cos h_6$$
$$+ \frac{x}{2}\left(\frac{\cos a_1 \cos h_1}{p_1} + \dots + \frac{\cos a_6 \cos h_6}{p_6}\right);$$

but from what we have just proved, the coefficient of x is zero, and hence we see that

$$\alpha_1 \cos h_1 + \dots + \alpha_6 \cos h_6$$

remains unchanged by any alteration in the pitch of α.

If we take three mutually rectangular screws, α, β, γ, then we have the three pitch invariants

$$L = \theta_1 \cos a_1 + \dots + \theta_6 \cos a_6,$$
$$M = \theta_1 \cos b_1 + \dots + \theta_6 \cos b_6,$$
$$N = \theta_1 \cos c_1 + \dots + \theta_6 \cos c_6.$$

It is obvious that any linear function of L, M, N, such as

$$fL + gM + hN,$$

is a pitch invariant.

We can further show that this is the most general type of linear pitch invariant.

For the conditions under which the general linear function

$$A_1\theta_1 + \dots + A_n\theta_n$$

shall be a pitch invariant are that equations of the type

$$\frac{A_1 \cos a_1}{p_1} + \dots + \frac{A_6 \cos a_6}{p_6} = 0; \text{ \&c.}$$

shall be satisfied for all possible rays.

Though these equations are infinite in number, yet they are only equivalent to three independent equations; in other words, if these equations are satisfied for three rays, a, b, c, which, for convenience, we may take to be rectangular, then they are satisfied for every ray.

For, take a ray ϵ, which makes direction-angles λ, μ, ν with a, b, c, then we have

$$\cos \epsilon_1 = \cos \lambda \cos a_1 + \cos \mu \cos b_1 + \cos \nu \cos c_1,$$

$$\cdots\cdots\cdots\cdots\cdots\cdots\cdots\cdots\cdots\cdots\cdots\cdots\cdots\cdots$$

$$\cos \epsilon_6 = \cos \lambda \cos a_6 + \cos \mu \cos b_6 + \cos \nu \cos c_6.$$

Hence

$$\frac{A_1 \cos \epsilon_1}{p_1} + \ldots + \frac{A_6 \cos \epsilon_6}{p_6} = \cos \lambda \Sigma \frac{A_1 \cos a_1}{p_1} + \cos \mu \Sigma \frac{A_1 \cos b_1}{p_1} + \cos \nu \Sigma \frac{A_1 \cos c_1}{p_1}.$$

If, therefore, we have

$$\Sigma \frac{A_1 \cos a_1}{p_1} = 0; \quad \Sigma \frac{A_1 \cos b_1}{p_1} = 0; \quad \Sigma \frac{A_1 \cos c_1}{p_1} = 0,$$

then, for every ray, we shall have

$$\Sigma \frac{A_1 \cos \epsilon_1}{p_1} = 0.$$

It thus follows that the coefficients of a linear function which possesses the property of a pitch invariant must be subjected to three conditions. There are accordingly only three coefficients left disposable in the most general type of linear pitch invariant. Now,

$$fL + gM + hN$$

is a pitch invariant which contains three disposable quantities, f, g, h; it therefore represents the most general form of linear function which possesses the required property.

We have thus solved the problem of finding a perfectly general expression for the linear pitch invariant function of the co-ordinates of a screw.

It is convenient to take the three fundamental rays as mutually rectangular; but it is, of course, easy to show that any linear pitch invariant can be expressed in terms of three pitch invariants unless their determining rays are coplanar. We may express the result thus:—Let L, M, N, O be four linear pitch invariants, no three of which have coplanar determining rays. Then it is always possible to find four parameters, λ, μ, ν, ρ, such that the following equation shall be satisfied identically:—

$$\lambda L + \mu M + \nu N + \rho O = 0.$$

275. Geometrical meaning.

The nature of the pitch invariant function can be otherwise seen. It is well known that in the composition of two or more twist velocities we discover the direction of the resultant screw and the magnitude of the resultant twist velocity by proceeding as if the twist velocities were vectors. Neither the pitches of the component screws nor their situations affect the magnitude of the resultant twist velocity or the direction of the resultant screw. This principle is, of course, an immediate consequence of the law of composition of twist velocities by the cylindroid.

Let any ray α make an angle λ with the ray θ, and angles $a_1, \ldots a_6$ with the six screws of reference. The twist velocity $\dot{\theta}$ on θ if resolved on α has a component $\dot{\theta}\cos\lambda$. This must be equal to the sum of the several components $\dot{\theta}_1$, $\dot{\theta}_2$, ... resolved on α; whence we have

$$\dot{\theta}\cos\lambda = \dot{\theta}_1 \cos a_1 + \ldots + \dot{\theta}_6 \cos a_6.$$

If we make $\dot{\theta} =$ unity, we obtain

$$\cos\lambda = \theta_1 \cos a_1 + \ldots + \theta_6 \cos a_6.$$

This gives a geometrical meaning to the pitch invariant. It is simply the cosine of the angle between the screws θ and α. As, of course, the pitch is not involved in the notion of this angle, it is, indeed, obvious that the expression for any function of the angle must be a pitch invariant.

We now see the meaning of the equation obtained by equating the pitch invariant to zero. If we make

$$\theta_1 \cos a_1 + \ldots + \theta_6 \cos a_6 = 0$$

it follows that α and θ must be at right angles. The equation therefore signifies the locus of all the screws that are at right angles to α.

The two equations

$$\theta_1 \cos a_1 + \ldots + \theta_6 \cos a_6 = 0,$$

$$\theta_1 \cos b_1 + \ldots + \theta_6 \cos b_6 = 0,$$

denote the screws perpendicular to the two directions of α and β. In other words, these two equations define all the screws perpendicular to a given plane.

276. Screws at infinity.

Let us now take the case where α, β, γ are three rectangular screws, and examine the conditions satisfied by $\theta_1, \ldots, \theta_6$ when subjected to the three following equations:

$$\theta_1 \cos a_1 + \ldots + \theta_6 \cos a_6 = 0,$$

$$\theta_1 \cos b_1 + \ldots + \theta_6 \cos b_6 = 0,$$

$$\theta_1 \cos c_1 + \ldots + \theta_6 \cos c_6 = 0.$$

The screws which satisfy these conditions must all be perpendicular to the three directions of α, β, γ. For real and finite rays this is impossible; for real and finite rays could not be perpendicular to each of three rays which were themselves mutually rectangular. This is only possible if the rays denoted by $\theta_1, \ldots \theta_6$ are lines at infinity.

It follows that the three equations, $L=0$; $M=0$; $N=0$, obtained by equating the three fundamental pitch invariants to zero, must in general express the collection of screws that are situated in the plane at infinity.

We can write the three equations in an equivalent form by the six equations

$$\theta_1 = f\frac{\cos a_1}{p_1} + g\frac{\cos b_1}{p_1} + h\frac{\cos c_1}{p_1},$$

$$\cdots\cdots\cdots\cdots\cdots\cdots\cdots\cdots\cdots\cdots$$

$$\theta_6 = f\frac{\cos a_6}{p_6} + g\frac{\cos b_6}{p_6} + h\frac{\cos c_6}{p_6},$$

where f, g, h are any quantities whatever; for it is obvious that, by substituting these values for $\theta_1, \ldots \theta_6$ in either L, or M, or N, these quantities are made to vanish by the formulæ of the type

$$\frac{\cos a_1 \cos b_1}{p_1} + \ldots + \frac{\cos a_6 \cos b_6}{p_6} = 0.$$

We have, consequently, in the expressions just written for $\theta_1, \ldots \theta_6$, the values of the co-ordinates of a screw which lies entirely in the plane at infinity.

277. Expression for the Pitch.

It is known that if α, β, γ be the direction-angles of a ray, and if P, Q, R be its shortest perpendicular distances from three rectangular axes, then

$$P \sin\alpha \cos\alpha + Q \sin\beta \cos\beta + R \sin\gamma \cos\gamma = 0.$$

Let η, ξ, ζ be three screws of zero pitch, which intersect at right angles, and let θ be another screw, then, if $\varpi_{\eta\theta}$ be the virtual coefficients of η and θ,

$$2\varpi_{\eta\theta} = p_\theta \cos\alpha - P \sin\alpha,$$

whence, by the theorem just mentioned, we have

$$p_\theta = 2\varpi_{\eta\theta}\cos\alpha + 2\varpi_{\xi\theta}\cos\beta + 2\varpi_{\zeta\theta}\cos\gamma.$$

Let $a_1, \ldots a_6$ be the angles made by η with the six co-reciprocal screws of reference, then

$$\cos\alpha = L = \theta_1 \cos a_1 + \ldots + \theta_6 \cos a_6,$$

and, similarly, for the two other angles,

$$\cos\beta = M = \theta_1 \cos b_1 + \ldots + \theta_6 \cos b_6,$$

$$\cos\gamma = N = \theta_1 \cos c_1 + \ldots + \theta_6 \cos c_6,$$

and $L^2 + M^2 + N^2 = 1$, whence we have for the pitch the homogeneous expression

$$p_\theta = \frac{2L\varpi_{\eta\theta} + 2M\varpi_{\xi\theta} + 2N\varpi_{\zeta\theta}}{L^2 + M^2 + N^2}.$$

It appears from this that the three equations,

$$\varpi_{\eta\theta} = 0; \quad \varpi_{\xi\theta} = 0; \quad \varpi_{\zeta\theta} = 0,$$

indicate that θ must be one of a pencil of rays of zero pitch radiating from a point.

The equations $L = 0$; $M = 0$; $N = 0$, define a screw of indeterminate pitch.

Why the screws in the plane at infinity (§ 46) should in general present themselves with indeterminate pitch is a point which requires some explanation. The twist about such a screw, as around any other, consists, of course, of a rotation and a translation. If, however, the finite parts of the body are only to be moved through a finite distance, the amplitude of the twist must be infinitely small, for a finite rotation around an axis at infinity would, of course, imply an infinitely great displacement of parts of the body which were at finite distances. The amplitude of the rotation is therefore infinitely small, so that, if the pitch is finite, the displacement parallel to the axis of the screw is infinitely small also. It thus appears that the effect of a small twist about a screw of any finite pitch at infinity is to give the finite parts of the body two displacements, one of which is infinitely insignificant as regards the other. We can therefore overlook the displacement due to the pitch, and consequently the pitch of the screw unless infinite is immaterial; in other words, in so far as the screw is the subject of our investigation, its pitch is indeterminate.

In like manner we can prove that a screw in the plane at infinity, when regarded as the seat of a wrench, must, when finite forces are considered, be regarded as possessing an indeterminate pitch. For, let the force appertaining to the wrench be of finite magnitude, then its effect on bodies at finite distances would involve a couple of infinite moment. It therefore follows that the force on the screw at infinity must be infinitely small if the effects of the wrench are finite. The moment of the couple on the screw of finite pitch is therefore infinitely small, nor is its magnitude increased by importation from infinity; therefore, at finite distances, the effect of the couple part of the wrench may be neglected in comparison with that of the force part of the wrench. But the pitch of the screw is only involved so far as the couple is concerned; and hence whatever be the pitch of the screw lying in the plane at infinity, its effect is inoperative so far as finite operations are concerned.

There is here a phenomenon of duality which, though full of significance in non-Euclidian space, merely retains a shred of its importance in the space of ordinary conventions. A displacement, such as we have been considering, may of course arise either from a twist about a screw of infinite pitch at an indefinite distance, or a twist about a screw of indefinite pitch at an infinite distance.

278. A System of Emanants which are Pitch Invariants*.

From the formula

$$2\varpi_{\alpha\beta} = (p_\alpha + p_\beta)\cos(\alpha\beta) - d_{\alpha\beta}\sin(\alpha\beta),$$

we obtain

$$\begin{aligned} d_{\alpha\beta}\sin(\alpha\beta) &= \tfrac{1}{2}(p_\alpha + p_\beta)\left(\alpha_1\frac{d}{d\beta_1} + \dots + \alpha_6\frac{d}{d\beta_6}\right)R_\beta \\ &\qquad - \left(\alpha_1\frac{d}{d\beta_1} + \dots + \alpha_6\frac{d}{d\beta_6}\right)p_\beta \\ &= (p_\alpha + p_\beta)\left(\alpha_1\frac{d}{d\beta_1} + \dots + \alpha_6\frac{d}{d\beta_6}\right)\sqrt{R_\beta} \\ &\qquad - \left(\alpha_1\frac{d}{d\beta_1} + \dots + \alpha_6\frac{d}{d\beta_6}\right)(p_\alpha + p_\beta) \\ &= \quad - \left(\alpha_1\frac{d}{d\beta_1} + \dots + \alpha_6\frac{d}{d\beta_6}\right)\left(\frac{p_\alpha + p_\beta}{\sqrt{R_\beta}}\right), \end{aligned}$$

or from symmetry

$$= \quad - \left(\beta_1\frac{d}{d\alpha_1} + \dots + \beta_6\frac{d}{d\alpha_6}\right)\left(\frac{p_\alpha + p_\beta}{\sqrt{R_\alpha}}\right).$$

We thus obtain an emanant function of the co-ordinates of α and β which expresses the product of the shortest distance between α and β into the sine of the angle between them. The evanescence of this emanant is of course the condition (§ 228) that α and β intersect.

This emanant is obviously a pitch invariant for each of the two screws involved. It will be a pitch invariant for α whatever be the screw β. Let us take for β the first screw of reference so that

$$\beta_1 = 1; \quad \beta_2 = 0 \ \dots \ \beta_6 = 0.$$

Then

$$\frac{d}{d\alpha_1}\left(\frac{p_\alpha + p_1}{\sqrt{R_\alpha}}\right)$$

must be a pitch invariant. It may be written

$$\frac{d}{d\alpha_1}\left(\frac{p_\alpha}{\sqrt{R_\alpha}}\right) + \frac{p_1}{2}\cdot\frac{dR_\alpha}{d\alpha_1},$$

* This article is due to Mr A. Y. G. Campbell.

but we know that the last term is itself a pitch invariant, and hence we have the following result.

If p_α be the pitch of a screw α expressed in terms of the co-ordinates, and if R_α denote the function $\alpha_1^2 + \ldots + \alpha_6^2 + 2\alpha_1\alpha_2 \cos(12) + \ldots = 1$, then the several functions

$$\frac{d}{d\alpha_1}\left(\frac{p_\alpha}{\sqrt{R_\alpha}}\right), \ldots \frac{d}{d\alpha_6}\left(\frac{p_\alpha}{\sqrt{R_\alpha}}\right)$$

remain unaltered if instead of $\alpha_1 \ldots \alpha_6$ the co-ordinates of any other screw on the same straight line as α should be substituted.

This is easily verified by the known formulæ that if α be any screw and θ another screw on the same axis as α whose pitch is $p_\alpha + x$, then

$$\theta_1 = \alpha_1 + \frac{x}{4p_1} \cdot \frac{dR}{d\alpha_1}, \ldots \theta_6 = \alpha_6 + \frac{x}{4p_6} \cdot \frac{dR}{d\alpha_6},$$

whence

$$\frac{d}{d\theta_1}\left(\frac{p_\theta}{\sqrt{R_\theta}}\right) = 2p_1\theta_1 - p_\theta \cos(\theta 1) = 2p_1\alpha_1 + \frac{x}{2} \cdot \frac{dR}{d\alpha_1} - (p_\alpha + x)\cos(\alpha 1)$$

$$= 2p_1\alpha_1 - p_\alpha \cos(\alpha 1) = \frac{d}{d\alpha_1}\left(\frac{p_\alpha}{\sqrt{R_\alpha}}\right).$$

CHAPTER XXI.

DEVELOPMENTS OF THE DYNAMICAL THEORY.

279. Expression for the Kinetic Energy.

Let us suppose that a body of mass M is twisting around a screw α with the twist velocity $\dot{\alpha}$. It is obvious that the kinetic energy of the body must be the product of $M\dot{\alpha}^2$ and some expression which has the dimensions of the square of a linear magnitude. This expression has a particular geometrical significance in the Theory of Screws, and the symbols of the theory afford a representation of the expression in an extremely concise manner.

Let η be the impulsive screw which corresponds to α as an instantaneous screw, the body being supposed to be perfectly unconstrained.

As usual p_α is the pitch of α and $(\alpha\eta)$ is the angle between α and η.

From the formulæ of § 80 we have, where H is a common factor,

$$H\eta_1 = +a\alpha_1; \qquad H\eta_2 = -a\alpha_2;$$
$$H\eta_3 = +b\alpha_3; \qquad H\eta_4 = -b\alpha_4;$$
$$H\eta_5 = +c\alpha_5; \qquad H\eta_6 = -c\alpha_6;$$

whence

$$H\left[(\eta_1+\eta_2)(\alpha_1+\alpha_2)+(\eta_3+\eta_4)(\alpha_3+\alpha_4)+(\eta_5+\eta_6)(\alpha_5+\alpha_6)\right]$$
$$= a(\alpha_1^2-\alpha_2^2)+b(\alpha_3^2-\alpha_4^2)+c(\alpha_5^2-\alpha_6^2)=p_\alpha$$

and we obtain

$$H = \frac{p_\alpha}{\cos(\alpha\eta)}.$$

The kinetic energy is

$$M\dot{\alpha}^2(a^2\alpha_1^2+a^2\alpha_2^2+b^2\alpha_3^2+b^2\alpha_4^2+c^2\alpha_5^2+c^2\alpha_6^2)$$
$$= M\dot{\alpha}^2\frac{p_\alpha}{\cos(\alpha\eta)}\left[a\alpha_1\eta_1-a\alpha_2\eta_2+b\alpha_3\eta_3-b\alpha_4\eta_4+c\alpha_5\eta_5-c\alpha_6\eta_6\right]$$
$$= M\dot{\alpha}^2\frac{p_\alpha}{\cos(\alpha\eta)}\varpi_{\alpha\eta}\text{*},$$

* *Trans. Roy. Irish Acad.*, Vol. XXXI. p. 99 (1896).

which is the required expression for the kinetic energy. It is remarkable that the co-ordinates of the rigid body are introduced by the medium of the impulsive screw alone.

280. Expression for the Twist Velocity.

If an impulsive wrench of unit intensity on a screw η be applied to a quiescent rigid body of unit mass which in consequence commences to twist about an instantaneous screw α, it is required to find the initial twist velocity $\dot{\alpha}$.

The impulsive wrench may be replaced by component impulsive forces $\eta_1, \ldots \eta_6$ on the six principal screws of inertia and component impulsive couples with moments $a\eta_1$, $-a\eta_2$, $b\eta_3$, $-b\eta_4$, $c\eta_5$, $-c\eta_6$ about those screws of inertia.

The force η_1 is expressed by the velocity it produces in the unit mass parallel to the direction of η. The component twist velocity of α is $\dot{\alpha}\alpha_1$ about the first principal screw, and accordingly the velocity of translation parallel to that screw is $a\dot{\alpha}\alpha_1$. Hence we have

$$a\dot{\alpha}\alpha_1 = \eta_1,$$

but

$$a\alpha_1 = \frac{p_\alpha}{\cos(\alpha\eta)}\eta_1,$$

whence we obtain*

$$\dot{\alpha} = \frac{\cos(\alpha\eta)}{p_\alpha}.$$

It will be noted that in this expression the co-ordinates of the rigid body are introduced through the medium of the impulsive screw alone.

A special case arises when the impulsive wrench is a couple, in which case of course $p_\eta = \infty$. As the effect of an impulsive couple is to produce a pure rotation only, we must under these circumstances have $p_\alpha = 0$. Poinsot's well-known construction exhibits the axis of the initial rotation as the diameter of the momental ellipsoid conjugate to the plane of the impulsive couple. As three conjugate diameters of an ellipsoid could not lie in the same plane, it follows that in the case of $p_\eta = \infty$ we can never have α and η at right angles. As p_α is zero while $\cos(\alpha\eta)$ is not zero, we must have $\dot{\alpha}$ infinite.

This might have been inferred from the fact that as the intensity of the impulsive wrench was not zero while the pitch of the screw on which it lay was infinite, the moment of the impulsive couple was infinite and consequently the initial twist velocity must be infinite.

* *Trans. Roy. Irish Acad.*, Vol. XXXI. p. 100 (1896).

Unless in this exceptional case where p_η is infinite it is always true that when p_α is zero, α and η are at right angles.

It is universally true that when the impulsive screw and the instantaneous screw are at right angles (the body being quite free), the pitch of the instantaneous screw must be zero.

For if p_α were not zero when $\cos(\alpha\eta)$ was zero then $\dot{\alpha}$ must be zero. As some motion must result from the impulse (the mass of the body being finite) we must have p_α infinite. The initial motion is thus a translation. Therefore the impulse must have been merely a force through the centre of gravity; α and η must be parallel and $\cos(\alpha\eta)$ could not be zero.

The expression for the kinetic energy in § 279,

$$M\dot{\alpha}^2 \frac{p_\alpha}{\cos(\alpha\eta)} \varpi_{\alpha\eta},$$

assumes an indeterminate form when the impulsive wrench reduces to a couple. For we then have $p_\alpha = 0$, but as $\cos(\alpha\eta)$ is not zero the expression for $\varpi_{\alpha\eta}$, i.e.

$$\tfrac{1}{2}\{(p_\alpha + p_\eta)\cos(\alpha\eta) - d_{\alpha\eta}\sin(\alpha\eta)\},$$

becomes infinite.

The expression for the kinetic energy arising from an impulsive wrench of unit intensity on a screw η applied to a free body of unit mass which thereupon begins to twist with an instantaneous movement about a screw α has the concise form

$$\frac{\cos(\alpha\eta)}{p_\alpha} \varpi_{\alpha\eta}.$$

281. Conditions to be fulfilled by two pairs of Impulsive and Instantaneous Screws.

Let α be a screw about which a free rigid body is made to twist in consequence of an impulsive wrench administered on some other screw η. Let β be another instantaneous screw corresponding in like manner to ξ as an impulsive screw. Then we have to prove that the two following formulæ are satisfied*:

$$\frac{p_\alpha}{\cos(\alpha\eta)} \cos(\beta\eta) + \frac{p_\beta}{\cos(\beta\xi)} \cos(\alpha\xi) = 2\varpi_{\alpha\beta},$$

$$\frac{p_\alpha}{\cos(\alpha\eta)} \varpi_{\beta\eta} = \frac{p_\beta}{\cos(\beta\xi)} \varpi_{\alpha\xi}.$$

* *Proceedings of the Camb. Phil. Soc.*, Vol. IX. Part iii. p. 193.

To demonstrate the first of these formulæ. Expand the left-hand side and it becomes

$$+\frac{p_\alpha}{\cos(\alpha\eta)}\{(\beta_1+\beta_2)(\eta_1+\eta_2)+(\beta_3+\beta_4)(\eta_3+\eta_4)+(\beta_5+\beta_6)(\eta_5+\eta_6)\}$$

$$+\frac{p_\beta}{\cos(\beta\xi)}\{(\alpha_1+\alpha_2)(\xi_1+\xi_2)+(\alpha_3+\alpha_4)(\xi_3+\xi_4)+(\alpha_5+\alpha_6)(\xi_5+\xi_6)\}.$$

But, as already shown,

$$\frac{p_\alpha}{\cos(\alpha\eta)}\eta_1=+a\alpha_1,\quad \frac{p_\alpha}{\cos(\alpha\eta)}\eta_2=-a\alpha_2,\ \ldots,$$

$$\frac{p_\beta}{\cos(\beta\xi)}\xi_1=+a\beta_1,\quad \frac{p_\beta}{\cos(\beta\xi)}\xi_2=-a\beta_2,\ \ldots;$$

whence, by substitution, the expression reduces to

$$\begin{aligned}&+a(\beta_1+\beta_2)(\alpha_1-\alpha_2)+a(\alpha_1+\alpha_2)(\beta_1-\beta_2)\\&+b(\beta_3+\beta_4)(\alpha_3-\alpha_4)+b(\alpha_3+\alpha_4)(\beta_3-\beta_4)\\&+c(\beta_5+\beta_6)(\alpha_5-\alpha_6)+c(\alpha_5+\alpha_6)(\beta_5-\beta_6)\\&=2a\alpha_1\beta_1-2a\alpha_2\beta_2+2b\alpha_3\beta_3-2b\alpha_4\beta_4+2c\alpha_5\beta_5-2c\alpha_6\beta_6\\&=2\varpi_{\alpha\beta}.\end{aligned}$$

To prove the second formula it is only necessary to note that each side reduces to

$$+a^2\alpha_1\beta_1+a^2\alpha_2\beta_2+b^2\alpha_3\beta_3+b^2\alpha_4\beta_4+c^2\alpha_5\beta_5+c^2\alpha_6\beta_6.$$

It will be observed that these two theorems are quite independent of the particular screws of reference which have been chosen.

282. Conjugate Screws of Inertia.

We have already made much use of the important principle that is implied in the existence of conjugate screws of inertia. If α be reciprocal to ξ then must η be reciprocal to β. This theorem implied the existence of some formula connecting $\varpi_{\alpha\xi}$ and $\varpi_{\beta\eta}$. We see this formula to be

$$\frac{p_\alpha}{\cos(\alpha\eta)}\varpi_{\beta\eta}=\frac{p_\beta}{\cos(\beta\xi)}\varpi_{\alpha\xi}.$$

We have now to show that if $\varpi_{\beta\eta}=0$, then must $\varpi_{\alpha\xi}=0$.

Let us endeavour to satisfy this equation when $\varpi_{\beta\eta}$ is zero otherwise than by making $\varpi_{\alpha\xi}$ zero. Let us make p_α infinite, then $\varpi_{\alpha\xi}$ will reduce to $\frac{1}{2}p_\alpha\cos(\alpha\xi)$ (for we may exclude the case in which p_ξ is also infinite because in that case $\varpi_{\alpha\xi}=0$, inasmuch as any two screws of infinite pitch are necessarily reciprocal).

The formula becomes, when p_α is very large,

$$\frac{p_\alpha}{\cos(\alpha\eta)}\varpi_{\beta\eta}=\frac{p_\beta}{\cos(\beta\xi)}\tfrac{1}{2}p_\alpha\cos(\alpha\xi).$$

In this case as the twist about α is merely a translation, we must have $\cos\alpha\eta=1$, so that

$$\varpi_{\beta\eta}=\frac{1}{2}\frac{p_\beta}{\cos(\beta\xi)}\cos(\alpha\xi).$$

$\varpi_{\beta\eta}$ is to vanish, but this cannot be secured by making p_β zero, because that cannot happen without $\cos(\beta\xi)$ being zero, except the pitch of ξ be infinite (§ 280) which is the case already excluded. It is therefore necessary that $\cos(\alpha\xi)$ be zero, but this requires that α and ξ be reciprocal, i.e. that $\varpi_{\alpha\xi}=0$.

Let us now suppose that we try to satisfy the original equations by making $\varpi_{\beta\eta}=0$, $p_\beta=0$. Here again we find that $p_\beta=0$ entails $\cos(\beta\xi)$ zero, except $p_\xi=\infty$. This in general makes $\varpi_{\alpha\xi}$ infinite so that the equation is not satisfied. If α and ξ were at right angles then no doubt the equation would be satisfied, but then $\varpi_{\alpha\xi}$ is zero. We thus see that notwithstanding the special form of the fundamental equation (§ 281) it implies no departure from the complete generality of the principle that whenever $\varpi_{\beta\eta}$ is zero then must $\varpi_{\alpha\xi}$ be also zero.

283. A Fundamental Theorem.

Let us suppose that a rigid body is either entirely free or constrained in any manner whatever. Let η be an impulsive screw whose pitch p_η is not infinite. Let η''' be the intensity of an impulsive wrench on that screw, it being understood that η''' is to be neither zero nor infinity. Let α be the instantaneous screw about which the body, having been previously quiescent, will commence to twist with an instantaneous twist velocity $\dot{\alpha}$. It is also supposed that p_α is neither zero nor infinity.

Let ξ be the impulsive screw similarly related to β, and let the affiliated symbols have the corresponding significations and limitations.

Let ζ be the impulsive screw similarly related to γ, and let the affiliated symbols have the corresponding significations and limitations.

The instantaneous movement of the body must necessarily be the same as if it had been quite free and had received in addition to the impulsive wrench of intensity η''' on the screw η, an impulsive wrench of intensity ρ''' situated on some screw ρ belonging to the system of screws reciprocal to the freedom of the body.

Let these two wrenches compound into a single wrench of intensity ω''' on a screw ω.

Then we have (§ 279),

$$\dot{\alpha} = \omega''' \frac{\cos(\alpha\omega)}{p_\alpha},$$

and also (§ 278),

$$\omega_1 = \frac{p_1\alpha_1}{p_\alpha}\cos(\alpha\omega), \quad \omega_2 = \frac{p_2\alpha_2}{p_\alpha}\cos(\alpha\omega), \; \ldots \; \omega_6 = \frac{p_6\alpha_6}{p_\alpha}\cos(\alpha\omega).$$

But from the fact that ω''' is the resultant of η''' and ρ''' we must have by resolving along the screws of reference

$$\omega'''\omega_1 = \eta'''\eta_1 + \rho'''\rho_1, \quad \omega'''\omega_2 = \eta'''\eta_2 + \rho'''\rho_2, \ldots \omega'''\omega_6 = \eta'''\eta_6 + \rho'''\rho_6 \ldots\ldots \text{(i)},$$

whence we obtain by substitution,

$$\dot{\alpha}p_1\alpha_1 = \eta'''\eta_1 + \rho'''\rho_1, \quad \dot{\alpha}p_2\alpha_2 = \eta'''\eta_2 + \rho'''\rho_2, \ldots \dot{\alpha}p_6\alpha_6 = \eta'''\eta_6 + \rho'''\rho_6 \ldots \text{(ii)}.$$

If we multiply the first of these equations by $p_1\beta_1$, the second by $p_2\beta_2$, &c., and then add, we obtain

$$\dot{\alpha}\Sigma p_1{}^2\alpha_1\beta_1 = \eta'''\varpi_{\beta\eta} + \rho'''\varpi_{\beta\rho};$$

as however ρ is on the reciprocal system we must have, except when $\rho''' = \infty$, to be subsequently considered,

$$\dot{\alpha}\Sigma p_1{}^2\alpha_1\beta_1 = \eta'''\varpi_{\beta\eta}.$$

In like manner,

$$\dot{\alpha}\Sigma p_1{}^2\alpha_1\gamma_1 = \eta'''\varpi_{\gamma\eta}.$$

We shall similarly find

$$\left.\begin{array}{ll} \dot{\beta}\Sigma p_1{}^2\beta_1\gamma_1 = \xi'''\varpi_{\gamma\xi}, & \dot{\beta}\Sigma p_1{}^2\beta_1\alpha_1 = \xi'''\varpi_{\alpha\xi} \\ \dot{\gamma}\Sigma p_1{}^2\gamma_1\alpha_1 = \zeta'''\varpi_{\alpha\zeta}, & \dot{\gamma}\Sigma p_1{}^2\gamma_1\beta_1 = \zeta'''\varpi_{\beta\zeta} \end{array}\right\} \ldots\ldots\ldots\ldots\ldots \text{(iii)},$$

whence by multiplication

$$\eta'''\xi'''\zeta'''\varpi_{\eta\beta}\varpi_{\xi\gamma}\varpi_{\zeta\alpha} = \eta'''\xi'''\zeta'''\varpi_{\eta\gamma}\varpi_{\xi\alpha}\varpi_{\zeta\beta}.$$

But we have chosen the intensities η''', ξ''', ζ''' so that no one of them is either zero or infinity, whence

$$\varpi_{\eta\beta}\varpi_{\xi\gamma}\varpi_{\zeta\alpha} = \varpi_{\eta\gamma}\varpi_{\xi\alpha}\varpi_{\zeta\beta} \ldots\ldots\ldots\ldots\ldots\ldots\ldots\ldots \text{(iv)}.$$

It remains to see whether this formula will continue to be satisfied in the cases excepted from this demonstration.

Let us take the case in which p_η is infinite, which makes $\eta_1 \ldots$ infinite. We have in the case of p_η very large,

$$\eta_1 = \frac{p_\eta \cos(\eta 1)}{2p_1},$$

the equations (ii) become

$$\dot{\alpha}p_1\alpha_1 = \frac{\eta'''p_\eta\cos(\eta 1)}{2p_1} + \rho''\rho_1; \; \ldots \; \dot{\alpha}p_6\alpha_6 = \frac{\eta'''p_\eta\cos(\eta 6)}{2p_6} + \rho''\rho_6;$$

multiplying the equations severally by $p_1, p_2, \ldots$ and adding, we get

$$\dot{\alpha}\Sigma p_1^2\alpha_1\beta_1 = \tfrac{1}{2}\eta''' p_\eta (\beta_1 \cos(\eta 1) + \beta_2 \cos(\eta 2) + \ldots + \beta_6 \cos(\eta 6))$$
$$= \tfrac{1}{2}\eta''' p_\eta \cos(\beta\eta)$$
$$= \eta''' \varpi_{\beta\eta} \text{ (since } p_\eta \text{ is indefinitely large)},$$

whence we proceed as before and we see that the theorem (iii) remains true, even if p_η or p_ξ or p_ζ be infinite.

If p_α be zero, then in general $\cos\alpha\omega$ is zero. But in this case $p_\alpha \div \cos\alpha\omega$ becomes d_α the length of the perpendicular from the centre of gravity upon α. Hence we have

$$\omega_1 = \frac{p_1\alpha_1}{d_\alpha}, \quad \omega_2 = \frac{p_2\alpha_2}{d_\alpha}, \ldots \omega_6 = \frac{p_6\alpha_6}{d_\alpha}$$

and the proof proceeds as before so that in this case also the theorem holds good.

Finally, let p_α be infinite, ω must then be of zero pitch and pass through the centre of gravity and

$$\dot{\alpha}p_\alpha = \omega'''.$$

We have

$$\omega_1 = \tfrac{1}{2}\cos(\alpha 1), \quad \omega_3 = \tfrac{1}{2}\cos(\alpha 3), \quad \omega_5 = \tfrac{1}{2}\cos(\alpha 5), \ldots$$

so that the equations (i) become

$$\tfrac{1}{2}\omega''' \cos(\alpha 1) = \eta'''\eta_1 + \rho'''\rho_1, \qquad \tfrac{1}{2}\omega''' \cos(\alpha 1) = \eta'''\eta_2 + \rho'''\rho_2,$$
$$\tfrac{1}{2}\omega''' \cos(\alpha 3) = \eta'''\eta_3 + \rho'''\rho_3, \qquad \tfrac{1}{2}\omega''' \cos(\alpha 3) = \eta'''\eta_4 + \rho'''\rho_4,$$
$$\tfrac{1}{2}\omega''' \cos(\alpha 5) = \eta'''\eta_5 + \rho'''\rho_5, \qquad \tfrac{1}{2}\omega''' \cos(\alpha 5) = \eta'''\eta_6 + \rho'''\rho_6.$$

Multiplying these equations by $+a\beta_1$, $-a\beta_2$, $+b\beta_3$, $-b\beta_3, \ldots$ and adding, we have

$$\tfrac{1}{2}\omega''' [a(\beta_1 - \beta_2)\cos(\alpha 1) + b(\beta_3 - \beta_4)\cos(\alpha 3) + c(\beta_5 - \beta_6)\cos(\alpha 5)] = \eta'''\varpi_{\eta\beta}.$$

Let σ be the screw belonging to the reciprocal system on which there is an impulsive wrench of intensity σ''' due to the reactions when an impulsive wrench is administered on ξ. Then we have

$$\dot{\beta}a\beta_1 = \xi'''\xi_1 + \sigma'''\sigma_1; \quad -\dot{\beta}a\beta_2 = \xi'''\xi_2 + \sigma'''\sigma_2,$$

whence

$$\dot{\beta}a(\beta_1 - \beta_2)\cos(\alpha 1) = \xi'''\cos(\xi 1)\cos(\alpha 1) + \sigma'''\cos(\sigma 1)\cos(\alpha 1),$$

with similar expressions for the two other pairs, whence by addition

$$\dot{\beta}\{a(\beta_1 - \beta_2)\cos(\alpha 1) + b(\beta_3 - \beta_4)\cos(\alpha 3) + c(\beta_5 - \beta_6)\cos(\alpha 5)\} = \xi'''\cos(\alpha\xi),$$

for since α and σ are reciprocal and $p_\alpha = \infty$ we must have $\cos(\alpha\sigma) = 0$.

We thus obtain

$$\frac{1}{2}\frac{\omega'''}{\dot{\beta}}\xi'''\cos(\alpha\xi) = \eta'''\varpi_{\eta\beta},$$

and similarly

$$\frac{1}{2}\frac{\omega'''}{\dot{\gamma}}\zeta'''\cos(\alpha\zeta) = \eta'''\varpi_{\eta\gamma},$$

whence

$$\frac{\dot{\gamma}\xi'''}{\dot{\beta}\zeta'''}\frac{\cos(\alpha\xi)}{\cos(\alpha\zeta)} = \frac{\varpi_{\eta\beta}}{\varpi_{\eta\gamma}},$$

or remembering that p_α is infinite,

$$\frac{\dot{\gamma}\xi'''}{\dot{\beta}\zeta'''}\frac{\varpi_{\alpha\xi}}{\varpi_{\alpha\zeta}} = \frac{\varpi_{\eta\beta}}{\varpi_{\eta\gamma}}.$$

But we had already from (iii),

$$\frac{\dot{\gamma}\xi'''}{\dot{\beta}\zeta'''} = \frac{\varpi_{\beta\zeta}}{\varpi_{\gamma\xi}},$$

whence we deduce that in this case also the formula remains true. We thus obtain the following general theorem.

If η, ξ, ζ be three impulsive screws and α, β, γ the three corresponding instantaneous screws, then in all cases, no matter how the movements of the body may be limited by constraints, the following formula holds good:*

$$\varpi_{\eta\beta}\varpi_{\xi\gamma}\varpi_{\zeta\alpha} = \varpi_{\eta\gamma}\varpi_{\xi\alpha}\varpi_{\zeta\beta}.$$

It is easily shown that this relation subsists when the correspondence between η and α is of the more general type implied by the equations

$$\eta_1 = \frac{1}{p_1}\frac{dU}{d\alpha_1}; \quad \eta_2 = \frac{1}{p_2}\frac{dU}{d\alpha_2}; \ \ldots \ \eta_6 = \frac{1}{p_6}\frac{dU}{d\alpha_6},$$

where U is any homogeneous function of the second degree in the co-ordinates.

284. Case of a Constrained Rigid Body.

Let η and ξ be, as before, a pair of impulsive screws, and let α and β be the corresponding pair of instantaneous screws. Let ρ be the screw on which a reaction is contributed by the constraints at the moment when the impulsive wrench is applied on η.

The movement of the body twisting about α is therefore the same as if it had been free, and one impulsive wrench had been imparted about η and another simultaneously about ρ, so that the following conditions are satisfied:

$$\dot{\alpha}p_1\alpha_1 = \eta'''\eta_1 + \rho'''\rho_1,$$
$$\cdots\cdots\cdots\cdots$$
$$\dot{\alpha}p_6\alpha_6 = \eta'''\eta_6 + \rho'''\rho_6.$$

* *Trans. Roy. Irish Acad.*, Vol. XXX. p. 575 (1894).

Multiplying by $p_1\alpha_1, \ldots p_6\alpha_6$, respectively, and adding, we have

$$\dot{\alpha} u_{\alpha\alpha} = \eta''' \varpi_{\eta\alpha},$$

where
$$u_{\alpha\alpha} = p_1^2\alpha_1^2 + \ldots + p_6^2\alpha_6^2;$$

because, as ρ belongs to the reciprocal system, we must have

$$\varpi_{\rho\alpha} = 0.$$

Similarly, if we multiplied the six equations by $p_1\beta_1, \ldots p_6\beta_6$, respectively, and added, we should get, since ρ is reciprocal to β also,

$$\dot{\alpha} u_{\alpha\beta} = \eta''' \varpi_{\eta\beta},$$

where
$$u_{\alpha\beta} = p_1^2\alpha_1\beta_1 + \ldots + p_6^2\alpha_6\beta_6.$$

Eliminating $\dot{\alpha}$ and η''' we have the concise result*,

$$u_{\alpha\alpha}\varpi_{\beta\eta} = u_{\alpha\beta}\varpi_{\alpha\eta}.$$

In a similar way we can deal with the pair of screws, β and ξ, and by eliminating σ, the reaction of the constraints in this case, we obtain the result

$$u_{\beta\beta}\varpi_{\alpha\xi} = u_{\beta\alpha}\varpi_{\beta\xi}.$$

Finally, from these two equations we can eliminate $u_{\alpha\beta}$, and we obtain

$$\frac{\varpi_{\alpha\eta}\varpi_{\alpha\xi}}{u_{\alpha\alpha}} = \frac{\varpi_{\beta\eta}\varpi_{\beta\xi}}{u_{\beta\beta}}.$$

This formula is a perfectly general relation, connecting any two pairs of impulsive and instantaneous screws η, α and ξ, β. It holds whether the body be free or constrained in any way whatever. If the body be perfectly free, then it is easy to show that it reduces to the result already found, viz.

$$\frac{p_\alpha}{\cos(\alpha\eta)}\varpi_{\beta\eta} = \frac{p_\beta}{\cos(\beta\xi)}\varpi_{\alpha\xi}.$$

285. Another Proof.

From the theory of impulsive and instantaneous screws in an n-system we know (§ 97) that if $\alpha_1, \ldots \alpha_n$ be the co-ordinates of an instantaneous screw, then the co-ordinates $\eta_1, \ldots \eta_n$ of the reduced impulsive screw may be determined as follows:

$$H\eta_1 = \frac{u_1^2}{p_1}\alpha_1,$$

$$\ldots\ldots\ldots\ldots\ldots$$

$$H\eta_n = \frac{u_n^2}{p_n}\alpha_n.$$

Multiplying severally by $p_1\alpha_1, \ldots p_n\alpha_n$, and adding, we have

$$H\varpi_{\eta\alpha} = u_{\alpha\alpha}.$$

* *Trans. Roy. Irish Acad.*, Vol. XXX. p. 573 (1894).

Multiplying similarly by $p_1\beta_1, \ldots p_n\beta_n$, and adding,

$$H\varpi_{\eta\beta} = u_{\alpha\beta}.$$

Eliminating H, we find

$$u_{\alpha\alpha}\varpi_{\beta\eta} - u_{\alpha\beta}\varpi_{\alpha\eta} = 0.$$

We may also prove this formula by physical consideration. Let α, β be the two screws which correspond, as instantaneous screws, to η and ξ, as impulsive screws.

Let us take on the cylindroid α, β, a screw θ, which is conjugate to α with respect to inertia (§ 81). Then, by known principles, the screw θ so defined must be reciprocal to η.

Hence
$$p_1\eta_1\theta_1 + \ldots + p_n\eta_n\theta_n = 0.$$

As, however, α and θ are conjugate, we have

$$u_1^2\alpha_1\theta_1 + \ldots + u_n^2\alpha_n\theta_n = 0;$$

also, since θ is co-cylindroidal with α and β, there must be relations of the kind

$$\theta_1 = \lambda\alpha_1 + \mu\beta_1; \ \ldots \ \theta_n = \lambda\alpha_n + \mu\beta_n.$$

Substituting these in the two previous equations, we get

$$\lambda\varpi_{\eta\alpha} + \mu\varpi_{\eta\beta} = 0;$$

$$\lambda u_{\alpha\alpha} + \mu u_{\alpha\beta} = 0;$$

whence, as before,

$$u_{\alpha\alpha}\varpi_{\beta\eta} - u_{\alpha\beta}\varpi_{\alpha\eta} = 0.$$

286. Twist Velocity acquired by an Impulse.

From the fact that the twist velocity $\dot{\alpha}$ acquired by a free body in consequence of an impulsive wrench of unit intensity on a screw η is expressed (§ 280) by the equation

$$\dot{\alpha} = \frac{\cos(\alpha\eta)}{p_\alpha}$$

we see that the second of the two formulae of § 281 may be expressed thus:—

$$\dot{\beta}\varpi_{\beta\eta} = \dot{\alpha}\varpi_{\alpha\xi}.$$

The proof thus given of this expression has assumed that the body is quite free.

It is however a remarkable fact that this formula holds good whatever be the constraints to which the body is submitted. If the body receive the unit impulsive wrench on a screw η, the body will commence to twist about a screw α. But the initial velocity of the body in this case will not generally be $\cos(\alpha\eta) \div p_\alpha$. It may be easily shown to be

$$\dot{\alpha} = \frac{\varpi_{\beta\eta}}{\Sigma p_1^2 \alpha_1 \beta_1}.$$

But we have also

$$\dot{\beta} = \frac{\varpi_{\alpha\xi}}{\Sigma p_1^2 \beta_1 \alpha_1},$$

whence in all cases

$$\dot{\beta}\varpi_{\beta\eta} = \dot{\alpha}\varpi_{\alpha\xi}.$$

This formula is therefore much more general besides being more concise than that of § 281.

287. System with Two Degrees of Freedom.

Let A, B, C, X, &c., and A', B', C', X', &c., be two homographic systems of points on a circle. These correspond respectively to two homographic systems of screws on the cylindroid according to the method of representation in Chap. XII. Then it is known, from geometrical principles, that if any two pairs, such as A, A' and B, B', be taken, the lines AB', BA' intersect on a definite straight line, which is the axis of the homography.

In general this axis may occupy any position whatever; if, however, it should pass through O, the pole of the axis of pitch, then the homography will assume a special type which it is our object to investigate.

In the first place, we may notice that under these circumstances the homography possesses the following characteristic:—

Let A, B be two screws, and A', B' their two correspondents; then, if A be reciprocal to B', B must be reciprocal to A'.

For in this case AB' must pass through O, and therefore BA' must pass through O also, i.e. B and A' must be reciprocal.

This cross relation suggests a name for the particular species of homography now before us. The form of the letter χ indicates so naturally the kind of relation, that I have found it convenient to refer to this type of homography as *Chiastic*. No doubt, in the present illustration I am only discussing the case of two degrees of freedom, but we shall presently see that chiastic homography is significant throughout the whole theory.

288. A Geometrical Proof.

It is known that in the circular representation the virtual coefficient of two screws is proportional to the perpendicular distance of their chord from the pole of the axis of pitch (§ 61).

Let α, β, γ be three screws of one system, and let η, ξ, ζ be the three corresponding screws, and, as usual, let $\varpi_{\alpha\xi}$ represent the virtual coefficient of α and ξ. Then whenever the homography is chiastic:—

$$\varpi_{\alpha\xi}\varpi_{\beta\zeta}\varpi_{\gamma\eta} = \varpi_{\alpha\zeta}\varpi_{\beta\eta}\varpi_{\gamma\xi}.$$

This is geometrically demonstrated when the following theorem is proved:—

If six points be inscribed on a circle, then the continued product of the three perpendiculars let fall from any point in a Pascal line formed from these six points upon three alternate sides of the corresponding hexagon is equal to the continued product of the three perpendiculars let fall from the same point on the other three sides.

Let $\alpha\alpha'$, $\beta\beta'$, $\gamma\gamma'$ be the three pairs of sides, and write the equation

$$\alpha\beta\gamma = \alpha'\beta'\gamma',$$

then this represents a cubic curve through the nine points $\alpha\alpha'$, $\alpha\beta'$, ..., and this cubic can only be the circle and the Pascal line.

289. Construction of Chiastic Homography on the Cylindroid.

It is first obvious that, if two corresponding pairs of screws be arbitrarily selected, it will always be possible to devise one chiastic homography of which those two pairs are corresponding members. The circular construction shows this at once for, join AB' and $A'B$, they intersect at T, then the line TO is the homographic axis, and the correspondent to X is found by drawing $A'X$, and then AX' through the intersection of $A'X$ and OT.

290. Homographic Systems on Two Cylindroids.

The fundamental theorem for the two cylindroids is thus expressed:—

Take any two screws, α and β, on one cylindroid, and any two screws, η and ξ, on the other, it will then be possible to inscribe one, and in general only one chiastic homography on the two surfaces, such that α and η shall be correspondents, and also β and ξ.

For, write the general equation

$$\varpi_{\alpha\zeta}\varpi_{\beta\eta}\varpi_{\gamma\xi} = \varpi_{\alpha\xi}\varpi_{\beta\zeta}\varpi_{\gamma\eta}.$$

If, then, α, β, η, ξ are known, and if γ be chosen arbitrarily on the first cylindroid, it will then be always possible to find one, but only one, screw ζ on the second cylindroid which satisfies the required condition.

If a body had two degrees of freedom expressed by a cylindroid A, and if an arbitrary cylindroid B were taken, then an impulsive wrench administered by any screw on B would make the body commence to twist about some corresponding screw on A, and the two systems of screws would have chiastic homography. If the body were given both in constitution and in position, then, of course, there would be nothing arbitrary in the choice of the corresponding screws. Suppose, however, that a screw η had been chosen arbitrarily on B to correspond to a screw α on A, it would then be generally possible to design and place a rigid body so that it should begin to twist about α in consequence of the impulse on η. There would, however, be no arbitrary element remaining in the homography. Thus, we see that, while for homography, in general, three pairs of correspondents can be arbitrarily assigned, there can only be two pairs so assigned for chiastic homography, while for such a particular type as that which relates to impulsive screws and the corresponding instantaneous screws, only one pair can be arbitrarily chosen.

291. Case of Normal Cylindroids.

We have already had occasion (§ 118), to remark on the curious relationship of two cylindroids when a screw can be found on either cylindroid which is reciprocal to all the screws on the other. If, for the moment, we speak of two such cylindroids as "normal," then we have the following theorem:—

Any homography of the screws on two cylindroids must be chiastic if the two cylindroids are normal.

Let α, β, γ be any three screws on one cylindroid, and η, ξ, ζ any three screws on the other; then, since the cylindroids are normal, we have

$$\varpi_{\beta\eta}\varpi_{\gamma\xi} = \varpi_{\beta\xi}\varpi_{\gamma\eta},$$

$$\varpi_{\alpha\zeta}\varpi_{\beta\xi} = \varpi_{\alpha\xi}\varpi_{\beta\zeta},$$

whence we obtain

$$\varpi_{\beta\xi}(\varpi_{\alpha\zeta}\varpi_{\beta\eta}\varpi_{\gamma\xi} - \varpi_{\alpha\xi}\varpi_{\beta\zeta}\varpi_{\gamma\eta}) = 0\,;$$

unless therefore ξ is reciprocal to β, we must have

$$\varpi_{\alpha\zeta}\varpi_{\beta\eta}\varpi_{\gamma\xi} - \varpi_{\alpha\xi}\varpi_{\beta\zeta}\varpi_{\gamma\eta} = 0.$$

If, however, ξ had been reciprocal to β, then one of these screws (suppose β) must have been the screw on its cylindroid reciprocal to the entire group of screws on the other cylindroid. In this case we must have

$$\varpi_{\beta\eta} = 0\,; \quad \varpi_{\beta\zeta} = 0,$$

so that even in this case it would still remain true that

$$\varpi_{\alpha\zeta}\varpi_{\beta\eta}\varpi_{\gamma\xi} - \varpi_{\alpha\xi}\varpi_{\beta\zeta}\varpi_{\gamma\alpha} = 0.$$

It is, indeed, a noteworthy circumstance that, for any and every three pairs of screws on two normal cylindroids, the relation just written must be fulfilled.

In general, when two pairs of screws are given on two cylindroids, the chiastic homography between the surfaces is determined. If, however, it were possible to determine two chiastic homographies having two pairs in common, then every homography is chiastic, and the cylindroids are normal.

Let α, η and β, ξ be the two pairs of correspondents, and let γ have the correspondents ζ and ζ', then we have

$$\varpi_{\alpha\zeta}\varpi_{\beta\eta}\varpi_{\gamma\xi} = \varpi_{\alpha\xi}\varpi_{\beta\zeta}\varpi_{\gamma\eta},$$

$$\varpi_{\alpha\zeta'}\varpi_{\beta\eta}\varpi_{\gamma\xi} = \varpi_{\alpha\xi}\varpi_{\beta\zeta'}\varpi_{\gamma\eta},$$

whence

$$\varpi_{\alpha\zeta}\varpi_{\beta\zeta'} = \varpi_{\alpha\zeta'}\varpi_{\beta\zeta},$$

i.e. the two cylindroids are normal.

292. General Conditions of Chiastic Homography.

We shall now discuss the relations of chiastic homography between two systems of screws in the same n-system. The first point to be demonstrated is, that in such a case every pair of the double screws are reciprocal.

Take α and β as two of the double screws, and η and ξ will coincide with them; whence the general condition,

$$\varpi_{\alpha\zeta}\varpi_{\beta\eta}\varpi_{\gamma\xi} = \varpi_{\alpha\xi}\varpi_{\beta\zeta}\varpi_{\gamma\eta}$$

becomes

$$\varpi_{\alpha\zeta}\varpi_{\alpha\beta}\varpi_{\beta\gamma} = \varpi_{\alpha\beta}\varpi_{\beta\zeta}\varpi_{\alpha\gamma},$$

$$\varpi_{\alpha\beta}(\varpi_{\alpha\zeta}\varpi_{\beta\gamma} - \varpi_{\alpha\gamma}\varpi_{\beta\zeta}) = 0.$$

One or other of these factors must be zero. We have to show that in general it is impossible for

$$\varpi_{\alpha\zeta}\varpi_{\beta\gamma} - \varpi_{\alpha\gamma}\varpi_{\beta\zeta}$$

to vanish.

For, take γ reciprocal to α but not to β, then $\varpi_{\alpha\gamma} = 0$; but $\varpi_{\beta\gamma}$ is not zero, and therefore $\varpi_{\alpha\zeta}$ would have to be zero; in other words ζ must be reciprocal to α. But this cannot generally be the case, and hence the other factor must vanish, that is

$$\varpi_{\alpha\beta} = 0.$$

In like manner it can be shown that every pair of the double screws must be reciprocal.

Conversely it can be shown that if the double screws of two homographic systems are co-reciprocal, then the homography is chiastic.

Let the n-double screws of the two systems be taken as the screws of reference; then if one screw in one system be denoted by the co-ordinates

$$\alpha_1, \ldots \alpha_n,$$

its correspondent in the other system will be

$$\lambda h_1 \alpha_1, \ldots \lambda h_n \alpha_n.$$

Similarly, the correspondent to

$$\beta_1, \ldots \beta_n$$

will have for its co-ordinates

$$\mu h_1 \beta_1, \ldots \mu h_n \beta_n,$$

and the correspondent to

$$\gamma_1, \ldots \gamma_n$$

will have for its co-ordinates

$$\nu h_1 \gamma_1, \ldots \nu h_n \gamma_n,$$

where λ, μ, ν, are the constants requisite to make the co-ordinates fulfil the fundamental conditions as to dimensions.

We thus compute

$$\varpi_{\alpha\xi} = \mu (p_1 h_1 \alpha_1 \beta_1 + \ldots + p_n h_n \alpha_n \beta_n);$$

and similarly for the other terms.

Whence, by substitution, we find the following equation identically satisfied:—

$$\varpi_{\alpha\xi} \varpi_{\beta\zeta} \varpi_{\gamma\eta} = \varpi_{\alpha\zeta} \varpi_{\beta\eta} \varpi_{\gamma\xi}.$$

It may be noted that, in a three-system, two homographies are chiastic when, in the plane representation by points, the double points of the two systems form a triangle which is self-conjugate with respect to the pitch conic.

293. Origin of the formulae of § 281*.

Let α be a screw about which a *free* rigid body is made to twist in consequence of an impulsive wrench administered on some other screw η. Except in the case where α and η are reciprocal, it will always be possible (in many different ways) to design and place a rigid body so that two arbitrarily chosen screws α and η will possess the required relation.

Let now β and ξ be two other screws (not reciprocal): we may consider the question as to whether a rigid body can be designed and placed so that α shall be the instantaneous screw corresponding to η as an impulsive screw, while β bears the same relation to ξ.

It is easy to see that it will not generally be possible for α, β, η, ξ to stand in the required relations. For, taking α and β as given, there are five

* *Proceedings of the Cambridge Phil. Soc.* Vol. IX. Part iii. p. 193.

disposable quantities in the choice of η, and five more in the choice of ξ. We ought, therefore, to have ten disposable co-ordinates for the designing and the placing of the rigid body. But there are not so many. We have three for the co-ordinates of its centre of gravity, three for the direction of its principal axes, and three more for the radii of gyration. The other circumstances of the rigid body are of no account for our present purpose.

It thus appears that if the four screws had been chosen arbitrarily we should have ten conditions to satisfy, and only nine disposable co-ordinates. It is hence plain that the four screws cannot be chosen quite arbitrarily. They must be in some way restricted. We can show as follows that these restrictions are not fewer than two.

Draw a cylindroid A through α, β, and another cylindroid P through $\eta\xi$. Then an impulsive wrench about any screw ω on P will make the body twist about some screw θ on A. As ω moves over P, so will its correspondent θ travel over A. It is shown in § 125 that any four screws on P will be equianharmonic with their four correspondents on A, and that consequently the two systems are homographic.

In general, to establish the homography of two cylindroids, three corresponding pairs of screws must be given; and, of course, there could be a triply infinite variety in the possible homographies. It is, however, a somewhat remarkable fact that in the particular homography with which we are concerned there is no arbitrary element. The fact that the rigid body is supposed quite free distinguishes this special case from the more general one of § 290. Given the cylindroids A and P, then, without any other considerations whatever, all the corresponding pairs are determined. This is first to be proved.

If the mass be one unit, and the intensity of the impulsive wrench on ω be one unit, then the twist velocity acquired by θ is (§ 280)

$$\frac{\cos(\theta\omega)}{p_\theta},$$

where $\cos(\theta\omega)$ denotes the cosine of the angle between the two screws θ and ω, and where p_θ is the pitch of θ. If, therefore, p_θ be zero, then $\cos(\theta\omega)$ must be zero. In other words, the two impulsive screws ω_1, ω_2 on P, which correspond to the two screws of zero pitch θ_1, θ_2 on A, must be at right angles to them, respectively. This will in general identify the correspondents on P to two known screws on A.

We have thus ascertained two pairs of correspondents, and we can now determine a third pair. For if ω_3 be a screw on P reciprocal to θ_2, then its correspondent θ_3 will be reciprocal to ω_2. Thus we have three pairs θ_1, θ_2, θ_3 on A, and their three correspondents ω_1, ω_2, ω_3 on P. This establishes the homography, and the correspondent θ to any other screw ω is assigned by the condition that the anharmonic ratio of $\omega_1\omega_2\omega_3\omega$ is the same as that of $\theta_1\theta_2\theta_3\theta$.

Reverting to our original screws α and η, β and ξ, we now see that they must fulfil the conditions

$$(\omega_1\omega_2\omega_3\eta) = (\theta_1\theta_2\theta_3\alpha), \qquad (\omega_1\omega_2\omega_2\xi) = (\theta_1\theta_2\theta_3\beta)$$

when the quantities in the brackets denote the anharmonic ratios.

It can be shown that these equations lead to the formulae of § 281.

294. Exception to be noted.

We have proved in the last article an instructive theorem which declares that when two cylindroids are given it is generally possible in one way, but in only one way, to correlate the several pairs of screws on the two surfaces, so that when a certain free rigid body received an impulse about the screw on one cylindroid, movement would commence by a twisting of the body about its correspondent on the other cylindroid. It is, however, easily seen that in one particular case the construction for correlation breaks down. The exception arises whenever the principal planes of the two cylindroids are at right angles.

The two correspondents on P to the zero-pitch screws on A had been chosen from the property that when p_α is zero the impulsive wrench must be perpendicular to α. We thus take the two screws on P which are respectively perpendicular to the two zero-pitch screws. But suppose there are *not* two screws on P which are perpendicular to the two zero-pitch screws on A. Suppose in fact that there is one screw on P which is parallel to the nodal axis of A, then the construction fails. We would thus have a single screw on P with *two* corresponding instantaneous screws for the same body. This is of course impossible, and accordingly in this particular case, which happens when the principal planes of P and A are rectangular, it is impossible to adjust the correspondence.

295. Impulsive and Instantaneous Cylindroids.

Let λ, λ' be two screws on a cylindroid whereof α and β are the two principal screws.

Let θ, θ' be the angles which λ and λ' respectively make with α.

We shall take the six absolute screws of inertia as the screws of reference and we have as the co-ordinates of λ

$$\alpha_1 \cos\theta + \beta_1 \sin\theta, \ldots \alpha_6 \cos\theta + \beta_6 \sin\theta,$$

and of λ'

$$\alpha_1 \cos\theta' + \beta_1 \sin\theta', \ldots \alpha_6 \cos\theta' + \beta_6 \sin\theta'.$$

In like manner, let ρ and ρ' be two screws on a cylindroid, of which the two principal screws are η and ξ.

Let ϕ, ϕ' be the angles which ρ and ρ' make respectively with ξ.

Then the co-ordinates of ρ are

$$\eta_1 \cos\phi + \xi_1 \sin\phi, \ldots \eta_6 \cos\phi + \xi_6 \sin\phi,$$

and of ρ'

$$\eta_1 \cos\phi' + \xi_1 \sin\phi', \ldots \eta_6 \cos\phi' + \xi_6 \sin\phi', \text{ \&c.}$$

We shall now suppose that the two cylindroids α, β and η, ξ are so circumstanced that the latter is the locus of the impulsive wrenches corresponding to the several instantaneous screws on the former with respect to the rigid body which is to be regarded as absolutely free. We shall further assume that ρ is the impulsive screw which has λ as its instantaneous screw, and that the relation of ρ' to λ' is of the same nature.

If, however, the four screws λ, λ', ρ, ρ' possess the relations thus indicated, it is necessary that they satisfy the conditions already proved (§ 281). These are two-fold, and they are expressed by the following equations, as already shown:—

$$\frac{p_\lambda}{\cos(\lambda\rho)} \cos(\lambda\rho') + \frac{p_{\lambda'}}{\cos(\lambda'\rho')} \cos(\lambda\rho') = 2\varpi_{\lambda\lambda'},$$

$$\frac{p_\lambda}{\cos(\lambda\rho)} \varpi_{\lambda'\rho} = \frac{p_\lambda}{\cos(\lambda'\rho')} \varpi_{\lambda\rho'}.$$

We shall arbitrarily choose λ' and ρ', so as to satisfy the conditions

$$\varpi_{\lambda'\rho} = 0, \quad \varpi_{\lambda\rho'} = 0,$$

and thus the second of the two equations is satisfied. These two equations will give θ' as a function of ϕ, and ϕ' as a function of θ. We can thus eliminate θ' and ϕ' from the first of the two equations, and the result will be a relation connecting θ and ϕ. This equation will exhibit the relation between any instantaneous screw θ on one cylindroid, and the corresponding impulsive screw ϕ on the other.

It will be observed that when the two cylindroids are given, the required equation is completely defined. The homographic relations of ρ and λ is thus completely determined by the geometrical relations of the two cylindroids.

The calculation* presents no difficulty and the result is as follows:—

$$
\begin{aligned}
0 = \cos\theta\cos\phi & \left[\begin{aligned} &+p_\alpha \cos(\beta\eta)\,[\varpi_{\alpha\eta}\cos(\beta\xi) - \varpi_{\alpha\xi}\cos(\beta\eta)] \\ &+p_\beta \cos(\alpha\eta)\,[\varpi_{\alpha\eta}\cos(\alpha\xi) - \varpi_{\alpha\xi}\cos(\alpha\eta)] \\ &-p_\alpha \varpi_{\beta\eta}\,[\cos(\alpha\xi)\cos(\beta\eta) - \cos(\alpha\eta)\cos(\beta\xi)] \end{aligned}\right] \\
+\cos\theta\sin\phi & \left[\begin{aligned} &+p_\alpha \cos(\beta\xi)\,[\varpi_{\alpha\eta}\cos(\beta\xi) - \varpi_{\alpha\xi}\cos(\beta\eta)] \\ &+p_\beta \cos(\alpha\xi)\,[\varpi_{\alpha\eta}\cos(\alpha\xi) - \varpi_{\alpha\xi}\cos(\alpha\eta)] \\ &-p_\alpha \varpi_{\beta\xi}\,[\cos(\alpha\xi)\cos(\beta\eta) - \cos(\alpha\eta)\cos(\beta\xi)] \end{aligned}\right] \\
+\sin\theta\cos\phi & \left[\begin{aligned} &+p_\beta \cos(\alpha\eta)\,[\varpi_{\beta\eta}\cos(\alpha\xi) - \varpi_{\beta\xi}\cos(\alpha\eta)] \\ &+p_\alpha \cos(\beta\eta)\,[\varpi_{\beta\eta}\cos(\beta\xi) - \varpi_{\beta\xi}\cos(\beta\eta)] \\ &+p_\beta \varpi_{\alpha\eta}\,[\cos(\alpha\xi)\cos(\beta\eta) - \cos(\alpha\eta)\cos(\beta\xi)] \end{aligned}\right] \\
+\sin\theta\sin\phi & \left[\begin{aligned} &+p_\beta \cos(\alpha\xi)\,[\varpi_{\beta\eta}\cos(\alpha\xi) - \varpi_{\beta\xi}\cos(\alpha\eta)] \\ &+p_\alpha \cos(\beta\xi)\,[\varpi_{\beta\eta}\cos(\beta\xi) - \varpi_{\beta\xi}\cos(\beta\eta)] \\ &+p_\beta \varpi_{\alpha\xi}\,[\cos(\alpha\xi)\cos(\beta\eta) - \cos(\alpha\eta)\cos(\beta\xi)] \end{aligned}\right]
\end{aligned}
$$

296. An exceptional Case.

A few remarks should be made on the failure of the correspondence when the principal planes of the two cylindroids are at right angles (§ 294). It will be noted that though this equation suffers a slight reduction when the principal planes of the two cylindroids are at right angles yet it does not become evanescent or impossible. For any value of θ defining a screw on one cylindroid, the equation provides a value of ϕ for the correspondent on the other cylindroid. Thus we seem to meet with a contradiction, for while the argument of § 294 shows that in such a case the homography is impossible, yet the homographic equation seemed to show that it was possible and indeed fixed the pairs of correspondents with absolute definiteness.

It is certainly true that if two cylindroids A and P admit of the correlation of their screws into pairs whereof those on P are impulsive screws and those on A are instantaneous screws, the pairs of screws by which the homographic equation is satisfied will stand to each other in the desired relation. If, however, the screws on two cylindroids be correlated into pairs in accordance with the indications of the homographic equation, though it will generally be true that there may be corresponding impulsive screws and instantaneous screws, yet in the case where the principal planes of the cylindroids are at right angles no such inference can be drawn.

The case is a somewhat curious one. It will be seen that the calculation

* See *Trans. Roy. Irish Acad.* Vol. xxx. p. 112 (1894).

of the homographic equation is based on the fact that if λ, λ' be two instantaneous screws and ρ, ρ' the corresponding impulsive screws, then the formula

$$\frac{p_\lambda}{\cos(\lambda\rho)}\cos(\lambda'\rho)+\frac{p_{\lambda'}}{\cos(\lambda'\rho')}\cos(\lambda\rho')=2\varpi_{\lambda\lambda'}$$

must be satisfied.

And *generally* it is satisfied. In the case of two cylindroids with normal planes it is however easy to show that there are certain pairs of screws for which this formula cannot obtain.

For in such a case there is one screw λ on A which is perpendicular to every screw on P, so that whatever be the ρ corresponding to λ,

$$\cos(\lambda\rho)=0.$$

Since no other screw λ' can be perpendicular to any screw on P we cannot have either

$$\cos(\lambda'\rho), \text{ or } \cos(\lambda'\rho'), \text{ zero.}$$

Hence this equation cannot be satisfied and the argument that the homographic equation defines corresponding pairs is in this case invalid.

We might have explained the matter in the following manner.

When the principal planes of A and P are normal there is one screw λ on A which is perpendicular to all the screws on P. If therefore the two cylindroids were to be impulsive and instantaneous, there must be a screw θ on P which corresponds to λ. It can be shown in general (§ 301) that

$$d_\lambda=p_\lambda\tan(\lambda\theta)$$

when d_λ is the perpendicular from the centre of gravity on λ; it follows that when $(\lambda\theta)=90°$ we must have either p_λ zero or d_λ infinite.

But of course it will not generally be the case that λ happens to be one of the screws of zero pitch on A. Hence we are reduced to the other alternative

$$d_\lambda=\text{infinity}.$$

This means that the centre of gravity is to be at infinity.

But when the centre of gravity of the body is at infinity a remarkable consequence follows. All the instantaneous screws must be parallel.

For if θ be the impulsive wrench corresponding to λ as the instantaneous screw, then we know that

$$d_\lambda=p_\lambda\tan(\lambda\theta),$$

and that the centre of gravity lies in a right line parallel to λ and distant from it by d_λ. In like manner if ϕ be an impulsive screw corresponding to μ as instantaneous screw we have another locus parallel to μ for the centre of gravity.

But as the centre of gravity is at infinity these two loci must there intersect, i.e. they must be parallel and so must λ and μ, and hence all instantaneous screws must be parallel.

Thus we see that all the screws on A must be parallel, i.e. that A must have degraded into an extreme type of cylindroid.

297. Another extreme Case.

Given any two cylindroids A and P it is, as we have seen, generally possible to correlate in one way the several screws on A to those on P so that an impulsive wrench given to a certain rigid body about any screw on P would make that body commence to move by twisting about its correspondent on A. One case of failure has just been discussed. The case now to be considered is not indeed one of failure but one in which *any* two pairs of screws on A and P will stand to each other in the desired relations.

Suppose that A and P happened to fulfil the single condition that each of them shall contain one screw which is reciprocal to the other cylindroid. We have called the cylindroids so circumstanced "normal."

Let λ be the screw on A which is reciprocal to every screw on P. If then P and A are to stand to each other in the required relation, λ must be reciprocal to its impulsive screw. But this is only possible on one condition. The mass of the body must be zero. In that case, if there is *no* mass involved any one of the screws on P may be the impulsive screw corresponding to any one of the screws on A.

Here again the question arises as to what becomes of the homographic equation which defines so precisely the screw on P which corresponds to the screws on A (§ 295). It might have been expected that in the case of two normal cylindroids this homographic equation should become evanescent. But it does not do so.

But there is no real contradiction. The greater includes the less. If *every* screw on P will suit as correspondent *every* screw on A then *à fortiori* will the pairs indicated by the homography fulfil the conditions requisite.

That any two pairs of screws will be correspondents in this case is obvious from the following.

Let λ be the screw on A which is reciprocal to P,

θ .... P .... A.

Then *any* screw μ on A and *any* screw ϕ on P fulfil the conditions

$$\varpi_{\lambda\phi} = 0, \quad \varpi_{\mu\theta} = 0.$$

Hence ϕ is the impulsive screw corresponding to μ as the instantaneous screw.

298. Three Pairs of Correspondents.

Let α, η; β, ξ; γ, ζ be three pairs of impulsive and instantaneous screws; let θ, ϕ be another pair. Then, if we denote by $L_{\alpha\beta} = 0$, and $M_{\alpha\beta} = 0$, the two fundamental equations

$$\frac{p_\alpha}{\cos(\alpha\eta)}\cos(\beta\eta) + \frac{p_\beta}{\cos(\beta\xi)}\cos(\alpha\xi) = 2\varpi_{\alpha\beta},$$

$$\frac{p_\alpha}{\cos(\alpha\eta)}\varpi_{\beta\eta} = \frac{p_\beta}{\cos(\beta\xi)}\varpi_{\alpha\xi},$$

we shall obtain six equations of the type

$$L_{\theta\alpha} = 0, \qquad L_{\theta\beta} = 0, \qquad L_{\theta\gamma} = 0,$$

$$M_{\theta\alpha} = 0, \qquad M_{\theta\beta} = 0, \qquad M_{\theta\gamma} = 0.$$

From these six it might be thought that $\phi_1, \ldots \phi_6$ could be eliminated, and thus it would, at first sight, seem that there must be an equation for θ to satisfy. It is, however, obvious that there can be no such condition, for θ can of course be chosen arbitrarily. The fact is, that these equations have a peculiar character which precludes the ordinary algebraical inference.

Since α, η; β, ξ; γ, ζ; are three pairs of screws, fulfilling the necessary six conditions, a rigid body can be adjusted to them so that they are respectively impulsive and instantaneous. We take the six principal screws of inertia of this body as the screws of reference. We thus have, where $\rho_\alpha, \rho_\beta, \rho_\gamma$ are certain factors,

$$\rho_\alpha\eta_1 = a\alpha_1, \qquad \rho_\alpha\eta_2 = -a\alpha_2, \qquad \rho_\alpha\eta_3 = b\alpha_3, \qquad \rho_\alpha\eta_4 = -b\alpha_4,$$

$$\rho_\beta\xi_1 = a\beta_1, \qquad \ldots\ldots\ldots\ldots, \qquad \ldots\ldots\ldots\ldots, \qquad \ldots\ldots\ldots\ldots,$$

$$\rho_\gamma\zeta_1 = a\gamma_1, \qquad \ldots\ldots\ldots\ldots, \qquad \ldots\ldots\ldots\ldots, \qquad \ldots\ldots\ldots\ldots$$

By putting the co-ordinates in this form, we imply that they satisfy the six equations of condition above written.

Substituting the co-ordinates in $L_{\theta\alpha} = 0$, we get

$$\begin{aligned}0 = &+(\alpha_1+\alpha_2)(\rho_\theta\phi_1+\rho_\theta\phi_2)+(\alpha_3+\alpha_4)(\rho_\theta\phi_3+\rho_\theta\phi_4)+(\alpha_5+\alpha_6)(\rho_\theta\phi_5+\rho_\theta\phi_6)\\ &+(\theta_1+\theta_2)(a\alpha_1-a\alpha_2)+(\theta_3+\theta_4)(b\alpha_3-b\alpha_4)+(\theta_5+\theta_6)(c\alpha_5-c\alpha_6)\\ &-2(a\alpha_1\theta_1-a\alpha_2\theta_2)-2(b\alpha_3\theta_3-b\alpha_4\theta_4)-2(c\alpha_5\theta_5-c\alpha_6\theta_6).\end{aligned}$$

Let

$$\rho_\theta\phi_1 - a\theta_1 = X_1, \qquad \rho_\theta\phi_2 + a\theta_2 = X_2,$$
$$\rho_\theta\phi_3 - b\theta_3 = X_3, \qquad \rho_\theta\phi_4 + b\theta_4 = X_4,$$
$$\rho_\theta\phi_5 - c\theta_5 = X_5, \qquad \rho_\theta\phi_6 + c\theta_6 = X_6,$$

and the equation becomes

$$0 = (\alpha_1 + \alpha_2)(X_1 + X_2) + (\alpha_3 + \alpha_4)(X_3 + X_4) + (\alpha_5 + \alpha_6)(X_5 + X_6);$$

and the two other L equations give

$$0 = (\beta_1 + \beta_2)(X_1 + X_2) + (\beta_3 + \beta_4)(X_3 + X_4) + (\beta_5 + \beta_6)(X_5 + X_6),$$
$$0 = (\gamma_1 + \gamma_2)(X_1 + X_2) + (\gamma_3 + \gamma_4)(X_3 + X_4) + (\gamma_5 + \gamma_6)(X_5 + X_6).$$

If we eliminate $X_1 + X_2$, $X_3 + X_4$, $X_5 + X_6$ from these equations, we should have

$$0 = \begin{vmatrix} \alpha_1 + \alpha_2 & \alpha_3 + \alpha_4 & \alpha_5 + \alpha_6 \\ \beta_1 + \beta_2 & \beta_3 + \beta_4 & \beta_5 + \beta_6 \\ \gamma_1 + \gamma_2 & \gamma_3 + \gamma_4 & \gamma_5 + \gamma_6 \end{vmatrix}.$$

But this would only be the case if α, β, γ were parallel to a plane, which is not generally true. Therefore, we can only satisfy these equations, under ordinary circumstances, by the assumption

$$X_1 + X_2 = 0, \qquad X_3 + X_4 = 0, \qquad X_5 + X_6 = 0.$$

In like manner, the equations of the M type give

$$\rho_\theta \varpi_{\alpha\phi} - \rho_\alpha \varpi_{\theta\eta} = 0,$$
$$\rho_\theta \varpi_{\beta\phi} - \rho_\beta \varpi_{\theta\xi} = 0,$$
$$\rho_\theta \varpi_{\gamma\phi} - \rho_\gamma \varpi_{\theta\zeta} = 0.$$

Substituting, in the first of these, we get

$$+ \rho_\theta (a\alpha_1\phi_1 - a\alpha_2\phi_2 + b\alpha_3\phi_3 - b\alpha_4\phi_4 + c\alpha_5\phi_5 - c\alpha_6\phi_6),$$
$$- \rho_\alpha (a\eta_1\theta_1 - a\eta_2\theta_2 + b\eta_3\theta_3 - b\eta_4\theta_4 + c\eta_5\theta_5 - c\eta_6\theta_6) = 0;$$

which reduces to

$$a\alpha_1 X_1 - a\alpha_2 X_2 + b\alpha_3 X_3 - b\alpha_4 X_4 + c\alpha_5 X_5 - c\alpha_6 X_6 = 0;$$

but we have already seen that $X_1 + X_2 = 0$, &c., whence we obtain

$$X_1 (a\alpha_1 + a\alpha_2) + X_3 (b\alpha_3 + b\alpha_4) + X_5 (c\alpha_5 + c\alpha_6) = 0;$$

with the similar equations

$$X_1 (a\beta_1 + a\beta_2) + X_3 (b\beta_3 + b\beta_4) + X_5 (c\beta_5 + c\beta_6) = 0,$$
$$X_1 (a\gamma_1 + a\gamma_2) + X_3 (b\gamma_3 + b\gamma_4) + X_5 (c\gamma_5 + c\gamma_6) = 0.$$

These prove that, unless α, β, γ be parallel to a plane, we must have

$X_1=0$, $X_3=0$, $X_5=0$. Combining these conditions with the last, we draw the general conclusion that

$$X_1=0,\quad X_2=0,\quad X_3=0,\quad X_4=0,\quad X_5=0,\quad X_6=0;$$

or $$\rho_0\phi_1=a\theta_1;\qquad \rho_0\phi_2=-a\theta_2;\qquad \rho_0\phi_3=b\theta_3;\qquad \rho_0\phi_4=-b\theta_4\ \ \text{\&c.}$$

Thus we demonstrate that if a pair of screws θ, ϕ satisfy the six conditions, they stand to each other in the relation of impulsive screws and instantaneous screws.

299. Cylindroid Reduced to a Plane.

Suppose that the family of rigid bodies be found which make α, η and β, ξ impulsive and instantaneous. Let there be any third screw, γ, and let us seek for the locus of its impulsive screw, ζ, for all the different rigid bodies of the family.

ζ must satisfy the four equations

$$\frac{p_\alpha}{\cos(\alpha\eta)}\cos(\gamma\eta)+\frac{p_\gamma}{\cos(\gamma\zeta)}\cos(\alpha\zeta)=2\varpi_{\alpha\gamma},$$

$$\frac{p_\beta}{\cos(\beta\xi)}\cos(\gamma\xi)+\frac{p_\gamma}{\cos(\gamma\zeta)}\cos(\beta\zeta)=2\varpi_{\beta\gamma},$$

$$\frac{p_\alpha}{\cos(\alpha\eta)}\varpi_{\gamma\eta}=\frac{p_\gamma}{\cos(\gamma\zeta)}\varpi_{\alpha\zeta},$$

$$\frac{p_\beta}{\cos(\beta\xi)}\varpi_{\gamma\xi}=\frac{p_\gamma}{\cos(\gamma\zeta)}\varpi_{\beta\zeta}.$$

As there are four linear equations in the coordinates of ζ, we have the following theorem.

If α, η and β, ξ be given pairs of impulsive and instantaneous screws, then the locus of ζ, the impulsive screw corresponding to γ, as an instantaneous screw, is a cylindroid.

But this cylindroid is of a special type. It is indeed a plane surface rather than a cubic. The equations for ζ can have this form:—

$$\cos(\alpha\zeta)=A\cos(\gamma\zeta),\qquad \varpi_{\alpha\zeta}=C\cos(\gamma\zeta),$$

$$\cos(\beta\zeta)=B\cos(\gamma\zeta),\qquad \varpi_{\beta\zeta}=D\cos(\gamma\zeta),$$

in which A, B, C, D are constants.

The fact that $\cos(\alpha\zeta)$ and $\cos(\gamma\zeta)$ have one fixed ratio, and $\cos(\beta\zeta)$ and $\cos(\gamma\zeta)$ another, shows that the direction of ζ is fixed. The cylindroidal locus of ζ, therefore, degenerates to a system of parallel lines.

At first it may seem surprising to find that $\varpi_{\alpha\zeta}$ is constant. But the necessity for this arrangement is thus shown. If not constant, then there would generally have been some screw ζ, for which $\varpi_{\alpha\zeta}$ was zero. In this case, of course, $\varpi_{\gamma\eta}$ would be generally zero also. But γ and η being both given, this is of course not generally true. The only escape is for $\varpi_{\alpha\zeta}$ to be constant.

300. A difficulty removed.

Given α and η, β and ζ, and also γ, then the plane of ζ is determined from the equations of the last article.

As $\varpi_{\alpha\zeta}$ and $\varpi_{\beta\zeta}$ are constant, both α and β must be parallel to the plane already considered. But as an impulsive screw could not be reciprocal to an instantaneous screw, it would seem that $\varpi_{\gamma\zeta}$ must *never* be zero, but this condition can only be fulfilled by requiring that ζ must be parallel to the same plane. Whence α, β, γ must be parallel to the same plane. But these three screws are quite arbitrary. Here then would seem to be a contradiction.

The difficulty can be explained as follows:—

Each rigid body, which conformed to the condition that α, β and η, ξ shall be two pairs of corresponding impulsive and instantaneous screws, will have a different screw ζ corresponding to a given screw γ. Thus, among the various screws ζ, in the degraded cylindroid, each will correspond to one rigid body. In general, of course, it would be impossible for ζ to be reciprocal to γ. It would be impossible for an impulsive wrench to make a body twist about a screw reciprocal thereto. Nevertheless, it seemed certain that, in general, there must be a screw ζ reciprocal to γ. For otherwise, α, β, γ should be all parallel to a plane, which, of course, is not generally true. If, however, a, or b, or c were zero, then the body will have *no mass*; consequently no impulse would be necessary to set it in motion. This clearly is the case when ζ is reciprocal to γ. We have thus got over the difficulty. ζ and γ are reciprocal, in the case when the rigid body is such that a, or b, or c is zero.

301. Two Geometrical Theorems.

The perpendicular from the centre of gravity on any instantaneous screw is parallel to the shortest distance between that instantaneous screw and the corresponding impulsive screw.

The perpendicular from the centre of gravity on any instantaneous screw is equal to the product of the pitch of that screw, and the tangent of the angle between it and the corresponding impulsive screw.

Let α be the instantaneous screw and d_α the length of the perpendicular thereon from the centre of gravity. If $\cos\lambda$, $\cos\mu$, $\cos\nu$ be the direction cosines of d_α then

$$d_\alpha \cos\lambda = (\alpha_5 - \alpha_6)(\alpha_3 + \alpha_4)\,c - (\alpha_5 + \alpha_6)(\alpha_3 - \alpha_4)\,b,$$
$$d_\alpha \cos\mu = (\alpha_1 - \alpha_2)(\alpha_5 + \alpha_6)\,a - (\alpha_1 + \alpha_2)(\alpha_5 - \alpha_6)\,c,$$
$$d_\alpha \cos\nu = (\alpha_3 - \alpha_4)(\alpha_1 + \alpha_2)\,b - (\alpha_3 + \alpha_4)(\alpha_1 - \alpha_2)\,a.$$

But if η is the impulsive screw corresponding to α as the instantaneous screw we have

$$a\alpha_1 = \frac{p_\alpha}{\cos(\alpha\eta)}\eta_1; \quad -a\alpha_2 = \frac{p_\alpha}{\cos(\alpha\eta)}\eta_2, \text{ \&c., \&c.,}$$

whence

$$d_\alpha \cos\lambda = \frac{p_\alpha}{\cos(\alpha\eta)}((\eta_5 + \eta_6)(\eta_3 + \alpha_4) - (\alpha_5 + \alpha_6)(\eta_3 + \eta_4)),$$

$$d_\alpha \cos\mu = \frac{p_\alpha}{\cos(\alpha\eta)}((\eta_1 + \eta_2)(\alpha_5 + \alpha_6) - (\alpha_1 + \alpha_2)(\eta_5 + \eta_6)),$$

$$d_\alpha \cos\nu = \frac{p_\alpha}{\cos(\alpha\eta)}((\eta_3 + \eta_4)(\alpha_1 + \alpha_2) - (\alpha_3 + \alpha_4)(\eta_1 + \eta_2)).$$

But

$$(\eta_5 + \eta_6)(\alpha_3 + \alpha_4) - (\alpha_5 + \alpha_6)(\eta_3 + \eta_4) = \sin(\alpha\eta)\cos\lambda',$$

with similar expressions for $\sin(\alpha\eta)\cos\mu'$ and $\sin(\alpha\eta)\cos\nu'$ where $\cos\lambda'$, $\cos\mu'$, and $\cos\nu'$ are the direction cosines of the common perpendicular to α and η. We have therefore

$$d_\alpha \cos\lambda = \frac{p_\alpha}{\cos(\alpha\eta)}\sin(\alpha\eta)\cos\lambda',$$

$$d_\alpha \cos\mu = \frac{p_\alpha}{\cos(\alpha\eta)}\sin(\alpha\eta)\cos\mu',$$

$$d_\alpha \cos\nu = \frac{p_\alpha}{\cos(\alpha\eta)}\sin(\alpha\eta)\cos\nu',$$

whence

$$\cos\lambda = \cos\lambda'; \quad \cos\mu = \cos\mu'; \quad \cos\nu = \cos\nu';$$

and

$$d_\alpha = p_\alpha \tan(\alpha\eta),$$

which proves the theorems.

CHAPTER XXII.

THE GEOMETRICAL THEORY*.

302. Preliminary.

It will be remembered how Poinsot advanced our knowledge of the dynamics of a rigid system by a beautiful geometrical theory of the rotation of a rigid body about a fixed point. We now specially refer to the geometrical construction by which he determined the instantaneous axis about which the body commenced to rotate when the plane of the instantaneous couple was given.

We may enunciate with a generality, increasing in successive steps, the problem which, in its simplest form, Poinsot had made classical. From the case of a rigid body which is constrained to rotate about a fixed point we advance to the wider conception of a body which has three degrees of freedom of the most general type. We can generalize this again into the case in which the body, instead of having a definite number of degrees of freedom has any number of such degrees. The range extends from the first or lowest degree, where the only movement of which the body is capable is that of twisting about a single fixed screw, up to the case in which the body being perfectly free, or in other words, having six degrees of freedom, is able to twist about every screw in space. It will, of course, be borne in mind that only small movements are to be understood.

In a corresponding manner we may generalize the forces applied to the body. In the problem solved by Poinsot the effective forces are equivalent to a couple solely. For the reaction of the fixed point is capable of reducing any system of forces whatever to a couple. But in the more generalized problems with which the theory of screws is concerned, we do not restrict the forces to the specialized pair which form a couple. We shall assume that the forces are of the most general type and represented by a wrench upon a screw. Thus, by generalizing the freedom of the rigid body, as well as the forces which act upon it, we may investigate the geometrical theory of the motion when a rigid body of the most general type, possessing a certain number of degrees of freedom of the most general type, is disturbed from a

* *Trans. Royal Irish Acad.* Vol. XXI. (1897) p. 185.

position of rest by an impulsive system of forces of the most general type. This is the object of the present chapter.

303. One Pair of Impulsive and Instantaneous Screws.

Let it be supposed that nothing is known of the position, mass, or other circumstances of an unconstrained rigid body save what can be deduced from the fact that, when struck from a position of rest by an impulsive wrench on a specified screw η, the effect is to make the body commence to move by twisting around a specified screw α.

As α, like every other screw, is defined by five coordinates, the knowledge of this screw gives us five data towards the nine data that are required for the complete specification of the rigid body and its position.

We have first to prove that the five elements which can be thence inferred with respect to the rigid body are in general—

(1) *A diameter of the momental ellipsoid.*

This is clearly equivalent to *two* elements, inasmuch as it restricts the position of the centre of gravity to a determinate straight line.

(2) *The radius of gyration about this diameter.*

This is, of course, *one* element.

(3) *A straight line in the plane conjugate to that diameter.*

A point in the plane would have been *one* element, but a straight line in the plane is equivalent to *two*. If the centre of gravity were also known, we should at once be able to draw the conjugate plane.

Draw a plane through both the instantaneous screw α and the common perpendicular to α and η. Then the centre of gravity of the rigid body must lie in that plane (§ 301). It was also shown that if p_α be the pitch of α, and if $(\alpha\eta)$ represent the angle between α and η, then the perpendicular distance of the centre of gravity from α will be expressed by $p_\alpha \tan(\alpha\eta)$ (§ 301). This expression is completely known since α and η are known. Thus we find that the centre of gravity must lie in a determinate ray parallel to α. There will be no ambiguity as to the side on which this straight line lies if it be observed that it must pass between α and the point in which η is met by the common perpendicular to η and α. In this manner from knowing α and η we discover a diameter of the momental ellipsoid.

If $\dot{\alpha}$ be the twist velocity with which a rigid body of mass M is twisting about any screw α. If η be the corresponding impulsive screw, and if $\varpi_{\alpha\eta}$ denote as usual the virtual coefficient of α and η, then it is proved in § 279 that the kinetic energy of the body

$$M\dot{\alpha}^2 \frac{p_\alpha}{\cos(\alpha\eta)} \varpi_{\alpha\eta}.$$

We can now determine the value of ρ_α^2 where ρ_α is the radius of gyration about an axis parallel to α through the centre of gravity. For the kinetic energy is obviously

$$\tfrac{1}{2}M\dot{\alpha}^2(\rho_\alpha^2 + p_\alpha^2 + a_\alpha^2).$$

By equating the two expressions we have

$$\rho_\alpha^2 = 2\frac{p_\alpha}{\cos(\alpha\eta)}\varpi_{\alpha\eta} - p_\alpha^2 - a_\alpha^2.$$

But when α and η are known the three terms on the right-hand side of this equation are determined. Thus we learn the radius of gyration on the diameter parallel to α.

It remains to show how a certain straight line in the plane which is conjugate to this diameter in the momental ellipsoid is also determined. Let a screw θ, of zero pitch, be placed on that known diameter of the momental ellipsoid which is parallel to α. Draw a cylindroid through the two screws θ and η. Let ϕ be the other screw of the zero pitch, which will in general be found on the same cylindroid.

We could replace the original impulsive wrench on η by its two component wrenches on any two screws of the cylindroid. We choose for this purpose the two screws of zero pitch θ and ϕ. Thus we replace the wrench on η by two forces, whose joint effect is identical with the effect that would have been produced by the wrench on η.

As to the force along the line θ it is, from the nature of the construction, directed through the centre of gravity. Such an impulsive force would produce a velocity of translation, but it could have no effect in producing a rotation. The rotatory part of the initial twist velocity must therefore be solely the result of the impulsive force on ϕ.

But when an impulsive force is applied to a quiescent rigid body we know, from Poinsot's theorem, that the rotatory part of the instantaneous movement must be about an axis parallel to the direction which is conjugate in the momental ellipsoid to the plane which contains both the centre of gravity and the impulsive force. It follows that the ray ϕ must be situated in that plane which is conjugate in the momental ellipsoid to the diameter parallel to α. But, as we have already seen, the position of ϕ is completely defined on the known cylindroid on which it lies. We have thus obtained a fixed ray in the conjugate plane to a known diameter of the momental ellipsoid.

The three statements at the beginning of this article have therefore been established. We have, accordingly, ascertained five distinct geometrical data towards the nine which are necessary for the complete specification of the rigid body. These five data are inferred solely from our knowledge of a single pair of corresponding impulsive and instantaneous screws.

304. An Important Exception.

If $p_\alpha = 0$, then $(\alpha\eta)$ is 90°, and consequently $p_\alpha \tan(\alpha\eta)$ is indefinite. If, therefore, the pitch of the instantaneous screw be zero, then we are no longer entitled to locate the centre of gravity in a certain ray. All we know is that it lies in the plane through α perpendicular to η. In general the knowledge of the impulsive screw corresponding to a given instantaneous screw implies five data, yet this ceases to be the case if p_α is zero, for as η must then be perpendicular to α there are really only four independent data given when η is given. We have, therefore, in this case one element the less towards the determination of the rigid body.

305. Two Pairs of Impulsive and Instantaneous Screws.

Let us next suppose that we are given a second pair of corresponding impulsive and instantaneous screws. We shall examine how much further we are enabled to proceed by the help of this additional information towards the complete determination of the rigid body in its abstract form. Any data in excess of nine, if not actually impossible, would be superfluous. If, therefore, we are given a second pair of impulsive and instantaneous screws, the five data which they bring cannot be wholly independent of the five data brought by the preceding pair. It is therefore plain that the quartet of screws forming two pairs of corresponding impulsive screws and instantaneous screws cannot be chosen arbitrarily. They must submit to at least one purely geometrical condition, so that the number of data independent of each other shall not exceed nine.

It is, however, not so obvious, though it is certainly true, as we found in § 281, that the two pairs of screws must conform not merely to one, but to no less than two geometrical conditions. In fact, if η, ξ be two impulsive screws, and if α, β be the two corresponding instantaneous screws, then, when the body acted upon is perfectly free, the following two formulæ must be satisfied:

$$\frac{p_\alpha}{\cos(\alpha\eta)}\varpi_{\beta\eta} = \frac{p_\beta}{\cos(\beta\xi)}\varpi_{\alpha\xi},$$

$$2\varpi_{\alpha\beta} = \frac{p_\alpha}{\cos(\alpha\eta)}\cos(\beta\eta) + \frac{p_\beta}{\cos(\beta\xi)}\cos(\alpha\xi).$$

We can enunciate two geometrical properties of the two pairs of screws, which are equivalent to the conditions expressed by these equations.

In the first place, each of the pairs of screws determines a diameter of the momental ellipsoid. The fact that the two diameters, so found, must intersect each other, is obviously one geometrical condition imposed on the system α, η and β, ξ.

Let G be this intersection, and draw GP parallel to α and equal to the radius of gyration about GP, which we have shown to be known from the fact that α and η are known. Let X be the plane conjugate to GP in the momental ellipsoid, then this plane is also known.

In like manner, draw GQ parallel to β and equal to the radius of gyration about GQ. Let Y be the plane, conjugate to GQ, in the momental ellipsoid.

Let P_1 and P_2 be the perpendiculars from P, upon X and Y respectively.

Let Q_1 and Q_2 be the perpendiculars from Q, upon X and Y respectively.

Then, from the properties of the ellipsoid, it is easily shown that

$$P_1 . P_2 = Q_1 . Q_2.$$

This is the second geometrical relation between the two pairs of screws α, η and β, ξ. Subject to these two geometrical conditions or to the two formulæ to which they are equivalent the two pairs of screws might be chosen arbitrarily.

As these two relations exist, it is evident that the knowledge of a second pair of corresponding impulsive screws and instantaneous screws cannot bring five independent data as did the first pair. The second pair can bring no more than three. From our knowledge of the two pairs of screws together we thus obtain no more than eight data. We are consequently short by one of the number requisite for the complete specification of the rigid body in its abstract form.

It follows that there must be a singly infinite number of rigid bodies, every one of which will fulfil the necessary conditions with reference to the two pairs of screws. For every one of those bodies α is the instantaneous screw about which twisting motion would be produced by an impulsive wrench on η. For every one of those bodies β is the instantaneous screw about which twisting motion would be produced by an impulsive wrench on ξ.

306. A System of Rigid Bodies.

We have now to study the geometrical relations of the particular system of rigid bodies in singly infinite variety which stand to the four screws in the relation just specified.

Draw the cylindroid (α, β) which passes through the two screws α and β. Draw also the cylindroid (η, ξ) which passes through the two corresponding impulsive screws η and ξ. It is easily seen that every screw on the first of these cylindroids if regarded as an instantaneous screw, with respect to the same rigid body, will have its corresponding impulsive screw on the second cylindroid. For any impulsive wrench on (η, ξ) can be decomposed into impulsive wrenches on η and ξ. The first of these will generate a twist velocity about α. The second will generate a twist velocity about β. These two can only compound into a twist velocity about some other screw on the cylindroid (α, β). This must, therefore, be the instantaneous screw corresponding to the original impulsive wrench on (η, ξ).

It is a remarkable point about this part of our subject that, as proved in § 293, we can now, without any further attention to the rigid body, correlate definitely each of the screws on the instantaneous cylindroid with its correspondent on the impulsive cylindroid.

We thus see how, from our knowledge of two pairs of correspondents, we can construct the impulsive screw on the cylindroid (η, ξ) corresponding to every screw on the cylindroid (α, β).

It has been already explained in the last article how a single known pair of corresponding impulsive and instantaneous screws suffice to point out a diameter of the momental ellipsoid, and also give its radius of gyration. A second pair of screws will give another diameter of the momental ellipsoid, and these two diameters give, by their intersection, the centre of gravity. As we have an infinite number of corresponding pairs, we thus get an infinite number of diameters, all, however, being parallel to the principal plane of the instantaneous cylindroid. The radius of gyration on each of these diameters is known. Thus we get a section S of the momental ellipsoid, and we draw any pair of conjugate diameters in that section. These diameters, as well as the radius of gyration on each of them, are thus definitely fixed.

When we had only a single pair of corresponding impulsive and instantaneous screws, we could still determine one ray in the conjugate plane to the diameter parallel to the instantaneous screw. Now that we have further ascertained the centre of gravity, the conjugate plane to the diameter, parallel to the instantaneous axis, is completely determined. Every pair of corresponding impulsive and instantaneous screws will give a conjugate plane to the diameter parallel to the instantaneous screw. Thus we know the conjugate planes to all the diameters in the plane S. All these planes must intersect, in a common ray Q, which is, of course, the conjugate direction to the plane S.

This ray Q might have been otherwise determined. Take one of the two screws, of zero pitch, in the impulsive cylindroid (η, ξ). Then the plane, through this screw and the centre of gravity, must, by Poinsot's theorem already referred to, be the conjugate plane to some straight line in S. Similarly, the plane through the centre of gravity and the other screw of zero pitch, on the cylindroid (η, ξ), will also be the conjugate plane to some ray in S. Hence, we see that the ray Q must lie in each of the planes so constructed, and is therefore determined. In fact, it is merely the transversal drawn from the centre of gravity to intersect both the screws of zero pitch on the cylindroid (η, ξ).

We have thus proved that when two pairs of corresponding impulsive screws and instantaneous screws are given, we know the centre of the momental ellipsoid, we know the directions of three of its conjugate diameters, and we know the radii of gyration on two of those diameters. The radius of gyration on the third diameter remains arbitrary. Be that radius what it may, the rigid body will still fulfil the condition rendering α, η and β, ξ respective pairs of instantaneous screws and impulsive screws. We had from the first foreseen that the data would only provide eight coordinates, while the specification of the body required nine. We now learn the nature of the undetermined coordinate.

It appears from this investigation that, if two pairs of impulsive screws and the corresponding instantaneous screws are known, but that if there be no other information, the rigid body is indeterminate. It follows that, if an impulsive screw be given, the corresponding instantaneous screw will not generally be determined. Each of the possible rigid bodies will have a different instantaneous screw, though the impulsive screw may be the same. It was, however, shown (§ 299), that all the instantaneous screws which pertain to a given impulsive screw lie on the same cylindroid. It is a cylindroid of extreme type, possessing a screw of infinite pitch, and degenerating to a plane.

Even while the body is thus indeterminate, there are, nevertheless, a system of impulsive screws which have the same instantaneous screw for every rigid body which complies with the expressed conditions. Among these are, of course, the several screws on the impulsive cylindroid (η, ξ) which have each the same corresponding screw on the instantaneous cylindroid (α, β), whatever may be the body of the system to which the impulsive wrench is applied. But the pairs of screws on these two cylindroids are indeed no more than an infinitesimal part of the total number of pairs of screws that are circumstanced in this particular way. We have to show that there is a system of screws of the fifth order, such that an impulsive wrench on any one of those screws η will make *any* body of the system commence to twist about the same screw α.

As already explained, the system of rigid bodies have a common centre of gravity. Any force, directed through the centre of gravity, will produce a linear velocity parallel to that force. This will, of course, apply to every body of the system. All possible forces, which could be applied to one point, form a system of the third order of a very specialized type. Each one of the screws of this system will have, as its instantaneous screw, a screw of infinite pitch parallel thereto. We have thus a system of impulsive screws of the third order, and a corresponding system of instantaneous screws of the third order, the relation between each pair being quite independent of whatever particular rigid body of the group the impulsive wrench be applied to.

This system of the third order taken in conjunction with the cylindroid (η, ξ) will enable us to determine the total system of impulsive screws which possess the property in question. Take any screw θ, of zero pitch, passing through the centre of gravity, and any screw, ϕ, on the cylindroid (η, ξ). We know, of course, as already explained, the instantaneous screws corresponding to θ and ϕ. Let us call them λ, μ, respectively. Draw the cylindroid (θ, ϕ), and the cylindroid (λ, μ). The latter of these will be the locus of the instantaneous screws, corresponding to the screws on the former as impulsive screws. From the remarkable property of the two cylindroids, so related, it follows that every impulsive screw on (θ, ϕ) will have its corresponding instantaneous screw on (λ, μ) definitely fixed. This will be so, notwithstanding the arbitrary element remaining in the rigid body. From the way in which the cylindroid (θ, ϕ) was constructed, it is plain that the screws belonging to it are members of the system of the fifth order, formed by combinations of screws on the cylindroid (η, ξ) with screws of the special system of the third order passing through the centre of gravity. But all the screws of a five-system are reciprocal to a single screw. The five-system we are at present considering consists of the screws which are reciprocal to that single screw, of zero pitch, which passes through the centre of gravity and intersects both the screws, of zero pitch, on the impulsive cylindroid (η, ξ). The corresponding instantaneous screws will also form a system of the fifth order, but it will be a system of a specialized type. It will be the result of compounding all possible displacements by translation, with all possible twists about screws on the cylindroid (α, β). The resulting system of the fifth order consists of all screws, of whatsoever pitch, which fulfil the single condition of being perpendicular to the axis of the cylindroid (α, β). Hence we obtain the following theorem:—

If an impulsive cylindroid, and the corresponding instantaneous cylindroid, be known, we can construct, from these two cylindroids, and without any further information as to the rigid body, two systems of screws of the fifth order, such that an impulsive wrench on a given screw of one system will produce an instantaneous twist velocity about a determined screw on the other system.

It is interesting to note in what way our knowledge of but *two* corresponding pairs of impulsive screws and instantaneous screws just fails to give complete information with respect to every other pair. If we take any ray in space, and assign to it an arbitrary pitch, the screw so formed may be regarded as an impulsive screw, and the corresponding instantaneous screw will not, in general, be defined. There is, however, a particular pitch for each such screw, which will constitute it a member of the system of the fifth order. It follows that any ray in space, when it receives the proper pitch, will be such that an impulsive wrench thereon would set any one of the singly infinite system of rigid bodies twisting about the same screw α.

307. The Geometrical Theory of Three Pairs of Screws.

We can now show how, when three pairs of corresponding impulsive screws and instantaneous screws are given, the instantaneous screw, corresponding to any impulsive screw, is geometrically constructed.

The solution depends upon the following proposition, which I have set down in its general form, though the application to be made of it is somewhat specialized.

Given any two independent systems of screws of the third order, P and Q. Let ω be any screw which does not belong either to P or to Q, then it is possible to find in one way, but only in one, a screw θ, belonging to P, and a screw ϕ, belonging to Q, such that ω, θ and ϕ shall all lie on the same cylindroid. This is proved as follows.

Draw the system of screws of the third order, P', which is reciprocal to P, and the system Q', which is reciprocal to Q. The screws belonging to P', and which are at the same time reciprocal to ω, constitute a group reciprocal to four given screws. They, therefore, lie on a cylindroid which we call P_0. In like manner, since Q is a system of the third order, the screws that can be selected from it, so as to be at the same time reciprocal to ω, will also form a cylindroid which we call Q_0.

It is generally a possible and determinate problem to find, among the screws of a system of the third order, one screw which shall be reciprocal to every screw, on an arbitrary cylindroid. For, take three screws from the system reciprocal to the given system of the third order, and two screws on the given cylindroid. As a single screw can be found reciprocal to any five screws, the screw reciprocal to the five just mentioned will be the screw now desired.

We apply this principle to obtain the screw θ, in the system P, which is reciprocal to the cylindroid Q_0. In like manner, we find the screw ϕ, in the system Q, which is reciprocal to the cylindroid P_0.

From the construction it is evident that the three screws θ, ϕ, and ω are all reciprocal to the two cylindroids P_0 and Q_0. This is, of course, equivalent to the statement that θ, ϕ, ω are all reciprocal to the screws of a system of the fourth order. It follows that, θ, ϕ, ω must lie upon the same cylindroid. Thus, θ, ϕ are the two screws required, and the problem has been solved. It is easily seen that there is only one such screw θ, and one such screw ϕ.

Or we might have proceeded as follows:—Take any three screws on P, and any three screws on Q. Then by a fundamental principle a wrench on ω can be decomposed into six component wrenches on these six screws. But the three component wrenches on P will compound into a single wrench on some screw θ belonging to P. The three component wrenches on Q will compound into a single wrench on some screw ϕ belonging to Q. Thus the original wrench on ω may be completely replaced by single wrenches on θ and ϕ. But this proves that θ, ϕ, and ω are co-cylindroidal.

In the special case of this theorem which we are now to use one of the systems of the third order assumes an extreme type. It consists simply of all possible screws of infinite pitch. The theorem just proved asserts that in this case a twist velocity about any screw ω can always be replaced by a twist velocity about some one screw belonging to any given system of the third order P, together with a suitable velocity of translation.

In the problem before us we know three corresponding pairs of impulsive screws and instantaneous screws (η, α), (ξ, β), (ζ, γ), and we seek the impulsive screw corresponding to some fourth instantaneous screw δ.

It should be noticed that the data are sufficient but not redundant. We have seen how a knowledge of two pairs of corresponding impulsive screws and instantaneous screws provided eight of the coordinates of the rigid body. The additional pair of corresponding screws only bring one further co-ordinate. For, though the knowledge of γ appropriate to a given ζ might seem five data, yet it must be remembered that the two pairs (η, α) and (ζ, γ) must fulfil the two fundamental geometrical conditions, and so must also the two pairs (ξ, β) and (ζ, γ); thus, as γ has to comply with four conditions, it really only conveys a single additional coordinate, which, added to the eight previously given, make the nine which are required for the rigid body. We should therefore expect that the knowledge of three corresponding pairs must suffice for the determination of every other pair.

Let the unit twist velocity about δ be resolved by the principles explained in this section into a twist velocity on some screw δ_0 belonging to α, β, γ, and into a velocity of translation on a screw δ_1 of infinite pitch.

We have already seen that the impulsive screw corresponding to δ_0 must lie on the system of the third order defined by η, ξ, and ζ, and that it is definitely determined. Let us denote by ψ this known impulsive screw which would make the body commence to twist about δ_0.

Let the centre of gravity be constructed as in the last section; then an impulsive force through the centre of gravity will produce the velocity of translation on δ_1. Let us denote by χ the screw of zero pitch on which this force lies.

We thus have χ as the impulsive screw corresponding to the instantaneous screw δ_1, while ψ is the impulsive screw corresponding to the instantaneous screw δ_0.

Draw now the cylindroids (χ, ψ) and (δ_1, δ_0). The first of these is the locus of the impulsive screws corresponding to the instantaneous screws on the second. As already explained, we can completely correlate the screws on two such cylindroids. We can, therefore, construct the impulsive screw on (χ, ψ) which corresponds to any instantaneous screw on (δ_1, δ_0). It is, however, obvious, from the construction, that the original screw δ lies on the cylindroid (δ_1, δ_0). Hence we obtain the impulsive screw which corresponds to δ as the instantaneous screw, and the problem has been solved.

308. Another method.

We might have proceeded otherwise as follows:—From the three given pairs of impulsive screws and instantaneous screws $\eta\alpha$, $\xi\beta$, $\zeta\gamma$ we can find other pairs in various ways. For example, draw the cylindroids (α, β) and (ξ, ζ); then select, by principles already explained, a screw δ on the first cylindroid, and its correspondent θ on the second. In like manner, from the cylindroids (α, γ) and (η, ζ), we can obtain another pair (ϕ, ϵ). We have thus five pairs of correspondents, $\eta\alpha$, $\xi\beta$, $\zeta\gamma$, $\theta\delta$, $\phi\epsilon$. Each of these will give a diameter of the momental ellipsoid, and the radius of gyration about that diameter. Thus we know the centre of the momental ellipsoid and five points on its surface. The ellipsoid can be drawn accordingly. Its three principal axes give the principal screws of inertia. All other pairs of correspondents can then be determined by a construction given later on (§ 311).

309. Unconstrained motion in system of second order.

Suppose that a cylindroid be drawn through any two (not lying along the same principal axis) of the six principal screws of inertia of a free rigid body. If the body while at rest be struck by an impulsive wrench about any one of the screws of the cylindroid it will commence to move by twisting about a screw which also lies on the cylindroid. For the given impulsive wrench can be replaced by two component wrenches on any two screws of the cylindroid. We shall, accordingly, take the component wrenches of the given impulse on the two principal screws of inertia which, by hypothesis, are contained on the cylindroid. Each of those components will, by the property of a principal screw of inertia, produce an instantaneous twist velocity about the same screw. But the two twist velocities so generated can, of course, only compound into a single twist velocity on some other screw of the cylindroid. We have now to obtain the geometrical relations characteristic of the pairs of impulsive and instantaneous screws on such a cylindroid.

In previous chapters we have discussed the relations between impulsive screws and instantaneous screws, when the movements of the body are confined, by geometrical constraint, to twists about the screws on a cylindroid. The problem now before us is a special case, for though the movements are no other than twists about the screws on a cylindroid, yet this restriction, in the present case, is not the result of constraint. It arises from the fact that two of the six principal screws of inertia of the rigid body lie on the cylindroid, while the impulsive wrench is, by hypothesis, limited to the same surface.

To study the question we shall make use of the circular representation of the cylindroid, § 50. We have there shown how, when the several screws on the cylindroid are represented by points on the circumference of a circle, various dynamical problems can be solved with simplicity and convenience. For example, when the impulsive screw is represented on the circle by one point, and the instantaneous screw by another, we have seen how these points are connected by geometrical construction (§ 140).

In the case of the unconstrained body, which is that now before us, it is known that, whenever the pitch of an instantaneous screw is zero, the corresponding impulsive screw must be at right angles thereto (§ 301).

In the circular representation, the angle between any two screws is equal to the angle subtended in the representative circle by the chord whose extremities are the representatives of the two screws. Two screws, at right angles, are consequently represented by the extremities of a diameter of the representative circle. If, therefore, we take A, B, two points on the circle, to represent the two screws of zero pitch, then the two points, P and Q, diametrically opposite to them, are the points indicating the corresponding impulsive screws. It is plain from § 287 that AQ and BP must intersect in the homographic axis, and hence the homographic axis is parallel to AQ and EP, and as it must contain the pole of AB it follows that the homographic axis XY must be the diameter perpendicular to AB.

The two principal screws of the cylindroid X and Y are, in this case, the principal screws of inertia. Each of them, when regarded as an impulsive screw, coincides with its corresponding instantaneous screw. The diameter XY bisects the angle between AP and BQ.

It is shown (§ 137) that the points which represent the instantaneous screws, and the points which represent the corresponding impulsive screws, form two homographic systems. A well-known geometrical principle asserts (§ 146), that if each point on a circle be joined to its homographic correspondent, the chord will envelop a conic which has double contact with the circle. It is easily seen that, in the present case, the conic must be the hyperbola which touches the circle at the ends of the diameter XY, and has the rays AP and BQ for its asymptotes. The hyperbola is completely defined by these conditions, so that the pairs of correspondents are uniquely determined.

Every tangent, IST, to this hyperbola will cut the circle in two points, I and S, such that S is the point corresponding to the impulsive screw, and I is the point which marks out the instantaneous screw. We thus obtain a concise geometrical theory of the connexion between the pairs of corresponding impulsive screws and instantaneous screws on a cylindroid which contains two of the principal screws of inertia of a free rigid body.

For completeness, it may be necessary to solve the same problem when the cylindroid is defined by two principal screws of inertia lying along the same principal axis of the rigid body. It is easy to see that if, on the principal axis, whose radius of gyration was a, there lay any instantaneous screw whose pitch was p_α, then the corresponding impulsive screw would be also on the same axis, and its pitch would be p_η where $p_\eta \times p_\alpha = a^2$.

310. Analogous Problem in a Three-system.

Let us now take the case of the system of screws of the third order, which contains three of the six principal screws of inertia of a free rigid body.

Any impulsive wrench, which acts on a screw of a system of the third order, can be decomposed into wrenches on any three screws of that system, and consequently, on the three principal screws of inertia, which in the present case the three-system has been made to contain. Each of these component wrenches will, from the property of a principal screw of inertia, generate an initial twist velocity of motion around the same screw. The three twist velocities, thence arising, can be compounded into a single twist velocity about some other screw of the system. We desire to obtain the geometrical relation between each such resulting instantaneous screw and the corresponding impulsive screw.

It has been explained in Chap. XV. how the several points in a plane are in correspondence with the several screws which constitute a system of the third order. It was further shown, that if by the imposition of geometrical constraints, the freedom of a rigid body was limited to twisting about the several screws of the system of the third order, a geometrical construction could be obtained for determining the point corresponding to any instantaneous screw, when the point corresponding to the appropriate impulsive screw was known.

We have now to introduce the simplification of the problem, which results when three of the principal screws of inertia of the body belong to the system. But a word of caution, against a possible misunderstanding, is first necessary. It is of course a fundamental principle, that when a rigid system has freedom of the nth order, there will always be, in the system of screws expressing that freedom, n screws such that an impulsive wrench administered on any one of those screws will immediately make the body begin to move by twisting about the same screw. These are the n *principal screws of inertia.*

But in the case immediately under consideration the rigid body is supposed to be free, and it has, therefore, six principal screws of inertia. The system of the third order, at present before us, is one which contains three of these principal screws of inertia of the free body. Such a system of screws possesses the property, that an impulsive wrench on any screw belonging to it will set the body twisting about a screw which also belongs to the same system. This is the case even though, in the total absence of constraints, there is no kinematical difficulty about the body twisting about any screw whatever.

As there are no constraints, we know that each instantaneous screw, of zero pitch, must be at right angles to the corresponding impulsive screw (§ 301). This condition will enable us to adjust the particular homography in the plane wherein each pair of correspondents represents an impulsive screw and the appropriate instantaneous screw.

The conic, which is the locus of points corresponding to the screws of a given pitch p, has as its equation (§ 204)

$$p_1\theta_1^2 + p_2\theta_2^2 + p_3\theta_3^2 - p\,(\theta_1^2 + \theta_2^2 + \theta_3^2) = 0.$$

The families of conics corresponding to the various values of p have a common self-conjugate triangle. The vertices of that triangle correspond to the three principal screws of inertia.

The three points just found are the double points of the homography which correlate the points representing the impulsive screws with those representing the instantaneous screws. Let us take the two conics of the system, corresponding to $p = 0$ and $p = \infty$. They are

$$p_1\theta_1^2 + p_2\theta_2^2 + p_3\theta_3^2 = 0 \quad \text{.........................(i)},$$

$$\theta_1^2 + \theta_2^2 + \theta_3^2 = 0 \quad \text{.........................(ii)}.$$

Two conjugate points to conic (i) denote two reciprocal screws. Two conjugate points to conic (ii) denote two screws at right angles.

Let A be any point representing an instantaneous screw. Take the polar of A, with respect to conic (i). Let P be the pole of this ray, with respect to conic (ii).

Then P will correspond to the impulsive screw, while A corresponds to the appropriate instantaneous screw. For this is clearly a homography of which A and P are two correspondents. Further, the double points of this homography are the vertices of the common conjugate triangle to conics (i) and (ii). If A lie on (i), then its polar is the tangent to (i); and as every point on this polar will be conjugate to P, with respect to conic (ii), it follows that A and P are conjugate, with respect to (ii)—that is, A and P are correspondents of a pair of screws at right angles. As the pitch of the screw, corresponding to A, is zero, we have thus obtained the solution of our problem.

311. Fundamental Problem with Free Body.

We now give the geometrical solution of the problem so fundamental in this present theory which may be thus stated:

A perfectly free body at rest is struck by an impulsive wrench upon a given screw. It is required to construct the instantaneous screw about which the body will commence to twist.

The rigid body being given, its three principal axes are to be drawn through its centre of gravity. The radii of gyration a, b, c about these axes are to be found. On the first principal axis two screws of pitches $+a$ and $-a$ respectively are to be placed. Similarly screws of pitches $+b, -b$, and $+c, -c$ are to be placed on the other two principal axes. These are, of course, the six principal screws of inertia: call them A_0, A_1, A_2, A_3, A_4, A_5. We then draw the five cylindroids

$$A_0A_1,\ A_0A_2,\ A_0A_3,\ A_0A_4,\ A_0A_5.$$

It is always possible to find one screw on a cylindroid reciprocal to any given screw. In certain cases, however, of a special nature, more than a single screw can be so found. Under such circumstances the present process is inapplicable, but the exceptional instances will be dealt with presently.

Choose on the cylindroid A_0A_1 a screw θ_1 which is reciprocal to the given impulsive screw η, which is, of course, supposed to lie anywhere and be of any pitch.

In like manner, choose on the other four cylindroids screws θ_2, θ_3, θ_4, θ_5, respectively, all of which are also reciprocal to η.

Let us now think of θ_1 as an instantaneous screw; it lies on the cylindroid A_0A_1, and this cylindroid contains two principal screws of inertia. It follows from § 309 that the corresponding impulsive screw ϕ_1 lies on the same cylindroid. That screw ϕ_1 can be determined by the construction there given. In like manner we construct on the other four cylindroids the screws ϕ_2, ϕ_3, ϕ_4, ϕ_5, which are the impulsive screws corresponding respectively to θ_2, θ_3, θ_4, θ_5, as instantaneous screws.

Consider then the two pairs of corresponding impulsive screws and instantaneous screws (η, α) and (ϕ_1, θ_1). We have arranged the construction so that θ_1 is reciprocal to η. Hence, by the fundamental principle so often employed, α and θ_1 are conjugate screws of inertia, so that α must be reciprocal to ϕ_1.

In like manner it can be proved that the instantaneous screw α for which we are in search must be reciprocal to ϕ_2, ϕ_3, ϕ_4, ϕ_5. We have thus discovered five screws, ϕ_1, ϕ_2, ϕ_3, ϕ_4, ϕ_5, to each of which the required screw α must be reciprocal. But it is a fundamental point in the theory that the single screw reciprocal to five screws can be constructed geometrically (§ 25). Hence α is determined, and the geometrical solution of the problem is complete.

It remains to examine the failure in this construction which arises when any one or more of the five screws $\phi_1 \dots \phi_5$ becomes indeterminate. This happens when η is reciprocal to *two* screws on the cylindroid in question. In this case η is reciprocal to every screw on the cylindroid. Any one of such screws might be taken as the corresponding ϕ, and, of course, θ would have been also indefinite, and α could not have been found. In this case η would have been reciprocal to the two principal screws of inertia, suppose A_0, A_1 which the cylindroid contained. Of course still more indeterminateness would arise if η had been also reciprocal to other screws of the series A_0, A_1, A_2, A_3, A_4, A_5. No screw could, however, be reciprocal to all of them. If η had been reciprocal to five, namely, A_1, A_2, A_3, A_4, A_5, then η could be no screw other than A_0, because the six principal screws of inertia are co-reciprocal; η would then be its own instantaneous screw, and the problem would be solved.

We may therefore, under the most unfavourable conditions, take η to be reciprocal to four of the principal screws of inertia A_0, A_1, A_2, A_3, but not to A_4 or A_5. We now draw the five cylindroids, A_0A_4, A_1A_4, A_2A_4, A_3A_4, A_0A_5. We know that η is reciprocal to no more than a single screw on each cylindroid. We therefore proceed to the construction as before, first finding $\theta_1 \dots \theta_5$, one on each cylindroid; then deducing $\phi_1 \dots \phi_5$, and thus ultimately obtaining α.

Thus the general problem has been solved.

312. Freedom of the First or Second Order.

If the rigid body have only a single degree of freedom, then the only movements of which it is capable are those of twisting to and fro on a single screw α. If the impulsive wrench η which acted upon the body happened to be reciprocal to α, then no movement would result. The forces would be destroyed by the reactions of the constraints. In general, of course, the impulsive screw η will not be reciprocal to α. A twisting motion about α will therefore be the result. All that can be said of the instantaneous screw is that it can be no possible screw but α.

In the next case the body has two degrees of freedom which, as usual, we consider to be of the most general type. It is required to obtain a construction for the instantaneous screw α about which a body will commence to twist in consequence of an impulsive wrench η.

The peculiarity of the problem when the notion of constraint is introduced depends on the circumstance that, though the impulsive screw may be situated anywhere and be of any pitch, yet that as the body is restrained to only two degrees of freedom, it can only move by twisting about one of the screws on a certain cylindroid. We are, therefore, to search for the instantaneous screw on the cylindroid expressing the freedom.

Let A be the given cylindroid. Let B be the system of screws of the fourth order reciprocal to that cylindroid. If the body had been free it would have been possible to determine, in the manner explained in the last section, the impulsive screw corresponding to each screw on the cylindroid A. Let us suppose that these impulsive screws are constructed. They will all lie on a cylindroid which we denote as P. In fact, if any two of such screws had been found, P would of course have been defined by drawing the cylindroid through those two screws.

Let Q be the system of screws of the fourth order which is reciprocal to P. Select from Q the system of the third order Q_1 which is reciprocal to η. We can then find one screw η_1 which is reciprocal to the system of the fifth order formed from A and Q_1. It is plain that η_1 must belong to B, as this contains every screw reciprocal to A.

Take also the one screw on the cylindroid A which is reciprocal to η, and find the one screw η_2 on the cylindroid P which is reciprocal to this screw on A.

Since η, η_1 and η_2 are all reciprocal to the system of the fourth order formed by A_1 and Q_1, it follows that η, η_1, and η_2 must all lie on the same cylindroid. We can therefore resolve the original wrench on η into two component wrenches on η_1 and η_2.

But it is of the essence of the theory that the reactions of the constraints y which the motion of the body is limited to twists about screws on the ylindroid A must be wrenches on the reciprocal system B. So far, therefore, s the body thus constrained is concerned, the reactions of the constraints ill neutralize the wrench on η_1. Thus the wrench on η_2 is the only part of ıe impulsive wrench which need be considered.

But we already know from the construction that an impulsive wrench on η_2 ill produce an instantaneous twist velocity about a determined screw α on l. Thus we have found the initial movement, and the investigation is geo-ıetrically complete.

313. Freedom of the Third Order.

We next consider the case in which a rigid body has freedom of the third rder. We require, as before, to find a geometrical construction for the in-tantaneous screw α corresponding to a given impulsive screw η.

Let A be the system of screws of the third order about which the body s free to twist. Let B be the system of screws of the third order reciprocal o A. We must first construct the system of the third order P which onsists of the impulsive screws that would have made the body, if perfectly free, twist about the several screws of A.

As already explained (§ 307) we can, in one way, but only in one way, resolve the original wrench on η into wrenches η_1 on B, and η_2 on P. The former is destroyed by the reactions of the constants. The latter gives rise to a twist velocity about a determinate screw on A. Thus the problem has been solved.

314. General Case.

We can obviously extend a similar line of reasoning to the cases where the body had freedom of the remaining degrees. It will, however, be as simple to write the general case at once.

Let A be a system of screws of the nth order, about which a body is free to twist, any other movements being prevented by constraints. If the body receive an impulsive wrench, on any screw η, it is required to determine the instantaneous screw, of course belonging to A, about which the body will commence to twist.

Let B be the system of screws of the $(6-n)$th order, reciprocal to A. The wrenches arising from the reaction of the constraints must, of course, be situated on the screws of the system B.

Let P be the system of screws of the nth order, which, in case the body had been free, would have been the impulsive screws, corresponding to the instantaneous screws belonging to A.

Let Q be the system of screw of the $(6-n)$th order, which are reciprocal to P.

Take from A the system of the $(n-1)$th order, reciprocal to η, and call it A_1.

Take from Q the system of the $(5-n)$th order, reciprocal to η, and call it Q_1.

As A is of the nth order, and Q_1 of the $(5-n)$th, they together define a system of the fifth order. Let η_1 be the single screw, reciprocal to this system of the fifth order.

As A_1 is of the $(n-1)$th order, and Q is of the $(6-n)$th order, they together define a system of the fifth order. Let η_2 be the single screw, reciprocal to this system of the fifth order.

η_1 is reciprocal to A_1 of the $(n-1)$th order, because A_1 forms part of A. It is reciprocal to Q_1 of the $(5-n)$th order, because it was so made by construction. Thus η_1 is reciprocal to both A_1 and Q_1, that is to a system of the fourth order.

In like manner, it can also be shown that η_2 is reciprocal to both A_1 and Q_1.

A_1 and Q_1 were originally chosen so as to be reciprocal to η. It thus appears that the three screws, η, η_1, η_2, are all reciprocal to the same system of the fourth order. They are, therefore, co-cylindroidal.

The initial wrench on η can therefore be adequately replaced by two components on η_1 and η_2. The former of these is destroyed by the reaction of the constraints. The latter gives rise to an initial movement on a determined screw of A. Thus, the most general problem of the effect of an impulsive wrench on a constrained rigid body has been solved geometrically.

315. Freedom of the Fifth Order.

The special case, where the rigid body has freedom of the fifth order, may be viewed as follows.

Let ρ be the screw reciprocal to the screw system of the fifth order, about which the body is free to twist.

Let λ be the instantaneous screw, determined in the way already explained, about which the body, had it been free, would have commenced to twist in consequence of an impulsive wrench on ρ.

Let η be any screw on which an impulsive wrench is imparted, and let α be the corresponding instantaneous screw, about which the body would have begun to twist had it been free.

Draw the cylindroid through α and λ, and choose on this cylindroid the screw μ, which is reciprocal to ρ.

Then μ is the instantaneous screw about which the body commences to twist, in consequence of the impulsive wrench on η.

For (η, ρ) is a cylindroid of impulsive screws, and (α, λ) are the corresponding instantaneous screws. As μ is reciprocal to ρ, it belongs to the system of the fifth order. The corresponding impulsive screw must lie on (η, ρ). The actual instantaneous motion could therefore have been produced by impulsive wrenches on η and ρ. The latter would, however, be neutralized by the reactions of the constraints. We therefore find that η is the impulsive screw, corresponding to α as the instantaneous screw.

316. Principal Screws of Inertia of Constrained Body.

There is no more important theorem in this part of the Theory of Screws than that which affirms that for a rigid body, with n degrees of freedom, there are n screws, such that if the body when quiescent receives an impulsive wrench about one of such screws, it will immediately commence to move by twisting about the same screw.

We shall show how the principles, already explained, will enable us to construct these screws.

We commence with the case in which the body has two degrees of freedom. We take three screws, η, ξ, ζ, arbitrarily selected on the cylindroid, which expresses the freedom of the body. We can then determine, by the preceding investigation, the three instantaneous screws, α, β, γ, on the same cylindroid, which correspond, respectively, to the impulsive screws. Of course, if η happened to coincide with α, or ξ with β, or ζ with γ, one of the principal screws of inertia would have been found. But, in general, such pairs will not coincide. We have to show how, from the knowledge of three such pairs, in general, the two principal screws of inertia can be found.

We employ the circular representation of the points on the cylindroid, as explained in § 50. The impulsive screws are represented by one system of points, the corresponding instantaneous screws are represented by another system of points. It is an essential principle, that the two systems of points, so related, are homographic. The discovery of the principal screws of inertia is thus reduced to the well-known problem of the discovery of the double points of two homographic systems on a circle.

The simplest method of solving this problem is that already given in § 139, in which we regard the six points, suitably arranged, as the vertices of a hexagon; then the Pascal line of the hexagon intersects the circle in two points which are the points corresponding to the principal screws of inertia.

317. Third and Higher Systems.

We next investigate the principal screws of inertia of a body which has three degrees of freedom. We have first, by the principles already explained, to discover four pairs of correspondents. When four such pairs are known, the principal screws of inertia can be constructed. Perhaps the best method of doing so is to utilize the plane correspondence, as explained in Chap. XV. The corresponding systems of impulsive screws and instantaneous screws, in the system of the third order, are then represented by the homographic systems of points in the plane. When four pairs of such correspondents are known, we can construct as many additional pairs as may be desired.

Let α, β, γ, δ be four points in the plane, and let η, ξ, ζ, θ be the points corresponding, so that η represents the impulsive screw, and α the instantaneous screw, and similarly for the other pairs. Let it be required to find the impulsive screw ϕ, which corresponds to any fifth instantaneous screw ϵ. Since anharmonic ratios are the same in two corresponding figures, we have

$$\alpha(\beta, \gamma, \delta, \epsilon) = \eta(\xi, \zeta, \theta, \phi),$$

thus we get one ray $\eta\phi$, which contains ϕ. We have also

$$\beta(\alpha, \gamma, \delta, \epsilon) = \xi(\eta, \zeta, \theta, \phi),$$

which gives a second ray $\xi\phi$, containing ϕ, and thus ϕ is known.

A construction for the double points of two homographic systems of points in the same plane is as follows:—

Let O and O' be a pair of corresponding points. Then each ray through O will have, as its correspondent, a ray through O'. The locus of the intersection of these rays will be a conic S. This conic S must pass through the three double points, and also through O and O'.

Draw the conic S', which is the locus of the points in the second system corresponding to the points on S, regarded as in the first system. Then since O lies on S, we must have O' on S'. But S' must also pass through the three double points. O' is one of the four intersections of S and S', and the three others are the sought double points. Thus the double points are constructed. And in this manner we obtain the three principal screws of inertia in the case of the system of the third order.

If a rigid body be free to move about screws of any system of the fourth order, we may determine its four principal screws of inertia as follows.

We correlate the several screws of the system of the fourth order with the points in space. The points representing impulsive screws will be a system homographic with the points representing the corresponding instantaneous screws. If we have five pairs of correspondents (α, η), (β, ξ), (γ, ζ), (δ, θ), (ϵ, ϕ), in such a homography we can at once determine the correspondent ψ to any other screw λ.

Draw a pencil of four planes through α, β, and the four points γ, δ, ϵ, λ, respectively.

Draw also a pencil of four planes through η, ξ, and the four points ζ, θ, ϕ, ψ, respectively. These two pencils will be equianharmonic. Thus we discover one plane which contains ψ. In like manner we can draw a pencil of planes through α, γ, and β, δ, ϵ, λ, respectively, and the equianharmonic pencil through η, ζ, and ξ, θ, ϕ, ψ, respectively. Thus we obtain a second plane which passes through ψ. A third plane may be found by drawing the pencil of four planes through α, δ, and the four points β, γ, ϵ, λ, respectively, and then constructing the equianharmonic pencil through η, θ, and ξ, ζ, ϕ, ψ, respectively. From the intersection of these three planes, ψ is known.

In the case of two homographic systems in three dimensions, there are, of course, four double points. These may be determined as follows. Let O and O' be two corresponding rays. Then any plane through O will have, as its correspondent, a plane through O'. It is easily seen that these planes intersect on a ray which has for its locus a quadric surface S, of which O and O' are also generators. This quadric must pass through the four double points.

Let S' be the quadric surface which contains all the points in the second system corresponding to the points of S regarded as the first system. Then O' will lie on S', and the rest of the intersection of S and S' will be a twisted cubic C, which passes through the four double points.

Take any point P on C, and draw any plane through P. Then every ray of the first system of the pencil through P in this plane will have as its correspondent in the second system the ray in some other plane pencil L. One, at least, of the rays in the pencil L will cut the cubic C. Call this ray X', and draw its correspondent X in the first system passing through P.

We thus have a pair of corresponding rays X and X', each of which intersects the twisted cubic C.

Draw pairs of corresponding planes through X and X'. The locus of their intersection will be a quadric S'', which also contains the four double points.

S'' and C, being of the second and the third order respectively, will intersect in six points. Two of these are on X and X', and are thus distinguished. The four remaining intersections will be the required double points, and thus the problem has been solved.

These double points correspond to the principal screws of inertia, which are accordingly determined.

In the case of freedom of the fifth order, the geometrical analogies which have hitherto sufficed are not available. We have to fall back on the general fact that the impulsive screws and the corresponding instantaneous screws form two homographic systems. There are five double screws belonging to this homography. These are the principal screws of inertia.

318. Correlation of Two Systems of the Third Order.

It being given that a certain system of screws of the third order, P, is the locus of impulsive screws corresponding to another given system of the third order, A, as instantaneous screws, it is required to correlate the corresponding pairs on the two systems.

We have already had frequent occasion to use the result demonstrated in § 293, namely, that when two impulsive and instantaneous cylindroids were known, we could arrange the several screws in corresponding pairs without any further information as to the rigid body. We have now to demonstrate that when we are given an impulsive system of the third order, and the corresponding instantaneous system, there can also be a similar adjustment of the corresponding pairs.

It has first to be shown, that the proposed problem is a definite one. The data before us are sufficient to discriminate the several pairs of screws, that is to say, the data are sufficient to point out in one system the correspondent to any specified screw in the other system. We have also to show that there is no ambiguity in the solution. There is only one rigid body (§ 293) which will comply with the condition, and it is not possible that there could be more than one arrangement of corresponding pairs.

Let α, β, γ be three instantaneous screws from A, and let their corresponding impulsive screws be η, ξ, ζ, in P. In the choice of a screw from a system of the third order there are two disposable quantities, so that, in the selection of three correspondents in P, to three given screws in A, there would be, in general, six disposable coordinates. But the fact that α, η and β, ξ are two pairs of correspondents necessitates, as we know, the fulfilment of two identical conditions among their coordinates. As there are three pairs of correspondents, we see at once that there are six equations to be fulfilled. These are exactly the number required for the determination of η, ξ, ζ, in the system P.

To the same conclusion we might have been conducted by a different line of reasoning. It is known that, for the complete specification of a system of the third order, nine co-ordinates are necessary (§ 75). This is the same number as is required for the specification of a rigid body. If, therefore, we are given that P, a system of the third order, is the collection of impulsive screws, corresponding to the instantaneous screws in the system A, we are given nine data towards the determination of a rigid body, for which A and P would possess the desired relation. It therefore follows that we have nine equations, while the rigid body involves nine unknowns. Thus we are led to expect that the number of bodies, for which the arrangement would be possible, is finite. When such a body is determined, then of course the correlation of the screws on the two systems is immediately accomplished. It thus appears that the general problem of correlating the screws on any two given systems of the third order, A and P, into possible pairs of impulsive screws and instantaneous screws, ought not to admit of more than a finite number of solutions.

We are now to prove that this finite number of solutions cannot be different from unity.

For, let us suppose that a screw λ, belonging to A, had two screws θ and ϕ, as possible correspondents in P. This could, of course, in no case be possible with the same rigid body. We shall show that it could not even be possible with two rigid bodies, M_1 and M_2. For, if two bodies could do what is suggested, then it can be shown that there are a singly infinite number of possible bodies, each of which would afford a different solution of the problem.

We could design a rigid body in the following manner:—

Increase the density of every element of M_1 in the ratio of $\rho_1 : 1$, and call the new mass M_1'.

Increase the density of every element of M_2 in the ratio of $\rho_2 : 1$, and call the new mass M_2'.

Let the two bodies, so altered, be conceived bound rigidly together by bonds which are regarded as imponderable.

Let ψ be any screw lying on the cylindroid (θ, ϕ). Then the impulsive wrench of intensity, ψ''' on ψ, may be decomposed into components

$$\psi''' \frac{\sin(\psi - \phi)}{\sin(\theta - \phi)} \text{ on } \theta, \text{ and } \psi''' \frac{\sin(\theta - \psi)}{\sin(\theta - \phi)} \text{ on } \phi.$$

If the former had been applied to M_1' it would have generated about α (§ 280) a twist velocity represented by

$$\dot{\psi}''' \frac{\sin(\psi - \phi)}{\sin(\theta - \phi)} \frac{1}{M_1'} \frac{\cos(\theta\alpha)}{p_\alpha}.$$

If the latter had been applied to the body M_2', it would have generated a twist velocity about the same screw, α, equal to

$$\dot{\psi}''' \frac{\sin(\theta - \psi)}{\sin(\theta - \phi)} \frac{1}{M_2'} \frac{\cos(\phi\alpha)}{p_\alpha}.$$

Suppose that these two twist velocities are equal, it is plain that the original wrench on ψ would, if it had been applied to the composite rigid body, produce a twisting motion about α. The condition is

$$\frac{\sin(\psi - \phi)}{M_1' \cos(\phi\alpha)} = \frac{\sin(\theta - \psi)}{M_2' \cos(\theta\alpha)}.$$

We thus obtain $\tan\psi$ in terms of $M_1' : M_2'$. As the structure of the composite body changes by alterations of the relative values of ρ_1 and ρ_2, so will ψ move over the various screws of the cylindroid (θ, ϕ).

This result shows that, if three screws, α, β, γ be given, then the possible impulsive screws, η, ξ, ζ, which shall respectively correspond to α, β, γ in a given system of the third order P, are uniquely determined.

For, suppose that a second group of screws, η', ξ', ζ', could also be determined which fulfilled the same property. We have shown how another rigid body could be constructed so that another screw, ψ, could be found on the cylindroid (η, η'), such that an impulse thereon given would make the body twist about α. It is plain that, for this body also, the impulsive wrench, corresponding to β, would be some screw on the cylindroid (ξ, ξ'). But all screws on this cylindroid belong to the system P. In like manner, the instantaneous screw γ would correspond for the composite body to some screw on the cylindroid (ζ, ζ'). Hence it follows that, for each different value of the ratio $\rho_1 : \rho_2$, we would have a different set of impulsive screws for the instantaneous screws α, β, γ. We thus find that, if there were more than one set of such impulsive screws to be found in the system P, there would be an infinite number of such sets. But we have already shown that the number of sets must be finite. Hence there can only be one set of screws, η, ξ, ζ, in the system P, which could be impulsive screws corresponding to the instantaneous screws, α, β, γ. We are thus led to the following important theorem, which will be otherwise proved in the next chapter.

Given any two systems of screws of the third order. It is generally possible, in one way, but only in one, to design, and place in a particular position a rigid body such that, if that body, while at rest and unconstrained, receive an impulsive wrench about any screw of the first system, the instantaneous movement will be a twist about a screw of the second system.

The two systems of corresponding impulsive and instantaneous screws on the two systems of the third order, form two homographic systems. There are, of course, infinite varieties in the possible homographic correspondences between the screws of two systems of the third order. The number of such correspondences is just so many as the number of possible homographic correspondences of points in two planes. There is, however, only one correspondence which will fulfil the peculiar requirements when one of the systems expresses the instantaneous screws, and the other the impulsive screws severally corresponding thereto.

If we are given one pair of corresponding impulsive and instantaneous screws, the body is not by such data fully determined. We are only given five coordinates, and four more remain, therefore, unknown. If we are given two corresponding impulsive cylindroids and instantaneous cylindroids, the body is still not completely specified. We have seen how eight of its coordinates are determined, but there is still one remaining indeterminate. If we are given a system of the fourth order of impulsive screws, and the corresponding system of the fourth order of instantaneous screws, the body, as in the other cases, remaining perfectly free, there are also, as we shall see in the next section, a singly infinite number of rigid bodies which fulfil the necessary conditions. In like manner, it will appear that, if we are given a system of the fifth order consisting of impulsive screws, and a corresponding system of the fifth order consisting of instantaneous screws, the body has really as much indeterminateness as if we had only been given a single pair of corresponding screws.

But the case of two systems of the third order is exceptional, in that when it is known that one of these is the locus of the instantaneous screws, which correspond to the screws of the other system regarded as impulsive screws, the rigid body for which this state of things is possible is completely and uniquely specified as to each and every one of its nine coordinates.

319. A Property of Reciprocal Screw Systems.

Given a system of the fourth order A and another system of the fourth order P. If it be known that the latter is the locus of the screws on which must lie the impulsive wrenches which would, if applied to a free rigid body, cause instantaneous twist velocities about the several screws on A, let us consider what can be inferred as to the rigid body from this fact alone.

Let A' be the cylindroid which is composed of the screws reciprocal to A.

Let P' be the cylindroid which is composed of the screws reciprocal to P. Let P_1, P_2, P_3, P_4 be any four impulsive screws on P. Let A_1, A_2, A_3, A_4 be the four corresponding instantaneous screws on A.

Take any screw α on the cylindroid P'. Let η be the corresponding impulsive screw. Since α is reciprocal to all the screws on P it must be reciprocal to P_1. It follows from the fundamental property of conjugate screws of inertia, that η must be reciprocal to A_1. In like manner we can show that η is reciprocal to A_2, A_3, and A_4. It follows that η is reciprocal to the whole system A, and therefore must be contained in the reciprocal cylindroid A'. Hence we obtain the following remarkable result, which is obviously generally true, though our proof has been enunciated for the system of the fourth order only.

Let P and A be any two systems of screws of the nth order, and P' and A' their respective reciprocal systems of the $(6-n)$th order. If P be the collection of impulsive screws corresponding severally to the screws of A as the instantaneous screws for a certain free rigid body; then, for the same free rigid body A' will be the collection of impulsive screws which correspond to the screws of P' as instantaneous screws.

320. Systems of the Fourth Order.

Thus we see that when we are given two systems of the fourth order P and A as correspondingly impulsive and instantaneous, we can immediately infer that, for the same rigid body, the screws on the cylindroid A' are the impulsive screws corresponding to the instantaneous screws on the cylindroid P'.

We can now make use of that instructive theorem (§ 293) which declares that when two given cylindroids are known to stand to each other in this peculiar relation, we are then able, without any further information, to mark out on the cylindroids the corresponding pairs of screws. We can then determine the centre of gravity of the rigid body on which the impulsive wrenches act. We can find a triad of conjugate diameters of the momental ellipsoid, and the radii of gyration about two of those diameters. Hence we have the following result:—

If it be given that a certain system of the fourth order is the locus of the impulsive screws corresponding to the instantaneous screws on another given system of the fourth order, the body being quite unconstrained, we can then determine the centre of gravity of the body, we can draw a triad of the conjugate diameters of its momental ellipsoid, and we can find the radii of gyration about two of those diameters.

There is still one undetermined element in the rigid body, namely, the radius of gyration about the remaining conjugate diameter. The data before us are not adequate to the removal of this indefiniteness. It must be remembered that the data in such a case are just so many but no more than suffice for the specification of the n-system A. The number of data necessary to define an n-system is $n(6-n)$. If, as in the present case, $n=4$, the number of data is 8. We are thus one short in the number of data necessary to specify a rigid body. Thus we confirm the result previously obtained. We can assert that for any one of the singly infinite number of rigid bodies which fulfil the necessary conditions, the system A will be the locus of the instantaneous screws which correspond to the screws of the system P as impulsive screws.

Though in the two cylindroids A' and P' we are able to establish the several pairs of correspondents quite definitely, yet we must not expect, with the data before us, to be able to correlate the pairs of screws in A and P definitely. If this could be done then the rigid body would be quite determinate, which we know is not the case. There is, however, only a single indeterminate element left in the correlation of the screws in A with the screws of P. This we prove as follows:—

Let ϕ be any screw of P on which an impulsive wrench is to act. Let δ be the instantaneous screw in A about which the movement commences. We shall now show that though δ cannot be completely defined, in the absence of any further data, yet it can be shown to be limited to a certain cylindroid.

Let G be the centre of gravity. Then we know that an impulsive force directed through G will generate a movement of translation in a direction parallel to the force. Such a movement may, of course, be regarded as a twist about a screw of infinite pitch.

Draw through G a plane normal to ϕ. Any screws of infinite pitch in this plane will be reciprocal to ϕ. It follows from the laws of conjugate screws of inertia that the impulsive forces in this plane, by which translations could be produced, must lie on screws of zero pitch which are reciprocal to δ. Take any two of such screws: then we know that δ is reciprocal to these two screws and also to P'. It follows that δ is reciprocal to the screws of a determinate system of the fourth order, and therefore δ must lie on a determined cylindroid.

We may commence to establish the correspondence between P and A by choosing some arbitrary screw ϕ on P, and then drawing the cylindroid on A, on which we know that the instantaneous screw corresponding to P must lie. Any screw on this cylindroid may be selected as the instantaneous screw which corresponds to ϕ. Once that screw δ had been so chosen there can be no further ambiguity. The correspondent in A to every other screw in P is completely known. To show this it is only necessary to take two pairs from A' and P' and the pair just found. We have then three corresponding pairs. It has been shown in § 307 how the correspondence is completely determined in this case.

Of course the fact that δ may be any screw on a cylindroid is connected with the fact that in this case the rigid body had one indeterminate element. For each of the possible rigid bodies δ would occupy a different position on its cylindroidal locus.

321. Systems of the Fifth Order.

It remains to consider the case where two screw systems of the fifth order are given, it being known that one of them P is the locus of the impulsive screws which correspond to the several screws of the other system A regarded as instantaneous screws.

Let P' be the screw reciprocal to P, and A' the screw reciprocal to A. Then from the theorem of § 319 it follows that an impulsive wrench on A' would make the body commence to move by twisting about P'. We thus know five of the coordinates of the rigid body. There remain four indeterminate elements.

Hence we see that, when the only data are the two systems P and A, there is a fourfold infinity in the choice of the rigid body. There are consequently four arbitrary elements in designing the correspondence between the several pairs of screws in the two systems.

We may choose any two screws η, ξ, on P, and assume as their two correspondents in A any two arbitrary screws α and β, provided of course that the three pairs A', B', η, α, and ξ, β fulfil the six necessary geometrical conditions (§ 304). Two of these conditions are obviously already satisfied by the circumstance that A' and P' are the reciprocals to the systems A and P. This leaves four conditions to be fulfilled in the choice of α and β. As each of these belongs to a system of the fifth order there will be four coordinates required for its complete specification. Therefore there will be eight disposable quantities in the choice of α and β. Four of these will be utilized in making them fulfil the geometrical conditions, so that four others may be arbitrarily selected. When these are chosen we have four coordinates of the rigid body which, added to the five data provided by A' and P', completely define the rigid body.

322. Summary.

We may state the results of this discussion in the following manner:—

If we are given two systems of the first, or the second, or the third order of corresponding impulsive screws and instantaneous screws, all the corresponding pairs are determined. There is no arbitrary element in the correspondence. There is no possible rigid body which would give any different correspondence.

If we are given two systems of the fourth order of corresponding impulsive screws and instantaneous screws then the essential geometrical conditions (§ 281), not here making any restriction necessary, we can select *one* pair of correspondents arbitrarily in the two systems, and find one rigid body to fulfil the requirements.

If we are given two systems of the fifth order of corresponding impulsive screws and instantaneous screws then subject to the observance of the geometrical conditions we can select *two* pairs of correspondents arbitrarily in the two systems, and find one rigid body to fulfil the requirements.

If we are given two systems of the sixth order of corresponding impulsive screws and instantaneous screws then subject to the observance of the geometrical conditions we can select *three* pairs of correspondents arbitrarily in the two systems, and find one rigid body to fulfil the requirements.

The last paragraph is, of course, only a different way of stating the results of § 307.

323. Two Rigid Bodies.

We shall now examine the circumstances under which pairs of impulsive and instantaneous screws are common to two, or more, rigid bodies. The problem before us may, perhaps, be most clearly stated as follows:—

Let there be two rigid bodies, M and M'. If M be struck by an impulsive wrench on a screw θ, it will commence to twist about some screw λ. If M' had been struck by an impulsive wrench on the same screw θ, the body would have commenced to twist about some screw μ, which would of course be generally different from λ. If θ be supposed to occupy different positions in space (the bodies remaining unaltered), so will λ and μ move into correspondingly various positions. It is proposed to inquire whether, under any circumstances, θ could be so placed that λ and μ should coincide. In other words, whether both of the bodies, M and M', when struck with an impulsive wrench on θ, will respond by twisting about the same instantaneous screw.

It is obvious, that there is at least one position in which θ fulfils the required condition. Let $G_1\, G_2$ be the centres of gravity of M and M'. Then a force along the ray $G_1\, G_2$, if applied either to M or to M', will do no more than produce a linear velocity of translation parallel thereto. Hence it follows, that a wrench on the screw of zero pitch, which lies on the ray $G_1 G_2$, will have the same instantaneous screw whether that wrench be applied to M or to M'.

We have now to examine whether there can be any other pair of impulsive and instantaneous screws in the same circumstances. Let us suppose that when θ assumes a certain position η, we have λ and μ coalescing into the single screw α.

We know that the centre of gravity lies in a plane through α, and the shortest distance between α and η. We know, also, that $d_\alpha = p_\alpha \tan(\alpha\eta)$, where d_α is the distance of the centre of gravity from α. It therefore follows that α must be parallel to $G_1 G_2$. We have, however, already had occasion (§ 303) to prove that, if ρ_α be the radius of gyration of a body about a ray through its centre of gravity, parallel to α,

$$\rho_\alpha^2 = \frac{2p_\alpha \varpi_{\alpha\eta}}{\cos(\alpha\eta)} - d_\alpha^2 - p_\alpha^2.$$

Hence it appears that, for the required condition to be satisfied, each of the two bodies must have the same radius of gyration about the axis through its centre of gravity, which is parallel to α. Of course this will not generally be the case. It follows that, in general, there cannot be any such pair of impulsive screws and instantaneous screws, as has been supposed. Hence we have the following result:—

Two rigid bodies, with different centres of gravity, have, in general, no other common pair of impulsive screws and instantaneous screws than the screw, of zero pitch, on the ray joining the centres of gravity, and the screw of infinite pitch parallel thereto.

We shall now consider what happens when the exceptional condition, just referred to, is fulfilled, that is, when the radius of gyration of the ray $G_1 G_2$ is the same for each of the bodies.

In each of the momental ellipsoids about the centres of gravity of the two bodies, draw the plane conjugate to the ray $G_1 G_2$. Let these planes intersect in a ray T. Suppose that an impulsive force, directed along T, be made to act on the body whose centre of gravity is G_1. It is plain, from Poinsot's well-known theory, that the rotation produced by such an impulse will be about a ray parallel to $G_1 G_2$. If this impulsive wrench had been applied to the body whose centre of gravity is G_2, the instantaneous screw would also be parallel to $G_1 G_2$. If we now introduce the condition that the radius of gyration of each of the bodies, about $G_1 G_2$, is the same, it can be easily deduced that the two instantaneous screws are identical. Hence we see that T, regarded as an impulsive screw of zero pitch, will have the same instantaneous screw for each of the two bodies.

If we regard T and G_1G_2 as two screws of zero pitch, and draw the cylindroid through these two screws, then any impulsive wrench about a screw on this cylindroid will have the same instantaneous screw for either of the two bodies to which it is applied.

For such a wrench may be decomposed into forces on T and on G_1G_2; these will produce, in either body, a twist about α, and a translation parallel to α, respectively. We therefore obtain the following theorem:—

If two rigid bodies have different centres of gravity, G_1 and G_2, and if their radii of gyration about the ray G_1G_2 are equal, there is then a cylindroid of screws such that an impulsive wrench on any one of these screws will make either of the rigid bodies begin to twist about the same screw, and the instantaneous screws which correspond to the several screws on this cylindroid, all lie on the same ray G_1G_2, but with infinitely varied pitch.

It is to be remarked that under no other circumstances can any impulsive screw, except the ray G_1G_2, with zero pitch, have the same instantaneous screw for each of the two bodies, so long as their centres of gravity are distinct.

We might have demonstrated the theorem, above given, from the results of § 303. We have there shown that, when an impulsive screw and the corresponding instantaneous screw are given, the rigid body must fulfil five conditions, the nature of which is fully explained. If we take two bodies which comply with these conditions, it appears that the ray through their centre of gravity is parallel to the instantaneous screw, and we also find that their radii of gyration must be equal about the straight line through their centres of gravity.

If two rigid bodies have the same centre of gravity, then, of course, any ray through this point will be the seat of an impulsive wrench on a screw of zero pitch such that it generates a twist velocity on a screw of infinite pitch, parallel to the impulsive screw. This will be the case to whichever of the two bodies the force be applied. We have therefore a system of the third order (much specialized no doubt) of impulsive screws, each of which has the same instantaneous screw for each of the two bodies. In general there will be no other pairs of common impulsive and instantaneous screws beyond those indicated.

Under certain circumstances, however, there will be other screws possessing the same relation.

We may suppose the two momental ellipsoids to be drawn about the common centre of gravity. These ellipsoids will, by a well-known property, possess one triad of common conjugate diameters. In general, of course, the radii of gyration of the two bodies, around any one of these three diameters, will not be equal. If, however, it should happen that the radius of gyration of one body be equal to that of the other body about one OX of these three common conjugate diameters, it can be shown that any screw, parallel to OX, whatever be its pitch regarded as an instantaneous screw, will have the same impulsive screw for either of the two bodies.

If the radii of gyration about two of the common conjugate diameters were equal for the two bodies, it will then appear that any instantaneous screw which is parallel to the plane of the common conjugate diameters, will have the same impulsive screw for each of the two bodies. The corresponding impulsive screws belong to the system of the fifth order, which is defined by being reciprocal to a screw of zero pitch on the third of the three common conjugate diameters.

Of course, if the radii of gyration coincide on this third diameter, then the two rigid bodies, regarded from our present point of view, would be identical.

CHAPTER XXIII.

VARIOUS EXERCISES.

324. The Co-ordinates of a Rigid Body.

We have already explained (§ 302) how nine co-ordinates define a rigid body sufficiently for the present theory. One set of such co-ordinates with respect to any three rectangular axes may be obtained as follows.

Let the element dm have the co-ordinates x, y, z, then causing the integrals to extend over the whole mass, we compute the nine quantities

$$\int x\,dm = Mx_0\,; \qquad \int y\,dm = My_0; \qquad \int z\,dm = Mz_0\,;$$

$$\int yz\,dm = Ml_1^2\,; \qquad \int xz\,dm = Ml_2^2; \qquad \int xy\,dm = Ml_3^2;$$

$$\int (y^2 + z^2)\,dm = M\rho_1^2; \quad \int (x^2 + z^2)\,dm = M\rho_2^2; \quad \int (x^2 + y^2)\,dm = M\rho_3^2.$$

The nine quantities x_0, y_0, z_0, l_1^2, l_2^2, l_3^2, ρ_1^2, ρ_2^2, ρ_3^2 constitute an adequate system of co-ordinates of the rigid body.

If θ_1, θ_2, ... θ_6 be the canonical co-ordinates of a screw about which twists a rigid body whose co-ordinates are x_0, y_0, z_0, l_1^2, l_2^2, l_3^2, ρ_1^2, ρ_2^2, ρ_3^2 with respect to the associated Cartesian axes, then the kinetic energy is $Mu_\theta^2\dot{\theta}^2$, where M is the mass, $\dot{\theta}$ the twist velocity, and where

$$\begin{aligned}
u_\theta^2 = {} & a^2\theta_1^2 + a^2\theta_2^2 + b^2\theta_3^2 + b^2\theta_4^2 + c^2\theta_5^2 + c^2\theta_6^2 \\
& + bx_0(\theta_3 - \theta_4)(\theta_5 + \theta_6) - cx_0(\theta_5 - \theta_6)(\theta_3 + \theta_4) \\
& + cy_0(\theta_5 - \theta_6)(\theta_1 + \theta_2) - ay_0(\theta_1 - \theta_2)(\theta_5 + \theta_6) \\
& + az_0(\theta_1 - \theta_2)(\theta_3 + \theta_4) - bz_0(\theta_3 - \theta_4)(\theta_1 + \theta_2) \\
& + \tfrac{1}{2}(\rho_1^2 - a^2)(\theta_1 + \theta_2)^2 + \tfrac{1}{2}(\rho_2^2 - b^2)(\theta_3 + \theta_4)^2 + \tfrac{1}{2}(\rho_3^2 - c^2)(\theta_5 + \theta_6)^2 \\
& - l_1^2(\theta_3 + \theta_4)(\theta_5 + \theta_6) - l_2^2(\theta_5 + \theta_6)(\theta_1 + \theta_2) - l_3^2(\theta_1 + \theta_2)(\theta_3 + \theta_4).
\end{aligned}$$

325. A Differential Equation satisfied by the Kinetic Energy.

If the pitch p of the screw θ about which a body is twisting receive a small increment δp while the twist velocity $\dot{\theta}$ is unaltered the change in kinetic energy is

$$Mp\delta p\dot{\theta}^2.$$

But the addition of δp to p has the effect (§ 264) of changing each canonical co-ordinate θ_1 into

$$\theta_1 + \frac{\delta p}{2p_1}(\theta_1 + \theta_2).$$

The variation thus arising in the kinetic energy equated to that already found gives the following differential equation which must be satisfied by u_θ^2,

$$a(\theta_1^2 - \theta_2^2) + b(\theta_3^2 - \theta_4^2) + c(\theta_5^2 - \theta_6^2)$$
$$= \frac{\theta_1 + \theta_2}{2a}\left(\frac{du_\theta^2}{d\theta_1} - \frac{du_\theta^2}{d\theta_2}\right) + \frac{\theta_3 + \theta_4}{2b}\left(\frac{du_\theta^2}{d\theta_3} - \frac{du_\theta^2}{d\theta_4}\right) + \frac{\theta_5 + \theta_6}{2c}\left(\frac{du_\theta^2}{d\theta_5} - \frac{du_\theta^2}{d\theta_6}\right).$$

If we assume that u_θ^2 must be a rational homogeneous function of the second order in $\theta_1, \ldots \theta_6$ we can, by solution of this equation, obtain the value of u_θ^2 given in the last Article.

326. Co-ordinates of Impulsive Screw in terms of the Instantaneous Screw.

If $\alpha_1, \ldots \alpha_6$ be the canonical co-ordinates of an instantaneous screw and $\eta_1, \ldots \eta_6$ the corresponding co-ordinates of the impulsive screw, then we have (§ 99),

$$\epsilon\eta_1 = \frac{1}{a}\frac{du_\alpha^2}{d\alpha_1}, \qquad \epsilon\eta_2 = -\frac{1}{a}\frac{du_\alpha^2}{d\alpha_2}, \ \&c.,$$

and we obtain the following:

$$\begin{aligned}
+\epsilon a\eta_1 = {} & (+\rho_1^2 + a^2)\alpha_1 \quad (+\rho_1^2 - a^2)\alpha_2 \quad + (az_0 - bz_0 - l_3^2)\alpha_3 \\
& + (az_0 + bz_0 - l_3^2)\alpha_4 + (-ay_0 + cy_0 - l_2^2)\alpha_5 + (-ay_0 - cy_0 - l_2^2)\alpha_6 \\
-\epsilon a\eta_2 = {} & (+\rho_1^2 - a^2)\alpha_1 \quad (+\rho_1^2 + a^2)\alpha_2 + (-az_0 - bz_0 - l_3^2)\alpha_3 \\
& (-az_0 + bz_0 - l_3^2)\alpha_4 + (+ay_0 + cy_0 - l_2^2)\alpha_5 + (+ay_0 - cy_0 - l_2^2)\alpha_6 \\
+\epsilon b\eta_3 = {} & (+az_0 - bz_0 - l_3^2)\alpha_1 + (-az_0 - bz_0 - l_3^2)\alpha_2 + \quad (+\rho_2^2 + b^2)\alpha_3 \\
& (+\rho_2^2 - b^2)\alpha_4 + (+bx_0 - cx_0 - l_1^2)\alpha_5 + \quad (bx_0 + cx_0 - l_1^2)\alpha_6 \\
-\epsilon b\eta_4 = {} & (+az_0 + bz_0 - l_3^2)\alpha_1 + (-az_0 + bz_0 - l_3^2)\alpha_2 \quad (+\rho_2^2 - b^2)\alpha_3 \\
& (+\rho_2^2 + b^2)\alpha_4 + (-bx_0 - cx_0 - l_1^2)\alpha_5 + (-bx_0 + cx_0 - l_1^2)\alpha_6 \\
+\epsilon c\eta_5 = {} & (-ay_0 + cy_0 - l_2^2)\alpha_1 + (+ay_0 + cy_0 - l_2^2)\alpha_2 + (+bx_0 - cx_0 - l_1^2)\alpha_3 \\
& (-bx_0 - cx_0 - l_1^2)\alpha_4 + \quad (+\rho_3^2 + c^2)\alpha_5 \quad (+\rho_3^2 - c^2)\alpha_6 \\
-\epsilon c\eta_6 = {} & (-ay_0 - cy_0 - l_2^2)\alpha_1 + \quad (ay_0 - cy_0 - l_2^2)\alpha_2 + (+bx_0 + cx_0 - l_1^2)\alpha_3 \\
& (-bx_0 + cx_0 - l_1^2)\alpha_4 + \quad (+\rho_3^2 - c^2)\alpha_5 + \quad (\rho_3^2 + c^2)\alpha_6.
\end{aligned}$$

327. Another proof of article 303.

As an illustration of the formulæ just given we may verify a theorem of § 303 showing that when we know the instantaneous screw corresponding to a given impulsive screw, then a ray along which the centre of gravity must lie is determined.

For subtracting the second equation from the first and repeating the process with each of the other pairs, we have

$$\epsilon(\eta_1+\eta_2)=2a(\alpha_1-\alpha_2)+2z_0(\alpha_3+\alpha_4)-2y_0(\alpha_5+\alpha_6),$$
$$\epsilon(\eta_3+\eta_4)=2b(\alpha_3-\alpha_4)+2x_0(\alpha_5+\alpha_6)-2z_0(\alpha_1+\alpha_2),$$
$$\epsilon(\eta_5+\eta_6)=2c(\alpha_5-\alpha_6)+2y_0(\alpha_1+\alpha_2)-2x_0(\alpha_3+\alpha_4).$$

Eliminating ϵ we have two linear equations in $x_0 y_0 z_0$, thus proving the theorem.

If we multiply these equations by $\alpha_1+\alpha_2$, $\alpha_3+\alpha_4$, $\alpha_5+\alpha_6$ respectively and add, we obtain

$$\epsilon\cos(\alpha\eta)=2p_\alpha,$$

thus giving a value for ϵ.

328. A more general Theorem.

If an instantaneous screw be given while nothing further is known as to the rigid body except that the impulsive screw is parallel to a given plane A, then the locus of the centre of gravity is a determinate plane.

Let λ, μ, ν be the direction cosines of a normal to A, then

$$\lambda(\eta_1+\eta_2)+\mu(\eta_3+\eta_4)+\nu(\eta_5+\eta_6)=0,$$

whence by substitution from the equations of the last Article we have a linear equation for x_0, y_0, z_0.

329. Two Three-Systems.

We give here another demonstration of the important theorem of § 318, which states that when two arbitrary three-systems U and V are given, it is in general possible to design and place a rigid body in one way but only in one way, such that an impulsive wrench delivered on any screw η of V shall make the body commence to move by twisting about some screw α of U.

Let the three principal screws of the system U have pitches a, b, c and take on the same three axes screws with the pitches $-a$, $-b$, $-c$ respectively. These six screws lying in pairs with equal and opposite pitches form the canonical co-reciprocals to be used.

As η belongs to the three-system V we must have the six co-ordinates of η connected by three linear equations (§ 77); solving these equations we have

$$\eta_2 = A\eta_1 + B\eta_3 + C\eta_5,$$
$$\eta_4 = A'\eta_1 + B'\eta_3 + C'\eta_5,$$
$$\eta_6 = A''\eta_1 + B''\eta_3 + C''\eta_5.$$

The nine coefficients A, B, C, A', B', C', A'', B'', C'' are essentially the co-ordinates of the three-system V. We now seek the co-ordinates of the rigid body in terms of these quantities.

Take the particular screw of U which has co-ordinates

$$1,\ 0,\ 0,\ 0,\ 0,\ 0.$$

Then the co-ordinates of the corresponding impulsive screw are η_1, η_2, ... where

$$+\epsilon a\eta_1 = \rho_1^2 - a^2; \quad +\epsilon b\eta_3 = az_0 - bz_0 - l_3^2; \quad +\epsilon c\eta_5 = -ay_0 + cy_0 - l_2^2;$$
$$-\epsilon a\eta_2 = \rho_1^2 - a^2; \quad -\epsilon b\eta_4 = az_0 + bz_0 - l_3^2; \quad -\epsilon c\eta_6 = -ay_0 - cy_0 - l_2^2.$$

Since by hypothesis this is to belong to V, the following equations must be satisfied:

$$\frac{a^2 - \rho_1^2}{a} = A\,\frac{a^2 + \rho_1^2}{a} + B\,\frac{az_0 - bz_0 - l_3^2}{b} + C\,\frac{cy_0 - ay_0 - l_2^2}{c},$$

$$\frac{-az_0 - bz_0 + l_3^2}{b} = A'\,\frac{a^2 + \rho_1^2}{a} + B'\,\frac{az_0 - bz_0 - l_3^2}{b} + C'\,\frac{cy_0 - ay_0 - l_2^2}{c},$$

$$\frac{ay_0 + cy_0 + l_2^2}{c} = A''\,\frac{a^2 + \rho_1^2}{a} + B''\,\frac{az_0 - bz_0 - l_3^2}{b} + C''\,\frac{cy_0 - ay_0 - l_2^2}{c}.$$

In like manner by taking successively for α the screws with co-ordinates

$$0,\ 0,\ 1,\ 0,\ 0,\ 0$$

and

$$0,\ 0,\ 0,\ 0,\ 1,\ 0,$$

we obtain six more equations of a similar kind. As these equations are linear they give but a single system of co-ordinates x_0, y_0, z_0, l_1^2, l_2^2, l_3^2, ρ_1^2, ρ_2^2, ρ_3^2 for the rigid body. The theorem has thus been proved, for of course if three screws of U correspond to three screws of V then every screw in U must have its correspondent restricted to V.

330. Construction of Homographic Correspondents.

If the screws in a certain three-system U be the instantaneous screws whose respective impulsive screws form the three-system V, then when three pairs of correspondents are known the determination of every other pair of correspondents may be conveniently effected as follows.

We know (§ 279) that to generate the unit twist velocity on an instantaneous screw α an impulsive wrench on the screw η is required, of which the intensity is

$$\frac{p_\alpha}{\cos(\alpha\eta)},$$

the mass being for convenience taken as unity.

Let α, β, γ be three of the instantaneous screws in U, and let η, ξ, ζ be their respective impulsive screws in V.

Let $\dot{\alpha}$, $\dot{\beta}$, $\dot{\gamma}$ be the component twist velocities on α, β, γ of a twist velocity $\dot{\rho}$ on any other screw ρ, belonging to the system U.

Then the impulsive wrench on V, which has ρ as its instantaneous screw will have as its components on η, ξ, ζ the respective quantities

$$\frac{p_\alpha}{\cos(\alpha\eta)}\dot{\alpha}, \quad \frac{p_\beta}{\cos(\beta\xi)}\dot{\beta}, \quad \frac{p_\gamma}{\cos(\gamma\zeta)}\dot{\gamma}.$$

These are accordingly the co-ordinates of the required impulsive wrench.

331. Geometrical Solution of the same Problem.

When three pairs of correspondents in the two impulsive and instantaneous systems of the third order V and U are known we can, in general, obtain the impulsive screw in V corresponding to any instantaneous screw ρ in U as follows.

Choose any screw other than ρ in the three-system U and draw the cylindroid H through that screw and ρ. Every screw on a cylindroid thus obtained must of course belong to U. Then H must have a screw in common with the cylindroid $(\alpha\beta)$ drawn through α and β, for this is necessarily true of any two cylindroids which lie in the same three-system. In like manner H must also have a screw in common with the cylindroid $(\alpha\gamma)$ drawn through α and γ. But by the principle of § 292 the several pairs of correspondents on the instantaneous cylindroid $(\alpha\beta)$ and the impulsive cylindroid $(\eta\xi)$ are determined. Hence the impulsive screw corresponding to one of the screws in H is known. In like manner the known pairs on the two cylindroids $(\alpha\gamma)$ and $(\eta\zeta)$ will discover the impulsive screw corresponding to another instantaneous screw on H. As therefore we know the impulsive screws corresponding to two of the screws on H we know the cylindroid H' which contains all the impulsive screws severally corresponding to instantaneous screws on H, of which of course ρ is one. But by § 293 we can now correlate the pairs on H and H', and thus the required correspondent to ρ is obtained.

332. Co-reciprocal Correspondents in two Three-systems.

If U be an instantaneous three-system and V the corresponding impulsive three-system it is in general possible to select one set of three co-reciprocal screws in U whose correspondents in V are also co-reciprocal.

As a preliminary to the formal demonstration we may note that the number of available constants is just so many as to suggest that some finite number of triads in U ought to fulfil the required condition.

In the choice of a screw α in U we have, of course, two disposable quantities. In the choice of β which while belonging to U is further reciprocal to α there is only one quantity disposable. The screw belonging to U, which is reciprocal both to α and β, must be unique. It is in fact reciprocal to five independent screws, *i.e.* to three of the screws of the system reciprocal to U, and to α and β in addition.

We have thus, in general, neither more nor fewer than three disposable elements in the choice of a set of three co-reciprocal screws α, β, γ in U. This is just the number of disposables required for the adjustment of the three correspondents η, ξ, ζ in V to a co-reciprocal system. We might, therefore, expect to have the number of solutions to our problem finite. We are now to show that this number is unity.

Taking the six principal screws of inertia of the rigid body as the screws of reference, we have as the co-ordinates of any screw in U

$$\begin{gathered}\lambda\alpha_1 + \mu\beta_1 + \nu\gamma_1,\\ \lambda\alpha_2 + \mu\beta_2 + \nu\gamma_2,\\ \cdots\cdots\cdots\cdots\\ \lambda\alpha_6 + \mu\beta_6 + \nu\gamma_6,\end{gathered}$$

where λ, μ, ν are numerical parameters.

The co-ordinates of the corresponding screw in V are

$$\begin{gathered}p_1(\lambda\alpha_1 + \mu\beta_1 + \nu\gamma_1),\\ p_2(\lambda\alpha_2 + \mu\beta_2 + \nu\gamma_2),\\ \cdots\cdots\cdots\cdots\\ p_6(\lambda\alpha_6 + \mu\beta_6 + \nu\gamma_6),\end{gathered}$$

where for symmetry $p_1, \ldots p_6$ are written instead of $+a, -a, +b, -b$, &c.

Three screws in U are specified by the parameters

$$\lambda', \mu', \nu'; \quad \lambda'', \mu'', \nu''; \quad \lambda''', \mu''', \nu'''.$$

If these screws are reciprocal, we have

$$0 = \Sigma p_1(\lambda'\alpha_1 + \mu'\beta_1 + \nu'\gamma_1)(\lambda''\alpha_1 + \mu''\beta_1 + \nu''\gamma_1),$$

or
$$\begin{aligned}0 = \lambda'\lambda''p_\alpha + \mu'\mu''p_\beta + \nu'\nu''p_\gamma &+ (\lambda'\mu'' + \lambda''\mu')\varpi_{\alpha\beta}\\ &+ (\lambda'\nu'' + \lambda''\nu')\varpi_{\alpha\gamma} + (\mu'\nu'' + \mu''\nu)\varpi_{\beta\gamma},\end{aligned}$$

and two similar equations.

If the corresponding impulsive screws are reciprocal, then

$$0 = \lambda'\lambda''\Sigma p_1{}^3\alpha_1{}^2 + \mu'\mu''\Sigma p_1{}^3\beta_1{}^2 + \nu'\nu''\Sigma p_1{}^3\gamma_1{}^2 + (\lambda'\mu'' + \lambda''\mu')\,\Sigma p_1{}^3\alpha_1\beta_1$$
$$+ (\lambda'\nu'' + \lambda''\nu')\,\Sigma p_1{}^3\alpha_1\gamma_1 + (\mu'\nu'' + \nu'\mu'')\,\Sigma p_1{}^3\beta_1\gamma_1,$$

and two similar equations.

Take the two conics whose equations are

$$0 = p_\alpha x^2 + p_\beta y^2 + p_\gamma z^2 + 2\varpi_{\alpha\beta}xy + 2\varpi_{\alpha\gamma}xz + 2\varpi_{\beta\gamma}yz,$$
$$0 = x^2\Sigma p_1{}^3\alpha_1{}^2 + y^2\Sigma p_1{}^3\beta_1{}^2 + z^2\Sigma p_1{}^3\gamma_1{}^2 + 2xy\Sigma p_1{}^3\alpha_1\beta_1 + 2xz\Sigma p_1{}^3\alpha_1\gamma_1$$
$$+ 2yz\Sigma p_1{}^3\beta_1\gamma_1,$$

these conics will generally have a single common conjugate triangle. If the co-ordinates of the vertices of this triangle be λ', μ', ν'; λ'', μ'', ν''; $\lambda''', \mu''', \nu'''$; then the equations just given in these quantities will be satisfied; and as there is only one such triangle, the required theorem has been proved.

It can be easily proved that a similar theorem holds good for a pair of impulsive and instantaneous cylindroids.

333. Impulsive and Instantaneous Cylindroids.

If a given cylindroid U be the locus of the screws about which a free rigid body would commence to twist if it had received an impulsive wrench about any screw on another given cylindroid V, it is required to calculate so far as practicable the co-ordinates of the rigid body.

Let us take our canonical screws of reference so that the two principal screws of the cylindroid U have as co-ordinates

$$1,\quad 0,\quad 0,\quad 0,\quad 0,\quad 0,$$
$$0,\quad 0,\quad 1,\quad 0,\quad 0,\quad 0.$$

The co-ordinates of any other screw on U will be

$$\alpha_1,\quad 0,\quad \alpha_3,\quad 0,\quad 0,\quad 0.$$

The cylindroid V will be determined by four linear equations in the co-ordinates of η. These equations may with perfect generality be written thus (§ 77),

$$\eta_3 = A\eta_1 + B\eta_2;\quad \eta_4 = A'\eta_1 + B'\eta_2;\quad \eta_5 + A''\eta_1 + B''\eta_2;\quad \eta_6 = A'''\eta_1 + B'''\eta_2,$$

where $A, B, A', B', A'', B'', A''', B'''$ are equivalent to the eight co-ordinates defining the cylindroid V.

The screw on U with co-ordinates

$$1,\quad 0,\quad 0,\quad 0,\quad 0,\quad 0$$

will have its corresponding impulsive screw defined by the equations (§ 326)

$$+\epsilon a\eta_1 = \rho_1^2 + a^2 \quad ; \quad -\epsilon a\eta_2 = \rho_1^2 - a^2 ;$$
$$+\epsilon b\eta_3 = az_0 - bz_0 - l_3^2 ; \quad -\epsilon b\eta_4 = az_0 + bz_0 - l_3^2 ;$$
$$+\epsilon c\eta_5 = cy_0 - ay_0 - l_2^2 ; \quad -\epsilon c\eta_6 = -ay_0 - cy_0 - l_2^2 .$$

By substituting these in the equations just given, we obtain

$$\frac{az_0 - bz_0 - l_3^2}{b} = A \frac{a^2 + \rho_1^2}{a} + B \frac{a^2 - \rho_1^2}{a}$$

$$\frac{-az_0 - bz_0 + l_3^2}{b} = A' \frac{a^2 + \rho_1^2}{a} + B' \frac{a^2 - \rho_1^2}{a},$$

$$\frac{cy_0 - ay_0 - l_2^2}{c} = A'' \frac{a^2 + \rho_1^2}{a} + B'' \frac{a^2 - \rho_1^2}{a},$$

$$\frac{ay_0 + cy_0 + l_2^2}{c} = A''' \frac{a^2 + \rho_1^2}{a} + B''' \frac{a^2 - \rho_1^2}{a}.$$

In like manner from the screw on U

$$0, \ 0, \ 1, \ 0, \ 0, \ 0$$

we obtain

$$+a\eta_1 = az_0 - bz_0 - l_3^2 ; \quad -a\eta_2 = -az_0 - bz_0 - l_3^2 ;$$
$$+b\eta_3 = \rho_2^2 + b^2 \quad ; \quad -b\eta_4 = \rho_2^2 - b^2 ;$$
$$+c\eta_5 = bx_0 - cx_0 - l_1^2 ; \quad -c\eta^6 = bx_0 + cx_0 - l_1^2 .$$

Introducing these into the equations for η, we have

$$\frac{b^2 + \rho_2^2}{b} = A \frac{az_0 - bz_0 - l_3^2}{a} + B \frac{az_0 + bz_0 + l_3^2}{a},$$

$$\frac{b^2 - \rho_2^2}{b} = A' \frac{az_0 - bz_0 - l_3^2}{a} + B' \frac{az_0 + bz_0 + l_3^2}{a},$$

$$\frac{bx_0 - cx_0 - l_1^2}{c} = A'' \frac{az_0 - bz_0 - l_3^2}{a} + B'' \frac{az_0 + bz_0 + l_3^2}{a},$$

$$\frac{-bx_0 - cx_0 + l_1^2}{c} = A''' \frac{az_0 - bz_0 - l_3^2}{a} + B''' \frac{az_0 + bz_0 + l_3^2}{a}.$$

Thus we have eight equations while there are nine co-ordinates of the rigid body. This ambiguity was, however, to be expected because, as proved in § 306, there is a singly infinite number of rigid bodies which stand to the two cylindroids in the desired relation.

The equations, however, contain one short of the total number of co-ordinates; x_0, y_0, z_0, l_1^2, l_2^2, l_3^2, ρ_1^2, ρ_2^2 are all present but ρ_3^2 is absent.

Hence from knowing the two cylindroids eight of the co-ordinates of the rigid body are uniquely fixed while the ninth remains quite indeterminate. Every value for ρ_3^2 will give one of the family of rigid bodies for which the desired condition is fulfilled.

We have already deduced geometrically (§ 306) the relations of these rigid bodies. We now obtain the same results otherwise.

The momental ellipsoid around the centre of gravity has as its equation

$$(x-x_0)^2\rho_1^2+(y-y_0)^2\rho_2^2+(z-z_0)^2\rho_3^2-2(y-y_0)(z-z_0)l_1^2$$
$$-2(z-z_0)(x-x_0)l_2^2-2(x-x_0)(y-y_0)l_3^2-(yx_0-xy_0)^2$$
$$-(zy_0-yz_0)^2-(xz_0-zx_0)^2=k^4.$$

This may be written in the form

$$\rho_3^2(z-z_0)^2=R,$$

where ρ_3^2 does not enter into R.

As ρ_3^2 varies this equation represents a family of quadrics which have contact along the section of $R=0$ by the plane $z-z_0=0$. This proves that a plane through the common centre of gravity and parallel to the principal plane of the cylindroid U passes through the conic along which the momental ellipsoids of all the different possible bodies have contact. All these quadrics touch a common cylinder along this conic. The infinite point on the axis of this cylinder is the pole of the plane $z-z_0=0$ for each quadric. Every chord parallel to the axis of the cylinder passes through this pole and is divided harmonically by the pole and the plane $z-z_0=0$. As the pole is at infinity it follows that in every quadric of the system a chord parallel to the axis of the cylinder is bisected by $z-z_0$. Hence a diameter parallel to the axis of the cylinder is conjugate to the plane $z-z_0$ in every one of the quadrics. Thus by a different method we arrive at the theorems of § 306.

334. The Double Correspondents on Two Cylindroids.

Referring to the remarkable homography between the impulsive screws on one cylindroid V and the corresponding instantaneous screws on another cylindroid U we have now another point to notice.

If the screws on U were the impulsive screws, while those on V were the instantaneous screws, there would also have been a unique homography, the rigid bodies involved being generally distinct.

But of course these homographies are in general quite different, that is to say, if A be a screw in U the instantaneous cylindroid, and B be its correspondent in V the impulsive cylindroid, it will not in general be true that if

A be a screw in U the impulsive cylindroid, then B will be its instantaneous screw in V the instantaneous cylindroid.

It is however to be now shown that there are two screws H_1 and H_2 on U, and their correspondents K_1 and K_2 on V, which possess the remarkable characteristic that whether V be the impulsive cylindroid and U the instantaneous cylindroid or *vice versâ*, in either case H_1 and K_1 are a pair of correspondents, and so are H_2 and K_2.

Let B_1, B_2, B_3, &c. be the screws on U corresponding severally to the screws A_1, A_2, A_3, &c., on V when V is the impulsive cylindroid and U the instantaneous cylindroid.

Let C_1, C_2, C_3, &c., be the screws on U corresponding severally to the screws A_1, A_2, A_3, &c., on V when U is now the impulsive cylindroid, and V the instantaneous cylindroid.

The systems A_1, A_2, A_3, &c., and B_1, B_2, B_3, &c., are homographic.

The systems C_1, C_2, C_3, &c., and A_1, A_2, A_3, &c., are homographic.

Hence also,

The systems B_1, B_2, B_3, &c., and C_1, C_2, C_3, &c. are homographic.

Let H_1, H_2 be the two double screws on U belonging to this last homography, then their correspondents K_1, K_2 on V will be the same whether U be the impulsive cylindroid and V the instantaneous cylindroid or *vice versâ*.

There can be no other pairs of screws on the two cylindroids possessing the same property.

335. A Property of Co-reciprocals.

Let α, β, γ be any three co-reciprocal screws. If η, ξ, ζ are the three screws on which impulsive wrenches would cause a free rigid body to twist about α, β, γ respectively, then

$$\cos(\alpha\zeta)\cos(\beta\eta)\cos(\gamma\xi) + \cos(\alpha\xi)\cos(\beta\zeta)\cos(\gamma\eta) = 0.$$

We have from § 281 the general formula

$$2\varpi_{\alpha\beta} = \frac{p_\alpha}{\cos(\alpha\eta)}\cos(\beta\eta) + \frac{p_\beta}{\cos(\beta\xi)}\cos(\alpha\xi),$$

but as α and β are reciprocal each side of this equation must be zero. We thus have

$$\frac{p_\alpha}{\cos(\alpha\eta)}\cos(\beta\eta) = -\frac{p_\beta}{\cos(\beta\xi)}\cos(\alpha\xi),$$

and similarly,

$$\frac{p_\beta}{\cos(\beta\xi)}\cos(\gamma\xi) = -\frac{p_\gamma}{\cos(\gamma\zeta)}\cos(\beta\zeta),$$

$$\frac{p_\gamma}{\cos(\gamma\zeta)}\cos(\alpha\zeta) = -\frac{p_\alpha}{\cos(\alpha\eta)}\cos(\gamma\eta),$$

whence we obtain

$$\cos(\alpha\zeta)\cos(\beta\eta)\cos(\gamma\xi) + \cos(\alpha\xi)\cos(\beta\zeta)\cos(\gamma\eta) = 0,$$

for it is shown in § 283 that $p_\alpha \div \cos(\alpha\eta)$ or the other similar expressions can never be zero.

336. Instantaneous Screw of Zero Pitch.

Let α be an instantaneous screw of zero pitch. Let two of the canonical co-reciprocals lie on α, then the co-ordinates of α are

$$\tfrac{1}{2},\ \tfrac{1}{2},\ 0,\ 0,\ 0,\ 0.$$

The co-ordinates of the impulsive screw η are given by the formulæ of § 326 which show that

$$\eta_1 + \eta_2 = 0; \quad \eta_3 + \eta_4 = -\frac{2z_0}{\epsilon}; \quad \eta_5 + \eta_6 = +\frac{2y_0}{\epsilon}.$$

We thus have

$$(\alpha_1 + \alpha_2)(\eta_1 + \eta_2) + (\alpha_3 + \alpha_4)(\eta_3 + \eta_4) + (\alpha_5 + \alpha_6)(\eta_5 + \eta_6) = 0,$$

which proves what we already knew, namely, that α and η are at right angles (§ 293).

We also have

$$y_0(\eta_3 + \eta_4) + z_0(\eta_5 + \eta_6) = 0,$$

which proves the following theorem:

If the instantaneous screw have zero-pitch then the centre of gravity of the body lies in the plane through the instantaneous screw and perpendicular to the impulsive screw.

337. Calculation of a Pitch Quadric.

If α, β, γ be three instantaneous screws it is required to find with respect to the principal axes through the centre of gravity, the equation to the pitch quadric of the three-system which contains the three impulsive screws corresponding respectively to α, β, γ. The co-ordinates of these screws are expressed with reference to the six principal screws of inertia.

We make the following abbreviations:

$$A = a^3(\alpha_1^2 - \alpha_2^2) + b^3(\alpha_3^2 - \alpha_4^2) + c^3(\alpha_5^2 - \alpha_6^2),$$
$$B = a^3(\beta_1^2 - \beta_2^2) + b^3(\beta_3^2 - \beta_4^2) + c^3(\beta_5^2 - \beta_6^2),$$
$$C = a^3(\gamma_1^2 - \gamma_2^2) + b^3(\gamma_3^2 - \gamma_4^2) + c^3(\gamma_5^2 - \gamma_6^2);$$

$$P = + bcx\,((\beta_5 - \beta_6)(\gamma_3 - \gamma_4) - (\beta_3 - \beta_4)(\gamma_5 - \gamma_6)),$$
$$+ cay\,((\gamma_5 - \gamma_6)(\beta_1 - \beta_2) - (\gamma_1 - \gamma_2)(\beta_5 - \beta_6)),$$
$$+ abz\,((\gamma_1 - \gamma_2)(\beta_3 - \beta_4) - (\beta_1 - \beta_2)(\gamma_3 - \gamma_4));$$

$$Q = + bcx\,((\alpha_3 - \alpha_4)(\gamma_5 - \gamma_6) - (\alpha_5 - \alpha_6)(\gamma_3 - \gamma_4)),$$
$$+ acy\,((\gamma_1 - \gamma_2)(\alpha_5 - \alpha_6) - (\alpha_1 - \alpha_2)(\gamma_5 - \gamma_6)),$$
$$+ abz\,((\alpha_1 - \alpha_2)(\gamma_3 - \gamma_4) - (\alpha_3 - \alpha_4)(\gamma_1 - \gamma_2));$$

$$R = + bcx\,((\alpha_5 - \alpha_6)(\beta_3 - \beta_4) - (\alpha_3 - \alpha_4)(\beta_5 - \beta_6)),$$
$$+ acy\,((\alpha_1 - \alpha_2)(\beta_5 - \beta_6) - (\beta_1 - \beta_2)(\alpha_5 - \alpha_6)),$$
$$+ abz\,((\beta_1 - \beta_2)(\alpha_3 - \alpha_4) - (\alpha_1 - \alpha_2)(\beta_3 - \beta_4));$$

$$L_{\alpha\beta} = a^3(\alpha_1 + \alpha_2)(\beta_1 - \beta_2) + b^3(\alpha_3 + \alpha_4)(\beta_3 - \beta_4) + c^3(\alpha_5 + \alpha_6)(\beta_5 - \beta_6),$$
$$L_{\beta\alpha} = a^3(\beta_1 + \beta_2)(\alpha_1 - \alpha_2) + b^3(\beta_3 + \beta_4)(\alpha_3 - \alpha_4) + c^3(\beta_5 + \beta_6)(\alpha_5 - \alpha_6);$$

$$L_{\alpha\gamma} = a^3(\alpha_1 + \alpha_2)(\gamma_1 - \gamma_2) + b^3(\alpha_3 + \alpha_4)(\gamma_3 - \gamma_4) + c^3(\alpha_5 + \alpha_6)(\gamma_5 - \gamma_6),$$
$$L_{\gamma\alpha} = a^3(\gamma_1 + \gamma_2)(\alpha_1 - \alpha_2) + b^3(\gamma_3 + \gamma_4)(\alpha_3 - \alpha_4) + c^3(\gamma_5 + \gamma_6)(\alpha_5 - \alpha_6);$$

$$L_{\beta\gamma} = a^3(\beta_1 + \beta_2)(\gamma_1 - \gamma_2) + b^3(\beta_3 + \beta_4)(\gamma_3 - \gamma_4) + c^3(\beta_5 + \beta_6)(\gamma_5 - \gamma_6),$$
$$L_{\gamma\beta} = a^3(\gamma_1 + \gamma_2)(\beta_1 - \beta_2) + b^3(\gamma_3 + \gamma_4)(\beta_3 - \beta_4) + c^3(\gamma_5 + \gamma_6)(\beta_5 - \beta_6).$$

The required equation is then as follows:

$$0 = \begin{vmatrix} A, & -R + L_{\beta\alpha}, & Q + L_{\gamma\alpha} \\ R + L_{\alpha\beta}, & B, & -P + L_{\gamma\beta} \\ -Q + L_{\alpha\gamma}, & P + L_{\beta\gamma}, & C \end{vmatrix}.$$

CHAPTER XXIV.

THE THEORY OF SCREW-CHAINS*.

338. Introduction.

In the previous investigations of this volume the Theory of Screws has been applied to certain problems in the dynamics of *one* rigid body. I propose to show in the present chapter to what extent the conceptions and methods of the Theory of Screws may be employed to elucidate certain problems in the dynamics of any material system whatever.

By such a system I mean an arbitrary arrangement of μ rigid bodies of any form or construction, each body being either entirely free or constrained in any manner by relations to fixed obstacles or by connexions of any kind with one or more of the remaining $\mu - 1$ pieces.

For convenience we may refer to the various bodies in the system by the respective numerals $1, 2, \ldots \mu$. This numbering may be quite arbitrary, and need imply no reference whatever to the mechanical connexions of the pieces. The entire set of material parts I call for brevity a *mass-chain*, and the number of the bodies in a mass-chain may be anything from unity to infinity.

I write, as before, of only small movements, but even with this limitation problems of equilibrium, of small oscillations and of impulsive movements are included. By the *order of the freedom* of the mass-chain, I mean the *number* of generalized co-ordinates which would be required to specify a position which that mass-chain was capable of assuming. The order cannot be less than one (if the mass-chain be not absolutely fixed), while if each element of the mass-chain be absolutely free, the order will be as much as 6μ.

Starting from any arbitrary position of the mass-chain, let it receive a small displacement. Each element will be displaced from its original position to an adjacent position, compatible of course with the conditions

* *Transactions Royal Irish Acad.* Vol. XXVIII. p. 99 (1881).

imposed by the constraints. The displacement of each element could, however, have been effected by a twist of appropriate amplitude about a screw specially correlated to that element. The total effect of the displacement could, therefore, have been produced by giving each element a certain twist about a certain screw.

339. The Graphic and Metric Elements.

In the lowest type of freedom which the mass-chain can possess (short of absolute fixity) the freedom is of the first order, and any position of the mass-chain admits of specification by a single co-ordinate. In such a case the screw appropriate to each element is unique, and is completely determined by the constraints both in position and in pitch. The ratio of the amplitude of each twist to the amplitudes of all the other twists is also prescribed by the constraints. The one co-ordinate which is arbitrary may be conveniently taken to be the amplitude of the twist about the first screw. To each value of this co-ordinate will correspond a possible position of the mass-chain. As the ratios of the amplitudes are all known, and as the first amplitude is given, then all the other amplitudes are known, and consequently the position assumed by every element of the mass-chain is known.

The whole series of screws and the ratios of the amplitudes thus embody a complete description of the particular route along which the mass-chain admits of displacement. The actual position of the mass-chain is found by adding to the purely graphic element which describes the *route* a metric element, to indicate the *amplitude* through which the mass-chain has travelled along that route. This *amplitude* is the arbitrary co-ordinate.

340. The Intermediate Screw.

It will greatly facilitate our further progress to introduce a conventional process, which will clearly exhibit the determinate character of the ratios of the amplitudes in the screw series. Consider the two first screws, α_1 and α_2 of the series. Draw the cylindroid (α_1, α_2) which contains these two screws. Since α_1 and α_2 are appropriated to two different elements of the mass-chain, no kinematical significance can be attached to the composition of the two twists on α_1 and α_2. If, however, the two twists on α_1 and α_2, having the proper ratio of amplitudes, had been applied to a single rigid body, the displacement produced is one which could have been effected by a single twist about a single screw on the cylindroid (α_1, α_2). If this intermediate screw be given, the ratio of the amplitudes of the twists on the given screws is determined. It is in fact equal to the ratio of the sines of the angles into which the intermediate screw divides the angle between the two given screws. With a similar significance we may conceive an intermediate screw inserted between every consecutive pair of the μ original screws.

341. The definition of a Screw-chain.

It will be convenient to have a name which shall concisely express the entire series of μ original screws with the $\mu - 1$ intermediate screws whose function in determining the amplitudes has just been explained. We may call it a *screw-chain*. A *twist about a screw-chain* will denote a displacement of a mass-chain, produced by twisting each element about the corresponding screw, through an amplitude whose ratio to the amplitudes on the two adjacent screws is indicated by the intermediate screws. The amplitude of the entire twist will, as already mentioned, be most conveniently expressed by the twist about the first screw of the chain. We hence have the following statement:—

The most general displacement of which a mass-chain is capable can be produced by a twist about a screw-chain.

342. Freedom of the first order.

Given a material system of μ elements more or less connected *inter se*, or related to fixed points or supports: let it be required to ascertain the freedom which this material system or mass-chain enjoys. The freedom is to be tested by the capacity for displacement which the mass-chain possesses. As each such displacement is a twist about a screw-chain, a complete examination of the freedom of the mass-chain will require that a trial be made to twist the mass-chain about every screw-chain in space which contains the right number of elements μ. If in the course of these trials it be found that the mass-chain cannot be twisted about any screw-chain, then the system is absolutely rigid, and has no freedom whatever. If after all trials have been made, one screw-chain, and only one, has been discovered, then the mass-chain has freedom of the first order, and we have the result thus stated:—

When a mass-chain is so limited by constraints, that its position can be expressed by a single co-ordinate, then the mass-chain is said to have freedom of the first order, and its possible movements are solely those which could be accomplished by twisting about one definite screw-chain.

By this method of viewing the question we secure the advantage of eliminating, as it were, the special characters of the constraints. The essential moving parts of a steam-engine, for example, have but one degree of freedom. Each angular position of the fly-wheel necessarily involves a definite position of all the other parts. A small angular motion of the fly-wheel necessarily involves a definite small displacement of each of the other parts. Complicated as the mechanism may be, it is yet always possible to construct a screw-chain, a twist about which would carry each element from its original position into the position it assumes after the displacement has been effected.

343. Freedom of the second order.

Suppose that after one screw-chain has been discovered, about which the mass-chain can be twisted, the search is continued until another screw-chain is detected of which the same can be asserted. We are now able to show, without any further trials whatever, that there must be an infinite number of other screw-chains similarly circumstanced. For, compound a twist of amplitude α' on one chain, α, with the twist of amplitude β' on the other, β. The position thus attained could have been attained by a twist about some single chain γ. As α' and β' are arbitrary, it is plain that γ can be only one of a system of screw-chains at least singly infinite in number about which twisting must be possible.

The problem to be considered may be enunciated in a somewhat more symmetrical manner, as follows:—

To determine the relations of three screw-chains, α, β, γ, such that if a mass-chain be twisted with amplitudes α', β', γ', about each of these screw-chains in succession, the mass-chain will regain the same position after the last twist which it had before the first.

This problem can be solved by the aid of principles already laid down (Chap. II.). Each element of the mass-chain receives two twists about α and β; these two twists can be compounded into a single twist about a screw lying on the cylindroid defined by the two original screws. We thus have for each element a third screw and amplitude by which the required screw-chain γ and its amplitude γ' can be completely determined.

A mass-chain free to twist to and fro on the chains α and β will therefore be free to twist to and fro on the chain γ. These three chains being known, we can now construct an infinite number of other screw-chains about which the mass-chain must be also able to twist.

Let δ be a further screw-chain of the system, then the screws $\alpha_1, \beta_1, \gamma_1, \delta_1$ which are the four first screws of the four screw-chains must be co-cylindroidal; so must α_2, β_2, γ_2, δ_2 and each similar set. We thus have μ cylindroids determined by the two first chains, and each screw of every chain derived from this original pair will lie upon the corresponding cylindroid. We have explained (§ 125) that by the anharmonic ratio of four screws on a cylindroid we mean the anharmonic ratio of a pencil of four lines parallel to these screws. If we denote the anharmonic ratio of four screws such as α_1, β_1, γ_1, δ_1 by the symbol

$$[\alpha_1, \beta_1, \gamma_1, \delta_1],$$

then the first theorem to be now demonstrated is that

$$[\alpha_1, \beta_1, \gamma_1, \delta_1] = [\alpha_2, \beta_2, \gamma_2, \delta_2] = \&c. = [\alpha_\mu, \beta_\mu, \gamma_\mu, \delta_\mu],$$

or that the anharmonic ratio of each group is the same.

This important proposition can be easily demonstrated by the aid of fundamental principles.

The two first chains, α and β, will be sufficient to determine the entire series of cylindroids. When the third chain, γ, is also given, the construction of additional chains can proceed by the anharmonic equality without any further reference to the ratios of the amplitudes.

When any screw, δ_1, is chosen arbitrarily on the first cylindroid, then δ_2, δ_3, &c., ... δ_μ, are all determined uniquely; for a twist about δ_1 can be decomposed into twists about α_1 and β_1. The amplitudes of the twists on α_1 and β_1 determine the amplitudes on α_2 and β_2 by the property of the intermediate screws which go to make up the screw-chains, and by compounding the twists on α_2 and β_2 we obtain δ_2. If any other screw of the series, for example, δ_2, had been given, then it is easy to see that δ_1 and all the rest, $\delta_3, \ldots \delta_\mu$, are likewise determined. Thus for the two first cylindroids, we see that to any one screw on either corresponds one screw on the other.

If one screw moves over the first cylindroid then its correspondent will move over the second and it will now be shown that these two screws trace out two homographic systems. Let us suppose that each screw is specified by the tangent of the angle which it makes with one of the principal screws of its cylindroid. Let θ_1, ϕ_1 be the angles for two corresponding screws on the first and second cylindroids, then we must have some relation which connects $\tan\theta_1$ and $\tan\phi_1$. But this relation is to be consistent with the condition that in every case *one* value of $\tan\theta_1$ is to correspond to *one* value of $\tan\phi_1$, and *one* value of $\tan\phi_1$ to *one* value of $\tan\theta_1$.

If for brevity we denote $\tan\theta_1$ by x and $\tan\phi_1$ by x' then the geometrical conditions of the system will give a certain relation between x and x'. The one-to-one condition requires that this relation must be capable of being expressed in either of the forms

$$x = U'; \quad x' = U,$$

where U' is some function of x' and where U is a function of x. From the nature of the problem it is easily seen that these functions are algebraical and as they must be one valued they must be rational. If we solve the first of these equations for x' the result that we obtain cannot be different from the second equation. The first equation must therefore contain x' only in the first degree in the form (see Appendix, Note 7)

$$x = \frac{px' + q}{p'x' + q'}.$$

The relation between $\tan\theta_1$ and $\tan\phi_1$ will therefore have the form which may generally be thus expressed,

$$a\tan\theta_1\tan\phi_1 + b\tan\theta_1 + c\tan\phi_1 + d = 0.$$

Let θ_1, θ_2, θ_3, θ_4 be the angles of four screws on the first cylindroid, then the anharmonic ratio will be

$$\frac{\sin(\theta_4-\theta_2)\sin(\theta_3-\theta_1)}{\sin(\theta_4-\theta_3)\sin(\theta_2-\theta_1)}.$$

From the relation just given between $\tan\theta_1$, $\tan\phi_1$, which applies of course to the other corresponding pair, it will be easily seen that this anharmonic ratio is unaltered when the angles ϕ_1, ϕ_2, &c., are substituted for θ_1, θ_2, &c.

We have, therefore, shown that the anharmonic ratio of four screws on the first cylindroid is equal to that of the four corresponding screws on the second cylindroid, and so on to the last of the μ cylindroids.

As soon, therefore, as any arbitrary screw δ_1 has been chosen on the first cylindroid, we can step from one cylindroid to the next, merely guided in choosing δ_2, δ_3, &c., by giving a constant value to the anharmonic ratio of the screw chosen and the three other collateral screws on the same cylindroid. Any number of screw-chains belonging to the system may be thus readily constructed.

This process, however, does not indicate the amplitudes of the twists appropriate to δ_1, δ_2, δ_3, &c. One of these amplitudes may no doubt be chosen arbitrarily, but the rest must be all then determined from the geometrical relations. We proceed to show how the relative values of these amplitudes may be clearly exhibited.

The first theorem to be proved is that in the three screw-chains α, β, γ the screws intermediate to α_1 and α_2, to β_1 and β_2, to γ_1 and γ_2 are co-cylindroidal. This important step in the theory of screw-chains can be easily inferred from the fundamental property that three twists can be given on the screw-chains α, β, γ, which neutralize, and that consequently the three twists on the screws α_1, β_1, γ_1 will neutralize, as will also those on α_2, β_2, γ_2. These six twists must neutralize when compounded in any way whatever. We shall accordingly compound α_1 and α_2 into one twist on their intermediate screw, and similarly for β_1 and β_2, and for γ_1 and γ_2. We hence see that the three twists about the three intermediate screws must neutralize, and consequently the three intermediate screws must be co-cylindroidal.

We thus learn that in addition to the several cylindroids containing the primary screws of each of the system of screw-chains about which a mass-chain with two degrees of freedom can twist, there are also a series of secondary cylindroids, on which will lie the several intermediate screws of the system of screw-chains.

If δ_1 be given, then it is plain that the intermediate screw between δ_1 and δ_2, as well as all the other screws of the chain and their intermediate screws, can be uniquely determined. If, however, the intermediate screw between δ_1 and δ_2 be given, the entire chain δ is also determined, yet it is not immediately obvious that that determination is unique. We can, however, show as follows that this is generally the case.

Let δ_{12} denote the given intermediate screw, and let this belong, not only to the chain δ_1, δ_2, &c., but to another chain δ_1', δ_2', &c. We then have δ_1, δ_{12}, δ_2 co-cylindroidal, and also δ_1', δ_{12}, δ_2' co-cylindroidal. Decompose any arbitrary twist of amplitude θ on δ_{12} into twists on δ_1 and δ_2, and a twist of amplitude $-\theta$ on the same δ_{12} into twists on δ_1' and δ_2'. Then the four twists must neutralize; but the two twists on δ_1' and δ_1 compound into a twist on a screw on the first cylindroid of the system; and δ_2' and δ_2 into a twist on the second cylindroid of the system; and as these two resultant twists must be equal and opposite it follows that they must be on the same screw, and that, therefore, the cylindroids belonging to the first and second elements of the system must have a common screw. It is, however, not generally the case that two cylindroids have a common screw. It is only true when the two cylindroids are themselves included in a three-system, this could only arise under special circumstances, which need not be further considered in a discussion of the general theory.

It follows from the unique nature of the correspondence between the intermediate cylindroids and the primary cylindroids that one screw on any cylindroid corresponds uniquely to one screw on each of the other cylindroids; the correspondence is, therefore, homographic.

We have now obtained a picture of the freedom of the second order of the most general type both as to the material arrangement and the character of the oonstraints: stating summarily the results at which we have arrived, they are as follows:—

A mass-chain of any kind whatever receives a small displacement. This displacement is under all circumstances a twist about a screw-chain. If the mass-chain admits of a displacement by a twist about a second screw-chain, then twists about an infinite number of other screw-chains must also be possible. To find, in the first place, a third screw-chain, give the mass-chain a small twist about the first chain; this is to be followed by a small twist about the second chain: the position of the mass-chain thus attained could have been reached by a twist about a third screw-chain. The system must, therefore, be capable of twisting about this third screw-chain. When three of the chains have been constructed, the process of finding the remainder is greatly simplified. Each element of the mass-chain is, in each of the three displacements just referred to, twisted about a screw. These three screws lie on one cylindroid appropriate to the element, and there are just so many of these cylindroids as there are elements in the mass-chain.

Between each two screws of a chain lies an intermediate screw, introduced for the purpose of defining the ratio of the amplitudes of the two screws of the chain on each side of it. In the three chains two consecutive elements will thus have three intermediate screws. These screws are co-cylindroidal. We thus have two series of cylindroids: the first of these is equal in number to the elements of the mass-chain (μ), each cylindroid corresponding to one element. The second series of cylindroids consists of one less than the entire number of elements ($\mu - 1$). Each of these latter cylindroids corresponds to the intermediate screw between two consecutive elements. An entire screw-chain will consist of μ primary screws, and $\mu - 1$ intermediate screws. To form such a screw-chain it is only necessary to inscribe on each of the $2\mu - 1$ cylindroids a screw which, with the other three screws on that cylindroid, shall have a constant anharmonic ratio. Any one screw on any one of the $2\mu - 1$ cylindroids may be chosen arbitrarily; but then all the other screws of that chain are absolutely determined, as the anharmonic ratio is known. The mass-chain which is capable of twisting about two screw-chains cannot refuse to be twisted about any other screw-chain constructed in the manner just described. It may, however, refuse to be twisted about any screw-chains not so constructed; and if so, then the mass-chain has freedom of the second order.

344. Homography of Screw-systems.

Before extending the conception of screw-chains to the examination of the higher orders of freedom, it will be necessary to notice some extensions of the notions of homography to the higher orders of screw systems. On the cylindroid the matter is quite simple. As we have already had occasion to explain, we can conceive the screws on two cylindroids to be homographically related, just as easily as we can conceive the rays of two plane pencils. The same ideas can, however, be adapted to the higher systems of screws—the 3rd, the 4th, the 5th—while a case of remarkable interest is presented in the homography of two systems of the 6th order.

The homography of two three-systems is completely established when to each screw on one system corresponds one screw on the other system, and conversely. We can represent the screws in a three-system by the points in a plane (see Chap. XV.). We therefore choose two planes, one for each of the three-systems, and the screw correspondence of which we are in search is identical with the homographic point-correspondence between the two planes.

We have already had to make use in § 317 of the fundamental property that when four pairs of correspondents in the two planes are given then the correspondence between every other pair of points is determined by rigorous construction. Any fifth point in one plane being indicated, the fifth point corresponding thereto in the other plane can be determined. It therefore follows that when four given screws on one three-system are the correspondents of four indicated screws on the other system, then the correspondence is completely established, and any fifth screw on one system being given, its correspondent on the other is determined.

345. Freedom of the third order.

We are now enabled to study the small movements of any mass-chain which has freedom of the third order. Let such a mass-chain receive any three displacements by twists about three screw-chains. It will, of course, be understood that these three screw-chains are not connected in the specialized manner we have previously discussed in freedom of the second order. In such a case the freedom of the mass-chain would be of the second order only and not of the third. The three screw-chains now under consideration are perfectly arbitrary; they may differ in every conceivable way, all that can be affirmed with regard to them is that the number of primary screws in each chain must of course be equal to μ, *i.e.* to the number of material elements of which the mass-chain consists.

It may be convenient to speak of the screws in the different chains which relate to the same element (or in the case of the intermediate screws, the same pair of elements) as *homologous* screws. Each set of three homologous screws will define a three-system. Compounding together any three twists on the screw-chains, we have a resultant displacement which could have been effected by a single twist about a fourth screw-chain. The first theorem to be proved is, that each screw in this fourth screw-chain must belong to the three-system which is defined by its three homologous screws.

So far as the primary screws are concerned this is immediately seen. Each element having been displaced by three twists about three screws, the resultant twist must belong to the same three-system, this being the immediate consequence of the definition of such a system. Nor do the intermediate screws present much difficulty. It must be possible for appropriate twists on the four screw-chains to neutralize. The four twists which the first element receives must neutralize: so must also the four twists imparted to the second element. These eight twists must therefore neutralize, however they may be compounded. Taking each chain separately, these eight twists will reduce to four twists about the four intermediate screws: these four twists must neutralize; but this is only possible if the four intermediate screws belong to a three-system.

On each of μ primary three-systems, and on each of $\mu - 1$ intermediate three-systems four screws are now supposed to be inscribed. We are to determine a fifth screw about which the system even though it has only freedom of the third order, must still be permitted to twist.

To begin with we may choose an arbitrary screw in any one of the three-systems. In the exercise of this choice we have two degrees of latitude; but once the choice has been made, the remainder of the screw-chain is fixed by the following theorem:—

If each set of five homologous screws of five screw-chains lies on a three-system, and if a mass-chain be free to twist about four of these screw-chains, it will also be free to twist about the fifth, provided each set of homologous screws is homographic with every other set.

Let δ denote the fifth screw-chain. If δ_1 be chosen arbitrarily on the three-system which included the first element, then a twist about δ_1 can be decomposed into three twists on α_1, β_1, γ_1. By the intermediate screws these three twists will give the amplitudes of the twists on all the other screws of the chains α, β, γ, and each group of three homologous twists being compounded, will give the corresponding screws on the chain δ. We thus see that when δ_1 is given, δ_2, δ_3, &c., are all determinate. It is also obvious that if δ_2, or any other primary screw of the chain, were given, then all the other screws of the chain would be determined uniquely.

If, however, an intermediate screw, δ_{12}, had been given, then, although the conditions are, so far as number goes, adequate to the determination of the screw-chain, it will be necessary to prove that the determination is unique. This is proved in the same manner as for freedom of the second order (§ 343). If there were two screw-chains which had the same intermediate screw, then it must follow that the two primary three-systems must have a common screw, which is not generally the case.

We have thus shown that when any one screw of the chain δ, whether primary or intermediate, is given, then all the rest of the screws of the chain are uniquely determinate. Each group of five homologous screws must therefore be homographic.

It is thus easy to construct as many screw-chains as may be desired, about which a mass-chain which has freedom of the third order must be capable of twisting. It is only necessary, after four chains have been found, to inscribe an arbitrary screw on one of the three-systems, and then to construct the corresponding screw on each of the other homologous systems.

In choosing one screw of the chain we have two degrees of latitude: we may, for example, move the screw chosen over the surface of any cylindroid embraced in the three-system: the remaining screws of the screw-chain, primary and intermediate, will each and all move over the surface of corresponding cylindroids.

If the mass-chain cannot be twisted about any screw-chain except those we have now been considering, then the mass-chain is said to have freedom of the third order. If, however, a fourth screw-chain can be found, about which the system can twist, and if that screw-chain does not belong to the doubly infinite system just described, then the mass-chain must have freedom of at least the fourth order.

346. Freedom of the fourth order.

The homologous screws in the four screw-chains about which the mass-chain can twist form each a four-system. All the other chains which can belong to the system must consist of screws, one of which lies on each of the four-systems.

It will facilitate the study of the homography of two four-systems to make use of the analogy between the homography of two spaces and the homography of two four-systems as already we had occasion to do in § 317. A screw in a four-system is defined by four homogeneous co-ordinates whereof only the ratios are significant. Each screw of such a system can therefore be represented by one point in space. The homography of two spaces will be completely determined if five points, a, b, c, d, e in one space, and the five corresponding points in the other space, a', b', c', d', e' are given.

From the four original screw-chains we can construct a fifth by compounding any arbitrary twists about two or more of the given chains. When five chains have been determined, then, by the aid of the principle of homography, we can construct any number.

That each set of six homologous screws is homographic with every other set can be proved, as in the other systems already discussed. With respect to the intermediate screws a different proof is, however, needed to show that when one of these screws is given the rest of the chain is uniquely determined. The proof we now give is perhaps simpler than that previously used, while it has the advantage of applying to the other cases as well. Let α, β, γ, δ be four screw-chains, and let ϵ_{12}, an intermediate screw of the chain ϵ, be given. We can decompose a twist on ϵ_{12} into components of definite amplitude on α_{12}, β_{12}, γ_{12}, δ_{12}. The first of these can be decomposed into twists on α_1 and α_2; the second on β_1 and β_2, &c. Finally, the four twists on α_1, β_1, γ_1, δ_1 can be compounded into one twist, ϵ_1, and those on α_2, β_2, γ_2, δ_2 compounded into a twist on ϵ_2. In this way it is obvious that when ϵ_{12} is given, then ϵ_1 and ϵ_2 are uniquely determined, and of course the same reasoning applies to the whole of the chain. We thus see that when any screw of the chain is known, then all the rest are uniquely determined, and therefore the principle of homography is applicable.

In the choice of a screw-chain about which a mass-chain with four degrees of freedom can twist there are three arbitrary elements. We may choose as the first screw of the chain any screw from a given four-system.

If one screw of the chain be moved over a two-system, or a three-system included in the given four-system, then every other screw of the chain will also describe a corresponding two-system or three-system.

347. Freedom of the fifth order.

In discussing the movements of a system which has freedom of the fifth order, the analogies which have hitherto guided us appear to fail. Homographic pencils, planes, and spaces have exhibited graphically the relations of the lower degrees of freedom; but for freedom of the fifth degree these illustrations are inadequate. No real difficulty can, however, attend the extension of the principles we have been considering to the freedom of the fifth order. We can conceive that two five-systems are homographically related, such that to each screw on the one corresponds one screw on the other, and conversely. To establish the homography of the two systems it will be necessary to know the six screws on one system which correspond to six given screws on the other: the screw in either system corresponding to any seventh screw in the other is then completely determined.

In place of the methods peculiar to the lower degrees of freedom, we shall here state the general analytical process which is of course available in the lower degrees of freedom as well.

A screw θ in a five-system is to be specified by five co-ordinates θ_1, θ_2, θ_3, θ_4, θ_5. These co-ordinates are homogeneous; but their ratios only are concerned, so they are equivalent to four data. The five screws of reference may be any five screws of the system. Let ϕ be the screw of the second system which is to correspond to θ in the first system. The co-ordinates of ϕ may be referred to any five screws chosen in the second system. It will thus be seen that the five screws of reference for ϕ are quite different from those of θ.

The geometrical conditions expressing the connection between ϕ and θ will give certain equations of the type

$$\phi_1 = U_1; \quad \phi_2 = U_2; \quad \ldots; \quad \phi_5 = U_5,$$

where $U_1, \ldots, U_5$ are homogeneous functions of $\theta_1, \ldots, \theta_5$. These equations express that one θ determines one ϕ. As however one ϕ is to determine one θ we must have also equations of the type

$$\theta_1 = U_1'; \quad \theta_2 = U_2'; \quad \ldots; \quad \theta_5 = U_5,$$

where $U_1', \ldots, U_5'$ are functions of $\phi_1, \ldots, \phi_5$.

From the nature of the problem these functions are algebraical and as they must be one valued they must be rational functions. We have therefore a case of "Rational Transformation" (see Salmon's *Higher Plane Curves*, Chap. VIII.). The theory is however here much simplified. In this case none of the special solutions are admissible which produce the critical cases. Consider the equations $U_2=0, \ldots, U_5=0$. They will give a number of systems of values for $\theta_1, \ldots, \theta_5$ equal to the product of the degrees of $U_2, \ldots, U_5$. Each of these θ screws would be a correspondent to the same ϕ screw 1, 0, 0, 0, 0. But in the problems before us this ϕ as every other ϕ can have only one correspondent. Hence all the functions U_1, U_2, etc. U_1', U_2', etc. must be linear. We may express the first set of equations thus:

$$\phi_1 = (11)\,\theta_1 + (12)\,\theta_2 + (13)\,\theta_3 + (14)\,\theta_4 + (15)\,\theta_5,$$

$$\phi_2 = (21)\,\theta_1 + (22)\,\theta_2 + (23)\,\theta_3 + (24)\,\theta_4 + (25)\,\theta_5,$$

$$\phi_3 = (31)\,\theta_1 + (32)\,\theta_2 + (33)\,\theta_3 + (34)\,\theta_4 + (35)\,\theta_5,$$

$$\phi_4 = (41)\,\theta_1 + (42)\,\theta_2 + (43)\,\theta_3 + (44)\,\theta_4 + (45)\,\theta_5,$$

$$\phi_5 = (51)\,\theta_1 + (52)\,\theta_2 + (53)\,\theta_3 + (54)\,\theta_4 + (55)\,\theta_5.$$

For the screw ϕ to be known whenever θ is given, it will be necessary to determine the various coefficients (11), (12), &c. These are to be determined from a sufficient number of given pairs of corresponding screws. Of these coefficients there are in all twenty-five. If we substitute the co-ordinates of one given screw θ, we have five linear equations between the co-ordinates. Of these equations, however, we can only take the ratios, for each of the co-ordinates may be affected by an arbitrary factor. Each of the given pairs of screws will thus provide four equations to aid in determining the coefficients. Six pairs of screws being given, we have twenty-four equations between the twenty-five coefficients. These will be sufficient to determine the ratios of the coefficients. We thus see that by six pairs of screws the homography of two five-systems is to be completely defined. To any seventh screw on one system corresponds a seventh screw on the other system, which can be constructed accordingly.

348. Application of Parallel Projections.

It will, however, be desirable at this point to introduce a somewhat different procedure. We can present the subject of homography from another point of view, which is specially appropriate for the present theory. The notions now to be discussed might have been introduced at the outset. It was, however, thought advantageous to concentrate all the light that could be obtained on the subject; we therefore used the point-homography of lines, of planes, and of spaces, so long as they were applicable.

The method which we shall now adopt is founded on an extension of what are known as "parallel projections" in Statics. We may here recall the outlines of this theory, with the view of generalizing it into one adequate for our purpose.

We can easily conceive of two systems of corresponding forces in two planes. To each force in one plane will correspond one force in the other plane, and *vice versâ*. To any system of forces in one plane will correspond a system of forces in the other plane. We are also to add the condition that if one force x vanishes, the corresponding force x' will also vanish.

The fundamental theorem which renders this correspondence of importance is thus stated:—

If a group of forces in one of the planes would equilibrate when applied to a rigid body, then the corresponding group of forces in the other plane would also equilibrate when applied to a rigid body.

Draw any triangle in each of these planes, then any force can be decomposed into three components on the three sides of the triangle. Let x, y, z be the components of such a force in the first plane, and let x', y', z' be the components of the corresponding force in the second plane; we must then have equations of the form

$$x' = ax + by + cz,$$

$$y' = a'x + b'y + c'z,$$

$$z' = a''x + b''y + c''z,$$

where a, b, c, &c., are constants. These equations do not contain any terms independent of the forces, because x', y', z' must vanish when x, y, z vanish. They are linear in the components of the forces, because otherwise one force in one plane will not correspond uniquely with one force in the other.

Let x_1, y_1, z_1; x_2, y_2, z_2; ... x_n, y_n, z_n be the components of forces in the first plane.

Let x_1', y_1', z_1'; x_2', y_2', z_2'; ... x_n', y_n', z_n' be the components of the corresponding forces in the second plane. Then we must have

$$x_k' = ax_k + by_k + cz_k,$$

$$y_k' = a'x_k + b'y_k + c'z_k,$$

$$z_k' = a''x_k + b''y_k + c''z_k,$$

where k has every value from 1 to n. If therefore we write

$$\Sigma x = x_1 + x_2 + \ldots + x_n,$$

and $$\Sigma x' = x_1' + x_2' + \ldots + x_n',$$

with similar values for Σy, $\Sigma y'$, Σz, $\Sigma z'$ then the above equations give

$$\Sigma x' = a\ \Sigma x + b\ \Sigma y + c\ \Sigma z,$$
$$\Sigma y' = a'\Sigma x + b'\Sigma y + c'\Sigma z,$$
$$\Sigma z' = a''\Sigma x + b''\Sigma y + c''\Sigma z.$$

If the system of forces in the first plane equilibrate, the following conditions must be satisfied:

$$\Sigma x = 0, \qquad \Sigma y = 0, \qquad \Sigma z = 0,$$

and from the equations just written, these involve

$$\Sigma x' = 0, \qquad \Sigma y' = 0, \qquad \Sigma z' = 0,$$

whence the corresponding system in the other plane must also equilibrate.

To determine the correspondence it will be necessary to know only the three forces in the second plane which correspond to three given forces in the first plane. We shall then have the nine equations which will be sufficient to determine the nine quantities a, b, c, &c.

It appears, from the form of the equations, that the ratio of the intensity of a force to the intensity of the corresponding force is independent of those intensities, *i.e.* it depends solely upon the situation of the lines in which the forces act.

Take any four straight lines in one system, and let four forces, X_1, X_2, X_3, X_4, on these four straight lines equilibrate. It is then well known that each of these forces must be proportional to certain functions of the positions of these straight lines. We express these functions by A_1, A_2, A_3, A_4. The four corresponding forces will be X_1', X_2', X_3', X_4', and as they must equilibrate, they must also be in the ratio of certain functions A_1', A_2', A_3', A_4' of the positions.

We thus have the equations

$$\frac{X_1}{A_1} = \frac{X_2}{A_2} = \frac{X_3}{A_3} = \frac{X_4}{A_4},$$
$$\frac{X_1'}{A_1'} = \frac{X_2'}{A_2'} = \frac{X_3'}{A_3'} = \frac{X_4'}{A_4'}.$$

We can select the ratio of X_1 to X_1' arbitrarily: for example, let this ratio be μ; then

$$\frac{X_2}{X_2'} = \frac{A_2}{A_2'}\mu,$$

whence the ratio of X_2 to X_2' is known. Similarly the ratio of the other intensities $X_3 : X_3'$, and $X_4 : X_4'$ is known. And generally the ratio of every pair of corresponding forces will be determined.

It thus appears that four straight lines in one system may be chosen arbitrarily to correspond respectively with four straight lines in the other system; and that one force being chosen on one of these straight lines in one system, the corresponding force may be chosen arbitrarily on the corresponding straight line in the other system. This having been done, the relation between the two systems is completely defined.

From the case of parallel projections in two planes it is easy to pass to the case which will serve our present purpose. Instead of the straight lines in the two planes we shall take screws in two n-systems. Instead of the forces on the lines we may take either twists or wrenches on the screws. More generally it will be better to use Plucker's word "Dyname," which we have previously had occasion to employ (§ 260) in the sense either of a twist or a wrench, or even a twist velocity. We shall thus have a Dyname in one system corresponding to a Dyname in the other.

Let us suppose that a Dyname on a screw of one n-system corresponds uniquely to a different Dyname on a screw of another n-system. The two n-systems may be coincident but we shall treat of the general case.

In the first place it can be shown that if any number of Dynames in the first system neutralize, their corresponding Dynames in the second system must also neutralize. Take n screws of reference in one system, and also n screws of reference in the corresponding system. Let θ be the Dyname in one system which corresponds to ϕ in the other; θ can be completely resolved into component Dynames of intensities $\theta_1, \ldots \theta_n$ on the n screws of reference in the first system and in like manner ϕ can be resolved into n components of intensities $\phi_1, \ldots \phi_n$ on the screws of reference in the second system ($n = < 6$).

From the fact that the relation between θ and ϕ is of the one-to-one type the several components $\phi_1, \ldots \phi_n$ are derived from $\theta_1, \ldots \theta_n$ by n equations which may be written

$$\phi_1 = (11)\,\theta_1 + (12)\,\theta_2 \ldots + (1n)\,\theta_n,$$

$$\cdots\cdots\cdots\cdots\cdots\cdots\cdots\cdots\cdots$$

$$\phi_n = (n1)\,\theta_1 + (n2)\,\theta_2 \ldots + (nn)\,\theta_n,$$

in which (11), (12), &c. must be independent of both θ and ϕ, for otherwise the correspondence would not be unique.

If there be a number of Dynames in the first system the sums of the intensities of their components on the n screws of reference may be expressed as $\Sigma\theta_1, \ldots \Sigma\theta_n$ respectively. In like manner the sums of the intensities of the components of their correspondents on the screws of reference of the second system may be represented by $\Sigma\phi_1, \ldots \Sigma\phi_n$ respectively. We therefore obtain the following equations by simply adding the equations just written for each separate screw

$$\Sigma\phi_1 = (11)\,\Sigma\theta_1 + (12)\,\Sigma\theta_2 \ldots + (1n)\,\Sigma\theta_n,$$

$$\ldots\ldots\ldots\ldots\ldots\ldots\ldots\ldots\ldots\ldots\ldots\ldots\ldots$$

$$\Sigma\phi_n = (n1)\,\Sigma\theta_1 + (n2)\,\Sigma\theta_2 \ldots + (nn)\,\Sigma\theta_n.$$

If the Dynames in the first system neutralize then their components on the screws of reference must vanish or

$$\Sigma\theta_1 = 0, \ldots \Sigma\theta_n = 0.$$

But it is obvious from the equations just written that in this case

$$\Sigma\phi_1 = 0, \ldots \Sigma\phi_n = 0,$$

and therefore the corresponding Dynames will also neutralize.

Given n pairs of corresponding Dynames in the two systems, we obtain n^2 linear equations which will be adequate to determine uniquely the n^2 constants of the type (11), (12), &c. It is thus manifest that n given pairs of Dynames suffice to determine the Dyname in either system, corresponding to a given Dyname in the other. It is of course assumed that in this case the intensities of the two corresponding Dynames in each of the n-pairs are given as well as the screws on which they lie.

349. Properties of this correspondence.

To illustrate the distinction between this Dyname correspondence and the screw correspondence previously discussed, let us take the case of two cylindroids. We have already seen that, given any *three* pairs of corresponding screws, the correspondence is then completely defined (§ 343). Any fourth screw on one of the cylindroids will have its correspondent on the other immediately pointed out by the equality of two anharmonic ratios. The case of the Dyname correspondence is, however, different inasmuch as we require more than *two* pairs of corresponding Dynames on the two cylindroids, in order to completely define the correspondence. For any third Dyname θ on one of the cylindroids can be resolved into two Dynames θ_1 and θ_2 on the two screws containing the given Dynames. These components will determine the components ϕ_1, ϕ_2 on the corresponding cylindroid, which being compounded, will give ϕ the Dyname corresponding to θ.

It is remarkable that *two* pairs of Dynames should establish the correspondence as completely as *three* pairs of screws. But it will be observed that to be given a pair of corresponding screws on the two cylindroids is in reality only to be given one datum. For one of the screws may be chosen arbitrarily; and as the other only requires one parameter to fix it on the cylindroid to which it is confined its specification merely gives a single datum. To be given a pair of corresponding Dynames is, however, to be given really two data—one of these is for the screws themselves as before, while the other is derived from the ratio of the amplitudes. Thus while three pairs of corresponding screws amount to three data, two pairs of corresponding Dynames amount to no less than four data; the additional datum in this case enabling us to indicate the intensity of each correspondent as well as the screw on which it is situated.

It can further be shown in the most general case of the correspondence of the Dynames in two n-systems that the number of pairs of Dynames required to define the correspondence is one less than the number of pairs of screws which would be required to define merely a screw correspondence in the same two n-systems. In an n-system a screw has $n-1$ disposable co-ordinates. To define the correspondence we require $n+1$ pairs of screws. Of course those on the first system may have been chosen arbitrarily, so that the number of data required for the correspondence is

$$(n-1)(n+1) = n^2 - 1.$$

A Dyname in an n-system has n arbitrary data, viz., $n-1$ for the screw, and one for the intensity: hence when we are given n pairs of corresponding Dynames we have altogether n^2 data. We thus see that the n pairs of corresponding Dynames really contribute one more datum to the problem than do the $n+1$ pairs of corresponding screws. The additional datum is applied in allotting the appropriate intensity to the sought Dyname.

We can then use either the n pair of Dyname correspondents or the $(n+1)$ pairs of screw correspondents. In previous articles we have used the latter; we shall now use the former.

350. **Freedom of the fifth order.**

In the higher orders of freedom the screw correspondence does not indeed afford quite so simple a means of constructing the several pairs of corresponding screws as we obtain by the Dyname correspondence. In two five-systems the correspondence is complete when we are given five Dynames in one and the corresponding five Dynames in the other. To find the Dyname X in the second system, corresponding to any given Dyname A in the first system, we proceed as follows:—Decompose A into Dynames on the five screws which contain the five given Dynames on the first system. This is always possible, and the solution is unique. These components will correspond to determinate Dynames on the five corresponding screws: these Dynames compounded together will give the required Dyname X both in intensity and position.

In the general case where a mass-chain possesses freedom of the fifth order we may, by trial, determine five screw-chains about which the system can be twisted. Each set of five homologous screws will determine a five-system. In this method of proceeding we need not pay any attention to the intermediate screws: it will only be necessary to inscribe one Dyname (in this case a twist) in each of the homologous five-systems so that the group of six shall be homographic. The set of twists so found will form a displacement which the system must be capable of receiving. This is perhaps the simplest geometrical presentment of which the question admits.

One more illustration may be given. Suppose we have a series of planes, and three arbitrary forces in each plane. We insert in one of the planes any arbitrary force, and its parallel projection can then be placed in all the other planes. Suppose a mechanical system, containing as many distinct elements as there are planes, be so circumstanced that each element is free to accept a rotation about each of the three lines of force in the plane, and that the amplitude of the rotation is proportional to the intensity of the force; it must then follow that the system will be also free to accept rotations about any other chain formed by an arbitrary force in one plane and its parallel projections in the rest.

We may, however, also examine the case of a mass-chain with freedom of the fifth order by the aid of the screw correspondence without introduction of the Dyname. We find, as before, five independent screw-chains which will completely define all the other movements which the system can accept. To construct the subsequent screw-chains, which are quadruply infinite in variety, we begin by first finding any sixth screw-chain of the system by actual composition of any two or more twists about two or more of the five screw-chains. When a sixth chain has been ascertained the construction of the rest is greatly simplified. Each set of six homologous screws lie in a five-system. Place in each of these five-systems another screw which, with the remaining six, form a set which is homographic with the corresponding set in each of the other five-systems. These screws so determined then form another screw-chain about which the system must be free to twist.

In the choice of the first screw with which to commence the formation of any further screw-chains of the five-system we have only a single condition to comply with: the screw chosen must belong to a given five-system. This implies that the chosen screw must be reciprocal merely to one given screw. On any arbitrary cylindroid a screw can be chosen which is reciprocal to this screw, and consequently on any cylindroid one screw can always be selected wherewith to commence a screw-chain about which a mass-chain with freedom of the fifth order must be free to twist.

351. Freedom of the sixth order.

In freedom of the sixth order we select at random six displacements of which the mass-chain admits, and then construct the six corresponding screw-chains. The homologous screws in this case lie on six-systems, but a six-system means of course every conceivable screw. It is easily shown (§ 248) that if to one screw in space corresponds another screw, and conversely, then the homography is completely established when we are given seven screws in one system, and the corresponding seven screws in the other. Any eighth screw in the one system will then have its correspondent in the other immediately determined.

It is of special importance in the present theory to dwell on the type of homography with which we are here concerned. If on the one hand it seems embarrassing, from the large number of screws concerned, on the other hand we are to recollect that the question is free from the complication of regarding the screws as residing on particular n-systems. Seven screws may be drawn *anywhere, and of any pitch;* seven other screws may also be chosen anywhere, and of any pitch. If these two groups be made to correspond in pairs, then any other screw being given, its corresponding screw will be completely determined. Nor is there in this correspondence any other condition, save the simple one, that to one screw of one system one screw of the other shall correspond linearly.

Six screw-chains having been found, a seventh is to be constructed. This being done, the construction of as many screw-chains as may be desired is immediately feasible. From the homographic relations just referred to we have appropriate to each element of the system seven homologous screws, and also appropriate to each consecutive pair of elements we have the seven homologous intermediate screws. An eighth screw, appropriate to any element, may be drawn arbitrarily, and the corresponding screw being constructed on each of the other systems gives at once another screw-chain about which the system must be free to twist.

When a mass-chain has freedom of the sixth order we see that any one element may be twisted about any arbitrary chosen screw, but that the screw about which every other element twists is then determined, and so are also the ratios of the amplitudes of the twists, by the aid of the intermediate screws.

352. Freedom of the seventh order.

Passing from the case of six degrees of freedom to the case of seven degrees, we have a somewhat remarkable departure from the phenomena shown by the lower degrees of freedom. Give to the mass-chain any seven arbitrary displacements, and construct the seven screw-chains, α, β, γ, δ, ϵ,

ζ, η by twists about which those displacements could have been effected. In the construction of an eighth chain, θ, we may proceed as follows:—Choose any arbitrary screw θ_1. Decompose a twist on θ_1 into components on α_1, β_1, γ_1, δ_1, ϵ_1, ζ_1. This must be possible, because a twist about any screw can be decomposed into twists about six arbitrary screws, for we shall not discuss the special exception when the six screws belong to a system of lower order.

The twists on $\alpha_1, \ldots \zeta_1$, &c., determine the twists on the screw-chains $\alpha, \ldots \zeta$, and therefore the twists on the screws $\alpha_2, \ldots \zeta_2$, which compound into a twist on θ_2, similarly for θ_3, &c.; consequently a screw-chain of which θ_1 is the first screw, and which belongs to the system, has been constructed. This is, however, only one of a number of screw-chains belonging to the system which have θ_1 for their first screw. The twist on θ_1 might have been decomposed on the six screws, β_1, γ_1, δ_1, ϵ_1, ζ_1, η_1, and then the screws θ_2, &c., might have been found as before. These will of course not be identical with the corresponding screws found previously. Or if we take the whole seven screws, $\alpha_1, \ldots \eta_1$, we can decompose a twist on θ_1 in an infinite number of ways on these seven screws. We may, in fact, choose the amplitude of the twist on any one of the screws of reference, α_1, for example, arbitrarily, and then the amplitudes on all the rest will be determined. It thus appears that where θ_1 is given, the screw θ_2 is not determined in the case of freedom of the seventh order; it is only indicated to be any screw whatever of a singly infinite number. The locus of θ_2 is therefore a ruled surface; so will be the locus of θ_3, &c. and we have, in the first place, to prove that all these ruled surfaces are cylindroids.

Take three twists on θ_1, such that the arithmetic sum of their amplitudes is zero, and which consequently neutralize. Decompose the first of these into twists on α_1, β_1, γ_1, δ_1, ζ_1, η_1, the second on α_1, β_1, γ_1, δ_1, ϵ_1, η_1, and the third on α_1, β_1, γ_1, δ_1, ϵ_1, ζ_1. It is still open to make another supposition about the twists on θ_1; let us suppose that they are such as to make the two components on η_1 vanish. It must then follow that the total twists on each of the remaining six screws, viz. α_1, β_1, γ_1, δ_1, ϵ_1, ζ_1 shall vanish, for their resultant cannot otherwise be zero. All the amplitudes of the twists about the screw-chains of reference must vanish, and so must also the amplitudes of the resultant twists when compounded. We should have three different screws for θ_2 corresponding to the three different twists on θ_1; and as the twists on these screws must neutralize, the three screws must be co-cylindroidal.

We can, therefore, in constructing a screw-chain of this system, not only choose θ_1 arbitrarily, but we can then take for θ_2 any screw on a certain cylindroid: this being done, the rest of the screw-chain is fixed, including the intermediate screws.

353. Freedom of the eighth and higher orders.

If the freedom be of the eighth order, then it is easily shown that the first screw of any other chain may be taken arbitrarily, and that even the second screw may be chosen arbitrarily from a three-system. Passing on to the twelfth order of freedom, the two first screws of the chain, as well as the amplitudes of their twists, may be chosen quite arbitrarily, and the rest of the chain is fixed. In the thirteenth order of freedom we can take the two first twists arbitrarily, while the third may be chosen anywhere on a cylindroid. It will not now be difficult to trace the progress of the chain to that unrestrained freedom it will enjoy when the mass-chain has 6μ degrees of freedom, when it is able to accept any displacement whatever. In the last stage, prior to that of absolute freedom, the system will have its position defined by $6\mu - 1$ co-ordinates. A screw-chain can then be chosen which is perfectly arbitrary in every respect, save that one of its screws must be reciprocal to a given screw.

354. Reciprocal Screw-Chains.

We have hitherto been engaged with the discussion of the geometrical or kinematical relations of a mass-chain of μ elements: we now proceed to the dynamical considerations which arise when the action of forces is considered.

Each element of the mass-chain may be acted upon by one or more external forces, in addition to the internal forces which arise from the reaction of constraints. This group of forces must constitute a wrench appropriated to the particular element. For each element we thus have a certain wrench, and the entire action of the forces on the mass-chain is to be represented by a series of μ wrenches. Recalling our definition of a screw-chain, it will be easy to assign a meaning to the expression, *wrench on a screw-chain*. By this we denote a series of wrenches on the screws of the chain, and the ratio of two consecutive intensities is given by the intermediate screw, as before. We thus have the general statement:—

The action of any system of forces on a mass-chain may be represented by a wrench on a screw-chain.

Two or more wrenches on screw-chains will compound into one wrench on a screw-chain, and the laws of the composition are exactly the same as for the composition of twists, already discussed.

Take, for example, any four wrenches on four screw-chains. Each set of four homologous screws will determine a four-system; the resulting wrench-chain will consist of a series of wrenches on these four-systems, each being the "parallel projection" of the other.

Let α and β be two screw-chains, each consisting of μ screws, appropriated one by one to the μ elements of the mass-chain. If the system receive a twist about the screw-chain β, while a wrench acts on the screw-chain α, some work will usually be lost or gained; if, however, no work be lost or gained, then the same will be true of a twist around α acting on a wrench on β. In this case the screw-chains are said to be *reciprocal*. The relation may be expressed somewhat differently, as follows:—

If a mass-chain, only free to twist about the screw-chain α, be in equilibrium, notwithstanding the presence of a wrench on the screw-chain β, then, conversely, a mass-chain only free to twist about the screw-chain β will be in equilibrium, notwithstanding the presence of a wrench on the screw-chain α.

This remarkable property of two screw-chains is very readily proved from the property of two reciprocal screws, of which property, indeed, it is only an extension.

Let $\alpha_1 \ldots \alpha_\mu$ be the screws of one screw-chain, and $\beta_1 \ldots \beta_\mu$ those of the other. Let $\alpha_1', \alpha_2', \ldots \alpha_\mu'$ denote amplitudes of twists on α_1, α_2, &c., and let α_1'', α_2'', &c., denote the intensities of wrenches on α_1, α_2, &c. Then, from the nature of the screw-chain, we must have

$$\alpha_1' : \alpha_1'' = \alpha_2' : \alpha_2'' = \alpha_3' : \alpha_3'', \text{ \&c.},$$
$$\beta_1' : \beta_1'' = \beta_2' : \beta_2'' = \beta_3' : \beta_3'', \text{ \&c.};$$

for as twists and wrenches are compounded by the same rules, the intermediate screws of the chain require that the ratio of two consecutive amplitudes of the twists about the chain shall coincide with the ratio of the intensities of the two corresponding wrenches. Denoting the virtual coefficient of α_1 and β_1 by the symbol $\varpi_{\alpha_1\beta_1}$, we have for the work done by a twist about α, against the screw-chain β,

$$2\alpha_1'\beta_1''\varpi_{\alpha_1\beta_1} + 2\alpha_2'\beta_2''\varpi_{\alpha_2\beta_2} + \text{\&c.},$$

while for the work done by a twist about β against the screw-chain α we have the expression

$$2\alpha_1''\beta_1'\varpi_{\alpha_1\beta_1} + 2\alpha_2''\beta_1'\varpi_{\alpha_2\beta_2}, \text{ \&c.}$$

If the first of these expressions vanishes, then the second will vanish also.

It will now be obvious that a great part of the Theory of Screws may be applied to the more general conceptions of screw-chains. The following theorem can be proved by the same argument used in the case when only a single pair of screws are involved.

If a screw-chain θ be reciprocal to two screw-chains α and β, then θ will be reciprocal to every screw-chain of the system obtained by compounding twists on α and β.

A screw-chain is defined by $6\mu - 1$ data (§ 353). It follows that a finite number of screw-chains can be determined, which shall be reciprocal to $6\mu - 1$ given screw-chains. It is, however, easy to prove that *that number must be one.* If two chains could be found to fulfil this condition, then every chain formed from the system by composition of two twists thereon would fulfil the same condition. Hence we have the important result—

One screw-chain can always be determined which is reciprocal to $6\mu - 1$ *given screw-chains.*

This is of course only the generalization of the fundamental proposition with respect to a single rigid body, that one screw can always be found which is reciprocal to five given screws (§ 25).

355. Twists on $6\mu + 1$ screw-chains.

Given $6\mu + 1$ screw-chains, it is always possible to determine the amplitudes of certain twists about those chains, such that if those twists be applied in succession to a mass-chain of μ elements, the mass-chain shall, after the last twist, have resumed the same position which it had before the first. To prove this it is first necessary to show that from the system formed by composition of twists about two screw-chains, one screw-chain can always be found which is reciprocal to any given screw-chain. This is indeed the generalization of the statement that one screw can always be found on a cylindroid which is reciprocal to a given screw. The proof of the more general theorem is equally easy. The number of screw-chains produced by composition of twists about the screw-chains α and β is singly infinite. There can, therefore, be a finite number of screw-chains of this system reciprocal to a given screw-chain θ. But that number must be one; for if even two screw-chains of the system were reciprocal to θ, then every screw-chain of the system must also be reciprocal to θ. The solution of the original problem is then as follows:—Let α and β be two of the given $6\mu + 1$ chains, and let θ be the one screw-chain which is reciprocal to the remaining $6\mu - 1$ chains. Since the $6\mu + 1$ twists are to neutralize, the total quantity of work done against any wrench-chain must be zero. Take, then, any wrench-chain on θ. Since this is reciprocal to $6\mu - 1$ of the screw-chains, the twists about these screw-chains can do no work against a twist on θ. It follows that the amplitudes of the twists about α and β must be such that the total amount of work done must be zero. For this to be the case, the two twists on α and β must compound into one twist on the screw-chain ν, which belongs to the system $(\alpha\beta)$, and is also reciprocal to θ. This defines the ratio of the amplitudes of the two twists on α and β. We may in fact draw any cylindroid containing three homologous screws of α, β, and γ, then the ratio of the sines of the angles into which γ divides the angle between α and β is the ratio of the amplitudes of the twists on α and β. In a similar manner the ratio of the amplitudes of any other pair of twists can be found, and thus the whole problem has been solved.

We are now able to decompose any given twist or wrench on a screw-chain into 6μ components on any arbitrary 6μ chains. The amplitudes or the intensities of these 6μ components may be termed the 6μ co-ordinates of the given twist or wrench. If the amplitude or the intensity be regarded as unity, then the 6μ quantities may be taken to represent the co-ordinates of the screw-chain. In this case only the ratios of the co-ordinates are of consequence.

If the mass-chain have only n degrees of freedom where n is less than 6μ, then all the screw-chains about which the mass-chain can be twisted are so connected together, that if any $n+1$ of these chains be taken arbitrarily, the system can receive twists about these $n+1$ chains of such a kind, that after the last twist the system has resumed the same position which it had before the first. In this case n co-ordinates will be sufficient to express the twist or wrench which the system can receive, and n co-ordinates, whereof only the ratios are concerned, will be sufficient to define any screw-chain about which the system can be twisted.

$6\mu - n$ screw-chains are taken, each of which is reciprocal to n screw-chains about which a mass-chain with freedom of the nth order can twist. The two groups of n screw-chains on the one hand, and $6\mu - n$ on the other, may each be made the basis of a system of chains about which a mechanism could twist with freedom of the nth order, or of the $(6\mu - n)$th order, respectively. These two systems are so related that each screw-chain in the one system is reciprocal to all the screw-chains in the other. They may thus be called two *reciprocal systems of screw-chains.*

Whatever be the constraints by which the freedom is hampered, the reaction of the constraints upon the elements must constitute a wrench on a screw-chain. It is a fundamental point of the present theory that this screw-chain belongs to the reciprocal system. For, as no work is done against the constraints by any displacement which is compatible with the freedom of the mass-chain, it must follow, from the definition, that the wrench-chain which represents the reactions must be reciprocal to all possible displacement chains, and must therefore belong to the reciprocal system.

For a wrench-chain applied to the mass-chain to be in equilibrium it must, if not counteracted by some other external wrench-chain, be counteracted by the reaction of the constraints. Thus we learn that

Of two reciprocal screw-chain systems, each expresses the collection of wrench-chains of which each one will equilibrate when applied to a mass-chain only free to twist about all the chains of the other system.

This is, perhaps, one of the most comprehensive theorems on Equilibrium which could be enunciated.

356. Impulsive screw-chains and instantaneous screw-chains.

Up to the present we have been occupied with considerations involving kinematics and statics: we now show how the principles of kinetics can be illustrated by the theory sketched in this chapter.

We shall suppose, as before, that the mechanical arrangement which we call the mass-chain consists of μ elements, and that those elements are so connected together that the mass-chain has n degrees of freedom. We shall also suppose that the mass-chain is acted upon by a wrench about any screw-chain whatever. The first step to be taken is to show that the given wrench-chain may be replaced by another more conveniently circumstanced. Take any n chains of the given system, and $6\mu - n$ chains of the reciprocal system, then the given wrench-chain can be generally decomposed into components on the $n + (6\mu - n)$ chains here specified. The latter, being all capable of neutralization by the reaction of the constraints, may be omitted, while the former n wrench-chains admit of being compounded into a single wrench-chain. We hence have the following important proposition:—

Whatever be the forces which act on a mass-chain, their effect is in general equivalent to that of a wrench on a screw-chain which belongs to the system of screw-chains expressing the freedom of the mass-chain.

The application of this theorem is found in the fact that, while we still retain the most perfect generality, it is only necessary, either for twists or wrenches, to consider the system, defined by n chains, about which the mass-chain can be twisted.

Let us consider the mass-chain at rest in a specified position, and suppose it receives the impulsive action of any set of forces, it is required to determine the instantaneous motion which the system will acquire. The first operation is to combine all the forces into a wrench-chain, and then to transform that wrench-chain, in the manner just explained, into an equivalent wrench-chain on one of the screws of the system. Let θ be the screw-chain of the system so found. In consequence of this impulsive action the mass-chain, previously supposed to be at rest, will commence to move; that motion can, however, be nothing else than an instantaneous twist velocity about a screw-chain α. We thus have an impulsive screw-chain θ corresponding to an instantaneous screw-chain α. In the same way we shall have the impulsive screw-chains ϕ, ψ, &c., correlated to the instantaneous chains, β, γ, &c.

The first point to be noticed is, that the correspondence is unique. To the instantaneous chain α one impulsive screw-chain θ will correspond. There could not be two screw-chains θ and θ' which correspond to the same instantaneous screw-chain α. For, suppose this were the case, then the twist velocity imparted by the impulsive wrench on θ could be neutralized by the impulsive wrench on θ'. We thus have the mass-chain remaining at rest in spite of the impulsive wrenches on θ and θ'. These two wrenches must therefore neutralize, and as, by hypothesis, they are on different screw-chains, this can only be accomplished by the aid of the reactions of the constraints. We therefore find that θ and θ' must compound into a wrench-chain which is neutralized by the reactions of the constraints. This is, however, impossible, for θ and θ' can only compound into a wrench on a screw-chain of the original system, while all the reactions of the constraints form wrenches on the chains of the wholly distinct reciprocal system.

We therefore see that to each instantaneous screw-chain α only one impulsive screw-chain θ will correspond. It is still easier to show that to each impulsive screw-chain θ only one instantaneous screw-chain α will correspond. Suppose that there were two screw-chains, α and α', either of which would correspond to an impulsive wrench on θ. We could then give the mass-chain, first, an impulsive wrench on θ of intensity λ, and make the mass-chain twist about α, and we could simultaneously give it an impulsive wrench on the same screw-chain θ of intensity $-\lambda$, and make the mass-chain twist about α'. The two impulses would neutralize, so that as a matter of fact the mass-chain received no impulse whatever, but the two twist velocities could not destroy, as they are on different screw-chains. We would thus have a twist velocity produced without any expenditure of energy.

We have thus shown that in the n-system of screw-chains expressing the freedom of the mass-chain, one screw-chain, regarded as an instantaneous screw-chain, will correspond to one screw-chain, regarded as an impulsive screw-chain, and conversely, and therefore linear relations between the co-ordinates are immediately suggested. That there are such relations can be easily proved directly from the laws of motion (see Appendix, note 7). We therefore have established a case of screw-chain homography between the two systems, so that if $\theta_1, \ldots \theta_n$ denote the co-ordinates of the impulsive screw-chain, and if $\alpha_1, \ldots \alpha_n$ denote the co-ordinates of the corresponding instantaneous screw-chain, we must have n equations of the type

$$\theta_1 = (11)\,\alpha_1 + (12)\,\alpha_2 + (13)\,\alpha_3 \ldots + (1\,n)\,\alpha_n,$$

$$\ldots\ldots\ldots\ldots\ldots\ldots\ldots\ldots\ldots\ldots\ldots\ldots$$

$$\theta_n = (n\,1)\,\alpha_1 + (n\,2)\,\alpha_2 + \ldots\ldots\ldots\ldots + (nn)\,\alpha_n,$$

where (11), (12), &c., are n^2 coefficients depending on the distribution of the masses, and the other circumstances of the mass-chain and its constraints. The equations having this form, the necessary one-to-one correspondence is manifestly observed.

357. The principal screw-chains of Inertia.

We are now in a position to obtain a result of no little interest. Just as we have two double points in two homographic rows on a line, so we have n double chains in the two homographic chain systems. If we make, in the foregoing equations,

$$\theta_1 = \rho\alpha_1; \qquad \theta_2 = \rho\alpha_2, \text{ \&c.},$$

we obtain, by elimination of $\alpha_1, \ldots \alpha_n$, an equation of the nth degree in ρ. The roots of this equation are n in number, and each root substituted in the equations will enable the co-ordinates of each of the n double screw-chains to be discovered. The mechanical property of these double chains is to be found in the following statement:—

If any mass-chain have n degrees of freedom, then in general n screw-chains can always be found (but not more than n), such that if the mass-chain receive an impulsive wrench from any one of these screw-chains, it will immediately commence to move by twisting about the same screw-chain.

In the case where the mass-chain reduces to a single rigid body, free or constrained, the n screw-chains to which we have just been conducted reduce to what we have called the *n principal screws of inertia.* In the case, still more specialized, of a rigid body only free to rotate around a point, the theorem degenerates to the well-known property of the principal axes. We may thus regard the n principal chains now found as the generalization of the familiar property of the principal axes for any system anyhow constrained.

Considerable simplification is introduced into the equations when, instead of choosing the chains of reference arbitrarily, we select the n principal screw-chains for this purpose; we then have the very simple results,

$$\theta_1 = (11)\,\alpha_1; \quad \theta_2 = (22)\,\alpha_2; \quad \ldots\, \theta_n = (nn)\,\alpha_n.$$

This gives a method of finding the impulsive screw-chain corresponding to any instantaneous screw-chain. It is only necessary to multiply the co-ordinates of the instantaneous screw-chain α_1, α_2 by the constant factors (11), (12), &c., in order to find the co-ordinates of the impulsive screw-chain.

The general type of homography here indicated has to be somewhat specialized for the case of impulsive screw-chains and instantaneous screw-chains. The n double screw-chains are generally quite unconnected—we might, indeed, have exhibited the relation between the two homographic systems of screw-chains by choosing n screws quite arbitrarily as the double screws of the two systems, and then appropriating to them n factors (11), (22), (33), &c., also chosen arbitrarily. In the case of impulsive and instantaneous chains, the n double chains are connected together by the relation that *each pair of them are reciprocal*, so that the whole group of n chains form what may be called a set of co-reciprocals.

To establish this we may employ some methods other than those previously used. Let us take a set of n-co-reciprocal chains, and let the co-ordinates of any other two chains, θ and ϕ of the same system, be $\theta_1, \ldots \theta_n$ and $\phi_1, \ldots \phi_n$. Let $2p_1$, $2p_2$, &c., $2p_n$ be certain constant parameters appropriated to the screws of reference. $2p_1$ is, for example, the work done by a twist of unit amplitude on the first screw-chain of reference against a wrench of unit intensity on the same chain. The work done by a twist θ_1 against a wrench ϕ_1 on this chain is $2p_1\theta_1\phi_1$. As the chains of reference are co-reciprocal, the twist on θ_1 does no work against the wrenches $\phi_2, \phi_3, \ldots$ &c.; hence the total work done by a twist on θ against the wrench on ϕ is

$$2p_1\theta_1\phi_1 + \ldots + 2p_n\theta_n\phi_n;$$

and hence if θ and ϕ be reciprocal,

$$p_1\theta_1\phi_1 + \ldots + p_n\theta_n\phi_n = 0.$$

The quantities $p_1, \ldots p_n$ are linear magnitudes, and they bear to screw-chains the same relation which the pitches bear to screws. If we use the word *pitch* to signify half the work done by a unit twist on a screw-chain against the unit wrench on the screw-chain, then we have for the pitch p_θ of the chain θ the expression

$$p_1\theta_1^2 + p_2\theta_2^2 + \ldots + p_n\theta_n^2.$$

The kinetic energy of the mass-chain, when animated by a twist velocity of given amount, depends on the instantaneous screw-chain about which the system is twisting. It is proportional to a certain quadratic function of the n co-ordinates of the instantaneous screw. By suitable choice of the screw-chains of reference it is possible, in an infinite number of ways, to exhibit this function as the sum of n squares. It follows from the theory of linear transformations that it is generally possible to make one selection of the screw-chains of reference which, besides giving the energy function the required form, will also exhibit p_θ as the sum of n squares. This latter condition means that the screw-chains of reference are co-reciprocal. It only remains to show that the n screw-chains of reference thus ascertained must be the n principal screw-chains to which we were previously conducted.

We may show this most conveniently by the aid of Lagrange's equations of motion in generalized co-ordinates (§ 86).

Let $\theta_1, \ldots \theta_n$ represent the co-ordinates of the impulsive screw-chain, and let $\alpha_1, \ldots \alpha_n$ be the co-ordinates of the corresponding instantaneous screw-chain, reference being made to the screw-chains of reference just found.

Lagrange's equations have the form

$$\frac{d}{dt}\left(\frac{dT}{d\dot{\alpha}_1}\right) - \frac{dT}{d\alpha_1} = -P_1,$$

where T is the kinetic energy, and where $P_1\delta\alpha_1$ denotes the work done against the forces by a twist of amplitude $\delta\alpha_1$.

If θ''' denote the intensity of the impulsive wrench, then its component on the first screw of reference is $\theta'''\theta_1$, and the work done is $2p_1\theta'''\theta_1\delta\alpha_1$, while, since the chains are co-reciprocal, the work done by $\delta\alpha_1$ against the components of θ''' on the other chains of reference is zero, we therefore have

$$P_1 = 2p_1\theta'''\theta_1.$$

We have also

$$T = M(u_1^2\dot{\alpha}_1^2 + \ldots + u_n^2\dot{\alpha}_n^2),$$

when $u_1, \ldots u_n$ are certain constants.

We have, therefore, from Lagrange's equation,

$$\frac{d}{dt}(Mu_1^2\dot{\alpha}_1) = -p_1\theta'''\theta_1,$$

whence, integrating during the small time t, during which the impulsive force acts,

$$Mu_1^2\dot{\alpha}\alpha_1 = -\theta_1 p_1 \int \theta''' dt,$$

in which $\dot{\alpha}$ is the actual twist velocity about the screw-chain, so that $\dot{\alpha}_1 = \dot{\alpha}\alpha_1$, each being merely the expression for the component of that twist velocity about the screw-chain.

We hence obtain $\theta_1, \ldots \theta_n$, proportional respectively to

$$\frac{u_1^2\alpha_1}{p_1}, \ldots \frac{u_n^2\alpha_n}{p_n}.$$

If we make $\dfrac{u_1^2}{p_1} = (11)$, &c., we have the previous result,

$$\theta_1 = (11)\,\alpha_1,$$

$$\cdots\cdots\cdots\cdots$$

$$\theta_n = (nn)\,\alpha_n.$$

358. Conjugate screw-chains of Inertia.

From the results just obtained, which relate of course only to the chains of reference, we can deduce a very remarkable property connecting instantaneous chains, and impulsive chains in general. *Let α and β be two instantaneous chains, and let θ and ϕ be the two corresponding impulsive chains, then when α*

is reciprocal to ϕ, β will be reciprocal to θ. This, it will be observed, is a generalization of a property of which much use has been previously made (§ 81). The proof is as follows.

The co-ordinates of the instantaneous chains are

$$\alpha_1, \ldots \alpha_n,$$

$$\beta_1, \ldots \beta_n.$$

The co-ordinates of the corresponding impulsive chains are

$$\frac{u_1^2\alpha_1}{p_1}, \ldots \frac{u_n^2\alpha_n}{p_n},$$

and

$$\frac{u_1^2\beta_1}{p_1}, \ldots \frac{u_n^2\beta_n}{p_n}.$$

If the chain α be reciprocal to the impulsive chain which produces β, then we have

$$u_1^2\alpha_1\beta_1 + \ldots + u_n^2\alpha_n\beta_n = 0;$$

but this being symmetrical in α and β is precisely the same as the condition that the impulsive chain corresponding to α shall be reciprocal to β. Following the analogy of our previous language we may describe two screw-chains so related as *conjugate screw-chains of Inertia.*

359. Harmonic screw-chains.

We make one more application of the theory of screw-chains to the discussion of a kinetical problem. Let us suppose that we have any material system with n degrees of freedom in a position of stable equilibrium under the action of a conservative system of forces. If the system receive a small displacement, the forces will no longer equilibrate, but the system will be exposed to the action of a wrench on a screw-chain. We thus have two corresponding sets of screw-chains, one set being the chains about which the system is displaced, the other set for the wrenches which are evoked in consequence of the displacements.

By similar reasoning to that which we have already used, it can be shown that these two corresponding chain systems are homographic. We can therefore find n screw-chains about which, if the system be displaced, a wrench will be evoked on the same screw-chain, and (the forces having a potential) it can be shown that this set of n screw-chains are co-reciprocal.

If after displacement the system be released it will continue to make small oscillations. The nature of these oscillations can be completely exhibited by the screw-chains. To a chain α, regarded as an instantaneous screw-chain, will correspond the screw θ as an impulsive screw-chain. To the chain α, regarded as the seat of a displacing twist, will correspond a wrench

ϕ which is evoked by the action of the forces. It will, of course, generally happen that the chain θ is different from the chain ϕ. It can however be shown that θ and ϕ are not in every case distinct. There are n different screw-chains, each of which regarded as α will have the two corresponding screws θ and ϕ identical. Nor is it difficult to see what the effect of such a displacement must be on the small oscillations which follow. A wrench is evoked by the displacement, and since θ and ϕ coincide, that wrench is undistinguishable from an infinitely small impulsive wrench which will make the system commence to twist about α. We are thus led to the result that—

There are n screw-chains such that if the system be displaced by a twist about one of these screw-chains, and then released, it will continue for ever to twist to and fro on the same screw-chain.

Following the language previously used, we may speak of these as *harmonic screw-chains*, and it can be shown that whatever be the small displacement of the system, and whatever be the small initial velocities with which it is started, the small oscillations are merely compounded of twist vibrations about the n harmonic screw-chains.

CHAPTER XXV.

THE THEORY OF PERMANENT SCREWS*.

360. Introduction.

In commencing this chapter it will be convenient to recite a well-known dynamical proposition, and then to enlarge its enunciation by successive abandonment of restrictions.

Suppose a rigid body free to rotate around a fixed point. There are, as is well known, three rectangular axes about any one of which the body when once set in rotation will continue to rotate uniformly so long as the application of force is withheld. These axes are known as permanent axes. The freedom of the body in this case is of a particular nature, included in the more general type known as Freedom of the Third Order. The Freedom of the Third Order is itself merely one subdivision of the class which, including the six orders of freedom, embraces every conceivable form of constraint that can be applied to a rigid body. We propose to investigate what may be called the theory of *permanent* screws for a body constrained in the most general manner.

The movement of the body at each moment must be a twist velocity about some one screw θ belonging to the system of screws prescribed by the character of the constraints. In the absence of forces external to those arising from the reactions of the constraints, the movement will not, in general, persist as a twist about the same screw θ. The instantaneous screw will usually shift its position so as to occupy a series of consecutive positions in the system. It must, however, be always possible to compel the body to remain twisting about θ. For this purpose a wrench of suitable intensity on an appropriate screw η may have to be applied. Without sacrifice of generality we can in general arrange that η is one of the system of screws

* *Trans. Roy. Irish Acad.*, Vol. XXIX. p. 613 (1890).

which expresses the freedom of the body (§ 96). It may sometimes appear that the intensity of the necessary wrench on η vanishes. The body in such a case requires no coercion beyond that of the original constraints to preserve θ as the screw about which it twists, and when this is the case we shall describe θ as a *permanent* screw. This use of the word *permanent* does not imply that the body could remain for ever twisting about this screw, for the movement of the body to an appreciable distance will in general entail some change in its relation to the constraints. The characteristic of the permanent screw is *the absence of any acceleration* in the body twisting about it, using the word acceleration in its widest sense.

In the earlier parts of the chapter we shall discard the restrictions involved in the assumption that the material arrangement is only a single rigid body. The doctrine of screw-chains (Chap. XXIV.) enables us to extend a considerable portion of the present theory to any mass-chain whatever. Any number of material parts connected in any manner must still conform to the general law, that the instantaneous movements can always be represented by a twist about a certain screw-chain. In general the mass-chain will have a tendency to wander from twisting about the original screw-chain. In such cases the position of the instantaneous screw-chain cannot be maintained without the imposition of further coercion than that which the constraints supply. This additional set of forces may be applied by a restraining wrench-chain, the relation of which to the instantaneous screw-chain we shall have to consider. Sometimes it may appear that no restraining wrench-chain is necessary beyond one of those provided by the reaction of the constraints. The instantaneous screw-chain is then to be described as *permanent*.

361. Different properties of a Principal Axis.

Another preliminary matter should be also noticed, because it exhibits the relation of the subject discussed in this chapter to some other parts of the Theory of Screws. In the ordinary theory of the rigid body there are, as is well known, two distinct properties of a principal axis which possess dynamical significance. We may think of a principal axis as the axis of a couple which, when applied impulsively to the body, will set it rotating about this axis. We may also think of the principal axis as a direction about which, if a body be once set in rotation, it will continue to rotate. The first of these properties by suitable generalization opens up the theory of principal screw-chains of inertia, which we have already explained in previous chapters. It is from the other property of the principal axis that the present investigation takes its rise. It is important to note that two quite different departments in the Theory of Screws happen to coalesce in the very special case of a rigid body rotating around a point.

362. A Property of the Kinetic Energy of a System.

It is obvious that the mere alteration of the azimuth about a fixed axis from which a rigid body is set into rotation will not affect its kinetic energy, provided the position of the axis and the angular velocity both remain unaltered.

A moment's reflection will show that this principle may be extended to any movement whatever of a rigid body. At each instant the body is twisting about some instantaneous screw α with a twist velocity $\dot{\alpha}$. Let the body be stopped in a position which we call A. Let it receive a displacement by a twist of any amplitude about α and thus be brought to a position which we call B*. Finally, let the body be started from its new position B so as to twist again about α with the original twist velocity $\dot{\alpha}$, then it is plain that the kinetic energy of the body just before being stopped at the position A is the same as its kinetic energy just after it is started from the position B.

Enunciated in a still more general form the same principle is as follows:—

Any mass-chain in movement is necessarily twisting about some screw-chain. If we arrest the movement, displace the mass-chain to an adjacent position on the same screw-chain, and then start the mass-chain to twist again on the same screw-chain, with its original twist velocity, the kinetic energy must remain the same as it was before the interruption.

This principle requires that whatever be the symbols employed, the function T, which denotes the kinetic energy, must satisfy a certain identical equation. I propose to investigate this equation, and its character will perhaps be best understood by first discussing the question with co-ordinates of a perfectly general type. We shall suppose the mass-chain has n degrees of freedom.

Let the co-ordinates $x_1, \dots x_n$ represent the position of the mass-chain, and let its instantaneous motion be indicated by $\dot{x}_1, \dots \dot{x}_n$. Let O be the initial position of the mass-chain, then in the time δt it has reached the position O', whereof the co-ordinates are

$$x_1 + \dot{x}_1\delta t, \dots x_n + \dot{x}_n\delta t.$$

The movement from O to O' must, like every possible movement of a system, consist of a twist about a screw-chain. This is a kinematical fact, wholly apart from whatever particular system of co-ordinates may have

* We have supposed that the pitch of this displacement is the same as the pitch of α. This restriction is only introduced here because the constraints will generally forbid the body to make any other twist about the axis of α. If the body were quite free we might discard the restriction altogether as is in fact done later on (§ 376).

been adopted. We call this screw-chain θ, and $\dot{\theta}$ denotes the twist velocity with which the system moves round O.

Choose any n independent screw-chains, about each one of which the mass-chain is capable of twisting. Then $\dot{\theta}$ can be decomposed into n components which have twist velocities $\dot{\theta}_1, \ldots \dot{\theta}_n$ about the several screw-chains of reference (§ 355).

Since everything pertaining to the position or the movement of the mass-chain must necessarily admit of being expressed in terms of the co-ordinates of the mass-chain, and since the quantities $\dot{\theta}_1, \ldots \dot{\theta}_n$ are definitely determined by the position and movements of the mass-chain, it follows that these quantities must be given by a group of formulae which may be written

$$\dot{\theta}_1 = f_1(x_1, \ldots x_n, \quad \dot{x}_1, \ldots \dot{x}_n),$$

$$\ldots\ldots\ldots\ldots\ldots\ldots\ldots\ldots\ldots\ldots$$

$$\dot{\theta}_n = f_n(x_1, \ldots x_n, \quad \dot{x}_1, \ldots \dot{x}_n).$$

Let the mass-chain be stopped in the position A. Let it then be displaced to an adjacent position B defined by the following variations of the co-ordinates

$$\delta x_1 = \dot{x}_1 \delta\epsilon, \ldots \delta x_n = \dot{x}_n \delta\epsilon,$$

where $\delta\epsilon$ is a small quantity. From this new position B let the mass-chain be started into motion so that it shall have the same twist velocity as it had just before being stopped in A and about the same screw-chain. This condition requires that each of the quantities $\dot{\theta}_1, \ldots \dot{\theta}_n$ shall resume its original value unaltered by the stoppage and subsequent restarting of the mass-chain from a new position. There must accordingly be an adjustment of $\delta\dot{x}_1, \ldots \delta\dot{x}_n$ to satisfy the equations

$$\frac{df_1}{dx_1}\delta x_1 \ldots + \frac{df_1}{dx_n}\delta x_n + \frac{df_1}{d\dot{x}_1}\delta\dot{x}_1 + \ldots \frac{df_1}{d\dot{x}_n}\delta\dot{x}_n = 0,$$

$$\frac{df_n}{dx_1}\delta x_1 \ldots + \frac{df_n}{dx_n}\delta x_n + \frac{df_n}{d\dot{x}_1}\delta\dot{x}_1 + \ldots \frac{df_n}{d\dot{x}_n}\delta\dot{x}_n = 0.$$

Under these circumstances T has obviously not altered, so that we have also

$$\frac{dT}{dx_1}\delta x_1 + \ldots \frac{dT}{dx_n}\delta x_n + \frac{dT}{d\dot{x}_1}\delta\dot{x}_1 + \ldots \frac{dT}{d\dot{x}_n}\delta\dot{x}_n = 0.$$

Let us assume, for brevity, the symbol Δ, such that

$$\Delta = \dot{x}_1 \frac{d}{dx_1} + \ldots + \dot{x}_n \frac{d}{dx_n}.$$

Then we obtain, by elimination of $\delta\dot{x}_1, \ldots \delta\dot{x}_n$, and $\delta\epsilon$,

$$\begin{vmatrix} \Delta T, & \frac{dT}{d\dot{x}_1}, \ldots & \frac{dT}{d\dot{x}_n} \\ \Delta f_1, & \frac{df_1}{d\dot{x}_1}, \ldots & \frac{df_1}{d\dot{x}_n} \\ \Delta f_n, & \frac{df_n}{d\dot{x}_1}, \ldots & \frac{df_n}{d\dot{x}_n} \end{vmatrix} = 0.$$

Such is the general condition which must be satisfied by the kinetic energy of any material arrangement whatever. But the equation is so complicated when expressed in ordinary rectangular co-ordinates that there is but little inducement to discuss it.

363. The Identical Equation in Screw-chain Co-ordinates.

The Theory of Screw-chains exhibits this equation in a form of special simplicity. For, suppose that

$$\theta_1' = x_1, \quad \dot{\theta}_1 = \dot{x}_1,$$
$$\cdots\cdots\cdots\cdots$$
$$\theta_n' = x_n, \quad \dot{\theta}_n = \dot{x}_n,$$

then the equation of the last article reduces to

$$\Delta T = 0.$$

We thus have the following theorem:—

If the co-ordinates, $\theta_1', \ldots \theta_n'$, of a mass-chain be n twists about n screw-chains, belonging to the system of screw-chains which express the freedom of the mass-chain, and if $\dot{\theta}_1, \ldots \dot{\theta}_n$ be the twist velocities of the mass-chain about these same screw-chains, then the kinetic energy T satisfies the equation

$$\dot{\theta}_1 \frac{dT}{d\theta_1'} \cdots + \dot{\theta}_n \frac{dT}{d\theta_n'} = 0.$$

I have thought it instructive to exhibit the origin of this equation as a special deduction from the case of co-ordinates of the general type. For a brief demonstration the following simple argument suffices:—

If the mass-chain be displaced through $\delta\theta_1', \ldots \delta\theta_n'$ while the velocities are unaltered, the change of kinetic energy is

$$\frac{dT}{d\theta_1'}\delta\theta_1' + \ldots + \frac{dT}{d\theta_n'}\delta\theta_n'.$$

If the change of the position be due to a small twist $\delta\epsilon$ around the screw-chain with co-ordinates $\dot{\theta}_1, \ldots \dot{\theta}_n$, then

$$\dot{\theta}_1\delta\epsilon = \delta\theta_1',$$
$$\cdots\cdots\cdots$$
$$\dot{\theta}_n\delta\epsilon = \delta\theta_n',$$

but from the physical property of the kinetic energy already cited, it appears that this kind of displacement cannot change the kinetic energy, whence

$$\dot{\theta}_1 \frac{dT}{d\theta_1'} \dots + \dot{\theta}_n \frac{dT}{d\theta_n'} = 0.$$

364. The Converse Theorem.

Let us take the general case where the co-ordinates are $x_1, \dots x_n$ and $\dot{x}_1, \dots \dot{x}_n$. Suppose that $\dot{x}_2, \dots \dot{x}_n$ are all zero, then $\dot{x}_1$ is the velocity of the mass-chain. We shall also take $x_2, \dots x_n$ to be zero, so that we only consider the position of the mass-chain defined by x_1. Think now of the two positions for which $x_1 = 0$ and $x_1 = x_1'$ respectively. Whatever be the character of the constraints it must be possible for the mass-chain to pass from the position $x_1 = 0$ to the position $x_1 = x_1'$ by a twist about a screw-chain. The magnitude x_1' is thus correlated to the position of the mass-chain on a screw-chain about which it twists.

If the co-ordinates are of such a kind that the identical equation which T must necessarily satisfy has the form

$$\dot{x}_1 \frac{dT}{dx_1'} \dots + \dot{x}_n \frac{dT}{dx_n'} = 0,$$

then for the particular displacement corresponding to the first co-ordinate, $\dot{x}_2, \dots \dot{x}_n$ are all zero, and

$$\frac{dT}{dx_1'} = 0;$$

and as T must involve $\dot{x}_1$ in the second degree, we have

$$T = H\dot{x}_1^2$$

where H is independent of x_1'.

Let $\dot{\theta}_1$ be the twist velocity about the screw-chain corresponding to the first co-ordinate, then, of course, A being a constant,

$$T = A\dot{\theta}_1^2,$$

whence

$$A\dot{\theta}_1^2 = H\dot{x}_1^2,$$

$$\sqrt{A}\dot{\theta}_1 = \sqrt{H}\dot{x}_1,$$

and by integration and adjustment of units and origins

$$\theta_1' = x_1.$$

We thus see that while the displacement corresponding to the first co-ordinate must always be a twist about a screw-chain, whatever be the actual nature of the metric element chosen for the co-ordinate, yet that when the identical equation assumes the form

$$\dot{\theta}_1 \frac{dT}{d\theta_1'} \dots + \dot{\theta}_n \frac{dT}{d\theta_n'} = 0,$$

the metric element must be essentially the *amplitude of the twist* about the screw-chain. We have thus proved the following theorem :—

The co-ordinates must be twists about n screw-chains of reference whenever the identical equation, satisfied by T, assumes the form

$$\dot{\theta}_1 \frac{dT}{d\theta_1'} \dots + \theta_n \frac{dT}{d\theta_n'} = 0.$$

365. Transformation of the Vanishing Emanant.

Suppose that the position and movement of a mass-chain were represented by the co-ordinates $\theta_1', \theta_2', \dots \theta_n'$; $\dot{\theta}_1, \dot{\theta}_2, \dots \dot{\theta}_n$ when referred to one set of n screws of reference, and by $\phi_1', \phi_2', \dots \phi_n'$; $\dot{\phi}_1, \dot{\phi}_2, \dots \dot{\phi}_n$ when referred to another set of screws of reference. Then of course these sets of co-ordinates must be linearly connected.

We may write

$$\theta_1' = (11)\,\phi_1' \dots + (1n)\,\phi_n',$$
$$\theta_n' = (n1)\,\phi_1' \dots + (un)\,\phi_n'.$$

Then, by differentiation

$$\dot{\theta}_1 = (11)\,\dot{\phi}_1 \dots + (1n)\,\dot{\phi}_n.$$

Thus the two sets of variables are co-gredients, and by the theory of linear transformations we must have

$$\dot{\theta}_1 \frac{dT}{d\theta_1'} \dots + \dot{\theta}_n \frac{dT}{d\theta_n'} = \dot{\phi}_1 \frac{dT}{d\phi_1'} \dots + \dot{\phi}_n \frac{dT}{d\phi_n'}.$$

The expression on either side of the equation is of course known in algebra as an *emanant* (§ 261).

We could have foreseen this result from the fact that whatever set of n independent screw-chains belonging to the system was chosen, the identical equation must in each case assume the standard form.

366. The General Equations of Motion with Screw-chain Co-ordinates.

The screw-chain co-ordinates of a mass-chain with n degrees of freedom are $\theta_1', \dots \theta_n'$; the co-ordinates of the velocities are $\dot{\theta}_1, \dots \dot{\theta}_n$. Let η be the wrench-chain which acts on the system. Let the components of the wrench-chain, when resolved on the screw-chains of reference, have for intensities $\eta_1'', \dots \eta_n''$. Let $p_1, \dots p_n$ be the pitches of the chains of reference, by which is meant that $2p_1$ is the work done on that screw-chain by a twist of unit amplitude against a wrench of unit intensity on the same screw-chain. Then the screw-chains of reference being supposed to be co-reciprocal, we have, from Lagrange's equations,

$$\frac{d}{dt}\left(\frac{dT}{d\dot\theta_1}\right) - \frac{dT}{d\theta_1'} = 2p_1\eta_1'',$$

$$\cdots\cdots\cdots\cdots\cdots\cdots$$

$$\frac{d}{dt}\left(\frac{dT}{d\dot\theta_n}\right) - \frac{dT}{d\theta_n'} = 2p_n\eta_n''.$$

These equations admit of a transformation by the aid of the identity

$$\dot\theta_1\frac{dT}{d\theta_1'}\cdots + \dot\theta_n\frac{dT}{d\theta_n'} = 0.$$

Differentiating this equation by $\dot\theta_1$, we find

$$\frac{dT}{d\theta_1'} + \dot\theta_1\frac{d^2T}{d\dot\theta_1 d\theta_1'} + \ldots + \dot\theta_n\frac{d^2T}{d\dot\theta_1 d\theta_n'} = 0;$$

but

$$\frac{d}{dt}\left(\frac{dT}{d\dot\theta_1}\right) = \ddot\theta_1\frac{d^2T}{d\dot\theta_1^2} + \ddot\theta_2\frac{d^2T}{d\dot\theta_1 d\dot\theta_2} + \ldots + \ddot\theta_n\frac{d^2T}{d\dot\theta_1 d\dot\theta_n} + \dot\theta_1\frac{d^2T}{d\dot\theta_1 d\theta_1'} + \ldots + \dot\theta_n\frac{d^2T}{d\dot\theta_1 d\theta_n'},$$

whence, by substitution

$$\frac{d}{dt}\left(\frac{dT}{d\dot\theta_1}\right) = \ddot\theta_1\frac{d^2T}{d\dot\theta_1^2}\cdots + \ddot\theta_n\frac{d^2T}{d\dot\theta_1 d\dot\theta_n} - \frac{dT}{d\theta_1'}.$$

Hence when screw-chain co-ordinates are employed Lagrange's equations may be written in the form

$$\ddot\theta_1\frac{d^2T}{d\dot\theta_1^2} + \ddot\theta_2\frac{d^2T}{d\dot\theta_1 d\dot\theta_2}\cdots + \ddot\theta_n\frac{d^2T}{d\dot\theta_1 d\dot\theta_n} = 2p_1\left(\eta_1'' + \frac{1}{p_1}\frac{dT}{d\theta_1'}\right),$$

$$\cdots\cdots\cdots\cdots\cdots\cdots\cdots\cdots\cdots\cdots$$

$$\ddot\theta_1\frac{d^2T}{d\dot\theta_1 d\dot\theta_n} + \ddot\theta_2\frac{d^2T}{d\dot\theta_2 d\dot\theta_n}\cdots + \ddot\theta_n\frac{d^2T}{d\dot\theta_n^2} = 2p_n\left(\eta_n'' + \frac{1}{p_n}\frac{dT}{d\theta_n'}\right).$$

367. Generalization of the Eulerian Equations.

The equations just written can be further simplified by appropriate choice of the screw-chains of reference. We have already assumed the screw-chains of reference to be co-reciprocal. If, however, we select that particular group which forms the principal screw-chains of inertia (§ 357), then every pair are conjugate screw-chains of inertia besides being reciprocal. In this case T takes the form

$$T = M(u_1^2\theta_1^2 + \ldots u_n^2\theta_n^2) + \theta_1'\frac{dT}{d\theta_1'} + \ldots \theta_n'\frac{dT}{d\theta_n'} + \tfrac{1}{2}\theta_1'^2\frac{d^2T}{d\theta_1'^2} + \&c.$$

Neglecting the small quantities $\theta_1' \ldots$ &c. we have

$$\frac{d^2T}{d\dot{\theta}_1^2} = 2Mu_1^2, \ldots \frac{d^2T}{d\dot{\theta}_n^2} = 2Mu_n^2,$$

$$\frac{d^2T}{d\dot{\theta}_1 d\dot{\theta}_2} = 0, \text{ \&c.}$$

Introducing these values we obtain

$$Mu_1^2\ddot{\theta}_1 = p_1\left(\eta_1'' + \frac{1}{p_1}\frac{dT}{d\theta_1'}\right),$$

$$\cdots\cdots\cdots\cdots\cdots\cdots\cdots\cdots\cdots\cdots$$

$$Mu_n^2\ddot{\theta}_n = p_n\left(\eta_n'' + \frac{1}{p_n}\frac{dT}{d\theta_n'}\right).$$

These may be regarded as the generalization for any material arrangement whatever of the well-known Eulerian equations for the rotation of a rigid body around a fixed point. If there are no external forces then $\eta_1'', \ldots \eta_n''$ are all zero, and the equations of movement assume the simple form

$$Mu_1^2\ddot{\theta}_1 = \frac{dT}{d\theta_1'};$$

$$\cdots\cdots\cdots\cdots\cdots$$

$$Mu_n^2\ddot{\theta}_n = \frac{dT}{d\theta_n'}.$$

368. The Restraining Wrench-chain.

If a mass-chain be twisting about an instantaneous screw-chain θ, the mass-chain will, in general, presently forsake θ and gradually adopt one instantaneous screw-chain after another. It is however possible, by the application of a suitable wrench-chain, to compel the mass-chain to continue twisting about the same screw θ with unchanged twist velocity. We now proceed to the discovery of this restraining wrench-chain when no other external forces act on the mass-chain.

As all the accelerations of θ must vanish, the co-ordinates of the wrench-chain required are obtained by imposing the conditions

$$\ddot{\theta}_1 = 0; \quad \ddot{\theta}_2 = 0, \ldots \ddot{\theta}_n = 0.$$

We therefore infer from the general equations of § 366, that if $\eta_1'', \ldots \eta_n''$ are the co-ordinates of the restraining wrench-chain we must have

$$\eta_1'' + \frac{1}{p_1}\frac{dT}{d\theta_1'} = 0,$$

$$\cdots\cdots\cdots\cdots\cdots\cdots$$

$$\eta_n'' + \frac{1}{p_n}\frac{dT}{d\theta_n'} = 0,$$

whence we deduce the following theorem:—

If the position of a mass-chain be referred to co-reciprocal screw-chains of reference, then

$$-\frac{1}{p_1}\frac{dT}{d\theta_1'}, \ldots -\frac{1}{p_n}\frac{dT}{d\theta_n'}$$

are the co-ordinates of the restraining wrench-chain which would coerce the mass-chain into continuing to twist about the same screw-chain θ*.

369. Physical meaning of the Vanishing Emanant.

We may verify this theorem by the following method of viewing the subject. It must be possible to coerce the system to twist about θ by the imposition of special constraints. The reactions of these constraints will constitute, in fact, the restraining wrench-chain. It is, however, a characteristic feature that, as the system is, *ex hypothesi*, still at liberty to twist about θ, the reaction of any constraints which are consistent with this freedom must lie on a screw-chain reciprocal to θ.

The condition that two screw-chains, θ and η, shall be reciprocal (§ 354) is

$$+p_1\theta_1\eta_1, \ldots +p_n\theta_n\eta_n = 0,$$

but this is clearly satisfied if for $\eta_1, \ldots$ we substitute

$$-\frac{1}{p_1}\frac{dT}{d\theta_1'}, \ldots -\frac{1}{p_n}\frac{dT}{d\theta_n'};$$

for the equation then becomes

$$\theta_1\frac{dT}{d\theta_1'}, \ldots +\theta_n\frac{dT}{d\theta_n'} = 0,$$

which, when multiplied by $\dot{\theta}$, reduces to the known identity

$$\dot{\theta}_1\frac{dT}{d\theta_1'}, \ldots +\dot{\theta}_n\frac{dT}{d\theta_n'} = 0.$$

We thus obtain a physical meaning of this equation. It is no more than an expression of the fact that the restraining wrench-chain must be reciprocal o the instantaneous screw-chain.

370. A displacement without change of Energy.

It should also be noticed that provided the twist velocities remain unltered the kinetic energy will be unchanged by any small displacement f the mass-chain arising from a twist on any screw-chain reciprocal to the estraining screw-chain.

* A particular case of this, or what is equivalent thereto, is given in Williamson and Tarleton's *ynamics*, 2nd ed., p. 432.

For, if $\theta_1', \ldots \theta_n'$ be the co-ordinates of the displacement, the change in T is

$$\theta_1' \frac{dT}{d\theta_1'} + \ldots + \theta_n' \frac{dT}{d\theta_n'},$$

which may be written

$$p_1\theta_1' \frac{1}{p_1}\frac{dT}{d\theta_1'} + \ldots + p_n\theta_n' \frac{1}{p_n}\frac{dT}{d\theta_n'};$$

but this will be zero if, and only if, the screw-chain $\theta_1', \ldots \theta_n'$ be reciprocal to the screw-chain

$$\frac{1}{p_1}\frac{dT}{d\theta_1'}, \ldots \frac{1}{p_n}\frac{dT}{d\theta_n'}.$$

371. The Accelerating Screw-chain.

When the mass-chain has forsaken the instantaneous screw-chain θ, and is twisting about another instantaneous screw-chain ϕ, there must be a twist velocity about some screw-chain ρ, which, when compounded with the twist velocity about θ, gives the twist velocity about ϕ. When ϕ and θ are indefinitely close, then ρ is the accelerating screw-chain.

Taking the n principal screw-chains of inertia as the screws of reference and assuming that external forces are absent, we have

$$Mu_1^2\ddot{\theta}_1 = \frac{dT}{d\theta_1'},$$

$$\ldots\ldots\ldots\ldots\ldots$$

$$Mu_n^2\ddot{\theta}_n = \frac{dT}{d\theta_n'}.$$

It is plain that the co-ordinates of the accelerating screw-chain are $\ddot{\theta}_1, \ldots \ddot{\theta}_n$, whence we have the following theorem:—

If a mass-chain be twisting around a screw-chain θ, and if external forces are absent, the co-ordinates of the corresponding accelerating screw-chain are proportional to

$$\frac{1}{u_1^2}\frac{dT}{d\theta_1'}, \ldots \frac{1}{u_n^2}\frac{dT}{d\theta_n'}.$$

372. Another Proof.

It is known from the theory of screw-chains (§ 357) that if a quiescent mass-chain receive an impulsive wrench-chain with co-ordinates

$$\frac{u_1^2}{p_1}\rho_1, \ldots \frac{u_n^2}{p_n}\rho_n,$$

the mass-chain commences to twist about the screw-chain, of which the co-ordinates are

$$\rho_1, \ldots \rho_n.$$

If, by imposition of a restraining wrench-chain, the mass-chain continues to twist about the same screw-chain θ, the restraining wrench-chain has neutralized the acceleration. It follows that the restraining wrench-chain, regarded as impulsive, must have generated an instantaneous twist velocity on the accelerating screw-chain, equal and opposite to the acceleration that would otherwise have taken place. The co-ordinates of this impulsive wrench-chain are proportional to

$$\frac{1}{p_1}\frac{dT}{d\theta_1'}, \dots \frac{1}{p_n}\frac{dT}{d\theta_n'}.$$

The corresponding instantaneous screw-chain is obtained by multiplying these expressions severally by

$$\frac{p_1}{u_1^2}, \dots \frac{p_n}{u_n^2},$$

and thus we find, as before, for the co-ordinates of the accelerating screw-chain

$$\frac{1}{u_1^2}\frac{dT}{d\theta_1'}, \dots \frac{1}{u_n^2}\frac{dT}{d\theta_n'}.$$

373. Accelerating Screw-chain and instantaneous Screw-chain.

We have, from the expressions already given,

$$M(u_1^2\dot{\theta}_1\ddot{\theta}_1 + \dots + u_n^2\dot{\theta}_n\ddot{\theta}_n) = \dot{\theta}_1\frac{dT}{d\theta_1'} + \dots \dot{\theta}_n\frac{dT}{d\theta_n'}.$$

But the right-hand side is the emanant which we know to be zero, whence

$$u_1^2\dot{\theta}_1\ddot{\theta}_1 + \dots + u_n^2\dot{\theta}_n\ddot{\theta}_n = 0.$$

This shows that $\dot{\theta}_1, \dots \dot{\theta}_n$, and $\ddot{\theta}_1, \dots \ddot{\theta}_n$ are on conjugate screw-chains of inertia, and hence we deduce the following theorem:—

Whenever a mass-chain is moving without the action of external forces, other than from constraints restricting the freedom, the instantaneous screw-chain and the accelerating screw-chain are conjugate screw-chains of inertia.

374. Permanent Screw-chains.

Reverting to the general system of equations (§ 366) we shall now investigate the condition that θ may be a permanent screw-chain. It is obvious that if $\ddot{\theta}_1, \dots \ddot{\theta}_n$ are all zero, then

$$\frac{dT}{d\theta_1'}, \dots \frac{dT}{d\theta_n'}$$

must each be zero. If, conversely, the differential coefficients just written are all zero, then the quantities $\ddot{\theta}_1, \dots \ddot{\theta}_n$ must each also vanish.

This is obviously true unless it were possible for the determinant

$$\begin{vmatrix} \frac{d^2T}{d\dot\theta_1^2}, & \frac{d^2T}{d\dot\theta_1 d\dot\theta_2}, & \dots, & \frac{d^2T}{d\dot\theta_1 d\dot\theta_n} \\ \frac{d^2T}{d\dot\theta_1 d\dot\theta_2}, & & \dots\dots\dots\dots, & \frac{d^2T}{d\dot\theta_2 d\dot\theta_n} \\ \frac{d^2T}{d\dot\theta_1 d\dot\theta_n}, & & \dots\dots\dots\dots, & \frac{d^2T}{d\dot\theta_n^2} \end{vmatrix}$$

to become zero. Remembering that T is a homogeneous function of the quantities $\dot\theta_1, \dots \dot\theta_n$ in the second degree, the evanescence of the determinant just written would indicate that T admitted of expression by means of $n-1$ square terms, such as

$$\pm L_1^2 \pm L_2^2 \dots \pm L^2_{n-1}.$$

This vanishes if

$$L_1 = 0; \quad L_2 = 0, \text{ \&c.}; \quad L_{n-1} = 0;$$

each of these is a linear equation in $\dot\theta_1, \dots \dot\theta_n$, and consequently a real system of values for $\dot\theta_1, \dots \dot\theta_n$ must satisfy these equations, and render T zero. It would thus appear that a real motion of the mass-chain would have to be compatible with a state of zero kinetic energy. This is, of course, impossible; it therefore follows that the determinant must not vanish, and consequently we have the following theorem:—

If the screw-chains of reference be co-reciprocal, then the necessary and the sufficient conditions for θ to be a permanent screw are that its co-ordinates $\dot\theta_1, \dot\theta_2, \dots \dot\theta_n$ shall satisfy the equations

$$\frac{dT}{d\theta_1'} = 0; \ \dots \ \frac{dT}{d\theta_n'} = 0.$$

There are n of these equations, but they are not independent. The emanant identity shows that if $n-1$ of them be satisfied, the co-ordinates so found must, in general, satisfy the last equation also.

375. Conditions of a permanent Screw-chain.

As the quantities $\theta_1', \dots \theta_n'$ are small, we may generally expand T in powers, as follows:—

$$T = T_0 + \theta_1' T_1 + \dots \theta_n' T_n$$
$$+ \theta_1'^2 T_{11} + \dots + 2\theta_1'\theta_2' T_{12} + \dots .$$

The equation

$$\frac{dT}{d\theta_1'} = 0$$

therefore becomes

$$T_1 + 2\theta_1' T_{11} + 2\theta_2' T_{12} + \dots = 0,$$

and as θ_1', θ_2', &c., are indefinitely small, this reduces to

$$T_1 = 0,$$

where T_1 is a homogeneous function of $\dot{\theta}_1, \dot{\theta}_2, \ldots \dot{\theta}_n$ in the second degree.

For the study of the permanent screws we have, therefore, n equations of the second degree in the co-ordinates of the instantaneous screw-chain, and any screw-chain will be permanent if its co-ordinates render the several differential coefficients zero. We may write the necessary conditions that have to be fulfilled, as follows:—

Let us denote the several differential coefficients of T with respect to the variables by I, II, III, &c. Then the emanant identity is

$$\dot{\theta}_1 \text{I} + \dot{\theta}_2 \text{II} + \dot{\theta}_3 \text{III} + \ldots = 0,$$

and we may develop any single expression, such as III, in the following form:—

$$\text{III} = \text{III}_{11}\dot{\theta}_1^2 + \text{III}_{22}\dot{\theta}_2^2 + \text{III}_{33}\dot{\theta}_3^2 + 2\text{III}_{12}\dot{\theta}_1\dot{\theta}_2 + \ldots 2\text{III}_{14}\dot{\theta}_1\dot{\theta}_4.$$

As the emanant is to vanish identically, we must have the coefficients of the several terms, such as $\dot{\theta}_1^3$, $\dot{\theta}_1^2\dot{\theta}_2$, $\dot{\theta}_1\dot{\theta}_2\dot{\theta}_3$, &c., all zero, the result being three types of equation—

$$\begin{array}{lll}
\text{I}_{11} = 0, & \text{I}_{22} + \text{II}_{12} = 0, & \text{I}_{23} + \text{II}_{13} + \text{III}_{12} = 0, \\
\text{II}_{22} = 0, & \text{II}_{11} + \text{I}_{12} = 0, & \&c., \\
\text{III}_{33} = 0, & \text{II}_{33} + \text{III}_{23} = 0, & \&c., \\
\text{IV}_{44} = 0, & \&c., & \&c. \\
\&c. & &
\end{array}$$

Of the first of these classes of equations, $\text{I}_{11} = 0$, there are n, of the second there are $n(n-1)$, and of the third, $\frac{n(n-1)(n-2)}{1.2.3}$, in all, $\frac{n(n+1)(n+2)}{1.2.3}$.

376. Another identical equation.

Let T be the kinetic energy of a perfectly free rigid body twisting for the moment around a screw θ. It is obvious that T will be a function of the six co-ordinates, $\theta_1', \ldots \theta_6'$, which express the position of the body, and also of $\dot{\theta}_1, \ldots \dot{\theta}_6$, the co-ordinates of the twist velocity,

$$T = f(\theta_1', \ldots \theta_6', \dot{\theta}_1, \ldots \dot{\theta}_6).$$

We may now make a further application of the principle employed in § 362. The kinetic energy will be unaltered if the motion of the body be arrested, and if, after having received a displacement by a twist of amplitude ϵ about a screw of *any* pitch on the same axis as the instantaneous screw, the body be again set in motion about the original screw with the original twist velocity. This obvious property is now to be stated analytically.

It has been shown in § 265 that if $\dot\theta_1, \dot\theta_2, \ldots \dot\theta_6$ are the co-ordinates of a screw of zero pitch, then the co-ordinates of a screw of pitch p_x on the same axis are respectively

$$\dot\theta_1 + \frac{p_x}{4p_1}\frac{dR}{d\dot\theta_1}, \quad \dot\theta_2 + \frac{p_x}{4p_2}\frac{dR}{d\dot\theta_2}, \ldots \dot\theta_6 + \frac{p_x}{4p_6}\frac{dR}{d\theta_6}.$$

In these expressions p_x denotes an arbitrary pitch, while R is the function

$$\dot\theta_1^2 + \dot\theta_2^2 \ldots + \dot\theta_6^2 + 2\,(12)\,\dot\theta_1\dot\theta_2 + 2\,(23)\,\dot\theta_2\dot\theta_3, \text{ \&c.},$$

where (12) is the cosine of the angle between the first and second screws of reference, and similarly for (23), &c.

The principle just stated asserts that T must remain unchanged if we substitute for θ_1', θ_2', &c. the expressions

$$\theta_1' + \left(\dot\theta_1 + \frac{p_x}{4p_1}\frac{dR}{d\dot\theta_1}\right)\epsilon, \quad \theta_2' + \left(\dot\theta_2 + \frac{p_x}{4p_2}\frac{dR}{d\dot\theta_2}\right)\epsilon, \text{ \&c.}$$

We thus obtain the formula

$$\left(\dot\theta_1 + \frac{p_x}{4p_1}\frac{dR}{d\dot\theta_1}\right)\frac{dT}{d\theta_1'} + \ldots + \left(\dot\theta_6 + \frac{p_x}{4p_6}\frac{dR}{d\dot\theta_6}\right)\frac{dT}{d\theta_6'} = 0.$$

As this must be true for every value of p_x, we must have, besides the vanishing emanant, the condition

$$\frac{1}{p_1}\frac{dR}{d\dot\theta_1}\frac{dT}{d\theta_1'} + \ldots + \frac{1}{p_6}\frac{dR}{d\dot\theta_6}\frac{dT}{d\theta_6'} = 0.$$

It is plain that this is equivalent to the statement that the screw whose co-ordinates are

$$\frac{1}{p_1}\frac{dR}{d\dot\theta_1}, \quad \frac{1}{p_2}\frac{dR}{d\dot\theta_2}, \ldots \frac{1}{p_6}\frac{dR}{d\dot\theta_6},$$

must be reciprocal to the screw defined by the co-ordinates

$$\frac{1}{p_1}\frac{dT}{d\theta_1'}, \quad \frac{1}{p_2}\frac{dT}{d\theta_2'}, \ldots \frac{1}{p_6}\frac{dT}{d\theta_6'}.$$

The former denotes a screw of infinite pitch parallel to θ and hence it follows that the restraining screw must be perpendicular to θ. Remembering also that the restraining screw is reciprocal to θ, it follows that the restraining screw must intersect θ. We thus obtain the following result:

If θ be the screw about which a free rigid body is twisting, then to check the tendency of the body to depart from twisting about θ a restraining wrench on a screw which intersects θ at right angles must in general be applied.

377. Different Screws on the same axis.

Let the body be displaced from a standard position to another position defined by the co-ordinates $\theta_1', \theta_2', \ldots \theta_6'$, and let it then be set in rotation about a screw of zero pitch with a twist velocity whose co-ordinates are $\dot{\theta}_1, \dot{\theta}_2, \ldots \dot{\theta}_6$. Let the kinetic energy of the body in this condition be T.

Suppose that in addition to the rotation about θ the body of mass M also received a velocity v of translation parallel to θ. Then the kinetic energy of the body would be T', where

$$T' = T + \tfrac{1}{2} M v^2.$$

It is obvious that the position of the body, i.e. the co-ordinates $\theta_1', \theta_2', \ldots \theta_6'$, can have no concern in $\tfrac{1}{2} M v^2$, whence

$$\frac{dT'}{d\theta_1'} = \frac{dT}{d\theta_1'}, \text{ and similar equations.}$$

But a body rotating about θ with an angular velocity $\dot{\theta}$ and translated parallel to θ with the velocity v is really rotating about a screw on the same axis as θ and with a pitch $v \div p$. As v may have any value we obtain the following theorem :—

All instantaneous screws lying on the same axis have the same restraining screw.

378. Co-ordinates of the Restraining Wrench for a free rigid body.

Suppose the body to have a standard position from which we displace it by small twists $\theta_1', \ldots \theta_6'$ around the six principal screws of inertia. While the body is in its new position it receives a twist velocity of which the co-ordinates relatively to the six principal screws of inertia are $\dot{\theta}_1, \ldots \dot{\theta}_6$.

To compute the kinetic energy we proceed as follows :—Let a point lie initially at x, y, z, then, by the placing of the body at the starting position the point is moved to X, Y, Z, where

$$X = a(\theta_1' - \theta_2') + y(\theta_5' + \theta_6') - z(\theta_3' + \theta_4') + x,$$
$$Y = b(\theta_3' - \theta_4') + z(\theta_1' + \theta_2') - x(\theta_5' + \theta_6') + y,$$
$$Z = c(\theta_5' - \theta_6') + x(\theta_3' + \theta_4') - y(\theta_1' + \theta_2') + z,$$

in which a, b, c are the radii of gyration on the principal axes. The six principal screws of inertia lie, of course, two by two on each of the three principal axes, with pitches $+a, -a$ on the first, $+b, -b$ on the second, and $+c, -c$ on the third.

In consequence of the twist velocity with the components $\dot{\theta}_1, \ldots \dot{\theta}_6$, each point X, Y, Z receives a velocity of which the components are

$$a(\dot{\theta}_1 - \dot{\theta}_2) + Y(\dot{\theta}_5 + \dot{\theta}_6) - Z(\dot{\theta}_3 + \dot{\theta}_4),$$
$$b(\dot{\theta}_3 - \dot{\theta}_4) + Z(\dot{\theta}_1 + \dot{\theta}_2) - X(\dot{\theta}_5 + \dot{\theta}_6),$$
$$c(\dot{\theta}_5 - \dot{\theta}_6) + X(\dot{\theta}_3 + \dot{\theta}_4) - Y(\dot{\theta}_1 + \dot{\theta}_2).$$

Before substitution for X, Y, Z it will be convenient to use certain abbreviations,

$$\theta_1' - \theta_2' = \epsilon_1; \quad \dot{\theta}_1 - \dot{\theta}_2 = \rho_1; \quad \theta_1' + \theta_2' = \lambda_1, \quad \dot{\theta}_1 + \dot{\theta}_2 = \omega_1,$$
$$\theta_3' - \theta_4' = \epsilon_2; \quad \dot{\theta}_3 - \dot{\theta}_4 = \rho_2; \quad \theta_3' + \theta_4' = \lambda_2, \quad \dot{\theta}_3 + \dot{\theta}_4 = \omega_2,$$
$$\theta_5' - \theta_6' = \epsilon_3; \quad \dot{\theta}_5 - \dot{\theta}_6 = \rho_3; \quad \theta_5' + \theta_6' = \lambda_3, \quad \dot{\theta}_5 + \dot{\theta}_6 = \omega_3.$$

With these substitutions in v^2 the square of the velocity of the element we readily obtain after integration and a few reductions and taking the total mass as unity,

$$\begin{aligned} \tfrac{1}{2}\int v^2 dm = {} & a^2\dot{\theta}_1^2 + a^2\dot{\theta}_2^2 + b^2\dot{\theta}_3^2 + b^2\dot{\theta}_4^2 + c^2\dot{\theta}_5^2 + c^2\dot{\theta}_6^2 \\ & + ab\epsilon_2\rho_1\omega_3 - ac\epsilon_3\rho_1\omega_2 - \lambda_1\omega_2\omega_3(b^2 - c^2) \\ & + bc\epsilon_3\rho_2\omega_1 - ba\epsilon_1\rho_2\omega_3 - \lambda_2\omega_3\omega_1(c^2 - a^2) \\ & + ca\epsilon_1\rho_3\omega_2 - cb\epsilon_2\rho_3\omega_1 - \lambda_3\omega_1\omega_2(a^2 - b^2), \end{aligned}$$

whence we easily find

$$\frac{dT}{d\theta_1'} = + ac\rho_3\omega_2 - ab\rho_2\omega_3 - (b^2 - c^2)\,\omega_2\omega_3.$$

If $\eta_1'', \ldots \eta_6''$ be the co-ordinates of the restraining wrench, then, as shown in § 368,

$$\eta_1'' = -\frac{1}{p_1}\frac{dT}{d\theta_1'},$$

whence we deduce the following fundamental expressions for the co-ordinates of the restraining wrench:—

$$\begin{aligned} p_1\eta_1'' &= - ac\rho_3\omega_2 + ab\rho_2\omega_3 + (b^2 - c^2)\,\omega_2\omega_3, \\ p_2\eta_2'' &= + ac\rho_3\omega_2 - ab\rho_2\omega_3 + (b^2 - c^2)\,\omega_2\omega_3, \\ p_3\eta_3'' &= - ab\rho_1\omega_3 + cb\rho_3\omega_1 + (c^2 - a^2)\,\omega_3\omega_1, \\ p_4\eta_4'' &= + ab\rho_1\omega_3 - cb\rho_3\omega_1 + (c^2 - a^2)\,\omega_3\omega_1, \\ p_5\eta_5'' &= - bc\rho_2\omega_1 + ac\rho_1\omega_2 + (a^2 - b^2)\,\omega_1\omega_2, \\ p_6\eta_6'' &= + bc\rho_2\omega_1 - ac\rho_1\omega_2 + (a^2 - b^2)\,\omega_1\omega_2. \end{aligned}$$

As usual, we here write for symmetry

$$p_1 = +a; \quad p_2 = -a; \quad p_3 = +b; \quad p_4 = -b; \quad p_5 = +c; \quad p_6 = -c.$$

We verify at once that

$$p_1\eta_1''\dot{\theta}_1 + \ldots p_6\eta_6''\dot{\theta}_6 = 0,$$

but this is of course known otherwise to be true, because the restraining screw must be reciprocal to the instantaneous screw.

These equations enable us to study the correspondence between each instantaneous screw θ and the corresponding restraining screw η. It is to be noted that this correspondence is not of the homographic, or one-to-one type, such as we meet with in the study of the Principal Screws of Inertia, and in other parts of the Theory of Screws. The correspondence now to be considered has a different character.

379. Limitation to the position of the Restraining Screw.

If a particular screw θ be given, then no doubt, a corresponding screw η is given definitely, but the converse is not true. If η be selected arbitrarily there will not in general be any possible θ. If, however, there be any one θ, then every screw on the same axis as θ will also correspond to the same η.

From the equations in the last article we can eliminate the six quantities, $\theta_1, \dots \theta_6$; we can also write $\eta_1'' = \eta''\eta_1, \dots \eta_n'' = \eta''\eta_n$ where η'' is the intensity of the restraining wrench and $\eta_1, \dots \eta_6$ the co-ordinates of the screw on which it acts.

We have
$$a(\eta_1'' + \eta_2'') = 2ab\rho_2\omega_3 - 2ac\rho_3\omega_2,$$
$$a(\eta_1'' - \eta_2'') = (b^2 - c^2)\,\omega_2\omega_3,$$
whence
$$\frac{b^2 - c^2}{a}\frac{\eta_1 + \eta_2}{\eta_1 - \eta_2} = b\frac{\rho_2}{\omega_2} - c\frac{\rho_3}{\omega_3},$$
and from the two similar equations we obtain, by addition,
$$\frac{b^2 - c^2}{a}\frac{\eta_1 + \eta_2}{\eta_1 - \eta_2} + \frac{c^2 - a^2}{b}\frac{\eta_3 + \eta_4}{\eta_3 - \eta_4} + \frac{a^2 - b^2}{c}\frac{\eta_5 + \eta_6}{\eta_5 - \eta_6} = 0.$$

It might at first have been supposed that any screw might be the possible residence of a restraining wrench, provided the corresponding instantaneous screw were fitly chosen. It should however be remembered that to each restraining screw corresponds a singly infinite number of possible instantaneous screws. As the choice of an instantaneous screw has five degrees of infinity, it was to be presumed that the restraining screws could only have four degrees of infinity, i.e. that the co-ordinates of a restraining screw must satisfy some equation, or, in other words, that they must belong to a screw system of the fifth order, as we have now shown them to do.

380. A verification.

We confirm the expression for the co-ordinates of η in the following manner. It has been shown (§ 376) that so long as θ retains the same direction and situation, its pitch is immaterial so far as η is concerned. This might have been inferred from the consideration that a rigid body twisting about a screw has no tendency to depart from the screw in so far as its velocity of *translation* is concerned. It is the *rotation* which necessitates the restraining wrench if the motion is to be preserved about the same instantaneous screw. We ought, therefore, to find that the expressions for the co-ordinates of η remained unaltered if we substituted for $\dot{\theta}_1, \ldots \dot{\theta}_6$, the co-ordinates of any other screw on the same straight line as θ. These are (§ 47)

$$\dot{\theta}_1 + \frac{H}{a}(\dot{\theta}_1 + \dot{\theta}_2), \quad \dot{\theta}_2 - \frac{H}{a}(\dot{\theta}_1 + \dot{\theta}_2),$$

$$\dot{\theta}_3 + \frac{H}{b}(\dot{\theta}_3 + \dot{\theta}_4), \quad \dot{\theta}_4 - \frac{H}{b}(\dot{\theta}_3 + \dot{\theta}_4),$$

$$\dot{\theta}_5 + \frac{H}{c}(\dot{\theta}_5 + \dot{\theta}_6), \quad \dot{\theta}_6 - \frac{H}{c}(\dot{\theta}_5 + \dot{\theta}_6),$$

where H is arbitrary.

Introducing these into the values for η_1'', it becomes

$$-ac\omega_2\left(\rho_3 + \frac{2H}{c}\omega_3\right) + ab\omega_3\left(\rho_2 + \frac{2H}{b}\omega_2\right) + (b^2 - c^2)\,\omega_2\omega_3,$$

from which H disappears, and the required result is proved.

The restraining screw is always reciprocal to the instantaneous screw, and, consequently, if ϵ be the angle between the two screws, and d their distance apart,

$$(p_\eta + p_\theta)\cos\epsilon - d\sin\epsilon = 0.$$

We have seen that this must be true for every value of p_θ, whence

$$\cos\epsilon = 0; \quad d = 0;$$

i.e. the two screws must intersect at right angles as we have otherwise shown in § 376.

This also appears from the formulae

$$\eta_1'' + \eta_2'' = 2b\rho_2\omega_3 - 2c\rho_3\omega_2,$$
$$\eta_3'' + \eta_4'' = 2c\rho_3\omega_1 - 2a\rho_1\omega_3,$$
$$\eta_5'' + \eta_6'' = 2a\rho_1\omega_2 - 2b\rho_2\omega_1;$$

multiplying respectively by $\omega_1, \omega_2, \omega_3$, and adding, we get

$$(\eta_1 + \eta_2)(\theta_1 + \theta_2) + (\eta_3 + \eta_4)(\theta_3 + \theta_4) + (\eta_5 + \eta_6)(\theta_5 + \theta_6) = 0,$$

which proves that η and θ are rectangular; but we already know that they are reciprocal, and therefore they intersect at right angles.

381. A Particular Case.

The expressions for the restraining wrenches can be illustrated by taking as a particular case an instantaneous screw which passes through the centre of gravity.

The equations to the axis of the screw are

$$\frac{a\rho_1 + y\omega_3 - z\omega_2}{\omega_1} = \frac{b\rho_2 + z\omega_1 - x\omega_3}{\omega_2} = \frac{c\rho_3 + x\omega_2 - y\omega_1}{\omega_3}.$$

If x, y, z are all simultaneously zero, then

$$\frac{a\rho_1}{\omega_1} = \frac{b\rho_2}{\omega_2} = \frac{c\rho_3}{\omega_3},$$

and these are, accordingly, the conditions that the instantaneous screw passes through the centre of gravity.

With these substitutions the co-ordinates become

$$p_1\eta_1'' = (b^2 - c^2)\,\omega_2\omega_3; \quad p_3\eta_3'' = (c^2 - a^2)\,\omega_3\omega_1; \quad p_5\eta_5'' = (a^2 - b^2)\,\omega_1\omega_2,$$
$$p_2\eta_2'' = (b^2 - c^2)\,\omega_2\omega_3; \quad p_4\eta_4'' = (c^2 - a^2)\,\omega_3\omega_1; \quad p_6\eta_6'' = (a^2 - b^2)\,\omega_1\omega_2;$$

remembering that $p_1 = +a$; $p_2 = -a$, &c., we have

$$\eta_1'' + \eta_2'' = 0; \quad \eta_3'' + \eta_4'' = 0; \quad \eta_5'' + \eta_6'' = 0;$$

but these are the conditions that the pitch of η shall be infinite; in other words the restraining wrench is a couple, as should obviously be the case.

From the equations already given, we can find the co-ordinates of the instantaneous screw in terms of those of the restraining screw.

We have

$$H = \sqrt{\frac{abc\,(\eta_1'' - \eta_2'')(\eta_3'' - \eta_4'')(\eta_5'' - \eta_6'')}{2\,(b^2 - c^2)(c^2 - a^2)(a^2 - b^2)}},$$

and $\omega_1 = H\dfrac{b^2 - c^2}{a\,(\eta_1'' - \eta_2'')}$; $\quad \omega_2 = H\dfrac{c^2 - a^2}{b\,(\eta_3'' - \eta_4'')}$; $\quad \omega_3 = H\dfrac{a^2 - b^2}{c\,(\eta_5'' - \eta_6'')}$.

If we make

$$L\dot\theta^2 = a\omega_1^2 + b\omega_2^2 + c\omega_3^2, \quad h^3\dot\theta^2 = a^3\omega_1^2 + b^3\omega_2^2 + c^3\omega_3^2,$$

then we have

$$\theta_1 = \omega_1\left(\frac{p_1 + p_\theta}{2p_1} + \frac{La^2 - h^3}{2abc}\right), \quad \theta_2 = \omega_1\left(\frac{p_2 + p_\theta}{2p_2} - \frac{La^2 - h^3}{2abc}\right),$$
$$\theta_3 = \omega_2\left(\frac{p_3 + p_\theta}{2p_3} + \frac{Lb^2 - h^3}{2abc}\right), \quad \theta_4 = \omega_2\left(\frac{p_4 + p_\theta}{2p_4} - \frac{Lb^2 - h^3}{2abc}\right),$$
$$\theta_5 = \omega_3\left(\frac{p_5 + p_\theta}{2p_5} + \frac{Lc^2 - h^3}{2abc}\right), \quad \theta_6 = \omega_3\left(\frac{p_6 + p_\theta}{2p_6} - \frac{Lc^2 - h^3}{2abc}\right).$$

In these expressions, p_θ is the pitch of θ, and is, of course, an indeterminate quantity.

382. Remark on the General Case.

If the freedom of a body be restricted, then any screw will be permanent, provided its restraining screw belong to the reciprocal system. For the body will not depart from the original instantaneous screw except by an acceleration. This must be on a screw which stands to the restraining screw in the relation of instantaneous to impulsive, but in the case supposed these two screws are reciprocal, therefore they cannot be so related, and therefore there is no acceleration.

There is little to be said as to the restraining wrench when the freedom is of the first order. Of course, in this case, as every movement of the body can only be a twist about the screw which prescribes its freedom, the restraining wrench is provided by the reactions of the constraints. It is only where the body has liberty to abandon its original instantaneous screw that the theory of the restraining wrench becomes significant.

383. Two Degrees of Freedom.

If a rigid body has *two degrees of freedom*, then it is free to twist about every screw on a certain cylindroid. If the body be set initially in motion by a twist velocity about some one screw on the surface, then, in general, it will not remain twisting about this screw. A movement will take place by which the instantaneous axis gradually comes into coincidence with other screws on the cylindroid. If we impose a suitable restraining wrench η, then of course θ can be maintained as the instantaneous screw; η is reciprocal to θ. It may be compounded with any reactions of the constraints of the system. Thus, given θ, there is an entire screw system of the fifth order, consisting of all possible screws reciprocal to θ, any one screw of which may be taken as the restrainer. Of this system there is one, but only one, which lies on the cylindroid itself. There are many advantages in taking it as the restraining wrench, and it entails no sacrifice of generality; we therefore have the following statement:—To each screw on the cylindroid, regarded as an instantaneous screw, will correspond one screw, also on the cylindroid, as a restraining screw.

The position of this restraining screw is at once indicated by the property that it must be reciprocal to the instantaneous screw. If we employ the circular representation for the screws on the cylindroid (fig. 42), and if O be the pole of the axis of pitch, then it is known that the extremities of any chord, such as IR drawn through O, will correspond to two reciprocal screws (§ 58). If therefore I be the instantaneous screw, R must be the restraining screw. If a body free to twist about all the screws on the cylindroid be set in motion by a twist velocity about I, it will be possible, by a suitable wrench

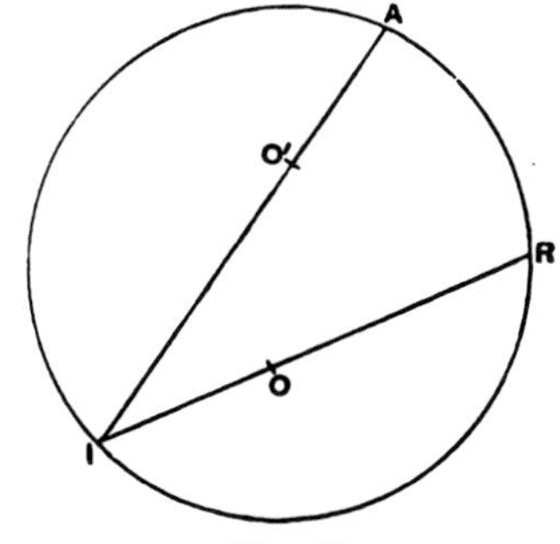

Fig. 42.

applied on the screw corresponding to R, to prohibit the body from changing its instantaneous screw.

Let O' be the pole of the axis of inertia, then, if IA be a chord drawn through O', the points I and A correspond to a pair of conjugate screws of inertia (§ 135). It further appears that A is the instantaneous screw corresponding to an impulsive wrench on R (§ 140). Therefore the effect of the wrench on R when applied to control the body twisting about I is to compound its movement with a nascent twist velocity about A. Therefore A must be the accelerating screw corresponding to I. We thus see that—

Of two conjugate screws of inertia, for a rigid body with two degrees of freedom, either is the accelerator for a body animated by a twist velocity about the other.

384. Calculation of T.

In the case of freedom of the second order we are enabled to obtain the form of T, from the fact that the emanant vanishes, that is,

$$\dot{\theta}_1 \frac{dT}{d\theta_1'} + \dot{\theta}_2 \frac{dT}{d\theta_2'} = 0.$$

If we assume that T is a homogeneous function of the second degree in $\dot{\theta}_1$ and $\dot{\theta}_2$, the solution of this equation must be

$$T = L\dot{\theta}_1^2 + 2S\dot{\theta}_1\dot{\theta}_2 + M\dot{\theta}_2^2 + H(\theta_1'\dot{\theta}_2 - \theta_2'\dot{\theta}_1)^2 + (\theta_1'\dot{\theta}_2 - \theta_2'\dot{\theta}_1)(A\dot{\theta}_1 + B\dot{\theta}_2),$$

in which L, S, M, H, A are constants. If we further suppose that θ_1' and θ_2' are so small that their squares may be neglected, then the term multiplied by H may be discarded, and we have

$$T = L\dot{\theta}_1^2 + 2S\dot{\theta}_1\dot{\theta}_2 + M\dot{\theta}_2^2 + (\theta_1'\dot{\theta}_2 - \theta_2'\dot{\theta}_1)(A\dot{\theta}_1 + B\dot{\theta}_2),$$

whence

$$\frac{dT}{d\theta_1'} = +\dot{\theta}_2(A\dot{\theta}_1 + B\dot{\theta}_2); \quad \frac{dT}{d\theta_2'} = -\dot{\theta}_1(A\dot{\theta}_1 + B\dot{\theta}_2).$$

Thus, for the co-ordinates of the restraining screw, supposing the screws of reference to be reciprocal, we have

$$\eta_1'' = \frac{1}{p_1}\frac{dT}{d\theta_1'} = +\frac{\dot{\theta}_2}{p_1}(A\dot{\theta}_1 + B\dot{\theta}_2); \quad \eta_2'' = \frac{1}{p_2}\frac{dT}{d\theta_2'} = -\frac{\dot{\theta}_1}{p_2}(A\dot{\theta}_1 + B\dot{\theta}_2);$$

from which it is evident that

$$p_1\eta_1''\dot{\theta}_1 + p_2\eta_2''\dot{\theta}_2 = 0,$$

which is, of course, merely expressing the fact that η and θ are reciprocal.

385. Another method.

It may be useful to show how the form of T, just obtained, can be derived from direct calculation. I merely set down here the steps of the work and the final result.

Let us take any two screws on the cylindroid α and β, and let their co-ordinates, when referred to the absolute screws of inertia, be

$$\alpha_1, \ldots \alpha_6, \text{ and } \beta_1, \ldots \beta_6.$$

Then any other screw on the cylindroid, about which the body has been displaced by a twist, by components θ_1' on α and θ_2' on β will have, for co-ordinates,

$$\alpha_1\theta_1' + \beta_1\theta_2', \ldots \alpha_6\theta_1' + \beta_6\theta_2',$$

and the screw about which the body is twisting, with a twist velocity $\dot{\theta}$, will have, for co-ordinates,

$$\alpha_1\dot{\theta}_1 + \beta_1\dot{\theta}_2, \ldots \alpha_6\dot{\theta}_1 + \beta_6\dot{\theta}_2.$$

It readily appears that, so far as the terms involving θ_1' and θ_2' are concerned, the kinetic energy is the expression

$$(\theta_1'\dot{\theta}_2 - \theta_2'\dot{\theta}_1)(A\dot{\theta}_1 + B\dot{\theta}_2),$$

where

$$\begin{aligned} A = &+ bc(\alpha_1 + \alpha_2)[(\alpha_5 - \alpha_6)(\beta_3 - \beta_4) - (\alpha_3 - \alpha_4)(\beta_5 - \beta_6)] \\ &\qquad + (b^2 - c^2)(\alpha_3 + \alpha_4)(\alpha_5 + \alpha_6)(\beta_1 + \beta_2) \\ &+ ca(\alpha_3 + \alpha_4)[(\alpha_1 - \alpha_2)(\beta_5 - \beta_6) - (\alpha_5 - \alpha_6)(\beta_1 - \beta_2)] \\ &\qquad + (c^2 - a^2)(\alpha_5 + \alpha_6)(\alpha_1 + \alpha_2)(\beta_3 + \beta_4) \\ &+ ab(\alpha_5 + \alpha_6)[(\alpha_3 - \alpha_4)(\beta_1 - \beta_2) - (\alpha_1 - \alpha_2)(\beta_3 - \beta_4)] \\ &\qquad + (a^2 - b^2)(\alpha_1 + \alpha_2)(\alpha_3 + \alpha_4)(\beta_5 + \beta_6), \end{aligned}$$

$$\begin{aligned} B = &+ bc(\beta_1 + \beta_2)[(\alpha_5 - \alpha_6)(\beta_3 - \beta_4) - (\alpha_3 - \alpha_4)(\beta_5 - \beta_6)] \\ &\qquad - (b^2 - c^2)(\alpha_1 + \alpha_2)(\beta_3 + \beta_4)(\beta_5 + \beta_6) \\ &+ ca(\beta_3 + \beta_4)[(\alpha_1 - \alpha_2)(\beta_5 - \beta_6) - (\alpha_5 - \alpha_6)(\beta_1 - \beta_2)] \\ &\qquad - (c^2 - a^2)(\alpha_3 + \alpha_4)(\beta_5 + \beta_6)(\beta_1 + \beta_2) \\ &+ ab(\beta_5 + \beta_6)[(\alpha_3 - \alpha_4)(\beta_1 - \beta_2) - (\alpha_1 - \alpha_2)(\beta_3 - \beta_4)] \\ &\qquad - (a^2 - b^2)(\alpha_5 + \alpha_6)(\beta_1 + \beta_2)(\beta_3 + \beta_4). \end{aligned}$$

386. The Permanent Screw.

We now write the equations of motion for a body which has two degrees of freedom, and is unacted upon by any force, the screws of reference being the two principal screws of inertia.

We have, from the general equations (§ 367)

$$Mu_1^2\ddot{\theta}_1 = \frac{dT}{d\theta_1'},$$

$$Mu_2^2\ddot{\theta}_2 = \frac{dT}{d\theta_2'}.$$

Introducing the value just obtained for T,

$$Mu_1^2\ddot{\theta}_1 = +\dot{\theta}_2(A\dot{\theta}_1 + B\dot{\theta}_2),$$

$$Mu_2^2\ddot{\theta}_2 = -\dot{\theta}_1(A\dot{\theta}_1 + B\dot{\theta}_2).$$

There must be one screw on the cylindroid, for which

$$A\dot{\theta}_1 + B\dot{\theta}_2 = 0.$$

This screw will have the accelerations $\ddot{\theta}_1$ and $\ddot{\theta}_2$, both zero, and thus we have the following theorem:—

If a rigid body has two degrees of freedom, then, among the screws about which it is at liberty to twist, there is one, and in general only one, which has the property of a permanent screw.

The existence of a single permanent screw in the case of freedom of the second order seems a noteworthy point. The analogy here ceases between the permanent screws and the principal screws of inertia. Of the latter there are two on the cylindroid (§ 84).

387. Geometrical Investigation.

Let N (fig. 43) be the critical point on the circle which corresponds to the permanent screw (§ 50). Let P be a screw θ, the twist velocity about

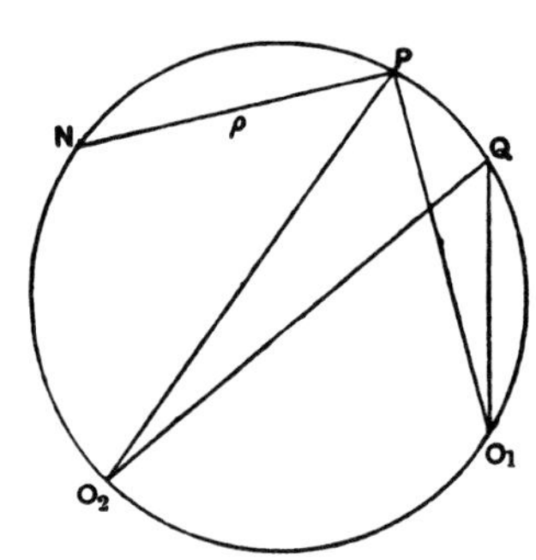

Fig. 43.

which is $\dot{\theta}$. Let u_θ be a linear parameter appropriate to the screw θ, such that $Mu_\theta^2\dot{\theta}^2$ is the kinetic energy.

Let O_1 and O_2 be the two screws of reference on the cylindroid and for convenience let the chord O_1O_2 be unity. Let the point Q correspond to another screw ϕ, then from § 57

$$\dot{\theta}_1 = \dot{\theta}PO_2, \qquad \dot{\theta}_2 = \dot{\theta}PO_1; \qquad \dot{\phi}_1 = \dot{\phi}QO_2, \qquad \dot{\phi}_2 = \dot{\phi}QO_1.$$

Ptolemy's theorem gives

$$PQ\dot{\theta}\dot{\phi} = \dot{\theta}_2\dot{\phi}_1 - \dot{\theta}_1\dot{\phi}_2.$$

Now let ϕ be the adjacent screw about which the body is twisting in a time t after it was twisting about θ.

Then

$$\phi_1 = \dot{\theta}_1 + \ddot{\theta}_1 \delta t,$$

$$\phi_2 = \dot{\theta}_2 + \ddot{\theta}_2 \delta t,$$

whence

$$\dot{\theta}^2 PQ = (\dot{\theta}_2 \ddot{\theta}_1 - \dot{\theta}_1 \ddot{\theta}_2)\, \delta t,$$

$$\frac{PQ}{\delta t} = \frac{\dot{\theta}_2 \ddot{\theta}_1 - \dot{\theta}_1 \ddot{\theta}_2}{\dot{\theta}^2},$$

which is, accordingly, the rate at which P will change its position. If we substitute for $\ddot{\theta}_1$ and $\ddot{\theta}_2$ their values already found in the last article, we obtain for the velocity of P the expression

$$\frac{1}{M} \frac{u_1^2 \dot{\theta}_1^2 + u_2^2 \dot{\theta}_2^2}{u_1^2 u_2^2 \dot{\theta}^2} (A\dot{\theta}_1 + B\dot{\theta}_2).$$

N being the position of the permanent screw, let ρ be the length of the chord PN, then the expression just written assumes the form

$$k\rho \dot{\theta} u_\theta^2$$

where k is a constant.

This expression illustrates the character of the screw corresponding to N. If ρ be zero, then the expression for this velocity vanishes. This means that P has no tendency to abandon N; in other words, that the screw corresponding to N is permanent.

388. Another method.

It is worth while to investigate the question from another point of view.

Let us think of any cylindroid S placed quite arbitrarily with respect to the position of the rigid body. A certain restraining screw η will correspond to each screw θ on S. As θ moves over the cylindroid, so must the corresponding screw η describe some other ruled surface S'. The two surfaces, S and S', will thus have two corresponding systems of screws, whereof every two correspondents are reciprocal. One screw can be discovered on S', which is reciprocal, not alone to its corresponding θ, but to all the screws on the cylindroid. A wrench on this η can be provided by the reactions of the constraints, and, consequently, the constraints will, in this case, arrest the tendency of the body to depart from θ as the instantaneous screw. It follows that this particular θ is the permanent screw.

The actual calculation of the relations between η and the cylindroid is as follows:—

A set of forces applied to a rigid system has components X, Y, Z at a point, and three corresponding moments F, G, H in the rectangular planes of reference.

Let p be the pitch of the screw on which the wrench thus represented lies, and let x, y, z be the co-ordinates of any point on this screw. Then, in the plane of Z the moments of the forces are $xY - yX$, and if to this be added pX, the whole must equal H.

Thus we have the three equations, so well known in statics,

$$F = pX + yZ - zY,$$
$$G = pY + zX - xZ,$$
$$H = pZ + xY - yX.$$

The centrifugal acceleration on a point P is, of course, $\omega^2 PH$, where ω is the angular velocity, and PH the perpendicular let fall on the axis. The three components of this force are X', Y', Z', where

$$X' = \omega^2 \sin\theta\,(x\sin\theta - y\cos\theta),$$
$$Y' = \omega^2 \cos\theta\,(y\cos\theta - x\sin\theta),$$
$$Z' = \omega^2 (z - m\sin 2\theta),$$

and the three moments are F', G', H', where

$$F' = \omega^2 \sin\theta\,(yz\sin\theta + xz\cos\theta - 2my\cos\theta),$$
$$G' = \omega^2 \cos\theta\,(-yz\sin\theta - xz\cos\theta + 2mx\sin\theta),$$
$$H' = \omega^2 \{(y^2 - x^2)\sin\theta\cos\theta + xy\cos 2\theta\}.$$

We are now to integrate these expressions over the entire mass, and we employ the following abbreviations (§ 324):—

$$\int x\,dm = Mx_0;\quad \int y\,dm = My_0;\quad \int z\,dm = Mz_0;$$
$$\int (y^2 - x^2)\,dm = (\rho_1^2 - \rho_2^2)\,M;$$
$$\int xy\,dm = Ml_3^2;\quad \int xz\,dm = Ml_2^2;\quad \int yz\,dm = Ml_1^2;$$
$$X = \int X'dm;\quad Y = \int Y'dm;\quad Z = \int Z'dm;$$
$$F = \int F'dm;\quad G = \int G'dm;\quad H = \int H'dm;$$

then, omitting the factor $M\omega^2$, we have

$$X = +(x_0\sin\theta - y_0\cos\theta)\sin\theta,$$
$$Y = -(x_0\sin\theta - y_0\cos\theta)\cos\theta,$$
$$Z = z_0 - m\sin 2\theta;$$

$$F = +\sin\theta\,(l_1^2\sin\theta + l_2^2\cos\theta) - 2my_0\sin\theta\cos\theta,$$
$$G = -\cos\theta\,(l_1^2\sin\theta + l_2^2\cos\theta) + 2mx_0\sin\theta\cos\theta,$$
$$H = (\rho_1^2 - \rho_2^2)\sin\theta\cos\theta + r^2\cos 2\theta.$$

We can easily verify that

$$FY - GX = 2mXY.$$

We now examine the points on the cylindroid intersected by the axis of the screw

$$F = pX + yZ - zY,$$
$$G = pY + zX - xZ,$$
$$H = pZ + xY - yX.$$

We write the equations of the cylindroid in the form

$$x = R\cos\phi; \quad y = R\sin\phi; \quad z = m\sin 2\phi;$$

then, eliminating p and R, and making

$$U = X^2 + Y^2 + Z^2,$$
$$V = FX + GY + HZ,$$

we find, after a few reductions,

$$\tan^3\phi\,(YV - GU) + \tan^2\phi\,(XV - FU + 2mXU) + \tan\phi\,(YV - GU - 2mYU) + XV - FU = 0.$$

This cubic corresponds, of course, to the three generators of the cylindroid which the ray intersects.

If we put

$$FY - GX = 2mXY,$$

then the cubic becomes, by eliminating m,

$$(Y\tan\phi + X)\{X(YV - GU)\tan^2\phi + (XV - FU)Y\} = 0.$$

The factor $Y\tan\phi + X$ simply means that the restraining screw cuts the instantaneous screw at right angles.

The two other screws in which η intersects the cylindroid are given by the equation

$$(XYV - XGU)\tan^2\phi + (XYV - FUY) = 0.$$

These two screws are of equal pitch, and the value of the pitch is

$$\frac{p_1(XYV - XGU) + p_2(FUY - XYV)}{U(FY - GX)};$$

where p_1 and p_2 are the pitches of the two principal screws on the cylindroid. After a few reductions the expression becomes

$$-\frac{V}{U} + \frac{(l_1^2 - x_0p_1)\sin\theta + (l_2^2 + y_0p_2)\cos\theta}{x_0\sin\theta - y_0\cos\theta}.$$

This is the pitch of the two equal pitch screws on the cylindroid which η intersects. If η is to be reciprocal to the cylindroid, then, of course, the pitch of η itself should be equal and opposite in value to this expression. Hence the permanent screw on the cylindroid is given by

$$(l_1^2 - x_0p_1)\sin\theta + (l_2^2 + y_0p_2)\cos\theta = 0.$$

We notice here the somewhat remarkable circumstance, that if

$$l_1^2 - x_0 p_1 = 0, \text{ and } l_2^2 + y_0 p_2 = 0,$$

then *all* the screws on the cylindroid are permanent screws.

It hence appears that if two screws on a cylindroid are permanent, then every screw on the cylindroid is permanent.

389. Three Degrees of Freedom.

Let us now specially consider the case of a rigid body which has freedom of the third order. On account of the evanescence of the emanant we have

$$\dot{\theta}_1 \frac{dT}{d\theta_1'} + \dot{\theta}_2 \frac{dT}{d\theta_2'} + \dot{\theta}_3 \frac{dT}{d\theta_3'} = 0.$$

It is well known that if U, V, W be three conics whose equations submit to the condition

$$xU + yV + zW = 0,$$

those conics must have three common intersections.

It therefore follows that the three equations

$$\frac{dT}{d\theta_1'} = 0, \quad \frac{dT}{d\theta_2'} = 0, \quad \frac{dT}{d\theta_3'} = 0,$$

must have three common screws. These are, of course, the permanent screws, and, accordingly, we have the theorem:—

A rigid system which has freedom of the third order has, in general, three permanent screws.

There will be a special convenience in taking these three screws as the screws of reference. We shall use the plane representation of the three-system, and the equations of the conics will be

$$A_1\dot{\theta}_2\dot{\theta}_3 + B_1\dot{\theta}_3\dot{\theta}_1 + C_1\dot{\theta}_1\dot{\theta}_2 = 0, \text{ or } U = 0,$$
$$A_2\dot{\theta}_2\dot{\theta}_3 + B_2\dot{\theta}_3\dot{\theta}_1 + C_2\dot{\theta}_1\dot{\theta}_2 = 0, \quad ,, \quad V = 0,$$
$$A_3\dot{\theta}_2\dot{\theta}_3 + B_3\dot{\theta}_3\dot{\theta}_1 + C_3\dot{\theta}_1\dot{\theta}_2 = 0, \quad ,, \quad W = 0;$$

but, as

$$\dot{\theta}_1 U + \dot{\theta}_2 V + \dot{\theta}_3 W = 0,$$

identically, we must have

$$B_1 = 0; \quad A_2 = 0; \quad A_3 = 0;$$
$$C_1 = 0; \quad C_2 = 0; \quad B_3 = 0;$$

and also

$$A_1 + B_2 + C_3 = 0.$$

or symmetry we may write

$$A_1 = \mu - \nu; \quad B_2 = \nu - \lambda; \quad C_3 = \lambda - \mu.$$

We thus find that when T is referred to the three permanent screws of the system, its expression must be

$$T = a\dot{\theta}_1^2 + b\dot{\theta}_2^2 + c\dot{\theta}_3^2 + 2f\dot{\theta}_2\dot{\theta}_3 + 2g\dot{\theta}_1\dot{\theta}_3 + 2h\dot{\theta}_1\dot{\theta}_2$$
$$+ (\mu - \nu)\,\theta_1'\dot{\theta}_2\dot{\theta}_3 + (\nu - \lambda)\,\theta_2'\dot{\theta}_3\dot{\theta}_1 + (\lambda - \mu)\,\theta_3'\dot{\theta}_1\dot{\theta}_2.$$

Let η'' be the intensity of any wrench acting on a screw η belonging to the system, and let $2\varpi_{1\eta}$ represent the virtual coefficient between η and the first of the three screws of reference.

Then, substituting for T in Lagrange's equations, we have

$$+ a\ddot{\theta}_1 + h\ddot{\theta}_2 + g\ddot{\theta}_3 - (\mu - \nu)\,\dot{\theta}_2\dot{\theta}_3 = \varpi_{1\eta}\eta'',$$
$$+ h\ddot{\theta}_1 + b\ddot{\theta}_2 + f\ddot{\theta}_3 - (\nu - \lambda)\,\dot{\theta}_3\dot{\theta}_1 = \varpi_{2\eta}\eta'',$$
$$+ g\ddot{\theta}_1 + f\ddot{\theta}_2 + c\ddot{\theta}_3 - (\lambda - \mu)\,\dot{\theta}_1\dot{\theta}_2 = \varpi_{3\eta}\eta''.$$

If η be the restraining screw, then an appropriate wrench η'' should be capable of annihilating the acceleration, i.e. of rendering

$$\ddot{\theta}_1 = 0; \quad \ddot{\theta}_2 = 0; \quad \ddot{\theta}_3 = 0;$$

whence the position of η, and the intensity η'' are indicated by the equations

$$(\nu - \mu)\,\dot{\theta}_2\dot{\theta}_3 = \varpi_{1\eta}\eta'',$$
$$(\lambda - \nu)\,\dot{\theta}_3\dot{\theta}_1 = \varpi_{2\eta}\eta'',$$
$$(\mu - \lambda)\,\dot{\theta}_1\dot{\theta}_2 = \varpi_{3\eta}\eta''.$$

We can now exhibit the nature of the correspondence between η and θ, for

$$\varpi_{1\eta} = p_1\eta_1 + \varpi_{12}\eta_2 + \varpi_{13}\eta_3,$$
$$\varpi_{2\eta} = \varpi_{12}\eta_1 + p_2\eta_2 + \varpi_{32}\eta_3,$$
$$\varpi_{3\eta} = \varpi_{13}\eta_1 + \varpi_{32}\eta_2 + p_3\eta_3.$$

If we make $H = \dot{\theta}_1\dot{\theta}_2\dot{\theta}_3 \div \eta''$, and omit the dots over θ_1, &c., we have

$$\theta_1(p_1\eta_1 + \varpi_{12}\eta_2 + \varpi_{13}\eta_3) = H(\gamma - \beta),$$
$$\theta_2(\varpi_{12}\eta_1 + p_2\eta_2 + \varpi_{32}\eta_3) = H(\alpha - \gamma),$$
$$\theta_3(\varpi_{13}\eta_1 + \varpi_{32}\eta_2 + p_3\eta_3) = H(\beta - \alpha).$$

We may reduce them to two homogeneous forms, viz.

$$\theta_1 L + \theta_2 M + \theta_3 N = 0,$$
$$\alpha\theta_1 L + \beta\theta_2 M + \gamma\theta_3 N = 0,$$

where $$L = \tfrac{1}{2}\frac{dp_\eta}{d\eta_1}; \quad M = \tfrac{1}{2}\frac{dp_\eta}{d\eta_2}; \quad N = \tfrac{1}{2}\frac{dp_\eta}{d\eta_3}.$$

390. Geometrical Construction for the Permanent Screws.

We see that η must lie on the polar of the point θ_1, θ_2, θ_3 with respect to the pitch conic (§ 201) or the locus of all the screws for which

$$p_\eta = 0.$$

We also see that η must lie on the polar of the point $\alpha\theta_1$, $\beta\theta_2$, $\gamma\theta_3$ with regard to the same conic.

We thus obtain a geometrical construction by which we discover the restraining screw when the instantaneous screw is given.

Two homographic systems are first to be conceived. A point of the first system, of which the co-ordinates are θ_1, θ_2, θ_3, has as its correspondent a point in the second system, with co-ordinates $\alpha\theta_1$, $\beta\theta_2$, $\gamma\theta_3$. The three double points of the homography correspond, of course, to the permanent screws.

To find the restraining screw η corresponding to a given instantaneous screw θ, we join θ to its homographic correspondent, and the pole of this ray, with respect to the pitch conic, is the position of η.

The pole of the same ray, with regard to the conic of inertia (§ 211), is the accelerator. It seems hardly possible to have a more complete geometrical picture of the relation between η and θ than that which these theorems afford.

391. Calculation of Permanent Screws in a Three-system.

When a three-system is given which expresses the freedom of a body we have seen how in the plane representation the knowledge of a conic (the conic of inertia) will give the instantaneous screw corresponding to any given impulsive screw. A conic is however specified completely by five data. The rigid body has nine co-ordinates. It therefore follows that there is a quadruply infinite system of rigid bodies which with respect to a given three-system will have the same conic of inertia. If in that three-system α be the instantaneous screw corresponding to η as the impulsive screw for any one body of the quadruply infinite system, then will η and α stand in the same relation to each other for every body of the system.

The point in question may be illustrated by taking the case of a four-system. The screws of such a system are represented by the points in space, and the equation obtained by equating the kinetic energy to zero indicates a quadric. For the specification of the quadric nine data are necessary. This is just the number of co-ordinates required for the specification of a rigid body. If therefore the inertia quadric in the space representation be assumed arbitrarily, then every instantaneous screw corresponding to a given impulsive screw will be determined; in this case there is only a finite number of rigid bodies and not an infinite system for which the correspondence subsists.

We thus note that there is a special character about the freedom of the fourth order which we may state more generally as follows. To establish a chiastic homography (§ 292) in an n-system requires $(n-1)(n+2)/2$ data.

If the restraints are such that the number of degrees of freedom is less than four, then an infinite number of rigid bodies can be designed, such that the impulsive screws and their corresponding instantaneous screws shall be represented by a given chiastic homography. If n exceed four then it will not in general be possible to design a rigid body such that its corresponding impulsive screws and instantaneous screws shall agree with a given chiastic homography. If, however, $n = 4$ then it is always possible to design one but only one rigid body so that its pairs of corresponding impulsive screws and instantaneous screws shall be represented by a given chiastic homography.

Returning to the three-system we may remark that, having settled the inertia conic in the plane representation we are not at liberty to choose three arbitrary points as representing the three permanent screws. For if these three points were to be chosen quite arbitrarily, then six relations among the co-ordinates of the rigid body would be given, and the conic of inertia would require five more conditions. Hence the co-ordinates of the rigid body would have in general to satisfy eleven conditions which, of course, is not generally possible, as there are only nine such co-ordinates. It is therefore plain that when the conic of inertia has been chosen at least two other conditions must necessarily be fulfilled by the three points which are to represent the permanent screw. This fact is not brought out by the method of § 389 in which, having chosen the three permanent screws arbitrarily, we have then written down the general equation of a conic as the inertia conic. This conic should certainly fulfil at least two conditions which the equations as there given do not indicate.

We therefore calculate directly the expression for the kinetic energy of a body in the position θ_1', θ_2', θ_3' twisting about a screw with twist velocities $\dot{\theta}_1$, $\dot{\theta}_2$, $\dot{\theta}_3$ when the screws of reference are the three principal screws of the three-system with pitches a, b, c, and when x_0, y_0, z_0, l_1^2, l_2^2, l_3^2, ρ_1^2, ρ_2^2, ρ_3^2 are the nine co-ordinates (§ 324) of the rigid body relative to these axes.

It is easily shown that we have for the kinetic energy the mass M of the body multiplied into the following expression where squares and higher powers of θ_1', θ_2', θ_3' are omitted:—

$$
\begin{aligned}
&\tfrac{1}{2}\left(a^2\dot{\theta}_1^2 + b^2\dot{\theta}_2^2 + c^2\dot{\theta}_3^2 + \rho_1^2\dot{\theta}_1^2 + \rho_2^2\dot{\theta}_2^2 + \rho_3^2\dot{\theta}_3^2\right) \\
&\quad + \dot{\theta}_2\dot{\theta}_3(c-b)\,x_0 + \dot{\theta}_3\dot{\theta}_1(a-c)\,y_0 + \dot{\theta}_1\dot{\theta}_2(b-a)\,z_0 \\
&\quad - \dot{\theta}_2\dot{\theta}_3 l_1^2 - \dot{\theta}_3\dot{\theta}_1 l_2^2 - \dot{\theta}_1\dot{\theta}_2 l_3^2 \\
&\quad + \theta_1'\begin{bmatrix} + a(\dot{\theta}_2^2 + \dot{\theta}_3^2)\,x_0 + (\dot{\theta}_3^2 - \dot{\theta}_2^2)\,l_1^2 \\ + \dot{\theta}_2\dot{\theta}_3(\rho_3^2 - \rho_2^2 + ac - ab) \\ + \dot{\theta}_3\dot{\theta}_1(l_3^2 - cz_0) - \dot{\theta}_1\dot{\theta}_2(l_2^2 + by_0) \end{bmatrix}
\end{aligned}
$$

$$+\theta_2'\begin{bmatrix}+b(\mathring{\theta}_3^2+\mathring{\theta}_1^2)\,y_0+(\mathring{\theta}_1^2-\mathring{\theta}_3^2)\,l_2^2\\ +\mathring{\theta}_3\mathring{\theta}_1(\rho_1^2-\rho_3^2+ba-bc)\\ +\mathring{\theta}_1\mathring{\theta}_2(l_1^2-ax_0)-\mathring{\theta}_2\mathring{\theta}_3(l_3^2+cz_0)\end{bmatrix}$$

$$+\theta_3'\begin{bmatrix}+c(\mathring{\theta}_1^2+\mathring{\theta}_2^2)\,z_0+(\mathring{\theta}_2^2-\mathring{\theta}_1^2)\,l_3^2\\ +\mathring{\theta}_1\mathring{\theta}_2(\rho_2^2-\rho_1^2+cb-ca)\\ +\mathring{\theta}_2\mathring{\theta}_3(l_2^2-by_0)-\mathring{\theta}_3\mathring{\theta}_1(l_1^2+ax_0)\end{bmatrix}$$

The coefficients of θ_1', θ_2', θ_3' respectively each equated to zero will give three conics $U=0$, $V=0$, $W=0$. These conics have three common points which are of course the three permanent screws.

If we introduce a new quantity Ω we can write the three equations

$$(bc-\rho_1^2+\Omega)\,\mathring{\theta}_1+(l_3^2+cz_0)\,\mathring{\theta}_2+(l_2^2-by_0)\,\mathring{\theta}_3=0,$$

$$(l_3^2-cz_0)\,\mathring{\theta}_1+(ac-\rho_2^2+\Omega)\,\mathring{\theta}_2+(l_1^2+ax_0)\,\mathring{\theta}_3=0,$$

$$(l_2^2+by_0)\,\mathring{\theta}_1+(l_1^2-ax_0)\,\mathring{\theta}_2+(ab-\rho_3^2+\Omega)\,\mathring{\theta}_3=0.$$

The elimination of Ω between each pair of these equations will produce the three equations $U=0$, $V=0$, $W=0$. If therefore we eliminate $\mathring{\theta}_1$, $\mathring{\theta}_2$, $\mathring{\theta}_3$ from the three equations just written the resulting determinant gives a cubic for Ω. The solution of this cubic will give three values for Ω which substituted in the three equations will enable the corresponding values of $\mathring{\theta}_1$, $\mathring{\theta}_2$, $\mathring{\theta}_3$ to be found. We thus express the co-ordinates of the three permanent screws in terms of the nine co-ordinates of the rigid body and their determination is complete.

It may be noted that the same permanent screws will be found for any one of the systems of rigid bodies whose co-ordinates are

$$x_0,\ y_0,\ z_0,\ l_1^2,\ l_2^2,\ l_3^2,\ \rho_1^2+h,\ \rho_2^2+h,\ \rho_3^2+h,$$

whatever h may be.

392. Case of Two Degrees of Freedom.

We have already shown that there is a single permanent screw in every case where the rigid body has two degrees of freedom. We can demonstrate this in a different manner as a deduction from the case of the three-system.

Consider a cylindroid in a three-system, that is of course a straight line in the plane representation (§ 200). Let this line be AB (fig. 44). If the movements of the body be limited to twists about the screws on the cylindroid, there may be reactions about the screw which corresponds to the pole P of this ray with respect to the pitch conic, in addition to the reactions of the three-system.

The permanent screw on this cylindroid will be one whereof the restraining screw coincides with P. In general, the points corresponding homo-

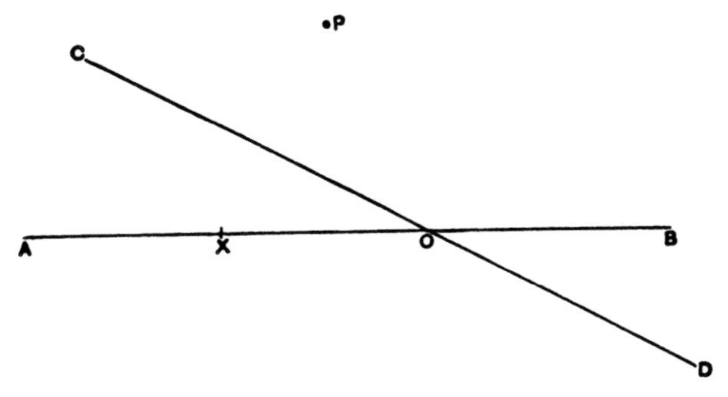

Fig. 44.

graphically to the points on the ray AB will form a ray CD. The intersection O, regarded as on CD, will be the correspondent of some point X on AB. The restraining screw corresponding to X will therefore lie at P, and will be provided by the constraints. Accordingly, X is a permanent screw on the cylindroid, and it is obvious from the construction that there can be no other screw of the same character.

We can also deduce the expression for T in the two-system from the expression of the more general type in the three-system; for we have

$$T = T_0 + (\mu - \nu)\,\theta_1'\dot{\theta}_2\dot{\theta}_3 + (\nu - \lambda)\,\theta_2'\dot{\theta}_3\dot{\theta}_1 + (\lambda - \mu)\,\theta_3'\dot{\theta}_1\dot{\theta}_2.$$

Consider any screw on the cylindroid defined by

$$\theta_1' = P\theta_2' + Q\theta_3',$$
$$\dot{\theta}_1 = P\dot{\theta}_2 + Q\dot{\theta}_3;$$

substituting, we obtain

$$(\theta_3'\dot{\theta}_2 - \theta_2'\dot{\theta}_3)\,[(\lambda - \mu)\,P\dot{\theta}_2 + (\lambda - \nu)\,Q\dot{\theta}_3],$$

which we already know to be the form of the function in the case of the two-system (§ 384).

393. Freedom of the Fourth Order.

The permanent screws in the case of a rigid body which has freedom of the fourth order may be investigated in the following manner:—If a screw θ be permanent, the corresponding restraining screw η must be provided by the reactions of the constraints. All the reactions in a case of freedom of the fourth order lie on the screws of a cylindroid. On a given cylindroid three possible η screws can be found. For, if we substitute $\alpha_1 + \lambda\beta_1$, $\alpha_2 + \lambda\beta_2$, &c., for η_1, η_2, &c., in the equation

$$\frac{b^2 - c^2}{a}\,\frac{\eta_1 + \eta_2}{\eta_1 - \eta_2} + \frac{c^2 - a^2}{b}\,\frac{\eta_3 + \eta_4}{\eta_3 - \eta_4} + \frac{a^2 - b^2}{c}\,\frac{\eta_5 + \eta_6}{\eta_5 - \eta_6} = 0,$$

we obtain a cubic for λ. The three roots of this cubic correspond to three η screws. Take the θ corresponding to one of the η screws, then, of course, θ will not, in general, belong to the four-system. We can, however, assign to θ any pitch we like, and as it intersects η at right angles, it must cut two other screws of equal pitch on the cylindroid (§ 22). Give to θ a pitch equal and opposite to that of the two latter screws, then θ is reciprocal to the cylindroid, and therefore it belongs to the four-system. We thus have a permanent screw of the system, and accordingly we obtain the following result:—

In the case of a rigid body with freedom of the fourth order there are, in general, three, and only three, permanent screws.

394. Freedom of the Fifth and Sixth Orders.

When a rigid body has freedom of the fifth order, the screws about which the body can be twisted are all reciprocal to a single screw ρ. In general, ρ does not lie on the system prescribed by the equation which the co-ordinates of all possible η screws have to satisfy. It is therefore, in general, not possible that the reaction of the constraints can provide an η. There are, however, three screws in any five-system which possess the property of permanent screws without however making any demand on the reaction of the constraints. The existence of these screws is thus demonstrated:— Through the centre of inertia of the body draw the three principal axes, then, on each of these axes one screw can always be found which is reciprocal to ρ. Each of these will belong to the five-system, and it is obvious from the property of the principal axes, that if the body be set twisting about one of these screws it will have no tendency to depart therefrom.

A body which has freedom of the sixth order is perfectly free. Any screw on one of the principal axes through the centre of inertia is a permanent screw, and, consequently, there is in this case a triply infinite number of permanent screws.

395. Summary.

The results obtained show that for a rigid body with the several degrees of freedom the permanent screws are as follows:—

	No. of Permanent Screws
Freedom I.	1
„ II.	1
„ III.	3
„ IV.	3
„ V.	3
„ VI.	Triply infinite

CHAPTER XXVI.

AN INTRODUCTION TO THE THEORY OF SCREWS IN NON-EUCLIDIAN SPACE.

396. Introduction.

The Theory of Screws in non-Euclidian space is a natural growth from some remarkable researches of Clifford* in further development of the Theory of Riemann, Cayley, Klein and Lindemann. I here give the investigation sufficiently far to demonstrate two fundamental principles (§§ 427, 434) which conduct the theory to a definite stage at which it seems convenient to bring this volume to a conclusion.

I have thought it better to develop from the beginning the non-Euclidian geometry so far as we shall at present require it. It is thus hoped to make it intelligible to readers who have had no previous acquaintance with this subject†. I give it as I have worked it out for my own instruction‡. It is indeed characteristic of this fascinating theory that it may be surveyed from many different points of view.

397. Preliminary notions.

Let x_1, x_2, x_3, x_4 be four numerical magnitudes of any description. We may regard these as the co-ordinates of an *object*. Let y_1, y_2, y_3, y_4 be the co-ordinates of another object, then we *premise* that the two objects will be identical if, and only if

$$\frac{x_1}{y_1}=\frac{x_2}{y_2}=\frac{x_3}{y_3}=\frac{x_4}{y_4}.$$

All possible objects may be regarded as constituting a *content*.

* "Preliminary Sketch of Biquaternions," *Proceedings of the London Mathematical Society*, Vol. IV. 381—395 (1873). See also "On the Theory of Screws in a Space of Constant Positive Curvature," *Mathematical Papers*, p. 402 (1876). Clifford's Theory was much extended by the labours of Buchheim and others; see the Bibliographical notes.

† We are fortunately now able to refer English readers to a Treatise in which the Theory of non-Euclidian space and allied subjects is presented in a comprehensive manner. Whitehead, *Universal Algebra*, Cambridge, 1898.

‡ *Trans. Roy. Irish Acad.*, Vol. XXVIII. p. 159 (1881), and Vol. XXIX. p. 123 (1887).

All objects whose co-ordinates satisfy one linear homogeneous equation we shall speak of as an *extent*.

All objects whose co-ordinates satisfy two linear homogeneous equations we shall speak of as a *range*.

It must be noticed that the content, with its objects, ranges, and extents, have no necessary connexion with *space*. It is only for the sake of studying the content with facility that we correlate its several objects with the points of space.

398. The Intervene.

In ordinary space the most important function of the co-ordinates of a pair of points is that which expresses their distance apart. We desire to create that function of a pair of objects which shall be homologous with the distance function of a pair of points in ordinary space.

The nature of this function is to be determined solely by the attributes which we desire it to possess. We shall take the most fundamental properties of distance in ordinary space. We shall then re-enunciate these properties in generalized language, and show how they suffice to determine a particular function of a pair of objects. This we shall call the *Intervene between the Two Objects*.

Let P, Q, R be three collinear points in ordinary space, Q lying between the other two; then we have, of course, as a primary notion of distance,

$$PQ + QR = PR.$$

In general, the distance between two points is not zero, unless the points are coincident. An exception arises when the straight line joining the points passes through either of the two circular points at infinity. In this case, however, the distance between every pair of points on the straight line is zero. These statements involve the second property of distance.

In ordinary geometry we find on every straight line one point which is at an infinite distance from every other point on the line. We call this the point at infinity. Sound geometry teaches us that this single point is properly to be regarded as a pair of points brought into coincidence by the assumptions made in Euclid's doctrine of parallelism. The existence of a pair of infinite points on a straight line is the third property which, by suitable generalization, will determine an important feature in the range. The fourth property of ordinary space is that which asserts that a point at infinity on a straight line is also at infinity on every other straight line passing through it. This obvious property is equivalent to a significant law of intervene which is vital in the theory. If we might venture to enunciate it in an epigrammatic fashion, we would say that there is no short cut to infinity.

The fifth property of common space which we desire to generalize is one which is especially obscured by the conventional coincidence of the two points at infinity on every straight line. We prefer, therefore, to adduce the analogous, but more perfect, theorem relative to two plane pencils of homographic rays in ordinary space, which is thus stated. If the two rays to the circular points at infinity in one pencil have as their correspondents the two rays to the circular points in the other pencil, then it is easily shown that the angle between any two rays equals that between their two correspondents.

We now write the five correlative properties which the intervene is to possess. They may be regarded as the axioms in the Theory of the Content. Other axioms will be added subsequently.

399. First Group of Axioms of the Content.

(I) If three objects, P, Q, R on a range be ordered in ascending parameter (§ 400), then the intervenes PQ, QR, PR are to be so determined that

$$PQ + QR = PR.$$

(II) The intervene between two objects cannot be zero unless the objects are coincident, or unless the intervene between every pair of objects on the same range is also zero.

(III) Of the objects on a range, two either distinct or coincident are at infinity, *i.e.* have each an infinite intervene with all the remainder.

(IV) An infinite object on any range has an infinite intervene from every object of the content.

(V) If the several objects on one range correspond one-to-one with the several objects on another, and if the two objects at infinity on one range have as their correspondents the two objects at infinity on the other. then the intervene between any two objects on the one range is equal to that between their correspondents on the other.

400. Determination of the Function expressing the Intervene between Two Objects on a Given Range.

Let x_1, x_2, x_3, x_4, and y_1, y_2, y_3, y_4 be the co-ordinates of the objects by which the range is determined. Then each remaining object is constituted by giving an appropriate value to ρ in the system,

$$x_1 + \rho y_1, \quad x_2 + \rho y_2, \quad x_3 + \rho y_3, \quad x_4 + \rho y_4.$$

Let λ and μ be the two values of ρ which produce the pair of objects of which the intervene is required. It is plain that the intervene, whatever it be, must be a function of x_1, x_2, x_3, x_4 and y_1, y_2, y_3, y_4, and also of λ and μ. So far as objects on the same range are concerned, we may treat the co-

28—2

ordinates of the originating objects as constant, and regard the intervene simply as a function of λ and μ, which we shall denote by $f(\lambda, \mu)$.

The form of this function will be gradually evolved, as we endow it with the attributes we desire it to possess. The first step will be to take a third object on the same range for which the parameter shall be ν, where λ, μ, ν are arranged in order of magnitude. Then, as we wish the intervene to possess the property specified in Axiom I., we have

$$f(\lambda, \mu) + f(\mu, \nu) = f(\lambda, \nu).$$

By the absence of μ from the right-hand side, we conclude that μ must disappear identically from the left-hand side. This must be the case whatever λ and ν may be. Hence, no term in which μ enters can have λ as a factor. It follows that $f(\lambda, \mu)$ must be simply the difference of two parts, one being a function of λ, and the other the same function of μ.

Accordingly, we write,

$$f(\lambda, \mu) = \phi(\lambda) - \phi(\mu).$$

The first step in the determination of the intervene function has thus been taken. But the form of ϕ is still quite arbitrary.

The rank of the objects in a range may be concisely defined by the magnitudes of their corresponding values of ρ. *Three objects are said to be ordered when the corresponding values of ρ are arranged in ascending or descending magnitude.*

Let P, Q, Q' be three ordered objects, then it is generally impossible that the intervenes PQ and PQ' shall be equal; for, suppose them to be so, then

$$PQ + QQ' = PQ' \text{ by Axiom I.};$$

but, by hypothesis, $$PQ = PQ',$$

and hence $$QQ' = 0.$$

But, from Axiom II., it follows that (Q and Q' being different) this cannot be true, unless in the very peculiar case in which the intervene between every pair of objects on the range is zero. Omitting this exception, to which we shall subsequently return, we see that PQ and PQ' cannot be equal so long as Q and Q' are distinct.

We hence draw the important conclusion that there is for each object P but a single object Q, which is at a stated intervene therefrom.

Fixing our attention on some definite value δ (*what* value it does not matter) of the intervene, we can, from each object λ, have an ordered equi-intervene object μ determined. Each λ will define one μ. Each μ will correspond to one λ. The values of λ with the correlated values of μ form two homographic systems. The relation between λ and μ depends, of course, upon the specific value of δ, but must be such that, when one of the quantities is given, the other shall be determined by a linear equation. It is therefore assumed that λ and μ must be related by the equation

$$A\lambda\mu + B\lambda + C\mu + D = 0,$$

where the ratios of the coefficients A, B, C, D shall depend, to some extent, upon δ. If λ' and μ' be a given pair of parameters belonging to objects at the required intervene, then

$$A\lambda'\mu' + B\lambda' + C\mu' + D = 0,$$

by which the disposable coefficients in the homographic equation are reduced to two.

The converse of Axiom II., though generally true, is not universally so. It will, of course, generally happen that when two objects coincide their intervene is zero. But on every range two objects can be found, each of which is truly to be regarded as two coincident objects of which the intervene is not zero.

Let us, for instance, make $\lambda = \mu$ in the above equation; then we have

$$A\lambda^2 + (B + C)\lambda + D = 0.$$

This equation has, of course, two roots, each of which points out an object of critical significance on the range. We shall denote these objects by O and O'. Each of them consists of a pair of objects which, though actually coincident, have the intervene δ. The fundamental property of O and O' is thus demonstrated.

Let X be any object on the range; then (Axiom I.)

$$XO + \delta = XO;$$

and as δ is not zero, we have

$$XO = \text{infinity}.$$

Therefore every object on the range is at an infinite intervene from O. A similar remark may be made with respect to O'; and hence we learn that *the two objects, O and O', are at infinity.*

We assume, in Axiom III., that there are not to be more than two objects on the range at infinity: these are, of course, O and O'. We must, therefore, be conducted to the same two objects at infinity, whatever be the value of the intervene δ, from which we started.* We thus see that while the original coefficients A, B, C, D do undoubtedly contain δ, yet that δ does not affect the equation

$$A\lambda^2 + (B + C)\lambda + D = 0.$$

* My attention was kindly directed to this point in a letter from Mr F. J. M'Aulay.

It follows that $D \div A$ and $(B + C) \div A$ must both be independent of δ. We may therefore make

$$A = A'; \quad B = B' + \Delta; \quad C = C' - \Delta; \quad D = D';$$

and thus the homographic equation becomes

$$A'\lambda\mu + (B' + \Delta)\lambda + (C' - \Delta)\mu + D' = 0,$$

where Δ is the only quantity which involves δ.

The equation can receive a much simpler form by taking the infinite objects as the two originating objects from which the range was determined. In this case the equation

$$A\lambda^2 + (B + C)\lambda + D = 0$$

must have as roots $\lambda = 0$ and $\lambda = \infty$, and therefore

$$A = 0; \quad D = 0,$$

hence the homographic equation reduces to

$$B\lambda + C\mu = 0;$$

since $B \div C$ is a function of the intervene δ, we may say, conversely, that

$$\delta = F\left(\frac{\lambda}{\mu}\right).$$

We have, however, already learned that the intervene is to have the form

$$\phi(\lambda) - \phi(\mu).$$

Now we find that it can also be expressed, with perfect generality, in the form

$$F\left(\frac{\lambda}{\mu}\right).$$

It follows that these two expressions must be equal, so that

$$\phi(\lambda) - \phi(\mu) = F\left(\frac{\lambda}{\mu}\right).$$

In this equation the particular value of δ does not appear, nor is it even implied. The formula must therefore represent an identical result true for all values of λ and all values of μ.

We may, therefore, differentiate the formula with respect both to λ and to μ, and thus we obtain

$$\phi'(\lambda) = \frac{1}{\mu} F'\left(\frac{\lambda}{\mu}\right)$$

$$-\phi'(\mu) = -\frac{\lambda}{\mu^2} F'\left(\frac{\lambda}{\mu}\right),$$

whence we deduce

$$\lambda\phi'(\lambda) = \mu\phi'(\mu).$$

As λ and μ are perfectly independent, this equation can only subsist by assuming for ϕ a form such that

$$\lambda\phi'(\lambda) = H,$$

where H is independent of λ. Whence we obtain

$$\phi'(\lambda) = \frac{H}{\lambda},$$

and, by integration,

$$\phi(\lambda) = H \log \lambda + \text{constant}.$$

The intervene is now readily determined, for

$$\phi(\lambda) - \phi(\mu) = H \log \lambda - H \log \mu = H \log \frac{\lambda}{\mu}.$$

We therefore obtain the following important theorem which is the well-known* basis of the mensuration of non-Euclidian space:—

Let x_1, x_2, x_3, x_4, and y_1, y_2, y_3, y_4 be the two objects at infinity on a range, and let $x_1 + \lambda y_1$, $x_2 + \lambda y_2$, $x_3 + \lambda y_3$, $x_4 + \lambda y_4$, and $x_1 + \mu y_1$, $x_2 + \mu y_2$, $x_3 + \mu y_3$, $x_4 + \mu y_4$ be any two other objects on the range, then their intervene will be expressed by

$$H (\log \lambda - \log \mu),$$

where H is a constant depending upon the adopted units of measurement.

It will be useful to obtain the expression for the intervene in a rather more general manner by taking the equation in λ and μ, for objects at the intervene δ, as

$$A\lambda\mu + B\lambda + C\mu + D = 0.$$

Let λ' and λ'' be the two roots of this equation when μ is made equal to λ. It follows that what we have just written may be expressed thus:—

$$\lambda\mu + \lambda(\theta - \tfrac{1}{2}\lambda' - \tfrac{1}{2}\lambda'') + \mu(-\theta - \tfrac{1}{2}\lambda' - \tfrac{1}{2}\lambda'') + \lambda'\lambda'' = 0.$$

For, if $\lambda = \mu$, this is satisfied by either λ' or λ'', while θ disappears. θ is, of course, a function of the intervene, and it is only through θ that the intervene comes into the equation. By solving for θ, we find

$$\theta = \frac{\lambda\mu - \tfrac{1}{2}(\lambda' + \lambda'')(\lambda + \mu) + \lambda'\lambda''}{\mu - \lambda}.$$

* Professor George Bruce Halsted remarks in *Science*, N. S., Vol. x., No. 251, pages 545—557, October 20, 1899, that "Roberto Bonola has just given in the *Bolletino di Bibliografia e Storia della Scienze Matematiche* (1899) an exceedingly rich and valuable Bibliografia sui Fondamenti, della Geometria in relazione alla Geometria non-Euclidea in which he gives 353 titles."

The intervene itself is $F(\theta)$, where F expresses some function; and, accordingly,

$$\phi(\lambda) - \phi(\mu) = F(\theta).$$

When we substitute in this expression the value of θ given above, we have an identity which is quite independent of the particular δ. We must, therefore, determine the functions so that this equation shall remain true for all values of λ, and all values of μ. The formulæ must therefore be true when differentiated—

$$\frac{d\theta}{d\lambda} = \frac{(\mu - \lambda')(\mu - \lambda'')}{(\mu - \lambda)^2}, \quad \frac{d\theta}{d\mu} = \frac{(\lambda - \lambda')(\lambda'' - \lambda)}{(\mu - \lambda)^2},$$

$$\phi'(\lambda) = \frac{dF}{d\theta}\frac{d\theta}{d\lambda}, \quad -\phi'(\mu) = \frac{dF}{d\theta}\frac{d\theta}{d\mu};$$

whence,

$$\frac{\phi'(\lambda)}{\phi'(\mu)} = \frac{(\mu - \lambda')(\mu - \lambda'')}{(\lambda - \lambda')(\lambda - \lambda'')},$$

or

$$(\lambda - \lambda')(\lambda - \lambda'')\,\phi'(\lambda) = (\mu - \lambda')(\mu - \lambda'')\,\phi'(\mu),$$

which has the form

$$\psi(\lambda) = \psi(\mu).$$

Considering the complete independence of both λ and μ, this equation requires that each of its members be independent alike of λ and μ. We shall denote them by $H(\lambda' - \lambda'')$ where H is a constant, whence

$$(\lambda - \lambda')(\lambda - \lambda'')\,\phi'(\lambda) = H(\lambda' - \lambda''),$$

or,

$$\phi'(\lambda) = \frac{H(\lambda' - \lambda'')}{(\lambda - \lambda')(\lambda - \lambda'')}$$

$$= H\left(\frac{1}{\lambda - \lambda'} - \frac{1}{\lambda - \lambda''}\right);$$

whence, integrating and denoting the arbitrary constant by C,

$$\phi(\lambda) = H[\log(\lambda - \lambda') - \log(\lambda - \lambda'')] + C;$$

similarly,

$$\phi(\mu) = H[\log(\mu - \lambda') - \log(\mu - \lambda'')] + C;$$

and, finally, we have for the intervene, or $\phi(\lambda) - \phi(\mu)$, the expression,

$$\delta = H\log\frac{(\lambda - \lambda')(\mu - \lambda'')}{(\lambda - \lambda'')(\mu - \lambda')}.$$

This expression discloses the intervene as the logarithm of a certain anharmonic ratio.

We may here note how a difficulty must be removed which is very likely to occur to one who is approaching the non-Euclidian geometry for the first time. No doubt we find the intervene to be the logarithm of an anharmonic ratio of four quantities, but these quantities are not distances nor are they quantities homologous with the intervene. They are simply numerical. The four numbers, λ, μ, λ', λ'' are merely introduced to define four objects, one of them being,

$$x_1 + \lambda y, \quad x_2 + \lambda y_2, \quad x_3 + \lambda y_3, \quad x_4 + \lambda y_4,$$

and the others are obtained by replacing λ by μ, λ', λ'', respectively. All we assert is, that if we choose to call the two objects defined by λ' and λ'' the objects at infinity, and that if we desire the intervene between the objects λ and μ to possess the properties that we have already specified, then the only function possible will be the logarithm of the anharmonic ratio of these four numbers.

The word anharmonic is ordinarily applied in describing a certain function of four collinear points. In the more general sense, in which we are at this moment using the word, it does not relate to any geometrical or spacial relation whatever; it is a purely arithmetical function of four abstract numbers.

We may also observe that the relation between θ and the intervene δ is given by the equation

$$\theta = \tfrac{1}{2}(\lambda'' - \lambda')\frac{e^{\frac{\delta}{H}} + 1}{e^{\frac{\delta}{H}} - 1},$$

and the expression of the intervene as a function of θ; that is, the expression $F(\theta)$ is

$$\delta = H \log \frac{\theta + \frac{1}{2}(\lambda'' - \lambda')}{\theta - \frac{1}{2}(\lambda'' - \lambda')}.$$

401. Another process.

We may also proceed in the following manner. Let us denote the values of λ for the infinite objects on the range by $\rho e^{i\theta}$ and $\rho e^{-i\theta}$.

If then λ, μ be two parameters for two objects at an intervene δ, we must have (p. 439)

$$\lambda\mu + \lambda(\epsilon - \rho\cos\theta) + \mu(-\epsilon - \rho\cos\theta) + \rho^2 = 0.$$

Solving for ϵ, we have

$$\epsilon = \frac{\lambda\mu - \rho\cos\theta(\lambda + \mu) + \rho^2}{\mu - \lambda}.$$

The intervene δ must be some function of ϵ, whence

$$\delta = \phi(\lambda) - \phi(\mu) = F(\epsilon),$$

whence

$$\phi'(\lambda) = \frac{dF}{d\epsilon}\frac{d\epsilon}{d\lambda} = \frac{dF}{d\epsilon}\frac{\mu^2 - 2\mu\cos\theta + \rho^2}{(\mu-\lambda)^2},$$

$$\phi'(\mu) = -\frac{dF}{d\epsilon}\frac{d\epsilon}{d\mu} = \frac{dF}{d\epsilon}\frac{\lambda^2 - 2\lambda\cos\theta + \rho^2}{(\mu-\lambda)^2},$$

$$(\lambda^2 - 2\lambda\cos\theta + \rho^2)\,\phi'(\lambda) = (\mu^2 - 2\mu\cos\theta + \rho^2)\,\phi'(\mu) = \rho\sin\theta \text{ suppose.}$$

Hence we have

$$\phi(\lambda) = \tan^{-1}\left(\frac{\rho\sin\theta}{\rho\cos\theta - \lambda}\right),$$

and thus we get for the intervene with a suitable unit

$$\delta = \tan^{-1}\left(\frac{\rho\sin\theta}{\rho\cos\theta - \lambda}\right) - \tan^{-1}\left(\frac{\rho\sin\theta}{\rho\cos\theta - \mu}\right).$$

402. On the Infinite Objects in an Extent.

On each range of the extent there will be two objects at infinity, by the aid of which the intervene between every two other objects on that range is to be ascertained. We are now to study the distribution of these infinite objects over the extent. Taking any range and one of its infinite objects, O, construct any other range in the same extent containing O as an object. This second range will also have two infinite objects. Is O to be one of them? Here we add another attribute to our, as yet, immature conception of the intervene.

In Euclidian space we cannot arrive at infinity except we take an infinitely long journey. This is because the point at infinity on one straight line is also the point at infinity on any other straight line passing through it. Were this not the case, then a finite journey to infinity could be taken by travelling along the two sides of a triangle in preference to the direct route *viâ* the third side. To develop the analogy between the conception of intervene and that of Euclidian distance, we therefore assume (in Axiom IV.) that an infinite object has an infinite intervene with every other object of the content. In consequence of this we have the general result, that

If O be an infinite object on one range, then it is an infinite object on every one of the ranges diverging from O.

The necessity for this assumption is made clear by the following consideration:—Suppose that O were an infinite object on one range containing the object A, but were not an infinite object on another range OB, diverging also from O; then, although the direct intervene OA is infinite, yet the intervenes from A to B and from B to O would be both finite. The only escape is by the assumption we have just italicised. Otherwise infinity could be reached by two journeys, each of finite intervene.

Take any infinite object O. Construct a series of ranges in the extent, each containing O. Each of these ranges will have another infinite object, O_1, O_2, O_3, &c. The values x_1, x_2, x_3, which define O_1, O_2, O_3, &c., must fulfil some general condition, which we may express thus:

$$f(x_1, x_2, x_3) = 0.$$

Form a range through O_1 and O_2. There must be two infinite objects on this range, and of course all other objects thereon will be defined by a linear equation $L=0$ in x_1, x_2, x_3.

Every object satisfying the condition $f(x_1, x_2, x_3) = 0$ is infinite, and therefore all the values of x_1, x_2, x_3 common to the two equations $L=0$ and $f(x_1, x_2, x_3) = 0$ must denote infinite objects. But we have already seen that there are only two infinite objects on one range; therefore there can be only two systems of values common to the two equations. In other words, $f(x_1, x_2, x_3)$ must be an algebraical function of the second degree. There can be no infinite object except those so conditioned; for, suppose that S were one, then any range through S would have two objects in common with f, and thus there would be three infinite objects on one range, which is contrary to Axiom III.

Hence we deduce the following important result:—

All the infinite objects in an extent lie on a range of the second degree.

We thus see that every range in an extent will have two objects in common with the infinite range of the second order. These are, of course, the two infinite objects on the range.

403. On the Periodic Term in the Complete Expression of the Intervene.

We have found for the intervene the general expression

$$H(\log \lambda - \log \mu).$$

We may, however, write instead of λ,

$$(\cos 2n\pi + i \sin 2n\pi)\lambda,$$

where n is any integer; but this equals

$$e^{2in\pi}\lambda;$$

hence,

$$\log \lambda = 2in\pi + \log \lambda;$$

and, consequently, the intervene is indeterminate to the extent of any number of integral multiples of

$$2iH\pi.$$

The expression just written is the intervene between any object and the same object, if we proceed round the entire circumference of the range. We may call it, in brief, the *circuit* of the range.

The intervene between the objects λ and $-\lambda$ is

$$iH\pi.$$

Nor is this inconsistent with the fact that $\lambda = $ zero denotes two coincident objects, as does also $\lambda = \pm$ infinity. In each of these cases the coincident objects are at infinity, and the intervene between two objects which coalesce into one of the objects at infinity has an indeterminate value, and may thus, of course, be $iH\pi$, as well as anything else.

404. Intervenes on Different Ranges in a Content.

Let us suppose any two ranges whatever. There are an infinite number of objects on one range, and an infinite number on the other. The well-known analogies of homographic systems on rays in space lead us to inquire whether the several objects on the two ranges can be correlated homographically. Each object in either system is to correspond definitely with a single object in the other system.

We determine an object on a range by its appropriate λ. Let the corresponding object on the other range be defined by λ'. The necessary conditions of homography demand that for each λ there shall be one λ', and *vice versâ*. Compliance with this is assured when λ and λ' are related by an equation of the form

$$P\lambda\lambda' + Q\lambda + R\lambda' + S = 0.$$

Let $\lambda_1, \lambda_2, \lambda_3, \lambda_4$ be any four values of λ, and let $\lambda_1', \lambda_2', \lambda_3', \lambda_4'$ be the corresponding four values of λ', then, by substitution in the equation just written, and elimination of P, Q, R, S, it follows that

$$\frac{\lambda_1 - \lambda_3}{\lambda_2 - \lambda_3} \times \frac{\lambda_2 - \lambda_4}{\lambda_1 - \lambda_4} = \frac{\lambda_1' - \lambda_3'}{\lambda_2' - \lambda_3'} \times \frac{\lambda_2' - \lambda_4'}{\lambda_1' - \lambda_4'}.$$

We now introduce the following important definition:—

By the expression, anharmonic ratio of four objects on a range, is meant the anharmonic ratio of the four values of the numerical parameter by which the objects are indicated.

We are thus enabled to enunciate the following theorem:—

When the objects on two ranges are ordered homographically, the anharmonic ratio of any four objects on one range equals the anharmonic ratio of their four correspondents on the other.

Three pairs of correspondents can be chosen arbitrarily, and then the equation last given will indicate the relation between every other λ and its corresponding λ'.

Among the different homographic systems there is one of special importance. It is that in which the intervene between any two objects in one range equals that between their correspondents in the other. But this homography is only possible when a critical condition is fulfilled.

In the first place, an infinite object on one range must have, as its correspondent, an infinite object on the other. For if X be an infinite object on one range, it has an infinite intervene with every other object on that range (Axiom IV.); therefore X', the correspondent of X, must have an infinite intervene with every other object on the second range.

If, then, X and Y are the infinite objects on one range, and X' and Y' the infinite objects on the other, and if A and B be two arbitrary objects on the first range, and A' and B' their correspondents on the other; then, using the accustomed notation for anharmonic ratio,

$$(ABXY) = (A'B'X'Y').$$

But, if H' be the factor (p. 440) for the second range, which H is for the first, we have, since the intervenes are equal,

$$H \log (ABXY) = H' \log (A'B'X'Y');$$

and, since the anharmonic ratios are equal, we obtain

$$H = H'.$$

If, then, it be possible to order two homographic systems of objects, so that the intervene between any two is equal to that between their correspondents, we must have H and H' equal; and conversely, when H and H' are equal, then equi-intervene homography is possible.

We have therefore assumed Axiom V. (§ 399) which we have now seen to be equivalent to the assumption that the metric constant H is to be the same for every range of the content.

Nor is there anything in Axiom V. which constitutes it a merely gratuitous or fantastic assumption. Its propriety will be admitted when we reduce our generalized conceptions to Euclidian space. It is an obvious notion that any two straight lines in space can have their several points so correlated that the distance between a pair on one line is the same as that between their correspondents on the other. In fact, this merely amounts to the statement that a straight line marked in any way can be conveyed, marks and all, into a different situation, or that a foot-rule will not change the length of its inches *because* it is carried about in its owner's pocket.

In a similar, but more general manner, we desire to have it possible, on any two ranges, to mark out systems of corresponding objects, such that the intervene between each pair of objects shall be equal to that between their correspondents. We have shown in this Article that such an arrangement is possible, when, and only when, the property V. is postulated. We may speak of such a pair of ranges as *equally graduated*.

The conception of *rigidity* involves the notion that it shall be possible to displace a system of points such that the distance between every pair of points in their original position equals that between the same pair after the displacement. We desire to have corresponding notions in the present Theory, which will only be possible when we have taken such a special view of the nature of the intervene as is implied in Axiom v.

405. Another Investigation of the possibility of equally Graduated Ranges.

The importance of the subject in the last Article is so great in the present Theory that I here give it from a different point of view.

Taking the infinite objects on a range as the originating objects, we have on the first range for the intervene between the objects λ and α,

$$H(\log\lambda - \log\alpha);$$

and for the second range for the intervene between μ and β, we have

$$H'(\log\mu - \log\beta).$$

Regarding α and β as fixed, and λ and μ as defining a pair of correlative objects, we get, as the relation between λ and μ for equally graduated ranges,

$$H\log\frac{\lambda}{\alpha} = H'\log\frac{\mu}{\beta};$$

whence the relation between λ and μ is thus given:

$$\frac{\lambda}{\alpha} = \left(\frac{\mu}{\beta}\right)^{\frac{H'}{H}}.$$

When this is the case, there are several values of λ corresponding to one value of μ, which may be thus found :—Let m be any integer; then, in the usual manner,

$$\frac{\lambda}{\alpha} = \left(\frac{\mu}{\beta}\right)^{\frac{H'}{H}}\left(\cos 2m\frac{H'}{H}\pi + i\sin 2m\frac{H'}{H}\pi\right);$$

and therefore

$$\frac{\lambda}{\alpha} = \left(\frac{\mu}{\beta}\right)^{\frac{H'}{H}},$$

or

$$= \left(\frac{\mu}{\beta}\right)^{\frac{H'}{H}}\left(\cos\frac{2H'}{H}\pi + i\sin\frac{2H'}{H}\pi\right),$$

or

$$=\left(\frac{\mu}{\beta}\right)^{\frac{H'}{H}}\left(\cos\frac{4H'}{H}\pi+i\sin\frac{4H'}{H}\pi\right),$$

&c. &c. &c.

We see that the correspondence between λ and μ cannot be of the homographic character, unless

$$H=H'.$$

The necessity for this condition may be otherwise demonstrated by considering the subject in the following manner:—

The intervene between any two objects on one range is, of course, ambiguous, to the extent of any integral number of the circuits on that range. Let C and C' be the circuits, and let δ be an intervene between two objects on the second range. If we try to determine two objects, α and λ, on the first range that shall have an intervene δ, we must also have another object λ', such that its intervene from α is $\delta+C'$. Similarly, there must be another object λ'' with the intervene $\delta+2C'$, &c. It is therefore impossible to have a single object at the intervene $\delta+mC'$ from α, unless it happened that

$$C=C',$$

or that

$$H=H'.$$

Thus, again, are we led to the conclusion that ranges cannot be equally graduated unless their circuits are the same.

The circuits on every range in the content being now taken to be equal, we can assume for the circuit any value we please. There are great advantages in so choosing our units that the circuit shall be π; but we have as its expression,

$$2iH\pi;$$

whence we deduce

$$2iH=1.$$

406. On the Infinite Objects in the Content.

Certain objects in the content are infinite, and it is proposed to determine the conditions imposed on x_1, x_2, x_3, x_4 when they indicate one of these. If an object O be infinite, then every range through that object will have one other infinite object. Let these be O_1, O_2, &c. These several objects will conform with the condition,

$$f(x_1,\ x_2,\ x_3,\ x_4)=0.$$

Every infinite object in the content must satisfy this equation; and, conversely, every object so circumstanced is infinite.

Two linear equations in x_1, x_2, x_3, x_4 determine a range, and the simultaneous solution of these equations with

$$f(x_1, x_2, x_3, x_4) = 0$$

gives the infinite objects on that range—but there can only be two—and hence we have the following important theorem:—

The co-ordinates x_1, x_2, x_3, x_4 of the infinite objects in a content satisfy an homogeneous equation of the second degree.

We denote this equation by

$$U = 0.$$

407. The Departure.

Let the ranges $x_1, x_2, 0, 0$ be formed when the parameter $x_1 \div x_2$ has every possible value, then the entire group of ranges produced in this way is called a *star*. In ordinary geometry the most important function of a pair of rays in a pencil is that which expresses their inclination. We have now to create, for our generalized conceptions, a function of two ranges in a star which shall be homologous with the notion of ordinary angular magnitude.

We shall call this function the *Departure*. Its form is to be determined by the properties that we wish it to possess. In the investigation of the departure between two ranges, we shall follow steps parallel to those which determined the intervene between two objects.

If OP, OQ, OR be three rays in an ordinary plane diverging from O, then

$$\angle POQ + \angle QOR = \angle POR.$$

In general the angle between two rays is not zero unless the rays are coincident; but this statement ceases to be true when the vertex of the pencil is at infinity. In this case, however, the angle between every pair of rays in the pencil is zero.

Every plane pencil has two rays (*i.e.* those to the circular points at infinity), which make an infinite angle with every other ray.

408. Second Group of Axioms of the Content.

We desire to construct a departure function which shall possess the following properties:—

(VI) If three ranges, P, Q, R, in a star, be ordered in ascending parameter, and if the departure between two ranges, for example, P and Q, be expressed by PQ, then

$$PQ + QR = PR.$$

(VII) The departure between two ranges cannot be zero unless the ranges are coincident, or unless the departure between every pair of ranges in that star is also zero.

(VIII) Of the ranges in a star, two (distinct or coincident) are at infinity, *i.e.* have each an infinite departure from all the remainder.

(IX) An infinite range has an infinite departure, not only with every range in its star, but with *every* range in the extent.

(X) If the several ranges in one star correspond one to one with the several ranges in another, and if the two infinite ranges in one star have as their correspondents the infinite ranges in the other; then the departure between any two ranges in one star is equal to that between the two corresponding ranges in the other.

409. The Form of the Departure Function.

The analogy of these several axioms to those which have guided us to the discovery of the intervene, shows that the investigation for the function of departure will be conducted precisely as that of the intervene has been; accordingly, we need not repeat the several steps of the investigation, but enunciate the general result, as follows:—

Let x_1, x_2, and y_1, y_2 be the co-ordinates of any two ranges in a star, and let λ_1, λ_2, and μ_1, μ_2 be the co-ordinates of the two infinite ranges in that star. Then the departure between (x_1, x_2) and (y_1, y_2) is

$$H \log \frac{(x_1\lambda_2 - x_2\lambda_1)(y_1\mu_2 - y_2\mu_1)}{(y_1\lambda_2 - y_2\lambda_1)(x_1\mu_2 - x_2\mu_1)}.$$

410. On the Arrangement of the Infinite Ranges.

Every star in the extent will have two infinite ranges, and we have now to see how these several infinite ranges in the extent can be compendiously organized into a whole.

To aid in this we have assumed Axiom IX., the effect of which is to render the following statement true. Let several objects on a range, O, be the vertices of a corresponding number of stars. If O be an infinite range in *any* one of the stars, then it is so in *every* one.

Let a_1, a_2, a_3 be any three ranges in an extent. Then every range in the same extent can be expressed by

$$x_1a_1 + x_2a_2 + x_3a_3,$$

where x_1, x_2, x_3 are the three co-ordinates of the range. It is required to determine the relation between x_1, x_2, x_3 if this range be infinite.

An object L will be defined by an equation of the form, where L_1, L_2, L_3 are numerical,

$$L_1x_1 + L_2x_2 + L_3x_3 = 0,$$

for any two sets of values, x_1, x_2, x_3, which satisfy this equation, will determine a pair of ranges which have the required object in common.

Let the relation between the co-ordinates of an infinite range be

$$f(x_1, x_2, x_3) = 0;$$

then the infinite ranges in the star, whose vertex is the object L, will be defined by co-ordinates obtained from the simultaneous solution of

$$L_1x_1 + L_2x_2 + L_3x_3 = 0,$$

$$f(x_1, x_2, x_3) = 0.$$

But there can only be two such ranges; and, accordingly, the latter of these equations must be of the second degree. We hence deduce the following important result:—

The infinite ranges in an extent may be represented by the different groups of values of the co-ordinates x_1, x_2, x_3 *which satisfy one homogeneous equation of the second degree.*

Remembering that the existence of zero intervene between every pair of objects on a range is a consequence of the coincidence of the two objects of infinite intervene on that range, we have the following result:—

The range through two consecutive objects of infinite intervene is a range of zero intervene.

And, similarly, we have the following:—

The object common to two consecutive ranges of infinite departure is the vertex of a star of zero departure.

411. Relations between Departure and Intervene.

With reference to the theory of Departure, we thus see that there is a system of critical ranges, and a system of critical objects in each extent. Every critical range has an infinite departure from every other range. Every critical object is the vertex of a star of which the departure between every pair of ranges is zero. It will be remembered that in the theory of the intervene we were conducted to the knowledge that every extent contained a system of objects and ranges, critical with regard to the intervene. Each critical object had an infinite intervene with every other object in the extent. Each critical range possessed the property that the intervene between any two objects thereon is zero. There are, thus, objects and ranges critical with repect to the intervene. There are also objects and ranges critical with respect to the departure. But nothing that we have hitherto assumed will entitle us to draw any inference as to the connexion, much less as to the actual identity, between the critical systems related to intervene, and those related to departure. We have already assumed five properties for the intervene, and five like properties for the departure. These are, in fact, the axioms by which alone the functions of intervene and of departure could be constructed. But another axiom of quite a distinct type has now to be introduced.

There are objects of infinite intervene, and objects of zero departure. There are ranges of infinite departure, and ranges of zero intervene. A range *generally* contains two objects of infinite intervene, and two of zero departure. A star *generally* contains two ranges of infinite departure, and two ranges of zero intervene. On a range of zero intervene the two objects of infinite intervene coalesce, and their intervene from other objects on the range becomes indeterminate. In a star of zero departure the two ranges of infinite departure coalesce, and their departure from other ranges in the same star becomes indeterminate. We have thus the following statement:—

On a range of zero intervene, the intervene between every pair of objects is zero, except where one particular object is involved, in which case the intervene is indeterminate.

In a star of zero departure, the departure between every pair of ranges is zero, except where one particular range is involved, in which case the departure is indeterminate.

The new axiom to be now introduced will be formed as the others have been by generalization from the conceptions of ordinary geometry. In that geometry we have two different aspects in which the phenomenon of parallelism may be presented. Two non-coincident lines are parallel when the angle between them is zero, or when their intersection is at an infinite distance. Without entering into any statement about parallel lines, we may simply say, that when two different straight lines are inclined at the angle zero, their point of intersection is at infinity. Generalizing this proposition, we assume the following axiom or property, which we desire that our systems of measurement shall possess.

412. The Eleventh Axiom of the Content.

This axiom, which is the first to bring together the notions of intervene and of departure, is thus stated:—

(xi) *If two ranges in the same extent have zero departure, their common object will be at infinity, and conversely.*

The vertex of every star of zero departure will thus be at infinity, and hence we deduce the important result that all the objects of infinite intervene are also objects of zero departure, and conversely. Thus we see that the two systems of critical objects in the extent coalesce into a single system in consequence of the assumption in Axiom XI.

Each consecutive pair of critical objects determine a range, which is a range of infinite departure as well as of zero intervene.

The expression *infinite objects* will then denote objects which possess the double property of having, in general, an infinite intervene from other objects in the extent, and of being also the vertices of stars of zero departure.

The expression *infinite ranges* will denote ranges which possess the double property of having, in general, an infinite departure with all other ranges, and which consist of objects, the intervene between any pair of which is, in general, zero.

There is still one more point to be decided. The measurement of departure, like that of intervene, is expressed by the product of a numerical factor with the logarithm of an anharmonic ratio. This factor is H for the intervene. Let us call it H' for the departure. What is to be the relation between H and H'? Here the analogy of geometry is illusory; for, owing to the coincidence between the points of infinity on a straight line, H has to be made infinite in ordinary geometry, while H' must be finite. But in the present more general theory H is finite, and we have found much convenience derived from making it equal to $-\frac{i}{2}$, for then the entire circuit of any range is π. We now stipulate that H' is also to be $-\frac{i}{2}$. The circuit of a star will then be π also.

With this assumption the theory of the metrics of an extent admits of a remarkable development.

Let x, y, z be any three objects. Let a, b, c denote the intervenes between y and z, z and x, x and y, respectively. Let the departure between the ranges from x to y and x to z be denoted by A, from y to z and y to x be denoted by B, and from z to x and z to y be denoted by C. Then,

$$\frac{\sin A}{\sin a} = \frac{\sin B}{\sin b} = \frac{\sin C}{\sin c};$$

$$\cos a = \cos b \cos c + \sin b \sin c \cos A,$$
$$\cos b = \cos c \cos a + \sin c \sin a \cos B,$$
$$\cos c = \cos a \cos b + \sin a \sin b \cos C.$$

Thus the formulæ of spherical trigonometry are generally applicable throughout the extent*.

* I learned this astonishing theorem from Professor Heath's very interesting paper, *Phil. Trans.* Part II. 1884.

413. Representation of Objects by Points in Space.

The several objects in a content are each completely specified when the four numbers, x_1, x_2, x_3, x_4, corresponding to each are known. It is only the ratios of these numbers that are significant. We may hence take them to be the four quadriplanar co-ordinates of a point in space. We are thus led to the construction of a system of one-to-one correspondence between the several points of an Euclidian space, and the several objects of a content. The following propositions are evident:

One object in a content has for its correspondent one point in space, and one point in space corresponds to one object in the content.

The several objects on a range correspond one to one with the several points on a straight line.

The several objects in an extent correspond one to one with the several points in a plane.

Since the objects at infinity are obtained by taking values of x_1, x_2, x_3, x_4, which satisfy a quadric equation, we find that—

The several objects at infinity in the content correspond with the several points of a quadric surface.

This surface we shall call the *infinite quadric*.

The following theorem in quadriplanar co-ordinates is the foundation of the metrics of the objects in the content by the points in space.

If x_1, x_2, x_3, x_4 and y_1, y_2, y_3, y_4 be the quadriplanar co-ordinates of two points P and Q respectively, and if O_1, O_2, O_3, O_4 be any other four points on the ray PQ whose co-ordinates are respectively

$$\begin{array}{cccc} x_1+\lambda_1 y_1, & x_2+\lambda_1 y_2, & x_3+\lambda_1 y_3, & x_4+\lambda_1 y_4, \\ x_1+\lambda_2 y_1, & x_2+\lambda_2 y_2, & x_3+\lambda_2 y_3, & x_4+\lambda_2 y_4, \\ x_1+\lambda_3 y_1, & x_2+\lambda_3 y_2, & x_3+\lambda_3 y_3, & x_4+\lambda_3 y_4, \\ x_1+\lambda_4 y_1, & x_2+\lambda_4 y_2, & x_3+\lambda_4 y_3, & x_4+\lambda_4 y_4, \end{array}$$

then we have the following identity

$$\frac{\lambda_1-\lambda_3}{\lambda_1-\lambda_4}\cdot\frac{\lambda_2-\lambda_4}{\lambda_2-\lambda_3}=\frac{O_1O_3\,.\,O_2O_4}{O_1O_4\,.\,O_2O_3}.$$

Remembering the definition of the anharmonic ratio of four objects on a range (§ 404), we obtain the following theorem:—

The anharmonic ratio of four objects on a range equals the anharmonic ratio of their four corresponding points on a straight line.

We hence deduce the following important result, that—

The intervene between any two objects is proportional to the logarithm of the anharmonic ratio in which the straight line joining the corresponding points is divided by the infinite quadric.

We similarly find that—

The departure between any two ranges in the same extent is proportional to the logarithms of the anharmonic ratio of the pencil formed by their two corresponding straight lines, and the two tangents in the same plane from their intersection to the infinite quadric.

414. Poles and Polars.

The point x_1, x_2, x_3, x_4 has for its polar, with regard to the infinite quadric, the plane,

$$x_1 \frac{dU}{dx_1} + x_2 \frac{dU}{dx_2} + x_3 \frac{dU}{dx_3} + x_4 \frac{dU}{dx_4} = 0.$$

Thus we see that an object corresponding to the point will have a polar extent corresponding to the polar of that point with regard to the infinite quadric. The following property of poles and polars follows almost immediately.

The intervene from an object to any object in its polar extent is equal to $\frac{\pi}{2}$

We have hitherto spoken of the departure between a pair of ranges which have a common object: we now introduce the notion of the departure between a pair of extents by the following definition:—

The departure between a pair of extents is equal to the intervene between their poles.

415. On the Homographic Transformation of the Content.

In our further study of the theory of the content we shall employ, instead of the objects themselves, their corresponding points in ordinary space. All the phenomena of the content can be completely investigated in this way. Objects, ranges, and extents, we are to replace by points, straight lines, and planes. Intervenes are to be measured, not, indeed, as distances, but as logarithms of certain anharmonic ratios obtained by ordinary distance measurement. Departures are to be measured, not, indeed, as angles, but as logarithms of anharmonic ratios of certain pencils obtained by ordinary angular measurement.

I now suppose the several objects of a content to be ordered in two homographic systems, A and B. Each object, X, in the content, regarded as belonging to the system A, will have another object, Y, corresponding thereto in the system B. The correspondence is to be simply of the one-to-one type. Each object in one system has one correspondent in the other system. But if X be regarded as belonging to the system B, its correspondent in A will not be Y, but some other object, Y'.

To investigate this correspondence we shall represent the objects by their correlated points in space. We take x_1, x_2, x_3, x_4 as the co-ordinates of a point corresponding to x, and y_1, y_2, y_3, y_4 as the co-ordinates of the point corresponding to y. We are then to have an unique correspondence between x and y, and we proceed to study the conditions necessary if this be complied with.

416. Deduction of the Equations of Transformation.

All the points in a plane L, taken as x points, must have as their correspondents the points also of a plane; for, suppose that the correspondents formed a surface of the nth degree, then three planes will have three surfaces of the nth degree as their correspondents, and all their n^3 intersections regarded as points in the second system will have but the single intersection of the three planes as their correspondent in the first system. But unless $n = 1$ this does not accord with the assumption that the correspondence is to be *universally* of the one-to-one type. Hence we see that to a plane of the first system must correspond a plane of the second system, and *vice versâ*.

Let the plane in the second system be

$$A_1y_1 + A_2y_2 + A_3y_3 + A_4y_4 = 0.$$

If we seek its corresponding plane in the first system, we must substitute for y_1, y_2, y_3, y_4 the corresponding functions of x_1, x_2, x_3, x_4. Now, unless these are homogeneous linear expressions, we shall not find that this remains a plane. Hence we see that the relations between x_1, x_2, x_3, x_4 and y_1, y_2, y_3, y_4 must be of the following type where (11), (12), &c., are constants:—

$$y_1 = (11)\,x_1 + (12)\,x_2 + (13)\,x_3 + (14)\,x_4,$$
$$y_2 = (21)\,x_1 + (22)\,x_2 + (23)\,x_3 + (24)\,x_4,$$
$$y_3 = (31)\,x_1 + (32)\,x_2 + (33)\,x_3 + (34)\,x_4,$$
$$y_4 = (41)\,x_1 + (42)\,x_2 + (43)\,x_3 + (44)\,x_4.$$

Such are the equations expressing the general homographic transformation of the objects of a content. From the general theory, however, we now proceed to specialize one particular kind of homographic transformation. It is suggested by the notion of a *displacement* in ordinary space. The displacement of a rigid system is only equivalent to a homographic transformation of all its points, conducted under the condition that the distance between every pair of points shall remain unaltered (see p. 2). In our extended conceptions we now study the possible homographic transformations of a content, conducted subject to the condition that the intervene between every pair of objects shall equal that between their correspondents.

417. On the Character of a Homographic Transformation which Conserves Intervene.

In all investigations of this nature the behaviour of the infinite objects is especially instructive. In the present case it is easily shown that every object O infinite before the transformation must be infinite afterwards when moved to O'. For, let X be any object which is not infinite before the transformation, nor afterwards, when it becomes X'. Then, by hypothesis, the intervene OX is equal to $O'X'$; but OX is infinite, therefore $O'X'$ must be also infinite, so that either O' or X' is infinite; but, by hypothesis, X' is finite, therefore O' must be infinite, so that *in a homographic transformation which conserves intervene, each object infinite before the transformation remains infinite afterwards.*

It follows that in the space representation each point, representing an infinite object, and therefore lying on the infinite quadric $U=0$ must, after transformation, be moved to a position which will also lie on the infinite quadric. Hence we obtain the following important result:—

In the space representation of a homographic transformation which conserves intervene, the infinite quadric $U=0$ is merely displaced on itself.

A homographic transformation of the points in space will not, *in general*, permit any quadric to remain unchanged. A certain specialization of the constants will be necessary. They must, in fact, satisfy a single condition, for which we shall presently find the expression.

Let x_1, x_2, x_3, x_4 be the quadriplanar co-ordinates of a point, and let us transform these to a new tetrahedron of which the vertices shall have as their co-ordinates with respect to the original tetrahedron

$$\begin{array}{cccc} x_1', & x_2', & x_3', & x_4', \\ x_1'', & x_2'', & x_3'', & x_4'', \\ x_1''', & x_2''', & x_3''', & x_4''', \\ x_1'''', & x_2'''', & x_3'''', & x_4''''. \end{array}$$

If then X_1, X_2, X_3, X_4 be the four co-ordinates of the point referred to the new tetrahedron

$$\begin{aligned} x_1 &= x_1'X_1 + x_1''X_2 + x_1'''X_3 + x_1''''X_4, \\ x_2 &= x_2'X_1 + x_2''X_2 + x_2'''X_3 + x_2''''X_4, \\ x_3 &= x_3'X_1 + x_3''X_2 + x_3'''X_3 + x_3''''X_4, \\ x_4 &= x_4'X_1 + x_4''X_2 + x_4'''X_3 + x_4''''X_4. \end{aligned}$$

For these equations must be linear and if X_2, X_3, X_4 are all zero then x_1, x_2, x_3, x_4 become x_1', x_2', x_3', x_4' as they ought to do, and similarly for the others, whence we get

$$\begin{vmatrix} x_1' & x_1'' & x_1''' & x_1'''' \\ x_2' & x_2'' & x_2''' & x_2'''' \\ x_3' & x_3'' & x_3''' & x_3'''' \\ x_4' & x_4'' & x_4''' & x_4'''' \end{vmatrix} X_1 = \begin{vmatrix} x_1 & x_1'' & x_1''' & x_1'''' \\ x_2 & x_2'' & x_2''' & x_2'''' \\ x_3 & x_3'' & x_3''' & x_3'''' \\ x_4 & x_4'' & x_4''' & x_4'''' \end{vmatrix}$$

We may write this result thus

$$HX_1 = \begin{vmatrix} x_1 & x_2 & x_3 & x_4 \\ x_1'' & x_2'' & x_3'' & x_4'' \\ x_1''' & x_2''' & x_3''' & x_4''' \\ x_1'''' & x_2'''' & x_3'''' & x_4'''' \end{vmatrix}$$

Let us now suppose that the vertices of this new tetrahedron are the double points of a homography defined by the equations

$$y_1 = (11)\,x_1 + (12)\,x_2 + (13)\,x_3 + (14)\,x_4,$$

$$y_2 = (21)\,x_1 + (22)\,x_2 + (23)\,x_3 + (24)\,x_4,$$

$$y_3 = (31)\,x_1 + (32)\,x_2 + (33)\,x_3 + (34)\,x_4,$$

$$y_4 = (41)\,x_1 + (42)\,x_2 + (43)\,x_3 + (44)\,x_4.$$

We have to solve the biquadratic

$$\begin{vmatrix} (11)-\rho & (12) & (13) & (14) \\ (21) & (22)-\rho & (23) & (24) \\ (31) & (32) & (33)-\rho & (34) \\ (41) & (42) & (43) & (44)-\rho \end{vmatrix} = 0.$$

Let the roots be ρ_1, ρ_2, ρ_3, ρ_4. Then we have

$$\rho_1 x_1' = (11)\,x_1' + (12)\,x_2' + (13)\,x_3' + (14)\,x_4',$$

$$\rho_1 x_2' = (21)\,x_1' + (22)\,x_2' + (23)\,x_3' + (24)\,x_4',$$

$$\rho_1 x_3' = (31)\,x_1' + (32)\,x_2' + (33)\,x_3' + (34)\,x_4',$$

$$\rho_1 x_4' = (41)\,x_1' + (42)\,x_2' + (43)\,x_3' + (44)\,x_4',$$

with similar equations for x_1'', x_1''', x_1'''', x_2'', &c.

We thus get

$$H\rho_1\rho_2\rho_3\rho_4 X_1 = \begin{vmatrix} x_1 & x_2 & x_3 & x_4 \\ x_1'' & x_2'' & x_3'' & x_4'' \\ x_1''' & x_2''' & x_3''' & x_4''' \\ x_1'''' & x_2'''' & x_3'''' & x_4'''' \end{vmatrix} \begin{vmatrix} 11 & 12 & 13 & 14 \\ 21 & 22 & 23 & 24 \\ 31 & 32 & 33 & 34 \\ 41 & 42 & 43 & 44 \end{vmatrix}$$

$$= \begin{vmatrix} y_1, & \rho_2 x_1'', & \rho_3 x_1''', & \rho_4 x_1'''' \\ y_2, & \rho_2 x_2'', & \rho_3 x_2''', & \rho_4 x_2'''' \\ y_3, & \rho_2 x_3'', & \rho_3 x_3''', & \rho_4 x_3'''' \\ y_4, & \rho_2 x_4'', & \rho_3 x_4''', & \rho_4 x_4'''' \end{vmatrix}$$

or

$$H\rho_1 X_1 = \begin{vmatrix} y_1 & y_2 & y_3 & y_4 \\ x_1'' & x_2'' & x_3'' & x_4'' \\ x_1''' & x_2''' & x_3''' & x_4''' \\ x_1'''' & x_2'''' & x_3'''' & x_4'''' \end{vmatrix}$$

But these determinants are the co-ordinates of y referred to the new tetrahedron, and omitting needless factors

$$Y_1 = \rho_1 X_1, \quad Y_2 = \rho_2 X_2, \quad Y_3 = \rho_3 X_3, \quad Y_4 = \rho_4 X_4.$$

We thus obtain the following theorem.

Let x_1, x_2, x_3, x_4 be the co-ordinates of a point with respect to any arbitrary tetrahedron of reference.

Let y_1, y_2, y_3, y_4 be the co-ordinates of the corresponding point in a homographic system defined by the equations

$$\begin{aligned} y_1 &= (11)\,x_1 + (12)\,x_2 + (13)\,x_3 + (14)\,x_4, \\ y_2 &= (21)\,x_1 + (22)\,x_2 + (23)\,x_3 + (24)\,x_4, \\ y_3 &= (31)\,x_1 + (32)\,x_2 + (33)\,x_3 + (34)\,x_4, \\ y_4 &= (41)\,x_1 + (42)\,x_2 + (43)\,x_3 + (44)\,x_4. \end{aligned}$$

If we transform the tetrahedron of reference to the four double points of the homography, and if X_1, X_2, X_3, X_4 be the co-ordinates of any point with regard to this new tetrahedron then the co-ordinates of its homographic correspondent are

$$\rho_1 X_1, \quad \rho_2 X_2, \quad \rho_3 X_3, \quad \rho_4 X_4,$$

where ρ_1, ρ_2, ρ_3, ρ_4 are the four roots of the equation,

$$\begin{vmatrix} (11)-\rho & (12) & (13) & (14) \\ (21) & (22)-\rho & (23) & (24) \\ (31) & (32) & (33)-\rho & (34) \\ (41) & (42) & (43) & (44)-\rho \end{vmatrix} = 0.$$

In general there are four, but only four double points, *i.e.* points which remain unaltered by the transformation. If however two of the roots of the biquadratic equation be equal, then every point on the ray connecting the two corresponding double points possesses the property of a double point.

For if $\rho_1 = \rho_2$, then

$$Y_1 : Y_2 :: X_1 : X_2,$$

and hence the point whose co-ordinates are

$$X_1,\ X_2,\ 0,\ 0,$$

being transformed into

$$Y_1,\ Y_2,\ 0,\ 0$$

remains unchanged.

Let us now suppose that a certain quadric surface is to remain unaltered by the homographic transformation.

At this point it seems necessary to choose the particular character of the quadric surface in the further developments to which we now proceed. The theory of any non-Euclidian geometry will of course depend on whether the surface adopted as the infinite be an ellipsoid or a double sheeted hyperboloid with no real generators or a single sheeted hyperboloid with real generators. We shall suppose the infinite, in the present theory, to be a single sheeted hyperboloid.

The homographic transformation which we shall consider will transform any generator of the surface into another generator of the same system, for if it transformed the generator into one of the other system, then the two rays would intersect, which is a special case that shall not be here further considered.

Let three rays R_1, R_2, R_3 be generators of the first system on the hyperboloid. After the transformation these rays will be transferred to three other positions R_1', R_2', R_3' belonging to the same system.

Let S_1, S_2 be two rays of the second system. Then the intersection of R_1, R_1', R_2, R_2' &c., with S_1 give two systems of homographic points. The two double points of these systems on S_1 give two points through which two rays of the first system must pass both before and after the transformation. Two similar points can also be found on S_2. These two pairs of Double points on S_1 and S_2 will fix a pair of generators of the first system which are unaltered by the transformation.

In like manner we find two rays of the second system which are unaltered. The four intersections of these rays must be the four double points of the system.

We can also prove in another way that in the homographic transformation which preserves intervene, the four double points must, in general, lie on the fundamental quadric.

For, suppose that one of the double points P was not on the quadric. Draw the tangent cone from P. The conic of contact will remain unaltered by the transformation. Therefore two points O_1 and O_2 on that conic will be unaltered (p. 2). So will R the intersection of the tangents to the conic at O_1 and O_2. The four double points will therefore be P, R, O_1 and O_2.

But PR cuts the quadric in two other points which cannot change. Hence PR will consist entirely of double points, and therefore the discriminant of the equation in ρ would have to vanish, which does not generally happen.

Of course, even in this case, there are still four double points on the quadric, *i.e.* O_1, O_2 and the two points in which PR cuts the quadric.

We may therefore generally assume that two pairs of opposite edges of the tetrahedron of double points are generators of the fundamental quadric, the latter must accordingly have for its equation

$$X_1X_2 + \lambda X_3X_4 = 0,$$

with the essential condition

$$\rho_1\rho_2 = \rho_3\rho_4.$$

Every point on any quadric of this family will remain upon that quadric notwithstanding the transformation.

Nor need we feel surprised, when in the attempt to arrange a homographic transformation which shall leave a *single* quadric unaltered, it appeared that if this was accomplished, then each member of a family of quadrics would be in the same predicament. Here again the resort to ordinary geometry makes this clear.

In the displacement of a rigid system in ordinary space one ray remains unchanged, and so does every circular cylinder of which this ray is the axis. Thus we see that there is a whole family of cylinders which remain unchanged; and if U be one of these cylinders, and V another, then all the cylinders of the type $U + \lambda V$ are unaltered, the plane at infinity being of course merely an extreme member of the series. More generally these cylinders may be regarded as a special case of a system of cones with a common vertex; and more generally still we may say that a family of quadrics remains unchanged.

Reverting, then, to a space of which the several points correspond to the objects of a content, we find, that for every homographic transformation which corresponds to a displacement in ordinary geometry a *singly* infinite family of quadrics is to remain unchanged, and the infinite quadric itself is to form one member of this family.

Let us now suppose a range in the content submitted to this description of homographic transformation. Let P, Q be two objects on the range, and let X, Y be the two infinite objects thereon. This range will be transformed to a new position, and the objects will now be P', Q', X', Y'. Since infinite objects must remain infinite, it follows that X' and Y' must be infinite, as well as X and Y. Also, since homographic transformation does not alter anharmonic ratio, we have

$$(PQXY) = (P'Q'X'Y');$$

whence, by Axiom v., we see that the intervene from P to Q equals the intervene from P' to Q'; in other words, that *all* intervenes remain unchanged by this homographic transformation.

Every homographic transformation which possesses these properties must satisfy a special condition in the coefficients. This may be found from the determinantal equation for ρ (p. 458), for then the following symmetric function of the four roots ρ_1, ρ_2, ρ_3, ρ_4 must vanish:

$$(\rho_1\rho_2 - \rho_3\rho_4)(\rho_1\rho_3 - \rho_2\rho_4)(\rho_1\rho_4 - \rho_2\rho_3).$$

418. The Geometrical Meaning of this Symmetric Function.

We may write the family of quadrics thus:

$$X_1X_2 + \lambda X_3X_4 = 0.$$

All these quadrics have two common generators of each kind:

$$\left\{\begin{matrix} X_1 = 0, & X_3 = 0 \\ & \text{and} \\ X_2 = 0, & X_4 = 0 \end{matrix}\right. \qquad \left\{\begin{matrix} X_1 = 0, & X_4 = 0 \\ & \text{and} \\ X_2 = 0, & X_3 = 0. \end{matrix}\right.$$

For the rays $X_1 = 0$, $X_3 = 0$, and $X_1 = 0$, $X_4 = 0$, are both contained in the plane X_1, and therefore intersect, and, accordingly, belong to the opposed system of generators.

The geometrical meaning of the equation

$$\rho_1\rho_2 - \rho_3\rho_4 = 0$$

can be also shown.

The tetrahedron formed by the intersection of the two pairs of generators just referred to remains unaltered by the transformation. Any point on the edge, $X_1 = 0$, $X_3 = 0$, of which the co-ordinates are

$$0,\ X_2,\quad 0,\ X_4,$$

will be transformed to

$$0,\ \rho_2 X_2,\ \ 0,\ \rho_4 X_4.$$

The question may be illustrated by Figure 45.

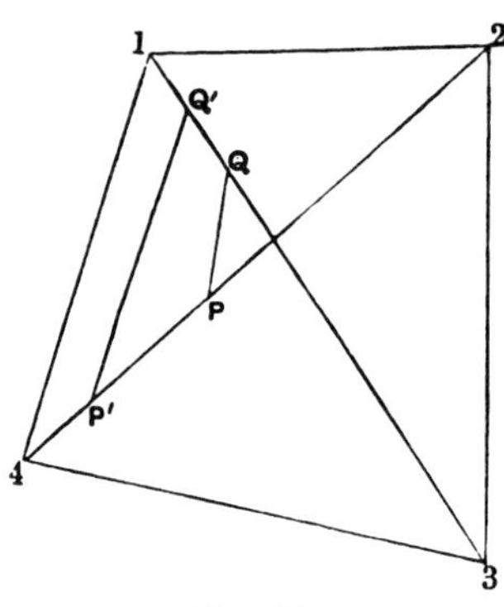

Fig. 45.

Let 1, 2, 3, 4 be the four corners of the tetrahedron. Let the transformation convey P to P' and Q to Q'. As P varies along the ray, so will P' vary, and the two will describe homographic systems, of which 2 and 4 are the double points. In a similar way, Q and Q' will trace out homographic systems on the ray 1 3. We shall write the points on 2 4, in the order,

$$2,\ 4,\ \ P,\ P'.$$

Through 2, the generator of the surface 2 3 can be drawn (1 2 is *not* a generator), and through 4 the generator 4 1 can be drawn (4 3 is not a generator); thus we have, for the corresponding order on 1 3.

$$3,\ \ 1,\ \ Q,\ \ Q'.$$

Points.	Co-ordinates.				Points.	Co-ordinates.			
2	0	1	0	0	3	0	0	1	0
4	0	0	0	1	1	1	0	0	0
P	0	X_2	0	X_4	Q	X_1'	0	X_3'	0
P'	0	$\rho_2 X_2$	0	$\rho_4 X_4$	Q'	$\rho_1 X_1'$	0	$\rho_3 X_3'$	0

The anharmonic ratio of the first set is that of $0,\ \infty,\ \dfrac{X_4}{X_2},\ \dfrac{\rho_4}{\rho_2}\dfrac{X_4}{X_2},$

" " second " $\infty,\ 0,\ \dfrac{X_3'}{X_1'},\ \dfrac{\rho_3}{\rho_1}\dfrac{X_3'}{X_1'};$

out since

$$\rho_1 \rho_2 = \rho_3 \rho_4,$$

hen the anharmonic ratios are equal.

The theorem can otherwise be shown by drawing Figure 46.

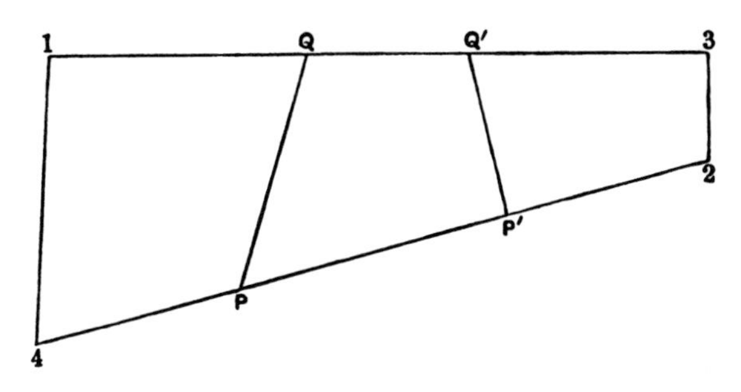

Fig. 46.

1 4 and 3 2 are to be generators of the infinite quadric. This will show that 4 (and not 2) is the correspondent to 1, and that 2 (and not 4) is the correspondent to 3, and thus the statement of anharmonic equality,

$$(1\ QQ'\ 3) = (4\ PP'\ 2),$$

becomes perfectly definite.

1 2 and 3 4 are, of course, not generators; they are two conjugate polars of the infinite quadric.

We can now see the reason of the anharmonic equality. Let PQ be a generator of the infinite quadric, as is clearly possible, for 1, 3 and 2, 4 are both generators of the opposite system. Then, since a generator of the infinite quadric must remain thereon after the displacement, it will follow that $P'Q'$, to which PQ is displaced, must also be a generator; and thus we have four generators, 4 1, PQ, $P'Q'$, 2 3, on a hyperboloid of one system intersecting the two generators of another, and by the well-known property of the surface,

$$(1\ QQ'\ 3) = (4\ PP'\ 2).$$

We also see why the infinite quadric is only one of a family which remains unaltered. For, if PQ be a generator of *any* quadric through the tetrahedron, 1, 2, 3, 4; then, since P and Q are conveyed to P' and Q', and since the anharmonic equality holds, it follows that $P'Q'$ will also be a generator of the quadric, *i.e.* a generator of the quadric will remain thereon after the displacement.

It is a remarkable fact that, when the linear transformation is given, the infinite quadric is not definitely settled. We have seen how, in the first place, the linear transformation must fulfil a fundamental condition; but when that condition is obeyed, then a whole family of quadrics present themselves, any one of which is equally eligible for the infinite.

419. On the Intervene through which each Object is Conveyed.

Given an object, X_1, X_2, X_3, X_4, find the intervene through which it is conveyed by the transformation, when

$$X_1X_2 + \lambda X_3X_4 = 0$$

is the infinite quadric.

In this equation substitute for X_1, $X_1 + \theta Y_1$, &c.; and, remembering that $Y_1 = \rho_1 X_1$, &c., we have,

$$(X_1 + \rho_1\theta X_1)(X_2 + \rho_2\theta X_2) + \lambda(X_3 + \rho_3\theta X_3)(X_4 + \rho_4\theta X_4) = 0,$$

or

$$\theta^2(\rho_1\rho_2 X_1X_2 + \lambda\rho_3\rho_4 X_3X_4)$$
$$+ \theta(\rho_1 X_1X_2 + \rho_2 X_1X_2 + \lambda\rho_3 X_3X_4 + \lambda\rho_4 X_3X_4)$$
$$+ X_1X_2 + \lambda X_3X_4 = 0.$$

We simplify this by introducing

$$\rho_1\rho_2 = \rho_3\rho_4,$$

and writing $\lambda X_3X_4 \div X_1X_2 = \phi$, whence the equation becomes

$$\theta^2\rho_1\rho_2(1+\phi) + \theta[\rho_1 + \rho_2 + \phi(\rho_3 + \rho_4)] + (1+\phi) = 0;$$

hence if δ be the intervene, we have,

$$\cos\delta = \frac{1}{2\sqrt{\rho_1\rho_2}}\,\frac{\rho_1 + \rho_2 + \phi(\rho_3 + \rho_4)}{1+\phi};$$

or, if we restore its value to ϕ,

$$\cos\delta = \frac{1}{2\sqrt{\rho_1\rho_2}}\,\frac{X_1X_2(\rho_1 + \rho_2) + (\rho_3 + \rho_4)\lambda X_3X_4}{X_1X_2 + \lambda X_3X_4}.$$

If

$$\rho_1 + \rho_2 = \rho_3 + \rho_4,$$

then

$$\cos\delta = \frac{\rho_1 + \rho_2}{2\sqrt{\rho_1\rho_2}};$$

i.e. all objects are translated through equal intervenes. This is the case which we shall subsequently consider under the title of the *vector*, as this remarkable conception of Clifford's is called. In this case, as

$$\rho_1 + \rho_2 = \rho_3 + \rho_4,$$

and also,

$$\rho_1\rho_2 = \rho_3\rho_4,$$

we must have

$$\rho_1 = \rho_3, \text{ and } \rho_2 = \rho_4,$$

or

$$\rho_1 = \rho_4, \text{ and } \rho_2 = \rho_3.$$

In either case the equation for ρ will become a perfect square.

In general let
$$X_1 = 0,$$
then
$$\cos\delta = \frac{\rho_3 + \rho_4}{2\sqrt{\rho_3\rho_4}};$$
whence we find that all objects in the extent X_1 are displaced through equal intervenes. This intervene can be readily determined for
$$e^{i\delta} + e^{-i\delta} = \sqrt{\frac{\rho_3}{\rho_4}} + \sqrt{\frac{\rho_4}{\rho_3}},$$
whence
$$i\delta = \log\sqrt{\frac{\rho_3}{\rho_4}},$$
or
$$\delta = \frac{1}{2i}(\log\rho_3 - \log\rho_4).$$

The intervene through which every object on X_2 is conveyed has the same value.

We could have also proved otherwise that objects on X_1 and X_2 are all displaced through equal intervenes, for the locus of objects so displaced is a quadric of the form
$$X_1X_2 + \lambda X_3X_4 = 0,$$
and, of course, for a special value of the distance this quadric becomes simply
$$X_1X_2 = 0.$$

If $X_1 = 0$, and $X_3 = 0$, then $\cos\delta$ becomes indeterminate; but this is as it should be, because all objects on X_1 and X_3 are at infinity.

420. The Orthogonal Transformation*.

The formulæ
$$y_1 = (11)\,x_1 + (12)\,x_2 + (13)\,x_3 + (14)\,x_4,$$
$$y_2 = (21)\,x_1 + (22)\,x_2 + (23)\,x_3 + (24)\,x_4,$$
$$y_3 = (31)\,x_1 + (32)\,x_2 + (33)\,x_3 + (34)\,x_4,$$
$$y_4 = (41)\,x_1 + (42)\,x_2 + (43)\,x_3 + (44)\,x_4,$$
denote the general type of transformation. The transformation is said to be *orthogonal* if when x_1, &c., are solved in terms of y_1, &c. we obtain as follows:—
$$x_1 = (11)\,y_1 + (21)\,y_2 + (31)\,y_3 + (41)\,y_4,$$
$$x_2 = (12)\,y_1 + (22)\,y_2 + (32)\,y_3 + (42)\,y_4,$$
$$x_3 = (13)\,y_1 + (23)\,y_2 + (33)\,y_3 + (43)\,y_4,$$
$$x_4 = (14)\,y_1 + (24)\,y_2 + (34)\,y_3 + (44)\,y_4.$$

* This is employed in Professor Heath's memoir, cited on p. 452.

From the first formulæ the equation for ρ is, as before, § 417

$$\begin{vmatrix} (11)-\rho & (12) & (13) & (14) \\ (21) & (22)-\rho & (23) & (24) \\ (31) & (32) & (33)-\rho & (34) \\ (41) & (42) & (43) & (44)-\rho \end{vmatrix} = 0.$$

From the second, the equation for ρ must be

$$\begin{vmatrix} (11)-\frac{1}{\rho} & (21) & (31) & (41) \\ (12) & (22)-\frac{1}{\rho} & (32) & (42) \\ (13) & (23) & (33)-\frac{1}{\rho} & (43) \\ (14) & (24) & (34) & (44)-\frac{1}{\rho} \end{vmatrix} = 0;$$

but we may interchange rows and columns in a determinant so that the last may be written,

$$\begin{vmatrix} (11)-\frac{1}{\rho} & (12) & (13) & (14) \\ (21) & (22)-\frac{1}{\rho} & (23) & (24) \\ (31) & (32) & (33)-\frac{1}{\rho} & (34) \\ (41) & (42) & (43) & (44)-\frac{1}{\rho} \end{vmatrix} = 0;$$

whence we see that the equation for ρ must be unaltered, if for ρ we substitute $\frac{1}{\rho}$. It must therefore be a reciprocal equation of the type

$$\rho^4 + 4A\rho^3 + 6B\rho^2 + 4A\rho + 1 = 0,$$

and the roots are of the form

$$\rho', \frac{1}{\rho'}, \quad \rho'', \frac{1}{\rho''};$$

and as

$$\rho' \frac{1}{\rho'} - \rho'' \frac{1}{\rho''} = 0,$$

this transformation fulfils the fundamental condition (§ 417).

421. Quadrics unaltered by the Orthogonal Transformation.

The special facilities of the orthogonal transformation in the present subject arise from the circumstance that it is the nature of this transformation to leave unaltered a certain family of quadrics. This is as we have seen the necessary characteristic of the homographic transformation which preserves intervene. The infinite quadric which the transformation fails to derange can be written at once, for we have

$$x_1^2 + x_2^2 + x_3^2 + x_4^2 = y_1^2 + y_2^2 + y_3^2 + y_4^2 = 0.$$

It is also easily seen that the expression

$$x_1 y_1 + x_2 y_2 + x_3 y_3 + x_4 y_4 = 0$$

is unchanged by the orthogonal transformation. We thus have the following quadric, which remains unaltered:—

$$x_1[(11)x_1+(12)x_2+(13)x_3+(14)x_4]+x_2[(21)x_1+(22)x_2+(23)x_3+(24)x_4]$$
$$+x_3[(31)x_1+(32)x_2+(33)x_3+(34)x_4]+x_4[(41)x_1+(42)x_2+(43)x_3+(44)x_4]=0;$$

or, writing it otherwise,

$$(11)x_1^2+(22)x_2^2+(33)x_3^2+(44)x_4^2+[(12)+(21)]x_1x_2+[(13)+(31)]x_1x_3$$
$$+[(14)+(41)]x_1x_4+[(23)+(32)]x_2x_3+[(24)+(42)]x_2x_4+[(43)+(34)]x_3x_4=0.$$

If this be denoted by U, and $x_1^2+x_2^2+x_3^2+x_4^2$ by Ω, then, more generally, $U-h\Omega$ is unaltered by the transformation.

We now investigate the intervene θ, through which every object on

$$U-h\Omega=0$$

is conveyed by the transformation.

If we substitute $x_1+\lambda y_1$ &c. for x_1 &c. in the infinite quadric we have

$$\Omega+2\lambda U+\lambda^2\Omega=0,$$

and, accordingly, the intervene θ, through which an object is conveyed by the orthogonal transformation is defined by the equation

$$\cos\theta=\frac{U}{\Omega};$$

hence the locus of objects moved through the intervene θ is simply

$$U-\Omega\cos\theta=0.$$

422. Proof that U and Ω have Four Common Generators.

The equation in ρ has four roots of the type

$$\rho',\ \frac{1}{\rho'},\ \rho'',\ \frac{1}{\rho''}.$$

These correspond to the vertices of the tetrahedron (fig. 47). Symmetry shows that the conjugate polars as distinguished from the generators will be the ray joining the vertices corresponding to

$$\frac{1}{\rho'} \text{ and } \rho',$$

and that joining

$$\frac{1}{\rho''} \text{ and } \rho''.$$

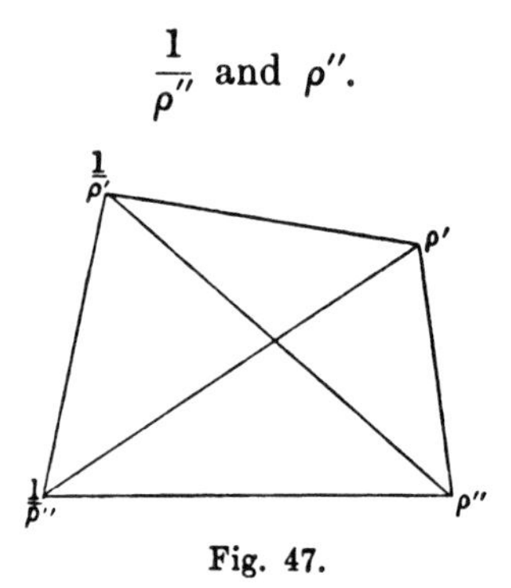

Fig. 47.

Let $\alpha_1, \alpha_2, \alpha_3, \alpha_4$ and $\beta_1, \beta_2, \beta_3, \beta_4$ be the co-ordinates of the corners at ρ' and ρ''. If we substitute $\alpha_1 + \lambda\beta_1$, $\alpha_2 + \lambda\beta_2 \ldots$ &c., for x_1, x_2 &c. in $\Omega = 0$,

$$2\lambda(\alpha_1\beta_1 + \alpha_2\beta_2 + \alpha_3\beta_3 + \alpha_4\beta_4) = 0.$$

Let us make the same substitution in U, we have, in general,

$$\begin{aligned} y_1 &= (11)\,x_1 + (12)\,x_2 + (13)\,x_3 + (14)\,x_4 \\ &= (11)(\alpha_1 + \lambda\beta_1) + (12)(\alpha_2 + \lambda\beta_2) + (13)(\alpha_3 + \lambda\beta_3) + (14)(\alpha_4 + \lambda\beta_4) \\ &= \rho'\alpha_1 + \lambda\rho''\beta_1\,; \end{aligned}$$

whence, remembering that

$$U = x_1y_1 + x_2y_2 + x_3y_3 + x_4y_4,$$

$$U = (\alpha_1 + \lambda\beta_1)(\rho'\alpha_1 + \lambda\rho''\beta_1) + (\alpha_2 + \lambda\beta_2)(\rho'\alpha_2 + \lambda\rho''\beta_2) + \&c.,$$

and as α and β are both on Ω, we have,

$$U = \lambda(\rho' + \rho'')(\alpha_1\beta_1 + \alpha_2\beta_2 + \alpha_3\beta_3 + \alpha_4\beta_4)\,;$$

but since the line joining ρ' and ρ'' is a generator of Ω, the last factor must vanish, and the line is therefore also a generator of U.

It is thus proved that U has four generators in common with Ω.

423. Verification of the Invariance of Intervene.

As an exercise in the use of the orthogonal system of co-ordinates, we may note the following proposition:—

Let x_1, x_2, x_3, x_4, and x_1', x_2', x_3', x_4', be two objects which are conveyed by the transformation to y_1, y_2, y_3, y_4, and y_1', y_2', y_3', y_4', respectively, it is desired to show that the intervene between the two original points is equal to that between the transformed. The expressions for the cosine of the intervene are

$$\frac{x_1x_1' + x_2x_2' + x_3x_3' + x_4x_4'}{(x_1^2 + x_2^2 + x_3^2 + x_4^2)^{\frac{1}{2}}\,(x_1'^2 + x_2'^2 + x_3'^2 + x_4'^2)^{\frac{1}{2}}},$$

and the similar one with y and y', instead of x and x'. The denominators are clearly equal, and we have only to notice that

$$x_1x_1' + x_2x_2' + x_3x_3' + x_4x_4' = y_1y_1' + y_2y_2' + y_3y_3' + y_4y_4';$$

as an immediate consequence of the formulæ connecting the orthogonal transformation.

424. Application of the Theory of Emanants.

We can demonstrate the same proposition in another manner by reverting to the general case.

Let $U=0$ be a function of x_1, x_2, x_3, x_4. Let x_1', x_2', x_3', x_4' be a system of variables cogredient with x_1, x_2, x_3, x_4, and let us substitute in U the expressions x_1+kx_1', x_2+kx_2', &c., for x_1, x_2. The value of U then becomes

$$U + k\Delta U + \frac{k^2}{1.2}\Delta^2 U + \text{\&c.},$$

where

$$\Delta = x_1'\frac{d}{dx_1} + x_2'\frac{d}{dx_2} + x_3'\frac{d}{dx_3} + x_4'\frac{d}{dx_4}.$$

If U be changed into V, a function of y, by the formulæ of transformation, we have, of course,

$$U = V;$$

but since y_1 is a linear function of x_1, &c., *i.e.*

$$y_1 = (11)x_1 + (12)x_2 + (13)x_3 + (14)x_4,$$

it follows that if we change x_1 into x_1+kx_1', &c., we simply change y_1 into y_1+ky_1'. Hence we deduce, that if U be transformed by writing x_1+kx_1', &c., for x, then V will be similarly transformed by writing y_1+ky_1' for y, and, of course, as the original U and V were equal, so will the transformed U and V be equal. It further follows that as k is arbitrary, the several coefficients will also be equal, and thus we have

$$x_1'\frac{dU}{dx_1} + \ldots = y_1'\frac{dU}{dy_1} + \ldots$$

Hence the intervene between two objects before displacement remains unaltered by that operation; for

$$\cos\delta = \frac{x_1'\dfrac{dU}{dx_1} + \text{\&c.}}{\sqrt{U_xU_x'}}:$$

and by what we have just proved, this expression will remain unaltered if y be interchanged with x.

425. The Vector in Orthogonal Co-ordinates.

Since, in general,

$$\cos\theta = \frac{U}{\Omega},$$

we have for the vector (§ 419) the following conditions:—

$$(11) = (22) = (33) = (44),$$

and also,

$$(12) + (21) = 0,$$

and the similar equations. In fact, U can only differ from Ω by a constant factor.

The orthogonal equations require the following conditions—

$$+(11).(12) - (12).(11) + (13).(23) + (14).(24) = 0,$$
$$+(11).(13) - (12).(23) - (13).(11) + (14).(34) = 0,$$
$$+(11).(14) - (12).(24) - (13).(34) - (14).(11) = 0,$$
$$+(12).(13) + (11).(23) - (23).(11) + (24).(34) = 0,$$
$$+(12).(14) + (11).(24) - (23).(34) - (24).(11) = 0,$$
$$+(13).(14) + (23).(24) + (11).(34) - (11).(34) = 0,$$
$$+(11)^2 + (12)^2 + (13)^2 + (14)^2 = 1,$$
$$+(12)^2 + (11)^2 + (23)^2 + (24)^2 = 1,$$
$$+(13)^2 + (23)^2 + (11)^2 + (34)^2 = 1,$$
$$+(14)^2 + (24)^2 + (34)^2 + (11)^2 = 1.$$

We now introduce the notation:—

$$(11) = \alpha; \quad (12) = \beta; \quad (13) = \gamma; \quad (14) = \delta,$$

and the equations give us

$$+\gamma(23) + \delta(24) = 0 \quad \text{.........................(i)},$$
$$-\beta(23) + \delta(34) = 0 \quad \text{......................... (ii)},$$
$$-\beta(24) - \gamma(34) = 0 \quad \text{......................... (iii)},$$
$$+\beta\gamma + (24)(34) = 0 \quad \text{......................... (iv)},$$
$$+\beta\delta - (23)(34) = 0 \quad \text{......................... (v)},$$
$$+\gamma\delta + (23)(24) = 0 \quad \text{......................... (vi)},$$

$$\left.\begin{aligned} +\alpha^2 + \beta^2 + \gamma^2 + \delta^2 &= 1 \\ +\beta^2 + \alpha^2 + (23)^2 + (24)^2 &= 1 \\ +\gamma^2 + (23)^2 + \alpha^2 + (34)^2 &= 1 \\ +\delta^2 + (24)^2 + (34)^2 + \alpha^2 &= 1 \end{aligned}\right\} \quad \text{.................. (vii)}.$$

From (iv), $\beta\gamma = -(24)(34)$.

From (v), $\beta\delta = +(23)(34)$;

by multiplication,

$$\beta^2\gamma\delta = -(23)(24)(34)^2;$$

but, from (vi),

$$\gamma\delta = -(23)(24);$$

whence, we deduce,

$$\beta^2 = (34)^2.$$

The significance of the double sign in the value of β will be afterwards apparent; for the present we take

$$\beta = +(34).$$

From (ii) $\delta = +(23)$,

From (iii) $\gamma = -(24)$,

while the group (vii) will be satisfied if

$$\alpha^2 + \beta^2 + \gamma^2 + \delta^2 = 1.$$

The scheme of orthogonal transformation for the *Right Vector* (for so we designate the case of $\beta = +(34)$,) is as follows:—

$$\begin{vmatrix} +\alpha & +\beta & +\gamma & +\delta \\ -\beta & +\alpha & +\delta & -\gamma \\ -\gamma & -\delta & +\alpha & +\beta \\ -\delta & +\gamma & -\beta & +\alpha \end{vmatrix}.$$

If we append the condition

$$\alpha^2 + \beta^2 + \gamma^2 + \delta^2 = 1,$$

then we have completely defined the *Right Vector*.

We now take the other alternative,

$$\beta = -(34);$$

then, from (ii), $\delta = -(23)$,

then, from (iii), $\gamma = +(24)$.

We thus have for the *Left Vector*, the form,

$$\begin{vmatrix} +\alpha & +\beta & +\gamma & +\delta \\ -\beta & +\alpha & -\delta & +\gamma \\ -\gamma & +\delta & +\alpha & -\beta \\ -\delta & -\gamma & +\beta & +\alpha \end{vmatrix},$$

with, as before, the condition,

$$\alpha^2 + \beta^2 + \gamma^2 + \delta^2 = 1.$$

If θ be the intervene through which the vector displaces an object then it is easily shown that $\cos\theta = \alpha$.

426. Parallel Vectors.

The several objects of a content are displaced by the same vector along ranges which are said to be *parallel*.

Taking the space representation, § 413, Clifford showed that all right vectors, which are parallel, intersect two generators of one system on the infinite quadric, while all left vectors, which are parallel, intersect two generators of the other system.

A generator intersected by two rays from a right vector may be defined by the points whose coordinates are

$$+\alpha', \quad -\beta, \quad -\gamma, \quad -\delta,$$

$$+\beta, \quad +\alpha', \quad -\delta, \quad +\gamma,$$

while a generator intersected by two rays from a left vector will be defined by

$$+\alpha_0', \quad -\beta_0, \quad -\gamma_0, \quad -\delta_0,$$

$$+\beta_0, \quad +\alpha_0', \quad +\delta_0, \quad -\gamma_0.$$

To prove the theorem, it is only necessary to show that these four points are coplanar, for then the two generators intersect, *i.e.* are of opposite systems. We have, then, only to show that the following determinant vanishes:—

$$\begin{vmatrix} \alpha' & -\beta & -\gamma & -\delta \\ \beta & \alpha' & -\delta & \gamma \\ \alpha_0' & -\beta_0 & -\gamma_0 & -\delta_0 \\ \beta_0 & \alpha_0' & \delta_0 & -\gamma_0 \end{vmatrix}.$$

This will be most readily shown by squaring, for with an obvious notation it then reduces to the simple form

$$\begin{vmatrix} 0 & 0 & [\alpha'\alpha_0'] & [\alpha'\beta_0] \\ 0 & 0 & [\beta\alpha_0'] & [\beta\beta_0] \\ [\alpha'\alpha_0'] & [\beta\alpha_0'] & 0 & 0 \\ [\beta_0\alpha'] & [\beta\beta_0] & 0 & 0 \end{vmatrix},$$

whence we see that the original determinant is simply

$$\begin{vmatrix} [\alpha'\alpha_0'] & [\beta\alpha_0'] \\ [\beta_0\alpha'] & [\beta\beta_0] \end{vmatrix};$$

which expanded, becomes

$$(\alpha'\alpha_0' + \beta\beta_0 + \gamma\gamma_0 + \delta\delta_0)(\beta\beta_0 + \alpha'\alpha_0' - \delta\delta_0 - \gamma\gamma_0)$$
$$-(\alpha'\beta_0 - \beta\alpha_0' - \gamma\delta_0 + \delta\gamma_0)(\beta\alpha_0' - \alpha'\beta_0 + \delta\gamma_0 - \gamma\delta_0)$$
$$= (\alpha'\alpha_0' + \beta\beta_0)^2 - (\gamma\gamma_0 + \delta\delta_0)^2 + (\alpha'\beta_0 - \alpha_0'\beta)^2 - (\delta\gamma_0 - \gamma\delta_0)^2$$
$$= \alpha'^2\alpha_0'^2 + \beta^2\beta_0^2 + \alpha'^2\beta_0^2 + \alpha_0'^2\beta^2 - \gamma^2\gamma_0^2 - \delta^2\delta_0^2 - \delta^2\gamma_0^2 - \gamma^2\delta_0^2$$
$$= \alpha'^2(\alpha_0'^2 + \beta_0^2) + \beta^2(\alpha_0'^2 + \beta_0^2) - \gamma^2(\gamma_0^2 + \delta_0^2) - \delta^2(\gamma_0^2 + \delta_0^2)$$
$$= (\alpha'^2 + \beta^2)(\alpha_0'^2 + \beta_0^2) - (\gamma^2 + \delta^2)(\gamma_0^2 + \delta_0^2);$$

but,

$$\alpha'^2 + \beta^2 + \gamma^2 + \delta^2 = 0;$$

whence this expression is

$$(\alpha'^2 + \beta^2)(\alpha_0'^2 + \beta_0^2 + \gamma_0^2 + \delta_0^2) = 0.$$

On the supposition that the vectors were homonymous, *i.e.* both right or both left, the corresponding determinant would have been

$$\begin{vmatrix} \alpha' & -\beta & -\gamma & -\delta \\ \beta & \alpha' & -\delta & \gamma \\ \alpha_0' & -\beta_0 & -\gamma_0 & -\delta_0 \\ \beta_0 & \alpha_0' & -\delta_0 & \gamma_0 \end{vmatrix}.$$

Squaring, we get, as before,

$$\begin{vmatrix} [\alpha'\alpha_0'] & [\beta\alpha_0'] \\ [\beta_0\alpha'] & [\beta\beta_0] \end{vmatrix};$$

but now,

$$[\alpha'\alpha_0'] = [\beta\beta_0],$$
$$[\beta_0\alpha'] = -[\beta\alpha_0'];$$

whence the determinant reduces to

$$[\alpha'\alpha_0']^2 + [\beta\alpha_0']^2,$$

a value very different from that in the former case.

427. The Composition of Vectors.

Let an object x be conveyed to y by the operation of a vector, and let the object y be then conveyed to z by the operation of a second vector, which we shall first suppose to be homonymous (*i.e.* both right or both left) with the preceding. Then we have, from the first, supposed right

$$y_1 = +\alpha x_1 + \beta x_2 + \gamma x_3 + \delta x_4,$$
$$y_2 = -\beta x_1 + \alpha x_2 + \delta x_3 - \gamma x_4,$$
$$y_3 = -\gamma x_1 - \delta x_2 + \alpha x_3 + \beta x_4,$$
$$y_4 = -\delta x_1 + \gamma x_2 - \beta x_3 + \alpha x_4,$$

and, from the second vector,

$$\begin{aligned} z_1 &= +\ \alpha' y_1 + \beta' y_2 + \gamma' y_3 + \delta' y_4, \\ z_2 &= -\ \beta' y_1 + \alpha' y_2 + \delta' y_3 - \gamma' y_4, \\ z_3 &= -\ \gamma' y_1 - \delta' y_2 + \alpha' y_3 + \beta' y_4, \\ z_4 &= -\ \delta' y_1 + \gamma' y_2 - \beta' y_3 + \alpha' y_4. \end{aligned}$$

Substituting for y_1, y_2, y_3, y_4, we obtain the following values for z_1, z_2, z_3, z_4.

The right vector, $\alpha, \beta, \gamma, \delta$, followed by the right vector, $\alpha', \beta', \gamma', \delta'$—

$$\begin{aligned} z_1 &= \overline{+\alpha\alpha' - \beta\beta' - \gamma\gamma' - \delta\delta'}\,x_1 + \overline{+\alpha\beta' + \beta\alpha' + \gamma\delta' - \delta\gamma'}\,x_2 + \overline{\alpha\gamma' - \beta\delta' + \gamma\alpha' + \delta\beta'}\,x_3 + \overline{+\alpha\delta' + \beta\gamma' - \gamma\beta' + \delta\alpha'}\,x_4, \\ z_2 &= \overline{-\alpha\beta' - \beta\alpha' - \gamma\delta' + \delta\gamma'}\,x_1 + \overline{+\alpha\alpha' - \beta\beta' - \gamma\gamma' - \delta\delta'}\,x_2 + \overline{\alpha\delta' + \beta\gamma' - \gamma\beta' + \delta\alpha'}\,x_3 + \overline{-\alpha\gamma' + \beta\delta' - \gamma\alpha' - \delta\beta'}\,x_4, \\ z_3 &= \overline{-\alpha\gamma' + \beta\delta' - \gamma\alpha' - \delta\beta'}\,x_1 + \overline{-\alpha\delta' - \beta\gamma' + \gamma\beta' - \delta\alpha'}\,x_2 + \overline{\alpha\alpha' - \beta\beta' - \gamma\gamma' - \delta\delta'}\,x_3 + \overline{+\alpha\beta' + \beta\alpha' + \gamma\delta' - \delta\gamma'}\,x_4, \\ z_4 &= \overline{-\alpha\delta' - \beta\gamma' + \gamma\beta' - \delta\alpha'}\,x_1 + \overline{+\alpha\gamma' - \beta\delta' + \gamma\alpha' + \delta\beta'}\,x_2 + \overline{-\alpha\beta' - \beta\alpha' - \gamma\delta' + \delta\gamma'}\,x_3 + \overline{+\alpha\alpha' - \beta\beta' - \gamma\gamma' - \delta\delta'}\,x_4. \end{aligned}$$

The right vector, $\alpha', \beta', \gamma', \delta'$, followed by the right vector, $\alpha, \beta, \gamma, \delta$—

$$\begin{aligned} z_1 &= \overline{+\alpha\alpha' - \beta\beta' - \gamma\gamma' - \delta\delta'}\,x_1 + \overline{+\alpha\beta' + \beta\alpha' - \gamma\delta' + \delta\gamma'}\,x_2 + \overline{+\alpha\gamma' + \beta\delta' + \gamma\alpha' - \delta\beta'}\,x_3 + \overline{+\alpha\delta' - \beta\gamma' + \gamma\beta' + \delta\alpha'}\,x_4, \\ z_2 &= \overline{-\alpha\beta' - \beta\alpha' + \gamma\delta' - \delta\gamma'}\,x_1 + \overline{+\alpha\alpha' - \beta\beta' - \gamma\gamma' - \delta\delta'}\,x_2 + \overline{+\alpha\delta' - \beta\gamma' + \gamma\beta' + \delta\alpha'}\,x_3 + \overline{-\alpha\gamma' - \beta\delta' - \gamma\alpha' + \delta\beta'}\,x_4, \\ z_3 &= \overline{-\alpha\gamma' - \delta\beta' - \gamma\alpha' + \delta\beta'}\,x_1 + \overline{-\alpha\delta' + \beta\gamma' - \gamma\beta' - \delta\alpha'}\,x_2 + \overline{+\alpha\alpha' - \beta\beta' - \gamma\gamma' - \delta\delta'}\,x_3 + \overline{+\alpha\beta' + \beta\alpha' - \gamma\delta' + \delta\gamma'}\,x_4, \\ z_4 &= \overline{-\alpha\delta' + \beta\gamma' - \gamma\beta' - \delta\alpha'}\,x_1 + \overline{+\alpha\gamma' + \beta\delta' + \gamma\alpha' - \delta\beta'}\,x_2 + \overline{-\alpha\beta' - \beta\alpha' + \gamma\delta' - \delta\gamma'}\,x_3 + \overline{+\alpha\alpha' - \beta\beta' - \gamma\gamma' - \delta\delta'}\,x_4. \end{aligned}$$

We thus learn the important truth, that when two or more *homonymous* vectors are compounded, the order of their application must be carefully specified. For example, if the object x be first transposed by the vector α and then by α', it attains a position different from that it would have gained if first transposed by α' and then by α.

We see, however, that in either case two homonymous vectors compound into a vector homonymous with the two components.

We now study two *heteronymous* vectors, *i.e.* one right and one left.

The right vector, $\alpha, \beta, \gamma, \delta$, followed by the left vector, $\alpha', \beta', \gamma', \delta'$—

$$\begin{aligned} z_1 &= \overline{+\alpha\alpha' - \beta\beta' - \gamma\gamma' - \delta\delta'}\,x_1 + \overline{+\alpha\beta' + \beta\alpha' + \gamma\delta' - \delta\gamma'}\,x_2 + \overline{+\alpha\gamma' - \beta\delta' + \gamma\alpha' + \delta\beta'}\,x_3 + \overline{+\alpha\delta' + \beta\gamma' - \gamma\beta' + \delta\alpha'}\,x_4, \\ z_2 &= \overline{-\alpha\beta' - \beta\alpha' + \gamma\delta' - \delta\gamma'}\,x_1 + \overline{+\alpha\alpha' - \beta\beta' + \gamma\gamma' + \delta\delta'}\,x_2 + \overline{-\alpha\delta' - \beta\gamma' - \gamma\beta' + \delta\alpha'}\,x_3 + \overline{+\alpha\gamma' - \beta\delta' - \gamma\alpha' - \delta\beta'}\,x_4, \\ z_3 &= \overline{-\alpha\gamma' - \beta\delta' - \gamma\alpha' + \delta\beta'}\,x_1 + \overline{+\alpha\delta' - \beta\delta' - \gamma\beta' - \delta\alpha'}\,x_2 + \overline{+\alpha\alpha' + \beta\beta' - \gamma\gamma' + \delta\delta'}\,x_3 + \overline{-\alpha\beta' + \beta\alpha' - \gamma\delta' - \delta\gamma'}\,x_4, \\ z_4 &= \overline{-\alpha\delta' + \beta\gamma' - \gamma\beta' - \delta\alpha'}\,x_1 + \overline{-\alpha\gamma' - \beta\delta' + \gamma\alpha' - \delta\beta'}\,x_2 + \overline{+\alpha\beta' - \beta\alpha' - \gamma\delta' - \delta\gamma'}\,x_3 + \overline{+\alpha\alpha' + \beta\beta' + \gamma\gamma' - \delta\delta'}\,x_4. \end{aligned}$$

The left vector, $\alpha', \beta', \gamma', \delta'$, followed by the right vector, $\alpha, \beta, \gamma, \delta$—

$$\begin{aligned} z_1 &= \overline{+\alpha\alpha' - \beta\beta' - \gamma\gamma' - \delta\delta'}\,x_1 + \overline{+\alpha\beta' + \beta\alpha' + \gamma\delta' - \delta\gamma'}\,x_2 + \overline{+\alpha\gamma' - \beta\delta' + \gamma\alpha' + \delta\beta'}\,x_3 + \overline{+\alpha\delta' + \beta\gamma' - \gamma\beta' + \delta\alpha'}\,x_4, \\ z_2 &= \overline{-\alpha\beta' - \beta\alpha' + \gamma\delta' - \delta\gamma'}\,x_1 + \overline{+\alpha\alpha' - \beta\beta' + \gamma\gamma' + \delta\delta'}\,x_2 + \overline{-\alpha\delta' - \beta\gamma' - \gamma\beta' + \delta\alpha'}\,x_3 + \overline{+\alpha\gamma' - \beta\delta' - \gamma\alpha' - \delta\beta'}\,x_4, \\ z_3 &= \overline{-\alpha\gamma' - \beta\delta' - \gamma\alpha' + \delta\beta'}\,x_1 + \overline{+\alpha\delta' - \beta\gamma' - \gamma\beta' - \delta\alpha'}\,x_2 + \overline{+\alpha\alpha' + \beta\beta' - \gamma\gamma' + \delta\delta'}\,x_3 + \overline{-\alpha\beta' + \beta\alpha' - \gamma\delta' - \delta\gamma'}\,x_4, \\ z_4 &= \overline{-\alpha\delta' + \beta\gamma' - \gamma\beta' - \delta\alpha'}\,x_1 + \overline{-\alpha\gamma' - \beta\delta' + \gamma\alpha' - \delta\beta'}\,x_2 + \overline{+\alpha\beta' - \beta\alpha' - \gamma\delta' - \delta\gamma'}\,x_3 + \overline{+\alpha\alpha' + \beta\beta' + \gamma\gamma' - \delta\delta'}\,x_4. \end{aligned}$$

We thus learn the remarkable fact, that if a right (left) vector be followed by a left (right) vector, the effect produced is the same as if the order of the two vectors had been interchanged.

This is *not true* for two right vectors or two left vectors.

The theorems at which we have arrived may be thus generally enunciated:—

In the composition of vectors the order of two heteronymous vectors does not affect the result, but that of two homonymous vectors does affect the result.

In the composition of two homonymous vectors the result is also an homonymous vector. In the composition of two heteronymous vectors the result is not a vector at all.

The theorems just established constitute the first of the fundamental principles relating to the Theory of Screws in non-Euclidian Space referred to in § 396. Their importance is such that it may be desirable to give a geometrical investigation.

428. Geometrical proof that two Homonymous Vectors compound into one Homonymous Vector.

Left vectors cannot disturb any right generators of the infinite quadric. Take two such generators, AB and $A'B'$ (Fig. 48). Let AA', BB', CC' be three left generators which the first vector conveys to A_1A_1', B_1B_1', C_1C_1', and the second vector further conveys to A_2A_2', B_2B_2', C_2C_2'. Let X and Y be the double points of the two homographic systems defined by A, B, C and A_2, B_2, C_2. Then we have

$$(ABCX) = (A_2B_2C_2X),$$

and

$$(ABCY) = (A_2B_2C_2Y).$$

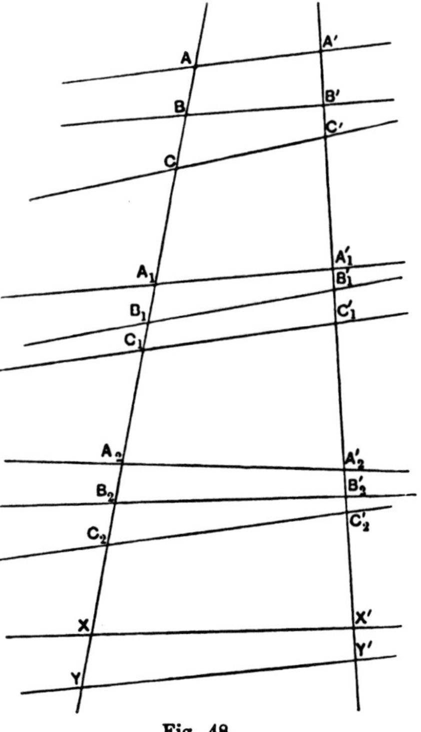

Fig. 48.

As anharmonic ratios cannot be altered by any rigid displacement, it follows that X and Y must each occupy the same position after the second vector which they had before the first, similarly, X' and Y' will remain unchanged, and as the two rays, AB and $A'B'$ are divided homographically, it follows that XX' and YY' are both generators. We therefore find that after the two vectors all the right generators remain as before, and so do also two left generators, *i.e.* the result of the two vectors could have been attained by a single vector homonymous therewith.

429. Geometrical proof of the Law of Permutability of Heteronymous Vectors.

Let AB and $A'B'$ be a pair of right generators (fig. 49), and CD and $C'D'$ a pair of left generators. Let the right vector convey P to Q, and

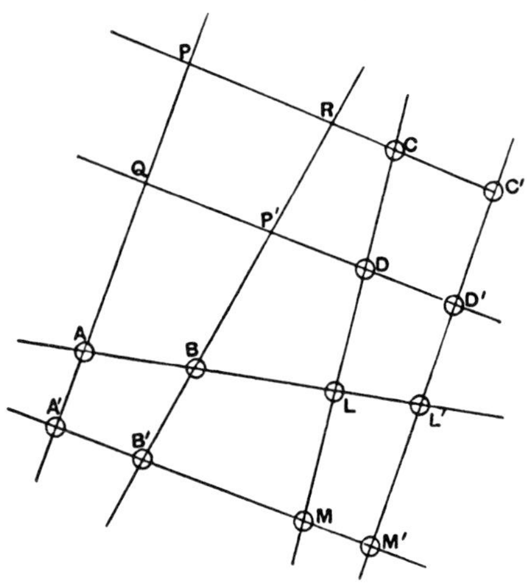

Fig. 49.

then let the left vector carry Q to the final position P'. We shall now show that P' would have been equally reached if P had gone first to R, so that intervene $PR = QP'$, and that then R was conveyed by the vector, $RB'B$, through a distance equal to PQ.

Draw through P the transversal $PRCC'$. Take R, so that

$$(PRCC') = (QP'DD');$$

but, because this relation holds,

$$PQ, \quad RP', \quad CD, \quad C'D'$$

must all lie on the same hyperboloid.

Therefore RP' must intersect AL and $A'M$, and therefore, also,

$$(PQAA') = (RP'BB').$$

Hence, finally, we have for the intervenes

$$PQ = RP' \text{ and } PR = QP'.$$

430. Determination of the Two Heteronymous Vectors equivalent to any given Motor.

If a right vector, α, β, γ, δ, be followed by a left vector, α', β', γ', δ', then the result obtained is a displacement of the most general type called a *motor*. We now prove Clifford's great theorem that a right vector and a left vector can be determined so as to form any motor, i.e. to accomplish any required homographic transformation that conserves intervene.

For, if we identify the several coefficients at the foot of p. 474 with those of § 420, we obtain equations of the type

$$\begin{aligned}(11) &= \alpha\alpha' - \beta\beta' - \gamma\gamma' - \delta\delta',\\ (21) &= -\alpha\beta' - \beta\alpha' + \gamma\delta' - \delta\gamma',\\ (31) &= -\alpha\gamma' - \beta\delta' - \gamma\alpha' + \delta\beta',\\ (41) &= -\alpha\delta' + \beta\gamma' - \gamma\beta' - \delta\alpha'.\end{aligned}$$

These can be simply reduced to a linear form; for multiply the first by α', and the second, third, and fourth by $-\beta'$, $-\gamma'$, $-\delta'$, respectively, and add, we obtain

$$(11)\,\alpha' - (21)\,\beta' - (31)\,\gamma' - (41)\,\delta' = \alpha;$$

for

$$\alpha'^2 + \beta'^2 + \gamma'^2 + \delta'^2 = 1.$$

In a similar manner we obtain a number of analogous equations, which are here all brought together for convenience—

$$\begin{aligned}
(11)\,\alpha' - (21)\,\beta' - (31)\,\gamma' - (41)\,\delta' &= \alpha,\\
-(21)\,\alpha' - (11)\,\beta' + (41)\,\gamma' - (31)\,\delta' &= \beta,\\
-(31)\,\alpha' - (41)\,\beta' - (11)\,\gamma' + (21)\,\delta' &= \gamma,\\
-(41)\,\alpha' + (31)\,\beta' - (21)\,\gamma' - (11)\,\delta' &= \delta.\\
+(22)\,\alpha' + (12)\,\beta' - (42)\,\gamma' + (32)\,\delta' &= \alpha,\\
+(12)\,\alpha' - (22)\,\beta' - (32)\,\gamma' - (42)\,\delta' &= \beta,\\
+(42)\,\alpha' - (32)\,\beta' + (22)\,\gamma' + (12)\,\delta' &- \gamma,\\
-(32)\,\alpha' - (42)\,\beta' - (12)\,\gamma' + (22)\,\delta' &= \delta.\\
+(33)\,\alpha' + (43)\,\beta' + (13)\,\gamma' - (23)\,\delta' &= \alpha,\\
+(43)\,\alpha' + (33)\,\beta' - (23)\,\gamma' - (13)\,\delta' &= \beta,\\
-(13)\,\alpha' - (23)\,\beta' - (33)\,\gamma' - (43)\,\delta' &= \gamma,\\
+(23)\,\alpha' + (13)\,\beta' - (43)\,\gamma' + (33)\,\delta' &= \delta.\\
+(44)\,\alpha' - (34)\,\beta' + (24)\,\gamma' + (14)\,\delta' &= \alpha,\\
+(34)\,\alpha' + (44)\,\beta' + (14)\,\gamma' - (24)\,\delta' &= \beta,\\
-(24)\,\alpha' - (14)\,\beta' + (44)\,\gamma' - (34)\,\delta' &= \gamma,\\
+(14)\,\alpha' - (24)\,\beta' - (34)\,\gamma' - (44)\,\delta' &= \delta.
\end{aligned}$$

These will enable α, β, γ, δ and α', β', γ', δ' to be uniquely determined.

431. The Pitch of a Motor.

Any small displacement of a rigid system in the content can be produced by a rotation (see § 417) α about one line followed by a rotation β about its conjugate polar with respect to the infinite quadric, the amplitudes of both rotations being small quantities. The two movements taken together constitute the motor. It will be necessary to set forth the conception in the theory of the motor, which is the homologue of the conception of *pitch* in the Theory of Screws in ordinary space. The pitch can most conveniently be expressed by the function

$$p = \frac{2\alpha\beta}{\alpha^2 + \beta^2}.$$

If either α or β vanish, then the pitch becomes zero. The motor then degenerates to a pure rotation about one or other of the two conjugate polars. This, of course, agrees with the ordinary conception of the pitch, which is zero whenever the general screw motion of the rigid body degrades to a pure rotation.

In ordinary space we have

$$p\alpha = d\beta,$$

where β is zero and where d is infinite. In this case

$$\frac{2\alpha\beta}{\alpha^2 + \beta^2} = 2\frac{\beta}{\alpha} = 2\frac{p}{d},$$

i.e. the pitch is proportional to the function now under consideration.

No generality will be sacrificed by the use of a single symbol to express the pitch. We may make $\alpha = \cos\theta$ and $\beta = \sin\theta$; the pitch then assumes the very simple form $\sin 2\theta$. We thus see that the pitch can never exceed unity.

If the motor be a vector, then we have $\beta = \pm\alpha$, or $\theta = \pm 45°$, and the pitch is simply ± 1.

It should be noticed that a rotation α about the line A, and a rotation β about its conjugate polar B, constitute a motor of the same pitch as a rotation β about A and α about B.

432. Property of Right and Left Vectors.

To take the next step it will be necessary to discuss some of the relations between right and left vectors. A right vector will displace any point P in a certain direction PA; a left vector will displace the same point in the direction PB. It will, of course, usually happen that the directions PA and PB are not identical. It is, however, necessary for us to observe that when P is situated on either of two rays then the directions of displacement are identical. To determine these two rays we draw the two pairs of generators corresponding to the two vectors. As these generators belong to opposite systems, they will form four edges of a tetrahedron. The two remaining edges are a pair of conjugate polars, and they form the two rays of which we are in search. The proof is obvious: a point P on one of these rays must be displaced along the same ray by either of the vectors, for this ray intersects both of the generators which define that vector.

Let a right vector consist of rotations $+\alpha$, $+\alpha$ about two conjugate polars, and let a left vector consist of rotations $+\alpha$, $-\alpha$, also about two conjugate polars. Without loss of generality we may take the two conjugate polars in both cases to be the pair just determined.

Let OO' and PP' be two conjugate polars (fig. 50). The right vector is appropriate to the generators OP and $O'P'$. The left vector to the generators

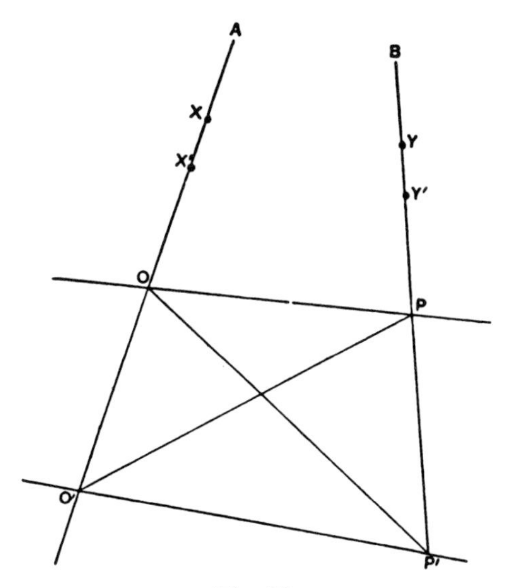

Fig. 50.

OP' and $O'P$. If we take the intersections with the quadric in the order OO' for A, then we must take them on B in the order PP' if we are considering a right vector, and in the order $P'P$ if we are considering a left vector. This is obvious, for in the first case we take the intersections of the conjugate polar with the generators OP and $O'P'$. In the second case we take the intersection of the conjugate polar with OP' and $O'P$.

If, therefore, the vector be right, we have for the displacements of X and Y,

$$H \log (XX'OO') = H \log (YY'PP').$$

If, however, the vector be left, then Y must be displaced to a distance Y_0, defined by

$$H \log (XX'OO') = H \log (YY_0P'P):$$

we therefore have

$$H \log (YY'PP') = H \log (YY_0P'P);$$

but from an obvious property of the logarithms,

$$H \log (YY_0P'P) = - H \log (YY_0PP');$$

whence, finally,

$$H \log (YY'PP') = - H \log (YY_0PP').$$

We hence have the important result, that the intervene through which a point on one of the common conjugate polars is displaced by one of two heteronymous vectors of equal amplitude, merely differs in sign from the displacement which the same point would receive from the other vector.

433. The Conception of Force in non-Euclidian Space.

In ordinary space we are quite familiar with the perfect identity which subsists between the composition of small rotations and the composition of forces. We shall now learn that what we so well know in ordinary space is but the survival, in an attenuated form, of a much more complete theory in non-Euclidian space. We have in non-Euclidian space force-motors and force-vectors, just as we have displacement-motors and displacement-vectors. We shall base the Dynamical theory on an elementary principle in the theory of Energy. Suppose that a force of intensity f act on a particle which is displaced in a direction directly opposed to the force through a distance δ, then the quantity of work done is denoted by $-f\delta$.

434. Neutrality of Heteronymous Vectors.

We are now able to demonstrate a very important theorem which lies at the foundation of all the applications of Dynamics in non-Euclidian space. The virtual moment of a force-vector and a displacement-vector will always vanish when the vectors are homonymous and at right angles. The analogies of ordinary geometry would have suggested this result, and it is easily shown to be true. If, however, the two vectors be not homonymous, the result is extremely remarkable. *The two vectors must then have their virtual moment zero under all circumstances.*

The proof of this singular proposition is very simple. Let the two vectors be what they may, we can always find one pair of conjugate polars which belong to them both. Let the two forces be λ, λ on the two conjugate polars, and let the displacements be μ, $-\mu$, then the work done is

$$\lambda\mu - \lambda\mu = 0.$$

It seems at first sight incredible that a theorem demonstrated with such simplicity should be of so much significance. It is not too much to say that the theory of Rigid Dynamics in non-Euclidian space depends, to a large extent, upon this result. This is the second of the two fundamental theorems referred to in § 396.

These two principles open up a geometrical Theory of Screws in non-Euclidian space. This is a subject too extensive to be here entered into any further. It is hoped that the present chapter will at least have conducted the reader to a point from which he can obtain a prospect of a great field of work. The few incursions that have as yet been made into this field (see the bibliographical notes) have shown the exceeding richness and interest of a region that still awaits a more complete exploration.

APPENDIX I.

NOTE I.

Another solution of the problem of § 28.

Let the intensities of the wrenches on α, β, ... η be as usual denoted by α'', β'', ... η'' respectively.

As the wrenches are to equilibrate we must have (§ 12)

$$\alpha''\varpi_{\alpha\lambda} + \beta''\varpi_{\beta\lambda} + \gamma''\varpi_{\gamma\lambda} + \delta''\varpi_{\delta\lambda} + \epsilon''\varpi_{\epsilon\lambda} + \zeta''\varpi_{\zeta\lambda} + \eta''\varpi_{\eta\lambda} = 0,$$

where λ is any screw whatever.

If six different but independent screws be chosen in succession for λ we have six independent linear equations, and thus $\alpha'' \div \beta''$ and the other ratios are known.

But the process will be much simplified by judicious choice of λ. If, for instance, we take as λ the screw ψ which is reciprocal to the five screws γ, δ, ϵ, ζ, η then we have

$$\varpi_{\gamma\psi} = 0, \quad \varpi_{\delta\psi} = 0, \quad \varpi_{\epsilon\psi} = 0, \quad \varpi_{\zeta\psi} = 0, \quad \varpi_{\eta\psi} = 0,$$

and we obtain

$$\alpha''\varpi_{\alpha\psi} + \beta''\varpi_{\beta\psi} = 0.$$

Let ρ be a screw on the cylindroid defined by α and β. Then wrenches on α, β, ρ will equilibrate (§ 14) provided their intensities are proportional respectively to

$$\sin(\beta\rho), \quad \sin(\rho\alpha), \quad \sin(\alpha\beta).$$

It follows that for any screw μ we must have

$$\sin(\beta\rho)\,\varpi_{\alpha\mu} + \sin(\rho\alpha)\,\varpi_{\beta\mu} + \sin(\alpha\beta)\,\varpi_{\rho\mu} = 0.$$

This is indeed a general relation connecting the virtual coefficients of three screws on a cylindroid with any other screw.

Let us now suppose that μ was the screw ψ just considered, and let us further take ρ to be that one screw on the cylindroid (α, β) which is reciprocal to ψ.

Then

$$\varpi_{\rho\psi} = 0,$$

and we have

$$\sin(\beta\rho)\,\varpi_{\alpha\psi} + \sin(\rho\alpha)\,\varpi_{\beta\psi} = 0.$$

But since the seven screws are independent both $\varpi_{\alpha\psi}$ and $\varpi_{\beta\psi}$ must be, in general, different from zero, whence by the former equation we have

$$\frac{\alpha''}{\sin(\beta\rho)} = \frac{\beta''}{\sin(\rho\alpha)}.$$

Thus we obtain the following theorem (§ 28).

If seven wrenches on seven given screws equilibrate and if the intensity α'' of one of the seven wrenches be given then the intensity of the wrench on any one β of the remaining six screws can be determined as follows.

Find the screw ψ reciprocal to the five screws remaining when α and β are excluded from the seven.

On the cylindroid $(\alpha\beta)$ find the screw ρ which is reciprocal to ψ.

Resolve the given wrench α'' on α into component wrenches on β and on ρ.

Then the intensity of the component wrench thus found on β is the required intensity β'' with its sign changed.

NOTE II.

Case of equal roots in the Equation determining Principal Screws of Inertia, § 86.

We have already made use of the important theorem that if U and V are both homogeneous quadratic functions of n variables, then the discriminant of $U + \lambda V$ when equated to zero must have n real roots for λ provided that *either* U or V admits of being expressed as the *sum* of n squares (§ 85).

The further important discovery has been made that whenever this determinantal equation has a repeated root, then every minor of the determinant vanishes (Routh, *Rigid Dynamics*, Part II. p. 51, 1892).

This theorem is of much interest in connection with the Principal Screws of Inertia. The result given at the end of § 86 is a particular case. It may be further presented as follows.

Taking the case of an n system each root of λ will give n equations

$$-2\lambda\theta_1 = \frac{1}{p_1}\frac{dT}{d\theta_1}, \ldots -2\lambda\theta_n = \frac{1}{p_n}\frac{dT}{d\theta_n}.$$

Of these $n-1$ are in general independent and these suffice to indicate the values of $\theta_1, \ldots \theta_n$.

But in the case of a root once repeated the theorem above stated shows that we have not more than $n-2$ independent equations in the series. The principal Screw of Inertia corresponding to this root is therefore indeterminate.

But it has a locus found from the consideration that besides these $n-2$ linear equations it must also satisfy the $6-n$ linear equations which imply that it belongs to the n-system.

In other words the co-ordinates satisfy $(n-2)+(6-n)$, *i.e.* four linear equations. But this is equivalent to saying that the screw lies on a cylindroid. We have thus the following result.

In the case when the equation for λ has two equal roots, there must be $n-2$ separate and distinct principal screws of inertia and also a cylindroid of which every screw is a principal Screw of Inertia.

For every value of n from 1 to 6 it is of course known that the celebrated harmonic determinantal equation of § 86 has n real roots.

But when the question arises as to the possibility of this equation as applied to our present problem having repeated roots, the several cases must be discriminated. It is to be understood that the body itself is to be of a general type without having *e.g.* two of the principal radii of gyration equal. The investigation relates to the possibility of a system of constraints which, while the body is still of the most general type, shall permit indeterminateness in the number of principal Screws of Inertia.

Of course if $n=1$ the equation is linear and has but a single root.

$n=2$. The equation under certain conditions may have two equal roots.

$n=3$. The equation under certain conditions may have two or three equal roots.

$n=4$. The equation under certain conditions may have two equal roots or three equal roots or four equal roots or two pairs of equal roots.

$n=5$. The equation can never have a repeated root.

$n=6$. The equation can never have a repeated root.

Here comes in the restriction that the body is of a general type, for of course the last statement could not be true if two of the radii of gyration are equal or if one of them was zero.

The curious contrast between the two last cases and those for the smaller values of n may be thus accounted for. The expression for T will contain $\frac{n(n+1)}{2}$ terms and the ratios only being considered T will contain

$$\frac{n(n+1)}{2}-1=\frac{(n+2)(n-1)}{2}$$

distinct parameters. As a rigid body is specified both as to position and character by 9 co-ordinates, it follows that the coefficients of T are not unrestricted if $\frac{(n+2)(n-1)}{2}$ is greater than 9. But this quantity is greater than 9 for the cases of $n=5$ and $n=6$.

We may put the matter in another way which will perhaps make it clearer. I shall take the two cases of $n=4$ and $n=5$.

In the case of $n=4$ the function T will consist of 10 terms such as

$$A_{11}\theta_1^2 + \ldots + 2A_{12}\theta_1\theta_2 + \ldots.$$

If any arbitrary values be assigned to A_{11}, A_{12} &c., it will still be possible to determine a rigid body such that this function shall represent u_θ^2 (to a constant factor), because we have 9 co-ordinates disposable in the rigid body. Hence for $n=4$ and *a fortiori* for any value of n less than four the function representing T will be a function in which the coefficients are perfectly unrestricted. Hence $n=$ or <4 the determinantal equation is in our theory of the most general type. The general theory while affirming that all the roots are real does not prohibit conditions arising under which roots are repeated. Hence Routh's important theorem becomes of significance in cases $n=2$, $n=3$, $n=4$ for in these equations the roots may be repeated.

But in the case of $n=5$ the function T consists of 15 terms. If arbitrary values could be assigned to the coefficients then of course the general theory would apply and cases of repeated roots might arise. But in our investigation the 15 coefficients are functions of the co-ordinates which express the most general place of a rigid body, and these co-ordinates are not more than nine. If these nine co-ordinates were eliminated we should have five conditions which must be satisfied by the coefficients of a general function before it could represent the T of our theory even to within a factor. The necessity that the coefficient of T shall satisfy these equations imports certain restrictions into the general theory of the determinantal equation based on T. One of these restrictions is that T shall have no repeated roots. The same conclusion applies *a fortiori* to the case of $n=6$.

The subject may also be considered as follows.

Let us first take the general theorem that when reference is made to n principal screws of inertia of an n-system the co-ordinates of the impulsive wrench corresponding to the instantaneous screw

$$\alpha_1 \ldots \alpha_n,$$

are (§ 97)

$$\frac{u_1^2}{p_1}\alpha_1 \ldots \frac{u_n^2}{p_n}\alpha_n.$$

For a principal screw of inertia the ratios must be severally equal or

$$\frac{u_1^2}{p_1}\frac{\alpha_1}{\alpha_1} = \frac{u_2^2}{p_2}\frac{\alpha_2}{\alpha_2} = \ldots = \frac{u_n^2}{p_n}\frac{\alpha_n}{\alpha_n}.$$

These equations can generally be only satisfied if $n-1$ of the quantities

$$\alpha_1 \ldots \alpha_n,$$

be zero, *i.e.* there are in general no more than the n principal screws of inertia.

If however

$$\frac{u_1^2}{p_1} = \frac{u_2^2}{p_2},$$

then though we must have

$$\alpha_3 = 0 \ldots \alpha_n = 0,$$

α_1 and α_2 remain arbitrary.

But $n-2$ linear equations in an n-system determine a cylindroid and hence we see that all the screws on this cylindroid will be principal Screws of Inertia.

In like manner if there be k repeated roots, *i.e.* if

$$\frac{u_1^2}{p_1} = \frac{u_2^2}{p_2} = \dots \frac{u_k^2}{p_k},$$

then $a_1, \dots a_k$ are arbitrary but $a_{k+1}, \dots a_n$ must be each zero. We have thus $n-k$ linear equations in the co-ordinations. They must also satisfy $6-n$ equations because they belong to the n-system and therefore they satisfy in all

$$6 - n + n - k = 6 - k \text{ equations},$$

whence we deduce that

If there be k repeated roots in the determinantal equation of § 86 then to those roots corresponds a k-system of screws each one of which is a principal screw of inertia and there are besides n − k additional principal Screws of Inertia.

So far as the cases of $n = 2$ and $n = 3$ are concerned the plane representations of Chaps. XII. and XV. render a complete account of the matter.

Let O (Fig. 10) be the pole of the axis of pitch, § 58, then O may lie either inside or outside the circle whose points represent the screws on the cylindroid.

Let O' (Fig. 22) be the pole of the axis of inertia, § 140, then O' must lie *inside* the circle, for otherwise the polar of O' would meet the circle, *i.e.* there would be one or two real screws about which the body could twist with a finite velocity but with zero kinetic energy.

We have seen that the two Principal Screws of Inertia are the points in which the chord OO' cuts the circle. If O' could be on the circle or outside the circle then we might have the two principal Screws of Inertia coalescing, or we might have them both imaginary. As however O' must be within the circle it is generally necessary that the two principal Screws of Inertia shall be both real and distinct.

But the points O and O' might have coincided. In this case every chord through O would have principal Screws of Inertia at its extremities. Thus every point on the circle is in this case a principal Screw of Inertia.

We thus see that with Freedom of the second order there are only two possible cases. Either every screw on the cylindroid is a principal Screw of Inertia or there are neither more nor fewer than two such screws, and both real.

If α and β be any two screws on the cylindroid then the conditions that all the screws are Principal Screws of Inertia are

$$p_\alpha u^2_{\alpha\beta} = \varpi_{\alpha\beta} u_\alpha^2,$$

$$p_\beta u^2_{\alpha\beta} = \varpi_{\alpha\beta} u_\beta^2.$$

With any rigid body in any position we can arrange any number of cylindroids which possess the required property. Choose *any* screw α and then take *any* screw β whose co-ordinates satisfy these two conditions.

We shall also use the plane representation of the 3-system.

Let $U=0$ be the pitch conic.

$V=0$ be the imaginary ellipse obtained by equating to zero the expression for the Kinetic Energy.

Then the vertices of a common conjugate triangle are of course the principal Screws of Inertia and generally there is only one such triangle.

It may however happen that U and V have more than a single common conjugate triangle, for let the cartesian co-ordinates of the four intersections of U and V be represented by

$$x_1,\ y_1;\ x_2,\ y_2;\ x_3,\ y_3;\ x_4,\ y_4.$$

As all the points on V are imaginary at least one co-ordinate of each intersection is imaginary. Suppose y_1 to be imaginary then it must be conjugate to y_2. If therefore the conic U touches V y_1 and y_3 must be respectively equal to y_2 and y_4. Hence we have only two values of y, and these are conjugate. Substituting these in U and V we see that there can only be two values of x, and consequently the intersections reduce to two pairs of coincident points.

Hence we see that V cannot touch U unless the two conics have double contact.

In this case the chord of contact possesses the property that each point on it is a principal Screw of Inertia while the pole of the chord with respect to either conic is also a principal Screw of Inertia.

If U and V coincided then every screw of the 3-system would be a principal Screw of Inertia.

The general theory on the subject is as follows.

Let $U=0$ be the quadratic relation among the co-ordinates of an n-system which expresses that its pitch is zero.

Let $V=0$ be the quadratic relation among the co-ordinates of a screw if a body twisting about that screw has zero kinetic energy.

The discriminant of $S=U-\lambda V$ equated to zero gives n real roots for λ. These roots substituted in the differential coefficients of S equated to zero give the corresponding principal Screws of Inertia. If however there be two equal roots for λ then for these roots every first minor of the discriminant vanishes. In this case S can be expressed as a function of $n-2$ linear quantities. Perhaps the most explicit manner of doing this is as follows.

Let $$S=a_{11}\theta_1^2-a_{22}\theta_2^2+2a_{12}\theta_1\theta_2+\ldots-a_{nn}\theta_n^2,$$

and let $$s_1=\frac{1}{2}\frac{dS}{d\theta_1},\ \ldots\ s_n=\frac{1}{2}\frac{dS}{d\theta_n}.$$

If all the first minors of the discriminant of S vanish we must have the following identity

$$-S\begin{vmatrix} a_{33}, & a_{34} \ldots a_{3n} \\ a_{43}, & a_{44} \ldots a_{4n} \\ \vdots & \\ \cdots & \cdots \\ a_{n}, & a_{n4} \ldots a_{nn} \end{vmatrix} = \begin{vmatrix} 0, & s_3, & s_4 \ldots s_n \\ s_3, & a_{33}, & a_{34} \ldots a_{3n} \\ s_4, & a_{43}, & a_{44} \ldots a_{4n} \\ \cdots & \cdots & \cdots \\ s_n, & a_{n3}, & a_{n4} \ldots a_{nn} \end{vmatrix},$$

by which we have

$$S = A_{33}s_3^2 + A_{44}s_4^2 + 2A_{34}s_3s_4 \ldots + A_{nn}s_n^2.$$

Hence $$U + \lambda V = A_{33}s_3^2 + A_{44}s_4^2 + 2A_{34}s_3s_4 \ldots + A_{nn}s_n^2.$$

In the case of $n=3$ we have

$$U + \lambda V = A_{33}s_3^2,$$

which proves that V and U have double contact as we already proved in a different manner.

In the general case all the differential coefficients of S will vanish if $s_3=0 \ldots s_n=0$, but these latter define a cylindroid and therefore whenever the discriminant of s has two equal roots, every screw on a certain cylindroid is a principal Screw of Inertia.

If the discriminant had three equal roots then S could be expressed in terms of $s_4, \ldots s_n$ and in this case every screw on a certain 3-system would be a principal Screw of Inertia.

If $n-1$ of the roots of the discriminant were equal, then every $(n-2)$nd minor would vanish, S would become the perfect square s_n^2 to a factor.

And we have

$$U + \lambda V = A_{nn}s_n^2.$$

In this case every screw of the $n-1$ system defined by $s_n = 0$ will be a principal Screw of Inertia.

NOTE III.

Twist velocity acquired by an impulsive wrench, § 90.

The problem solved in § 90 may be thus stated.

A body of mass M only free to twist about α is acted upon by a wrench of intensity η''' on a screw η. Find the twist velocity acquired.

From Lagrange's equations we have, § 86

$$\frac{d}{dt}\left(\frac{dT}{d\dot{\alpha}}\right) - \frac{dT}{d\alpha} = 2\eta''\varpi_{\eta\alpha}.$$

But as the wrench is very great the initial acceleration is great and consequently the second term on the left-hand side is negligible compared with the first.

Whence
$$\frac{dT}{d\dot{a}} = 2\varpi_{\eta a}\int \eta'' dt = 2\varpi_{\eta a}\eta''',$$

but
$$T = Mu_a^2\dot{a}^2 \text{ (§ 89)},$$

whence
$$2Mu_a^2\dot{a} = 2\varpi_{\eta a}\eta''',$$

or
$$\dot{a} = \frac{1}{M}\frac{\varpi_{\eta a}}{u_a^2}\eta'''.$$

The kinetic energy
$$T = Mu_a^2\dot{a}^2 = \frac{1}{M}\frac{\varpi^2_{\eta a}}{u_a^2}\eta'''^2, \quad \text{(§ 91)}.$$

NOTE IV.

Professor C. J. Joly's theory of the triple contact of conic and cubic.

Professor C. J. Joly has pointed out to me that the conics of § 162 and § 165 which have triple contact with the nodal cubic are but particular instances of the more general theory which he investigates as follows.

Let t be the parameter necessary to define a particular generator on a given cylindroid; we first show that the condition that a line, *i.e.* a screw θ of zero pitch should intersect this generator may be expressed in the form

$$at^3 + bt^2 + ct + d = 0,$$

where a, b, c, d are linear functions of the co-ordinates of θ and, of course, functions also of the constants defining the cylindroid.

For if α and β be the two principal screws on the cylindroid then the co-ordinates of the screw ϵ on the cylindroid making an angle λ with α are

$$\cos\lambda\alpha_1 + \sin\lambda\beta_1, \ \ldots \ \cos\lambda\alpha_6 + \sin\lambda\beta_6,$$

whence
$$\varpi_{\epsilon\theta} = \cos\lambda\varpi_{\epsilon\alpha} + \sin\lambda\varpi_{\epsilon\beta},$$
$$\cos(\epsilon\theta) = \cos\lambda\cos(\epsilon\alpha) + \sin\lambda\cos(\epsilon\beta),$$
$$p_\epsilon = \cos^2\lambda p_\alpha + 2\cos\lambda\sin\lambda\varpi_{\alpha\beta} + \sin^2\lambda p_\beta.$$

If ϵ and θ intersect then

$$2\varpi_{\epsilon\theta} = \cos(\epsilon\theta)\,p_\epsilon,$$

or putting
$$t = \tan\lambda,$$

$$\begin{aligned}
&(\cos(\epsilon\beta)\,p_\beta - 2\varpi_{\epsilon\beta})\,t^3 \\
&+ (\cos(\epsilon\alpha)\,p_\beta + 2\cos(\epsilon\beta)\,\varpi_{\alpha\beta} - 2\varpi_{\epsilon\alpha})\,t^2 \\
&+ (\cos(\epsilon\beta)\,p_\alpha + 2\cos(\epsilon\alpha)\,\varpi_{\alpha\beta} - 2\varpi_{\epsilon\beta})\,t \\
&+ (\cos(\epsilon\alpha)\,p_\alpha - 2\varpi_{\epsilon\alpha}) = 0,
\end{aligned}$$

which has the form just given.

Conversely if we are given a, b, c, d we have a cubic equation in t which on solution determines the three generators of the cylindroid which a given line intersects.

If the generators are connected in pairs by a one-to-one relation of the type

$$ltt' + m(t + t') + n = 0,$$

we may for convenience speak of the pairs of generators as being in "involution."

Suppose that two of the generators met by an arbitrary line are in "involution" we have two roots of the cubic

$$at^3 + bt^2 + ct + d = 0,$$

connected by the relation

$$lt_1t_2 + m(t_1 + t_2) + n = 0,$$

where t_1 and t_2 are the parameters of the two generators and of course roots of the cubic. Let the third root be t_3 and form the product P of the three factors

$$lt_1t_2 + m(t_1 + t_2) + n,$$

$$lt_2t_3 + m(t_2 + t_3) + n,$$

$$lt_3t_1 + m(t_3 + t_1) + n.$$

If we replace the symmetric functions of the roots by their values we find that P is a homogeneous function of a, b, c, d in the second degree.

The equation $P = 0$ represents the complex of transversals intersecting corresponding generators of the involution. This complex is of the second order and the transversals in a plane therefore envelop a conic and those through a point lie on a quadric cone.

In like manner the discriminant of the cubic itself when equated to zero represents a complex Q of the fourth order which consists of all the tangents to the cylindroid. The lines in a plane envelop a curve of the fourth class (the section of the cylindroid) and the lines through a point are generators of the tangent cone of the fourth order.

Let us now consider the lines common to the two complexes P and Q.

If we suppose two roots of the cubic equal, for example

$$t_3 = t_2,$$

then $$P = [lt_1t_2 + m(t_1 + t_2) + n]^2 [lt_2^2 + 2mt_2 + n].$$

The common lines fall into two groups (1) transversals of the united lines of the "Involution" where the parameters of these united lines satisfy $lt^2 + 2mt + n = 0$, and (2) where the odd point on the transversal coincides with one of the points in which the transversal meets the conjugate generators. The occurrence of the square factor shows that these latter lines are to be counted twice.

In any plane we have belonging to these complexes eight common lines which are the common tangents of a curve of the fourth and a curve of the second class. Two and only two of these intersect the united lines of the involution. The occurrence of the square factor indicates that the remaining six coincide in pairs and hence we have the general result that the conic has triple contact with the cubic.

NOTE V.

Remarks on § 210.

Professor C. J. Joly has communicated to me the following theorems with regard to the cubic which is the locus of the points corresponding to the screws of a 3-system which intersect a given screw of the system (§ 210).

Let O be the double point on the cubic and P_1, P_2 the two points corresponding to a pair of screws of equal pitch which intersect O. Then all the chords P_1P_2 for different pitches are concurrent.

For the cylindroid defined by the screws corresponding to P_1P_2 must be cut by the screw corresponding to O in a third point which lies on the generator O' of the cylindroid such that O and O' are at right angles (§ 22). As there is only one screw of the 3-system intersecting O at right angles it follows that all the chords P_1P_2 will be concurrent. The point corresponding to O' is that whose co-ordinates are given on p. 213, viz.

$$\frac{p_3 - p_2}{\alpha_1}, \quad \frac{p_1 - p_3}{\alpha_2}, \quad \frac{p_2 - p_1}{\alpha_3},$$

where α_1, α_2, α_3 are the co-ordinates of O.

There is also to be noted the construction for the tangents at the double point of the cubic. They are the lines to the points in which the pitch-conic through the double point is met by the polar of the double point with respect to the conic of infinite pitch.

Let $S_p = 0$ be the conic of pitch p. Let $P_p = 0$ be the polar of O with respect to the conic of pitch p and let $S_p' = 0$ be the result of substituting the co-ordinates of O in the equation of the conic $S_p = 0$.

As all the conics have four points common, suppose

$$S_0 = kS_p + lS_\infty,$$

where k and l are certain constants.

Likewise $$P_0 = kP_p + lP_\infty; \quad S_0' = kS_p' + lS'_\infty,$$

whence after a few steps (§ 210) we have the new form for the cubic

$$2P_pS_\infty S'_\infty - P_\infty (S_pS'_\infty + S_\infty S_p') = 0.$$

If $S_p = 0$ passes through the double point then $S_p' = 0$ and the cubic is

$$2P_pS_\infty - P_\infty S_p = 0,$$

we form the polar conic of the point O by changing separately S to P and P to S' and we have its equation as

$$2S_p' S_\infty + P_p P_\infty - S'_\infty S_p = 0,$$

which as $S_p' = 0$ becomes

$$P_p P_\infty - S'_\infty S_p = 0.$$

Repeating this operation we have for the equation of the polar of O or the tangent to the cubic the vanishing expression

$$P_p S'_\infty - S'_\infty P_p \equiv 0.$$

This proves the duplicity of the point.

Therefore $P_p P_\infty - S'_\infty S_p = 0$ represents the tangents at the point and these accordingly pass through the intersection of $P_\infty = 0$ and $S_p = 0$.

I may also add that the principles here laid down will enable us to investigate the various relations between the screws of a 3-system which intersect. Let us seek for example the number of screws of the system which are common transversals to two screws which also belong to the system and which are represented by O and O'. If we draw two cubics of the class just considered from O and O' as double points, they will in general intersect in nine distinct points. Of these, four will of course be the points common to all these cubics on the conic of infinite pitch. We have thus five remaining intersections each of which corresponds to a screw of the system, whence we deduce the theorem that any two screws of a 3-system will in general be both intersected by five other screws of the 3-system.

NOTE VI.

Remarks on § 224 by Professor C. J. Joly.

If there is no speciality the nodal curve of the sextic ruled surface of the quadratic 2-system is of the tenth degree with four triple points on the surface. Of course every generator of the surface meets four other generators; this follows from the plane representation. An arbitrary section is a unicursal sextic having therefore $\frac{1}{2}(6-1)(6-2) = 10$ double points. A section through a generator is the generator plus a unicursal quintic, and a section through two generators consists of the generators and a trinodal quartic. When the director cone of the surface breaks into a pair of planes, the nodal curve rises to the eleventh degree and consists of the two double lines, the common generator and the remaining curve of intersection of the two cylindroids into which the surface degrades. The four triple points are those in which the double lines of one cylindroid meets the other —not on the common generator. We should expect to find four triads of concurrent axes belonging to the quadratic system.

The locus of the feet of perpendiculars from an arbitrary origin is a twisted quartic. The quartic is not the intersection of two quadrics. Only one quadric can be drawn through it. This is borne out in the case of the two cylindroids. The two conics have but one point common, and the only quadric through both consists of their planes. The line of intersection of the planes intersects the conics in three distinct points, and hence another quadric cannot be drawn through the conics.

As regards the ruled surfaces generated by the axes of a three-system which are parallel to the edges of a cone of degree m, the degree of the surface is evidently $3m$. For the axes of the enclosing system which meet any assumed line are parallel to the edges of a cubic cone, and there are $3m$ directions common to this cone and the director cone. Again the locus of the feet of perpendiculars on the generators from any point is a curve of degree $2m$ which viewed from the point appears to have three multiple points of order m situated on the axes of the reciprocal three-system passing through the point. For if we take any plane and consider its intersections with the curve, we find easily that the axes of the enclosing system whose feet of perpendiculars from the point lie in the plane are parallel to the edges of a *quadric* cone. The theorem about the apparent multiple points follows from consideration of the cylindroids of the enclosing system whose double lines pass through the assumed point.

We also note this construction for Art. 180. Assume a radius of the pitch quadric, draw the tangent plane and let fall the central perpendicular on the plane. Measure off on the radius and on the perpendicular the reciprocals of their lengths, thus determining a triangle. Through the centre draw a normal to the plane of the triangle equal in length to double the area of the triangle multiplied by the product of the three principal pitches. This is the perpendicular to the required axis if we consider rotation round the line from the perpendicular to the radius as positive.

One more point may be mentioned. If we take the cone reciprocal to the director cone, that is the cone whose edges are perpendicular to the tangent planes, and if we use this new cone for selecting the generators of a ruled surface from the reciprocal three-system, the two ruled surfaces have a common line of striction and they touch one another along this line. This is the extension of the theorem that the reciprocal screw at right angles to the generators of a cylindroid coincides with the axis.

NOTE VII.

Note on homographic transformation, § 246.

That there must be in general linear relations between the co-ordinates of the screws of an instantaneous system, and the co-ordinates of the corresponding impulsive screws is proved as follows.

Let α be an impulsive screw and let β be the corresponding instantaneous screw with respect to a free body.

Let A, B, C, D, E, F be six independent screws which we shall take as screws of reference. Let $\alpha_1, \ldots \alpha_6$ and $\beta_1, \ldots \beta_6$ be the co-ordinates of α and β with respect to A, B, C, &c.

Let $A_1, \ldots A_6$ be the co-ordinates of A with respect to the six principal Screws of Inertia and similarly let $B_1, \ldots B_6$ be the co-ordinates of B and in like manner for C, D, E, &c.

An impulsive wrench on α of intensity α''' will have for components $\alpha'''\alpha_1$ on $A \ldots$ and $\alpha'''\alpha_6$ on F. These components on $A, \ldots F$ may each be resolved into six component wrenches on the principal screws of inertia, viz.

$$\alpha'''\alpha_1 A_1 + \alpha'''\alpha_2 B_1 \ldots + \alpha'''\alpha_6 F_1,$$

$$\alpha'''\alpha_1 A_2 + \alpha'''\alpha_2 B_2 \ldots + \alpha'''\alpha_6 F_2,$$

$$\ldots\ldots\ldots\ldots\ldots\ldots\ldots\ldots\ldots\ldots$$

$$\alpha'''\alpha_1 A_6 + \alpha'''\alpha_2 B_6 \ldots + \alpha'''\alpha_6 F_6.$$

But these impulsive wrenches give rise to an instantaneous twist velocity about α whence by § 80, we have, if h be a common factor, and a, b, c the principal radii of gyration

$$+ ha\beta_1 = \alpha_1 A_1 + \alpha_2 B_1 + \alpha_3 C_1 + \alpha_4 D_1 + \alpha_5 E_1 + \alpha_6 F_1,$$

$$- ha\beta_2 = \alpha_1 A_2 + \alpha_2 B_2 + \alpha_3 C_2 + \alpha_4 D_2 + \alpha_5 E_2 + \alpha_6 F_2,$$

$$+ hb\beta_3 = \alpha_1 A_3 + \alpha_2 B_3 + \alpha_3 C_3 + \alpha_4 D_3 + \alpha_5 E_3 + \alpha_6 F_3,$$

$$- hb\beta_4 = \alpha_1 A_4 + \alpha_2 B_4 + \alpha_3 C_4 + \alpha_4 D_4 + \alpha_5 E_4 + \alpha_6 F_4,$$

$$+ hc\beta_5 = \alpha_1 A_5 + \alpha_2 B_5 + \alpha_3 C_5 + \alpha_4 D_5 + \alpha_5 E_5 + \alpha_6 F_5,$$

$$- hc\beta_6 = \alpha_1 A_6 + \alpha_2 B_6 + \alpha_3 C_6 + \alpha_4 D_6 + \alpha_5 E_6 + \alpha_6 F_6.$$

Thus the linear relations are established.

NOTE VIII.

Remarks on § 268.

It ought to have been mentioned that the relation between four points on a sphere used in this article is a well known theorem, see Salmon, *Geometry of Three Dimensions*, § 56 and Casey's *Spherical Trigonometry*, § 111.

It is also worth while to add that $\sqrt{E_{123}}$ is the function which on other grounds has been called the sine of the solid angle formed by the straight lines 1, 2, 3 (Casey, *Spherical Trigonometry*, § 28). The three formulæ of this article have been proved as they stand for sets of six co-reciprocal screws. Mr J. H. Grace has however kindly pointed out to me (1898) that the second of the formulæ would be also true for any set of five co-reciprocal screws, and the third would be true for any set of four co-reciprocal screws. We thus have for a set of four co-reciprocals

$$p_1 \sin^2(234) + p_2 \sin^2(341) + p_3 \sin^2(412) + p_4 \sin^2(123) = 0,$$

where $\sin^2(234)$ is the square of the sine of the solid angle contained by the straight lines 2, 3, 4.

APPENDIX II.

ADDRESS TO THE MATHEMATICAL AND PHYSICAL SECTION OF THE BRITISH ASSOCIATION.

MANCHESTER, 1887.

A Dynamical Parable.

THE subject I have chosen for my address to you to-day has been to me a favourite topic of meditation for many years. It is that part of the science of theoretical mechanics which is usually known as the "Theory of Screws."

A good deal has been already written on this theory, but I may say with some confidence that the aspect in which I shall invite you now to look at it is a novel one. I propose to give an account of the proceedings of a committee appointed to undertake some experiments upon certain dynamical phenomena. It may appear to you that the experiments I shall describe have not as yet been made, that even the committee itself has not as yet been called together. I have accordingly ventured to call this address "A Dynamical Parable."

There was once a rigid body which lay peacefully at rest. A committee of natural philosophers was appointed to make an experimental and rational inquiry into the dynamics of that body. The committee received special instructions. They were to find out why the body remained at rest, notwithstanding that certain forces were in action. They were to apply impulsive forces and observe how the body would begin to move. They were also to investigate the small oscillations. These being settled, they were then to—— But here the chairman interposed; he considered that for the present, at least, there was sufficient work in prospect. He pointed out how the questions already proposed just completed a natural group. "Let it suffice for us," he said, "to experiment upon the dynamics of this body so long as it remains in or near to the position it now occupies. We may leave to some more ambitious committee the task of following the body in all conceivable gyrations through the universe."

The committee was judiciously chosen. Mr Anharmonic undertook the geometry. He was found to be of the utmost value in the more delicate parts of the work, though his colleagues thought him rather prosy at times. He was much aided by his two friends, Mr One-to-One, who had charge of the homographic department, and Mr Helix, whose labours will be seen to be of much importance. As a most respectable, if rather old-fashioned member, Mr Cartesian was added to the committee, but his antiquated tactics were quite out-manoeuvred by those of Helix and One-to-One. I need only mention two more names. Mr Commonsense was, of course, present as an *ex-officio* member, and valuable service was rendered even by Mr Querulous, who objected at first to serve on the committee at all. He said that the inquiry was all nonsense, because everybody knew as much as they wished to know about the dynamics of a rigid body. The subject was as old as the hills, and had all been settled long ago. He was persuaded, however, to look in occasionally. It will appear that a remarkable result of the labours of the committee was the conversion of Mr Querulous himself.

The committee assembled in the presence of the rigid body to commence their memorable labours. There was the body at rest, a huge amorphous mass, with no regularity in its shape—no uniformity in its texture. But what chiefly alarmed the committee was the bewildering nature of the constraints by which the movements of the body were hampered. They had been accustomed to nice mechanical problems, in which a smooth body lay on a smooth table, or a wheel rotated on an axle, or a body rotated around a point. In all these cases the constraints were of a simple character, and the possible movements of the body were obvious. But the constraints in the present case were of puzzling complexity. There were cords and links, moving axes, surfaces with which the body lay in contact, and many other geometrical constraints. Experience of ordinary problems in mechanics would be of little avail. In fact, the chairman truly appreciated the situation when he said, that the *constraints were of a perfectly general type*.

In the dismay with which this announcement was received Mr Commonsense advanced to the body and tried whether it could move at all. Yes, it was obvious that in some ways the body could be moved. Then said Commonsense, 'Ought we not first to study carefully the nature of the freedom which the body possesses? Ought we not to make an inventory of every distinct movement of which the body is capable? Until this has been obtained I do not see how we can make any progress in the dynamical part of our business.'

Mr Querulous ridiculed this proposal. 'How could you,' he said, 'make any geometrical theory of the mobility of a body without knowing all about the constraints? And yet you are attempting to do so with perfectly general constraints of which you know nothing. It must be all waste of time, for though I have read many books on mechanics, I never saw anything like it.'

Here the gentle voice of Mr Anharmonic was heard. 'Let us try, let us simply experiment on the mobility of the body, and let us faithfully record what we find.' In justification of this advice Mr Anharmonic made a remark which was new to most members of the committee: he asserted that, *though the constraints may be of endless variety and complexity, there can be only a very limited variety in the types of possible mobility*.

It was therefore resolved to make a series of experiments with the simple object of seeing how the body could be moved. Mr Cartesian, having a reputation for such work, was requested to undertake the inquiry and to report to the committee. Cartesian commenced operations in accordance with the well-known traditions of his craft. He erected a cumbrous apparatus which he called his three rectangular axes. He then attempted to push the body parallel to one of these axes, but it would not stir. He tried to move the body parallel to each of the other axes, but was again unsuccessful. He then attached the body to one of the axes and tried to effect a rotation around that axis. Again he failed, for the constraints were of too elaborate a type to accommodate themselves to Mr Cartesian's crude notions.

We shall subsequently find that the movements of the body are necessarily of an exquisitely simple type, yet such was the clumsiness and the artificial character of Mr Cartesian's machinery that he failed to perceive the simplicity. To him it appeared that the body could only move in a highly complex manner; he saw that it could accept a composite movement consisting of rotations about two or three of his axes and simultaneous translations also parallel to two or three axes. Cartesian was a very skilful calculator, and by a series of experiments even with his unsympathetic apparatus he obtained some knowledge of the subject, sufficient for purposes in which a vivid comprehension of the whole was not required. The inadequacy of Cartesian's geometry was painfully evident when he reported to the committee on the mobility of the rigid body. 'I find,' he said, 'that the body is unable to move parallel to x, or to y, or to z; neither can I make it rotate around x, or y, or z; but I could push it an inch parallel to x, provided that at the same time I pushed it a foot parallel to y and a yard backwards parallel to z, and that it was also turned a degree around x, half a degree the other way around y, and twenty-three minutes and nineteen seconds around z.'

'Is that all?' asks the chairman. 'Oh, no,' replied Mr Cartesian, 'there are other proportions in which the ingredients may be combined so as to produce a possible movement,' and he was proceeding to state them when Mr Commonsense interposed. 'Stop! stop!' said he, 'I can make nothing of all these figures. This jargon about x, y, and z may suffice for your calculations, but it fails to convey to my mind any clear or concise notion of the movements which the body is free to make.'

Many of the committee sympathised with this view of Commonsense, and they came to the conclusion that there was nothing to be extracted from poor old Cartesian and his axes. They felt that there must be some better method, and their hopes of discovering it were raised when they saw Mr Helix volunteer his services and advance to the rigid body. Helix brought with him no cumbrous rectangular axes, but commenced to try the mobility of the body in the simplest manner. He found it lying at rest in a position we may call A. Perceiving that it was in some ways mobile, he gave it a slight displacement to a neighbouring position B. Contrast the procedure of Cartesian with the procedure of Helix. Cartesian tried to force the body to move along certain routes which he had arbitrarily chosen, but which the body had not chosen; in fact the body would not take any one of his routes separately, though it would take all of them together in a most embarrassing manner. But Helix had no preconceived scheme as to the nature of the movements to be expected. He simply found the body in a certain position A, and then he coaxed the body to move, not in this particular way or in that particular way, but any way the body liked to any new position B.

Let the constraints be what they may—let the position B lie anywhere in the close neighbourhood of A—Helix found that he could move the body from A to B by an extremely simple operation. With the aid of a skilful mechanic he prepared a screw with a suitable pitch, and adjusted this screw in a definite position. The rigid body was then attached by rigid bonds to a nut on this screw, and it was found that the movement of the body from A to B could be effected by simply turning the nut on the screw. A perfectly definite fact about the mobility of the body had thus been ascertained. It was able to twist to and fro on a certain screw.

Mr Querulous could not see that there was any simplicity or geometrical clearness in the notion of a screwing movement; in fact he thought it was the reverse of simple. Did not the screwing movement mean a translation parallel to an axis *and* a rotation around that axis? Was it not better to think of the rotation and the translation separately than to jumble together two things so totally distinct into a composite notion?

But Querulous was instantly answered by One-to-One. 'Lamentable, indeed,' said he, 'would be a divorce between the rotation and the translation. Together they form the unit of rigid movement. Nature herself has wedded them, and the fruits of their union are both abundant and beautiful.'

The success of Helix encouraged him to proceed with the experiments, and speedily he found a second screw about which the body could also twist. He was about to continue when he was interrupted by Mr Anharmonic, who said, 'Tarry a moment, for geometry declares that a body free to twist about two screws is free to twist about a myriad of screws. These form the generators of a graceful ruled surface known as the cylindroid. There may be infinite variety in the conceivable constraints, but there can be no corresponding variety in the character of this surface. Cylindroids differ in size, they have no difference in shape. Let us then make a cylindroid of the right size, and so place it that two of its screws coincide with those you have discovered; then I promise you that the body can be twisted about every screw on the surface. In other words, if a body has two degrees of freedom the cylindroid is the natural and the perfectly general method for giving an exact specification of its mobility.'

A single step remained to complete the examination of the freedom of the body. Mr Helix continued his experiments and presently detected a third screw, about which the body can also twist in addition to those on the cylindroid. A flood of geometrical light then burst forth and illuminated the whole theory. It appeared that the body was free to twist about ranks upon ranks of screws all beautifully arranged by their pitches on a system of hyperboloids. After a brief conference with Anharmonic and One-to-One, Helix announced that sufficient experiments of this kind had now been made. By the single screw, the cylindroid, and the family of hyperboloids, every conceivable information about the mobility of the rigid body can be adequately conveyed. Let the body have any constraints, howsoever elaborate, yet the definite geometrical conceptions just stated will be sufficient.

With perfect lucidity Mr Helix expounded the matter to the committee. He exhibited to them an elegant fabric of screws, each with its appropriate pitch, and then he summarised his labours by saying, 'About every one of these screws you can displace the body by twisting, and, what is of no less importance, it will not admit of any movement which is not such a twist.' The committee expressed their satisfaction with this information. It was both clear and complete. Indeed, the chairman remarked with considerable force that *a more thorough method of specifying the freedom of the body was inconceivable.*

The discovery of the mobility of the body completed the first stage of the labours of the committee, and they were ready to commence the serious dynamical work. Force was now to be used, with the view of experimenting on the behaviour of the body under its influence. Elated by their previous success the committee declared that they would not rest satisfied until they had again obtained the most perfect solution of the most general problem.

'But what is force?' said one of the committee. 'Send for Mr Cartesian,' said the chairman, 'we will give him another trial.' Mr Cartesian was accordingly requested to devise an engine of the most ferocious description wherewith to attack the rigid body. He was promptly ready with a scheme, the weapons being drawn from his trusty but old-fashioned armoury. He would erect three rectangular axes, he would administer a tremendous blow parallel to each of these axes, and then he would simultaneously apply to the body a forcible couple around each of them; this was the utmost he could do.

'No doubt,' said the chairman, 'what you propose would be highly effective, but, Mr Cartesian, do you not think that while you still retained the perfect generality of your attack, you might simplify your specification of it? I confess that these three blows all given at once at right angles to each other, and these three couples which you propose to impart at the same time, rather confuse me. There seems a want of unity somehow. In short, Mr Cartesian, your scheme does not create a distinct geometrical image in my mind. We gladly acknowledge its suitability for numerical calculation, and we remember its famous achievements, but it is utterly inadequate to the aspirations of this committee. We must look elsewhere.'

Again Mr Helix stepped forward. He reminded the committee of the labours of Mathematician Poinsot, and then he approached the rigid body. Helix commenced by clearing away Cartesian's arbitrary scaffolding of rectangular axes. He showed how an attack of the most perfect generality could be delivered in a form that admitted of concise and elegant description. 'I shall,' he said, 'administer a blow upon the rigid body from some unexpected direction, and at the same instant I shall apply a vigorous couple in a plane perpendicular to the line of the blow.'

A happy inspiration here seized upon Mr Anharmonic. He knew, of course, that the efficiency of a couple is measured by its moment—that is, by the product of a force and a linear magnitude. He proposed, therefore, to weld Poinsot's force and couple into the single conception of a *wrench* on a screw. The force would be directed along the screw while the moment of the couple would equal the product of the force and the pitch of the screw. 'A screw,' he said, 'is to be regarded merely as a directed straight line with an associated linear magnitude called the pitch. The screw has for us a dual aspect of much significance. No small movement of the body is conceivable which does not consist of a twist about a screw. No set of forces could be applied to the body which were not equivalent to a wrench upon a screw. Everyone remembers the two celebrated rules that forces are compounded like rotations and that couples are compounded like translations. These may now be replaced by the single but far more compendious rule which asserts that wrenches and twists are to be compounded by identical laws. Would you unite geometry with generality in your dynamics? It is by screws that you are enabled to do so.'

These ideas were rather too abstract for Cartesian, who remarked that, as D'Alembert's principle provided for everything in dynamics, screws could not be needed. Mr Querulous sought to confirm him by saying that he did not see how screws helped the study either of Foucault's Pendulum or of the Precession of the Equinoxes.

Such absurd observations kindled the intellectual wrath of One-to-One, who rose and said, 'In the development of the natural philosopher two epochs may be noted. At the first he becomes aware that problems exist. At the second he discovers their solution. Querulous has not yet reached the first epoch; he cannot even conceive those problems which the "Theory of Screws" proposes to solve. I may, however, inform him that the "Theory of Screws" is not a general dynamical calculus. It is the discussion of a particular class of dynamical problems which do not admit of any other enunciation except that which the theory itself provides. Let us hope that ere our labours have ended Mr Querulous may obtain some glimmering of the subject.' The chairman happily assuaged matters. 'We must pardon,' he said, 'the vigorous language of our friend Mr One-to-One. His faith in geometry is boundless—in fact he is said to believe that the only real existence in the universe is anharmonic ratio.'

It was thus obvious that screws were indispensable alike for the application of the forces and for the observation of the movements. Special measuring instruments were devised by which the positions and pitches of the various screws could be carefully ascertained. All being ready the first experiment was commenced.

A screw was chosen quite at random, and a great impulsive wrench was administered thereon. In the infinite majority of cases this would start the body into activity, and it would commence to move in the only manner possible—*i.e.* it would begin to twist about some screw. It happened, however, that this first experiment was unsuccessful; the impulsive wrench failed to operate, or at all events the body did not stir. 'I told you it would not do,' shouted Querulous, though he instantly subsided when One-to-One glanced at him.

Much may often be learned from an experiment which fails, and the chairman sagaciously accounted for the failure, and in doing so directed the attention of the committee to an important branch of the subject. The mishap was due, he thought, to some reaction of the constraints which had neutralised the effect of the wrench. He believed it would save time in their future investigations if these reactions could be first studied and their number and position ascertained.

To this suggestion Mr Cartesian demurred. He urged that it would involve an endless task. 'Look,' he said, 'at the complexity of the constraints: how the body rests on these surfaces here; how it is fastened by links to those points there; how there are a thousand-and-one ways in which reactions might originate.' Mr Commonsense and other members of the committee were not so easily deterred, and they determined to work out the subject thoroughly. At first they did not see their way clearly, and much time was spent in misdirected attempts. At length they were rewarded by a curious and unexpected discovery, which suddenly rendered the obscure reactions perfectly transparent.

A trial was being made upon a body which had only one degree of freedom; was, in fact, only able to twist about a single screw, X. Another screw, Y, was speedily found, such that a wrench thereon failed to disturb the body. It now occurred to the committee to try the effect of interchanging the relation of these screws. They accordingly arranged that the body should be left only free to twist about Y, while a wrench was applied on X. Again the body did not stir. The importance of this fact immediately arrested the attention of the more intelligent observers, for it established the following general law: If a wrench on X fails to move a body only free to twist about Y, then a wrench on Y must be unable to move a body only free to twist about X. It was determined to speak of two screws when related in this manner as *reciprocal.*

Some members of the committee did not at first realise the significance of this discovery. Their difficulty arose from the restricted character of the experiments by which the law of reciprocal screws had been suggested. They said, 'You have shown us that this law is observed in the case of a body only free to twist about one screw at a time; but how does this teach anything of the general case in which the body is free to twist about whole shoals of screws?' Mr Commonsense immediately showed that the discovery could be enunciated in a quite unobjectionable form. 'The law of reciprocal screws,' he said, 'does not depend upon the constraints or the limitations of the freedom. It may be expressed in this way:—*Two screws are reciprocal when a small twist about either can do no work against a wrench on the other.*'

This important step at once brought into view the whole geometry of the reactions. Let us suppose that the freedom of the body was such that it could twist about all the screws of a system which we shall call U. Let all the possible reactions form wrenches on the screws of another system, V. It then appeared that every screw upon U is reciprocal to every screw upon V. A body might therefore be free to twist about every screw of V and still remain in equilibrium, notwithstanding the presence of a wrench on every screw of U. A body free to twist about all the screws of V can therefore be only partially free. Hence V must be one of those few types of screw system already discussed. It was accordingly found that the single screw, the cylindroid, and the set of hyperboloids completely described every conceivable reaction from the constraints just as they described every conceivable kind of freedom. The committee derived much encouragement from these discoveries; they felt that they must be following the right path, and that the bounty of Nature had already bestowed on them some earnest of the rewards they were ultimately to receive.

It was with eager anticipation that they now approached the great dynamical question. They were to see what would happen if the impulsive wrench were not neutralised by the reactions of the constraints. The body would then commence to move—that is, to twist about some screw which it would be natural to call the instantaneous screw. To trace the connection between the impulsive screw and the corresponding instantaneous screw was the question of the hour. Before the experiments were commenced, some shrewd member remarked that the issue had not yet been presented with the necessary precision. 'I understand,' he said, 'that when you apply a certain impulsive wrench, the body will receive a definite twist velocity about a definite screw; but the converse problem is ambiguous. Unless the body be quite free, there are myriads of impulsive screws corresponding to but one instantaneous screw.' The chairman perceived the difficulty, and not in vain did he appeal to the geometrical instinct of Mr One-to-One, who at once explained the philosophy of the matter, dissipated the fog, and disclosed a fresh beauty in the theory.

'It is quite true,' said Mr One-to-One, 'that there are myriads of impulsive screws, any one of which may be regarded as the correspondent to a given instantaneous screw, but it fortunately happens that among these myriads there is always one screw so specially circumstanced that we may select it as *the* correspondent, and then the ambiguity will have vanished.'

As several members were not endowed with the geometrical insight possessed by One-to-One, they called on him to explain how this special screw was to be identified; accordingly he proceeded:—'We have already ascertained that the constraints permit the body to be twisted about any screw of the system, U. Out of the myriads of impulsive screws, corresponding to a single instantaneous screw, it almost always happens that one, but never more than one, lies on U. This is the special screw. No matter where the impulsive wrench may lie throughout all the realms of space, it may be exchanged for a precisely equivalent wrench lying on U. Without the sacrifice of a particle of generality, we have neatly circumscribed the problem. For one impulsive there is one instantaneous screw, and for one instantaneous screw there is one impulsive screw.'

The experiments were accordingly resumed. An impulsive screw was chosen, and its position and its pitch were both noted. An impulsive wrench was administered, the body commenced to twist, and the instantaneous screw was ascertained by the motion of marked points. The body was brought to rest. A new impulsive screw was then taken. The experiment was again and again repeated. The results were tabulated, so that for each impulsive screw the corresponding instantaneous screw was shown.

Although these investigations were restricted to screws belonging to the system which expressed the freedom of the body, yet the committee became uneasy when they reflected that the screws of that system were still infinite in number, and that consequently they had undertaken a task of infinite extent. Unless some compendious law should be discovered, which connected the impulsive screw with the instantaneous screw, their experiments would indeed be endless. Was it likely that such a law could be found—was it even likely that such a law existed? Mr Querulous decidedly thought not. He pointed out how the body was of the most hopelessly irregular shape and mass, and how the constraints were notoriously of the most embarrassing description. It was, therefore, he thought, idle to search for any geometrical law connecting the impulsive screw and the instantaneous screw. He moved that the whole inquiry be abandoned. These sentiments seemed to be shared by other members of the committee. Even the resolution of the chairman began to quail before a task of infinite magnitude. A crisis was imminent—when Mr Anharmonic rose.

'Mr Chairman,' he said, 'Geometry is ever ready to help even the most humble inquirer into the laws of Nature, but Geometry reserves her most gracious gifts for those who interrogate Nature in the noblest and most comprehensive spirit. That spirit has been ours during this research, and accordingly Geometry in this our emergency places her choicest treasures at our disposal. Foremost among these is the powerful theory of homographic systems. By a few bold extensions we create a comprehensive theory of homographic screws. All the impulsive screws form one system, and all the instantaneous screws form another system, and these two systems are homographic. Once you have realised this, you will find your present difficulty cleared away. You will only have to determine a few pairs of impulsive and instantaneous screws by experiment. The number of such pairs need never be more than seven. When these have been found, the homography is completely known. The instantaneous screw corresponding to every impulsive screw will then be completely determined by geometry both pure and beautiful.' To the delight and amazement of the committee, Mr Anharmonic demonstrated the truth of his theory by the supreme test of fulfilled prediction. When the observations had provided him with a number of pairs of screws, one more than the number of degrees of freedom of the body, he was able to predict with infallible accuracy the instantaneous screw corresponding to any impulsive screw. Chaos had gone. Sweet order had come.

A few days later the chairman summoned a special meeting in order to hear from Mr Anharmonic an account of a discovery he had just made, which he believed to be of signal importance, and which he was anxious to demonstrate by actual experiment. Accordingly the committee assembled, and the geometer proceeded as follows:—

'You are aware that two homographic ranges on the same ray possess two double points, whereof each coincides with its correspondent; more generally when each point in space, regarded as belonging to one homographic system, has its correspondent belonging to another system, then there are four cases in which a point coincides with its correspondent. These are known as the four double points, and they possess much geometrical interest. Let us now create conceptions of an analogous character suitably enlarged for our present purpose. We have discovered that the impulsive screws and the corresponding instantaneous screws form two homographic systems. There will be a certain limited number (never more than six) of double screws common to these two systems. As the double points in the homography of point systems are fruitful in geometry, so the double screws in the homography of screw systems are fruitful in Dynamics.'

A question for experimental inquiry could now be distinctly stated. Does a double screw possess the property that an impulsive wrench delivered thereon will make the body commence to move by twisting about the same screw? This was immediately tested. Mr Anharmonic, guided by the indications of homography, soon pointed out the few double screws. One of these was chosen, a vigorous impulsive wrench was imparted thereon. The observations were conducted as before, the anticipated result was triumphantly verified, for the body commenced to twist about the identical screw on which the wrench was imparted. The other double screws were similarly tried, and with a like result. In each case the instantaneous screw was identical both in pitch and in position with the impulsive screw.

'But surely,' said Mr Querulous, 'there is nothing wonderful in this. Who is surprised to learn that the body twists about the same screw as that on which the wrench was administered? I am sure I could find many such screws. Indeed, the real wonder is not that the impulsive screw and the instantaneous screw are ever the same, but that they should ever be different.'

And Mr Querulous proceeded to illustrate his views by experiments on the rigid body. He gave the body all sorts of impulses, but in spite of all his endeavours the body invariably commenced to twist about some screw which was *not* the impulsive screw. 'You may try till Doomsday,' said Mr Anharmonic, 'you will never find any besides the few I have indicated.'

It was thought convenient to assign a name to these remarkable screws, and they were accordingly designated the *principal screws of inertia*. There are for example six principal screws of inertia when the body is perfectly free, and two when the body is free to twist about the screws of a cylindroid. The committee regarded the discovery of the principal screws of inertia as the most remarkable result they had yet obtained.

Mr Cartesian was very unhappy. The generality of the subject was too great for his comprehension. He had an invincible attachment to the x, y, z, which he regarded as the *ne plus ultra* of dynamics. 'Why will you burden the science,' he sighs, 'with all these additional names? Can you not express what you want without talking about cylindroids, and twists, and wrenches, and impulsive screws, and instantaneous screws, and all the rest of it?' 'No,' said Mr One-to-One, 'there can be no simpler way of stating the results than that natural method we have followed. You would not object to the language if your ideas of natural phenomena had been sufficiently capacious. We are dealing with questions of perfect generality, and it would involve a sacrifice of generality were we to speak of the movement of a body except as a twist, or of a system of forces except as a wrench.'

'But,' said Mr Commonsense, 'can you not as a concession to our ignorance tell us something in ordinary language which will give an idea of what you mean when you talk of your "principal screws of inertia"? Pray for once sacrifice this generality you prize so much and put the theory into some familiar shape that ordinary mortals can understand.'

Mr Anharmonic would not condescend to comply with this request, so the chairman called upon Mr One-to-One, who somewhat ungraciously consented. 'I feel,' said he, 'the request to be an irritating one. Extreme cases frequently make bad illustrations of a general theory. That zero multiplied by infinity may be anything is not surely a felicitous exhibition of the perfections of the multiplication table. It is with reluctance that I divest the theory of its flowing geometrical habit, and present it only as a stiff conventional guy from which true grace has departed.

'Let us suppose that the rigid body, instead of being constrained as heretofore in a perfectly general manner, is subjected merely to a special type of constraint. Let it in fact be only free to rotate around a fixed point. The beautiful fabric of screws, which so elegantly expressed the latitude permitted to the body before, has now degenerated into a mere horde of lines all stuck through the point. Those varieties in the pitches of the screws which gave colour and richness to the fabric have also vanished, and the pencil of degenerate screws have a monotonous zero of pitch. Our general conceptions of mobility have thus been horribly mutilated and disfigured before they can be adapted to the old and respectable problem of the rotation of a rigid body about a fixed point. For the dynamics of this problem the wrenches assume an extreme and even monstrous type. Wrenches they still are, as wrenches they ever must be, but they are wrenches on screws of infinite pitch; they have even ceased to possess definite screws as homes of their own. We often call them couples.

'Yet so comprehensive is the doctrine of the principal screws of inertia that even to this extreme problem the theory may be applied. The principal screws of inertia reduce in this special case to the three principal axes drawn through the point. In fact we see that the famous property of the principal axes of a rigid body is merely a very special application of the general theory of the principal screws of inertia. Every one who has a particle of mathematical taste lingers with fondness over the theory of the principal axes. Learn therefore,' says One-to-One in conclusion, 'how great must be the beauty of a doctrine which comprehends the theory of principal axes as the merest outlying detail.'

Another definite stage in the labours of the committee had now been reached, and accordingly the chairman summarised the results. He said that a geometrical solution had been obtained of every conceivable problem as to the effect of impulse on a rigid body. The impulsive screws and the corresponding instantaneous screws formed two homographic systems. Each screw in one system determined its corresponding screw in the other system, just as in two anharmonic ranges each point in one determines its correspondent in the other. The double screws of the two homographic systems are the principal screws of inertia. He remarked in conclusion that the geometrical theory of homography and the present dynamical theory mutually illustrated and interpreted each other.

There was still one more problem which had to be brought into shape by geometry and submitted to the test of experiment.

The body is lying at rest though gravity and many other forces are acting upon it. These forces constitute a wrench which must lie upon a screw of the reciprocal system, inasmuch as it is neutralised by the reaction of the constraints. Let the body be displaced from its initial position by a small twist. The wrench will no longer be neutralised by the reaction of the constraints; accordingly when the body is released it will commence to move. So far as the present investigations are concerned these movements are small oscillations. Attention was therefore directed to these small oscillations. The usual observations were made, and Helix reported them to be of a very perplexing kind. 'Surely,' said the chairman, 'you find the body twisting about some screw, do you not?' 'Undoubtedly,' said Helix; 'the body can only move by twisting about some screw; but, unfortunately, this screw is not fixed, it is indeed moving about in such an embarrassing manner that I can give no intelligible account of the matter.' The chairman appealed to the committee not to leave the interesting subject of small oscillations in such an unsatisfactory state. Success had hitherto guided their efforts. Let them not separate without throwing the light of geometry on this obscure subject.

Mr Querulous here said he must be heard. He protested against any further waste of time; it was absurd. Everybody knew how to investigate small oscillations; the equations were given in every book on mechanics. You had only to write down these equations, solve these equations again for the thousandth time and the thing was done. But the more intelligent members of the committee took the same view as the chairman. They did not question the truth of the formulæ which to Querulous seemed all sufficient, but they wished to see whether geometry could not illuminate the subject. Fortunately this view prevailed, and new experiments were commenced under the direction of Mr Anharmonic, who first quelled the elaborate oscillations which had so puzzled the committee, reduced the body to rest, and then introduced the discussion as follows:—

'The body now lies at rest. I displace it a little, and hold it in its new position. The wrench, which is the resultant of all the varied forces acting on the body, is no longer completely neutralised by the reactions of the constraints. Indeed, I can feel it in action. Our apparatus will enable us to measure the intensity of this wrench, and to determine the screw on which it acts.'

A series of experiments was then made, in which the body was displaced by a twist about a screw, which was duly noted, while the corresponding evoked wrench was determined. The pairs of screws so related were carefully tabulated. When we remember the infinite complexity of the forces, of the constraints and of the constitution of the body, it might seem an endless task to determine the connection between the two systems of screws. Mr Anharmonic pointed out how modern geometry supplied the wants of Dynamics. As in the previous case the two screw systems were homographic, and when a number of pairs, one more than the degrees of freedom of the body, had been found all was determined. This statement was put to the test. Again and again the body was displaced in some new fashion, but again and again did Mr Anharmonic predict the precise wrench which would be required to maintain the body in its new position.

'But,' said the chairman, 'are not these purely statical results? How do they throw light on those elaborate oscillations which seem at present so inexplicable?' 'This I shall explain,' said Anharmonic; 'but I beg of you to give me your best attention, for I think the theory of small oscillations will be found worthy of it.

'Let us think of any screw α belonging to the system U, which expresses the freedom of the body. If α be an instantaneous screw, there will of course be a corresponding impulsive screw θ also on U. If the body be displaced from a position of equilibrium by a small twist about α, then the uncompensated forces produce a wrench ϕ, which, without loss of generality, may also be supposed to lie on U. According as the screw α moves over U so will the two corresponding screws θ and ϕ also move over U. The system represented by α is homographic with both the systems of θ and of ϕ respectively. But two systems homographic with the same system are homographic with each other. Accordingly, the θ system and the ϕ system are homographic. There will therefore be a certain number of double screws (not more than six) common to the systems θ and ϕ. Each of these double screws will of course have its correspondent in the α system, and we may call them α_1, α_2, &c., their number being equal to the degrees of freedom of the body. These screws are most curiously related to the small oscillations. We shall first demonstrate by experiment the remarkable property they possess.'

The body was first brought to rest in its position of equilibrium. One of the special screws α having been carefully determined both in position and in pitch, the body was displaced by a twist about this screw and was then released. As the forces were uncompensated, the body of course commenced to move, but the oscillations were of unparalleled simplicity. With the regularity of a pendulum the body twisted to and fro on this screw, just as if it were actually constrained to this motion alone. The committee were delighted to witness a vibration so graceful, and, remembering the complex nature of the ordinary oscillations, they appealed to Mr Anharmonic for an explanation. This he gladly gave, not by means of complex formulæ, but by a line of reasoning that was highly commended by Mr Commonsense, and to which even Mr Querulous urged no objection.

'This pretty movement,' said Mr Anharmonic, 'is due to the nature of the screw α_1. Had I chosen any screw at random, the oscillations would, as we have seen, be of a very complex type; for the displacement will evoke an uncompensated wrench, in consequence of which the body will commence to move by twisting about the instantaneous screw corresponding to that wrench; and of course this instantaneous screw will usually be quite different from the screw about which the displacement was made. But you will observe that α_1 has been chosen as a screw in the instantaneous system, corresponding to one of the double screws in the θ and ϕ systems. When the body is twisted about α_1 a wrench is evoked on the double screw, but as α_1 is itself the instantaneous screw, corresponding to that double screw, the only effect of the wrench will be to make the body twist about α_1. Thus we see that the body will twist to and fro on α_1 for ever. Finally, we can show that the most elaborate oscillations the body can possibly have may be produced by compounding the simple vibrations on these screws α_1, α_2, &c.'

Great enlightenment was thus diffused over the committee, and now Mr Querulous began to think there must be something in it. Cordial unanimity prevailed among the members, and it was appropriately suggested that the screws of simple vibration should be called *harmonic screws*. This view was adopted by the chairman, who said he thought he had seen the word harmonic used in 'Thomson and Tait.'

The final meeting showed that real dynamical enthusiasm had been kindled in the committee. Vistas of great mathematical theories were opened out in many directions. One member showed how the theory of screws could be applied not merely to a single rigid body but to any mechanical system whatever. He sketched a geometrical conception of what he was pleased to call a *screw-chain*, by which he said he could so bind even the most elaborate system of rigid bodies that they would be compelled to conform to the theory of screws. Nay, soaring still further into the empyrean, he showed that all the instantaneous motions of every molecule in the universe were only a twist about one screw-chain while all the forces of the universe were but a wrench upon another.

Mr One-to-One expounded the 'Ausdehnungslehre' and showed that the theory of screws was closely related to parts of Grassmann's great work; while Mr Anharmonic told how Sir W. R. Hamilton, in his celebrated "*Theory of systems of rays*" had by his discovery of the cylindroid helped to lay the foundations of the Theory of Screws.

The climax of mathematical eloquence was attained in the speech of Mr Querulous, who, with newborn enthusiasm, launched into appalling speculations. He had evidently been reading his 'Cayley' and had become conscious of the poverty of geometrical conception arising from our unfortunate residence in a space of an arbitrary and unsymmetrical description.

'Three dimensions,' he said, 'may perhaps be enough for an intelligent geometer. He may get on fairly well without a four-dimensioned space, but he does most heartily remonstrate against a flat infinity. Think of infinity,' he cries, 'as it should be, perhaps even as it is. Talk not of your scanty straight line at infinity and your miserable pair of circular points. Boldly assert that infinity is an ample quadric, and not the mere ghost of one; and then geometry will become what geometry ought to be. Then will every twist resolve into a right vector and a left vector, as the genius of Clifford proved. Then will the theory of screws shed away some few adhering incongruities and fully develop its shapely proportions. Then will——' But here the chairman said he feared the discussion was beginning to wax somewhat transcendental. For his part he was content with the results of the experiments even though they had been conducted in the vapid old space of Euclid. He reminded them that their functions had now concluded, for they had ascertained everything relating to the rigid body which had been committed to them. He hoped they would agree with him that the enquiry had been an instructive one. They had been engaged in the study of Nature, they had approached the problems in the true philosophical spirit, and the rewards they had obtained proved that

'Nature never did betray
The heart that truly loved her.'

BIBLIOGRAPHICAL NOTES.

I HERE briefly refer to the principal works known to me which bear on the subject of the present volume.

POINSOT (L.)—*Sur la composition des moments et la composition des aires* (1804). Journal de l'École Polytechnique; vol. vi. (13 cah.), pp. 182–205 (1806).

In this paper the author of the conception of the couple, and of the laws of composition of couples, has demonstrated the important theorem that *any system of forces applied to a rigid body can be reduced to a single force, and a couple in a plane perpendicular to the force.*

CHASLES (M.)—*Note sur les propriétés générales du système de deux corps semblables entr'eux et placés d'une manière quelconque dans l'espace; et sur le déplacement fini ou infiniment petit d'un corps solide libre.* Férussac, Bulletin des Sciences Mathématiques, Vol. xiv., pp. 321–326 (1830).

The author shows that there always exists one straight line, about which it is only necessary to rotate one of the bodies to place it similarly to the other. Whence (p. 324) he is led to the following fundamental theorem:—

L'on peut toujours transporter un corps solide libre d'une position dans une autre position quelconque, déterminée par le mouvement continu d'une vis à laquelle ce corps serait fixé invariablement.

Three or four years later than the paper we have cited, Poinsot published his celebrated *Théorie Nouvelle de la Rotation des Corps* (Paris, 1834). In this he enunciates the same theorem without reference to Chasles, but that it is really due to Chasles there can be little doubt. He explicitly claims it in note 34 to the *Aperçu Historique.* Bruxelles Mém. Couronn. XI., 1837.

HAMILTON (W. R.)—*First supplement to an essay on the Theory of Systems of Rays.* Transactions of the Royal Irish Academy, Vol. xvi., pp. 4—62 (1830).

That conoidal cubic surface named the cylindroid which plays so fundamental a part in the Theory of Screws was first discovered by Sir William Rowan Hamilton.

In his celebrated memoir on the Theory of Systems of Rays he demonstrates the remarkable proposition which may be thus enunciated:—

The lines of shortest distance between any ray of the system and the other contiguous rays of the system have a surface for their locus, and that surface is a cylindroid.

We can illustrate this as follows by the methods of the present volume.

The Hamiltonian system of rays here considered form a congruency. If we except all rays save those contiguous to any one ray then the congruency may be regarded as *linear*. Hence any property of a linear congruency must apply to the Hamiltonian system as restricted in the proposition before us.

A linear congruency is constituted by those screws of zero pitch whose coordinates satisfy two linear equations. They are the screws which belong to a 4-system and which each have zero pitch (§ 76). But we know that each such screw must intersect both of the screws of zero pitch on the cylindroid reciprocal to the 4-system (§ 212). It has also been shown that any transversal meeting two screws of equal pitch on a cylindroid must *intersect at right angles* a third screw on that surface (§ 22). Hence the shortest distance from any ray of the congruency to the axis of the cylindroid must lie on a generator of the cylindroid. This is however only true for one particular ray.

Hamilton's most instructive theorem shows, more generally, that the shortest distances between *any* specified ray R of the congruency and all the other *contiguous* rays have a conoidal cubic as their locus, such as might be represented by the equation

$$z(x^2 + y^2) = Ax^2 + 2Bxy + Cy^2.$$

There are two disposable quantities in the selection of the origin and the axis of x. If these quantities be so taken as to render $A = 0$; $C = 0$, then the equation is at once shown to represent a cylindroid of which R is the axis. Of course all rays of this congruency intersect two fixed rays, and the axis of the cylindroid must also intersect both of these rays.

MÖBIUS (A. F.)—*Lehrbuch der Statik* (Leipzig, 1837).

This book is, we learn from the preface, one of the numerous productions to which the labours of Poinsot gave rise. The first part, pp. 1–355, discusses the laws of equilibrium of forces, which act upon a single rigid body. The second part, pp 1–313, discusses the equilibrium of forces acting upon several rigid bodies connected together. The characteristic feature of the book is its great generality. I here enunciate some of the principal theorems.

If a number of forces acting upon a free rigid body be in equilibrium, and if a straight line of arbitrary length and position be assumed, then the algebraic sum of the tetrahedra, of which the straight line and each of the forces in succession are pairs of opposite edges, is equal to zero (p. 94).

If four forces are in equilibrium they must be generators of the same hyperboloid (p. 177).

If five forces be in equilibrium they must intersect two common straight lines (p. 179).

If the *lines of action* of five forces be given, then a certain plane S through any point P is determined. If the five forces can be equilibrated by one force through P, then this one force must lie in S (p. 180).

To adopt the notation of Professor Cayley, we denote by 12 the perpendicular distance between two lines 1, 2, multiplied into the sine of the angle between them (Comptes Rendus, Vol. lxi., pp. 829–830 (1865)). Möbius shows (p. 189) that if forces along four lines 1, 2, 3, 4 equilibrate, the intensities of these forces are proportional to

$$\sqrt{23 \,.\, 24 \,.\, 34}, \quad \sqrt{13 \,.\, 14 \,.\, 34}, \quad \sqrt{12 \,.\, 14 \,.\, 24}, \quad \sqrt{12 \,.\, 13 \,.\, 23}.$$

It is also shown that the product of the forces on 1 and 2, multiplied by 12, is equal to the product of the forces on 3 and 4 multiplied by 34. He hence deduces Chasles' theorem (Liouville's Journal, 1st Ser., Vol. xii., p. 222 (1847)), that the volume of the tetrahedron formed by two of the forces is equal to that formed by the remaining two.

MÖBIUS (A. F.)—*Ueber die Zusammensetzung unendlich kleiner Drehungen.* Crelle's Journal; Vol. xviii., pp. 189–212 (1838).

This memoir contains many very interesting theorems, of which the following are the principal:—Any given small displacement of a rigid body can be effected by two small rotations. Two equal parallel and opposite rotations compound into a translation. Small rotations about intersecting axes are compounded like forces. If a number of forces acting upon a free body make equilibrium, then the final effect of a number of rotations (proportional to the forces) on the same axes will be zero. If a body can undergo small rotations about six independent axes, it can have any small movement whatever. He illustrates this by the case of a series of bodies of which each one is hinged to those on either side of it. If the first of the series be fixed then in general the seventh of the series will be perfectly free for small movements (see Rittershaus, p. 524).

RODRIGUES (O.)—*Des lois géométriques qui régissent les déplacements d'un système solide dans l'espace et de la variation des coordonnées provenant de ces déplacements considérés indépendamment des causes qui peuvent les produire.* Liouville's Journal Math.; Vol. v., pp. 380—440 (5th Dec., 1840).

This paper consists mainly of elaborate formulæ relating to displacements of finite magnitude. It has been already cited for an important remark (§ 9).

CHASLES (M.)—*Propriétés géométriques relatives au mouvement infiniment petit dans un corps solide libre dans l'espace.* Paris, Comptes Rendus; Vol. xvi., pp. 1420–1432 (1843).

A pair of "droites conjuguées" are two lines by rotations about which a given displacement can be communicated to a rigid body. Two pairs of "droites conjuguées" are always generators of the same hyperboloid.

HAMILTON (Sir W. R.)—*On some additional applications of the Theory of Algebraic Quaternions.* Royal Irish Academy Proceedings; Vol. iii. (1845–1847). Appendix No. 5, pp. li.–lx. (Communicated Dec. 8, 1845.)

On p. lvii. he states "the laws of equilibrium of several forces applied to various points of a solid body, are thus included in the two equations,

$$\Sigma\beta = 0\,;\quad \Sigma(\alpha\beta - \beta\alpha) = 0\,;$$

the vector of the point of application being α, and the vector representing the force applied at that point being β." On the same page he writes,

"Instead of the two equations of equilibrium, we may employ the single formula

$$\Sigma\,.\,\alpha\beta = -c,$$

c here denoting a scalar (or real) quantity, which is independent of the origin of vectors, and seems to have some title to be called the total *tension* of the system."

HAMILTON (Sir W. R.)—*Some applications of Quaternions to questions connected with the Rotation of a Solid Body.* Royal Irish Academy Proceedings; Vol. iv. (1847–1850) pp. 38–56. (Communicated Jan. 10, 1848.)

In this paper with the same notation as before, he takes the general case of a Rigid Body acted on by forces and considers the Quaternion

$$\frac{\Sigma V\alpha\beta}{\Sigma\beta} = w + ix + jy + kz.$$

"The number w" which he sees does not depend on the choice of origin "will denote the (real) quotient obtained by *dividing the moment of the principal resultant couple by the intensity of the resultant force*; with the known direction of which force the axis of this *principal* (and known) couple coincides, being the line which is known by the name of the *central axis* of the system." The vector part of the quaternion is the vector "perpendicular let fall from the assumed origin on the central axis of the system" (p. 40). It is interesting to note that the scalar w is what we would term the pitch of the Screw on which the wrench acts.

POINSOT (L.)—*Théorie nouvelle de la rotation des corps.* Liouville's Journal Math.; Vol. xvi., pp. 9–129, 289—336 (March, 1851).

This is Poinsot's classical memoir, which contains his beautiful geometrical theory of the *rotation of a rigid body about a fixed point.* In a less developed form the Theory had been previously published in Paris in 1834, as already mentioned.

SCHÖNEMANN (T.)—*Ueber die Construction von Normalen und Normalebenen gewisser krummer Flächen und Linien.* Monatsberichte der königlichen preussischen Akademie der Wissenchaften für das Jahr 1855, pp. 255–260.

Believing that this paper was but little known Herr Geiser reprinted it in Crelle's Journal, Vol. xc., pp. 44–48 (1881). Schönemann there gave the important theorem which has since been independently discovered by others, namely that whenever a rigid body is so displaced that four of its points, A, B, C, D move on fixed surfaces the normals to the surfaces which are the trajectories of all its points intersect two fixed rays. Herr Geiser gives an analytical proof (Crelle, Vol. xc., pp. 39–43, 1881). In our language the two rays are the two screws of zero pitch on the cylindroid reciprocal to the freedom of the body, and the cylindroid is itself determined by being reciprocal to four screws of zero pitch on the normals at A, B, C, D respectively to the four fixed surfaces. Another proof is given by Ribaucour, Comptes rendus, Vol. lxxvi., p. 1347 (2 June, 1873). See also Mannheim (A.), Liouville's Journal de Mathématiques, 2e Sér., Vol. xl., 1866.

WEIERSTRASS (C.). *Ueber ein die homogenen Functionen zweiten grades betreffendes. Theorem nebst Anwendung desselben auf die Theorie der kleinen Schwingungen.* Monatsberichte der k. preussischen Akademie der Wissenschaften, 1858, pp. 207–220; and Mathematische Werke, Vol. i. pp. 233–246.

Let ϕ, ψ be two homogeneous quadratic functions of n variables $x_1, \ldots x_n$ and let $f(s)$ be the discriminant of $s\phi - \psi$.

If the discriminant of one of the functions, say ϕ, does not vanish, and if further ϕ is essentially one-signed vanishing only when all the variables vanish, it can be shown that $s_1, s_2, \ldots s_n$ the roots of $f(s) = 0$ (assumed distinct) are all real and ϕ, ψ can then be reduced to the forms

$$\phi = \epsilon\,(y_1^2 + \ldots + y_n^2)$$

$$\psi = \epsilon\,(s_1 y_1^2 + \ldots s_n y_n^2)$$

where $y_1 \ldots y_n$ are all real linear functions of $x_1 \ldots x_n$ and ϵ is ± 1 according as ϕ is positive or negative. See § 86 and p. 484.

CAYLEY (A.)—*On a new analytical representation of curves in space.* Quarterly Mathematical Journal; Vol. iii., pp. 225–236 (1860). Vol. v., pp. 81–86 (1862). Coll. Math. Papers, Vol. iv. pp. 446–455, 490–494.

In this paper the *conception of the six co-ordinates of a line* is introduced for the first time. This is of importance in connection with our present subject because the six coordinates of a screw may be regarded as the generalization of the six coordinates of a straight line.

If $\alpha_1, \dots \alpha_6$ be the six coordinates of a screw then when we express that its pitch is zero by the condition

$$p_1\alpha_1^2 + \dots + p_6\alpha_6^2 = 0$$

we obtain the coordinates of a straight line. This is perhaps the most symmetrical form of the quadratic condition which must subsist between six quantities constituting the coordinates of a line. In Cayley's system it is given by equating the sum of three products to zero.

SYLVESTER (J. J.)—*Sur l'involution des lignes droites dans l'espace, considérées comme des axes de rotation.* Paris, Comptes Rendus; vol. lii., pp. 741–746 (April, 1861).

Any small displacement of a rigid body can *generally* be represented by rotations about six axes (Möbius). But this is not the case if forces can be found which equilibrate when acting along the six axes on a rigid body. The six axes in this case are in *involution.* The paper discusses the geometrical features of such a system, and shows, when five axes are given, how the locus of the sixth is to be found. Möbius had shown that through any point a plane of lines can be drawn in involution with five given lines. The present paper shows how the plane can be constructed. All the transversals intersecting a pair of *conjugate axes* are in involution with five given lines. Any two pairs of conjugate axes lie on the same hyperboloid. Two forces can be found on any pair of conjugate axes, which are statically equivalent to two given forces on any other given pair of conjugate axes. In presenting this paper M. Chasles remarks that Mr Sylvester's results lead to the following construction:—Conceive that a rigid body receives any small displacement, then lines drawn through any six points of the body perpendicular to their trajectories are in involution. M. Chasles also takes occasion to mention some other properties of the conjugate axes.

SYLVESTER (J. J.)—*Note sur l'involution de six lignes dans l'espace.* Paris, Comptes Rendus; vol. lii., pp. 815–817 (April, 1861).

The six lines are 1, 2, 3, 4, 5, 6. Let the line i be represented by the equations

$$a_i x + b_i y + c_i z + d_i u = 0,$$

$$\alpha_i x + \beta_i y + \gamma_i z + \delta_i u = 0,$$

and let i, j represent the determinant

$$\left|\begin{matrix} a_i & b_i & c_i & d_i \\ \alpha_i & \beta_i & \gamma_i & \delta_i \\ a_j & b_j & c_j & d_j \\ \alpha_j & \beta_j & \gamma_j & \delta_j \end{matrix}\right| .$$

Form now the determinant Δ_6—

$$\begin{vmatrix} & 1,2 & 1,3 & 1,4 & 1,5 & 1,6 \\ 2,1 & & 2,3 & 2,4 & 2,5 & 2,6 \\ 3,1 & 3,2 & & 3,4 & 3,5 & 3,6 \\ 4,1 & 4,2 & 4,3 & & 4,5 & 4,6 \\ 5,1 & 5,2 & 5,3 & 5,4 & & 5,6 \\ 6,1 & 6,2 & 6,3 & 6,4 & 6,5 & \end{vmatrix}$$

If $\Delta_6 = 0$, the lines are in involution. Considering only the figures 1, 2, 3, 4, 5, the determinant Δ_5 can be formed. If $\Delta_6 = 0$ and $\Delta_5 = 0$, the five lines 1, 2, 3, 4, 5 are in involution. If all the other minors are zero, the six lines will intersect a single transversal. If $\Delta_5 = 0$, without any other condition, the five lines 1, 2, 3, 4, 5 intersect a single transversal. If $\Delta_4 = 0$, without any other condition, the lines 1, 2, 3, 4 have but one common transversal (Cayley). A determinant can be found which is equal to the square root of Δ_6.

GRASSMANN (H.)—*Die Ausdehnungslehre.* Berlin (1862).

This remarkable work, a development of an earlier volume (1844), by the same author, contains much that is of instruction and interest in connection with the present theory.

A system of n, numerically equal, "Grossen erster Stufe," of which each pair are "normal," is discussed on p. 113. A set of co-reciprocal screws is a particular case of this very general conception.

The "inneres Produkt" of two "Grössen" divided by the product of their numerical values, is the cosine of the angle between the two "Grössen." If $a, b, c, \ldots$ be normal, and if k, l be any two other "Grössen," then

$$\cos \angle kl = \cos \angle ak \cos \angle al + \cos \angle bk \,.\, \cos \angle bl, + \&c. \text{ (p. 139).}$$

Here we have a very general theory, which includes screw co-ordinates as a particular case.

In a note on p. 222 the author states that the displacement of a body in space, or a general system of forces, form an "allgemeine räumliche Grösse zweiter Stufe."

The "kombinatorisches Produkt" (p. 41) of n screws will contain as a factor that single function whose evanescence would express that the n screws belonged to a screw system of the $(n-1)$th order.

PLÜCKER (J.)—*On a new geometry of space.* Phil. Trans., Vol. clv., pp. 725—791. 1865.

In this paper the *linear complex* is defined (p. 733). Some applications to optics are made (p. 760); the six co-ordinates of a line are considered (p. 774); and the applications to the geometry of forces (p. 786).

This is of importance for our purpose because the *linear complex* may be also defined with perfect generality as the axes of all the screws of any stated pitch which belong to a 5-system. The relation of the linear geometry to Dynamics is developed in the Theory of Screws.

HAMILTON (Sir W. R.)—*Elements of Quaternions.* Dublin, 1866.

In Art. 416 the equation $\Sigma V(\alpha - \gamma)\beta = 0$, is regarded as the single equation of equilibrium when it is satisfied for *all* values of γ, the vector to an arbitrary point C in space. In general if γ is not supposed to vary in this arbitrary manner, the equation is that of the central axis.

He then considers the quaternion $q = \dfrac{\Sigma V\alpha\beta}{\Sigma\beta}$, already mentioned (p. 512), and introduces the new quaternion $Q = \dfrac{\Sigma\alpha\beta}{\Sigma\beta} = c + \gamma$. The scalar c (the pitch) is independent of the assumed origin, and the vector γ is the vector to a definite point C on the central axis. This point does not vary with the position of the assumed origin, and is called the "Centre of the System of Forces." When the forces are all parallel C coincides with the centre of the parallel forces. In general

$$T\Sigma\alpha\beta = \sqrt{(c^2 + T\gamma^2)}\, T\Sigma\beta,$$

or the tensor of the total moment is constant for all points O situated on a sphere whose centre is C, and becomes a minimum when O coincides with C.

In Art. 396 Hamilton says "the *passage of a right line* from any *one* given position in space to any *other* may be conceived to be accomplished by a sort of *screw motion*," and on these kinematical lines he worked out his theory of the "Surface of Emanants," generated by a line moving according to some given law and constantly intersecting a given curve in space.

PLÜCKER (J.)—*Fundamental views regarding mechanics.* Phil. Trans. (1866); Vol. clvi., pp. 361—380.

The object of this paper is to "connect, in mechanics, translatory and rotatory movements with each other by a principle in geometry analogous to that of reciprocity." One of the principal theorems is thus enunciated:—"Any number of rotatory forces acting simultaneously, the co-ordinates of the resulting rotatory force, if there is such a force, if there is not, the co-ordinates of the resulting rotatory dyname, are obtained by adding the co-ordinates of the given rotatory forces. In the case of equilibrium the six sums obtained are equal to zero."

SPOTTISWOODE (W.)—*Note sur l'équilibre des forces dans l'espace.* Comptes Rendus; Vol. lxvi., pp. 97-103 (January, 1868).

If $P_0 \ldots P_{n-1}$ be n forces in equilibrium, and if $(0, 1)$ denote the moment of P_0, P_1, then the author proves* that

$$\begin{array}{llll} & P_1(0,1) + & P_2(0,2) + \ldots & = 0, \\ P_0(1,0) + & & + P_2(1,2) + \ldots & = 0, \\ P_0(2,0) + & P_1(2,1) + & + \ldots & = 0. \end{array}$$

As we have thus n equations to determine only the *relative values* of n quantities, the redundancy is taken advantage of to prove that

$$\frac{P_0^2}{[0,0]} = \frac{P_1^2}{[1,1]} = \&c.,$$

where $[0, 0]$, $[1, 1]$, &c., are the coefficients of $(0, 0)$, $(1, 1)$, &c., in the determinant

$$\begin{vmatrix} (0,0), & (0,1) & \ldots \\ (1,0), & (1,1) & \ldots \\ \ldots & \ldots & \ldots \end{vmatrix}$$

* We may remark that since the moment of two lines is the virtual coefficient of two screws of zero pitch, these equations are given at once by virtual velocities, if we rotate the body round each of the forces in succession.

When the forces are fewer than seven, the formulæ admit of a special transformation, which expresses certain further conditions which must be fulfilled.

This very elegant result may receive an extended interpretation. If P_0, P_1, P_2, &c., denote the intensities of wrenches on the screws 0, 1, 2, &c.; and if (12) denote the virtual coefficient of 1 and 2, then, when the formulæ of Mr Spottiswoode are satisfied, the n wrenches equilibrate, provided that the screws belong to a screw complex of the $(n-1)$th order and first degree.

PLÜCKER (J.)—*Neue Geometrie des Raumes gegründet auf die Betrachtung der geraden Linie als Raumelement.* Leipzig (B. G. Teübner, 1868–69), pp. 1–374.

This work is of course the principal authority on the theory of the linear complex. The subject here treated is essentially geometrical rather than dynamical, but there are a few remarks which are specially significant in our present subject; thus the author, on p. 24, introduces the word "Dyname":—"Durch den Ausdruck 'Dyname,' habe ich die Ursache einer beliebigen Bewegung eines starren Systems, oder, da sich die Natur dieser Ursache, wie die Natur einer Kraft überhaupt, unserem Erkennungsvermögen entzieht, die Bewegung selbst, statt der Ursache die Wirkung, bezeichnet." Although it is not very easy to see the precise meaning of this passage, yet it appears that a '*Dyname*' may be either a twist or a wrench (to use the language of the Theory of Screws).

On p. 25 we read:—"Dann entschwindet das specifisch Mechanische, und, um mich auf eine kurze Andeutung zu beschränken: es treten geometrische Gebilde auf, welche zu Dynamen in derselben Beziehung stehen, wie gerade Linien zu Kräften und Rotationen." There can be little doubt that the "geometrische Gebilde," to which Plücker refers, are what we have called screws.

As we have already stated (§ 13), we find in this book the discussion of the surface which we call the *cylindroid*, to which, as pointed out on p. 510, Sir W. R. Hamilton had been previously conducted.

Through any point a cone of the second degree can be drawn, the generators of which are lines belonging to a linear complex of the second degree. If the point be limited to a certain surface the cone breaks up into two planes. This surface is of the fourth class and fourth degree, and is known as Kummer's surface. See papers by Kummer in the Monatsberichte of the Berlin Academy, 1864, pp. 246–260, and 495–499. It has since been extensively studied from various points of view by many mathematicians. This theory is of interest for our purpose, because the locus of screws reciprocal to a cylindroid is a very special linear complex of the second degree, of which the cylindroid itself is the surface of singularities. Kummer's surface has in this case broken up into a plane and a cylindroid.

KLEIN (F.)—*Zur Theorie der Linien-Complexe des ersten und zweiten Grades.* Math. Ann.; Vol. II., pp. 198–226 (14th June, 1869).

The "simultaneous invariant" of two linear complexes is discussed. In our language this function is the virtual coefficient of the two screws reciprocal to the complexes. The six fundamental complexes are considered at length, and many remarkable geometrical properties proved. It is a matter of no little interest that these purely geometrical researches have a physical significance attached to them by the Theory of Screws.

This paper also contains the following proposition:—If $x_1, \ldots, x_6$ be the co-ordinates of a line, and $k_1, \ldots k_6$, be constants, then the family of linear complexes denoted by

$$\frac{x_1^2}{k_1-\lambda}+\ \ldots\ +\frac{x_6^2}{k_6-\lambda}=0$$

have a common surface of singularities where λ is a variable parameter. If the roots λ_1, &c. be known, we have a set of quasi elliptic co-ordinates for the line x. (Compare § 234.)

It is in this memoir that we find the enunciation of the remarkable *geometrical* principle which, when transformed into the language and conceptions of the Theory of Screws, asserts the existence of one screw reciprocal to five given screws. (§ 25.)

KLEIN (F.)—*Die allgemeine lineare Transformation der Linien Coordinaten.* Math. Ann.; Vol. II., pp. 366–371 (August 4, 1869).

Let $U_1, \ldots U_6$ denote six linear complexes. The moments of a straight line, with its conjugate polars with respect to $U_1, \ldots U_6$, are, when multiplied by certain constants, the homogeneous co-ordinates of the straight line, and are denoted by $x_1, \ldots x_6$. Arbitrary values of x_1, &c., do not denote a straight line, unless a homogeneous function of the second degree vanishes*. If this condition be not satisfied, then a linear complex is defined by the co-ordinates, and the function is called the *invariant* of the linear complex. The *simultaneous invariant* of two linear complexes is a function of the co-ordinates, and is equal to

$$\Delta \sin \phi - (K + K') \cos \phi,$$

where K and K' are the parameters of the linear complexes, Δ the perpendicular distance, and ϕ the angle between their principal axes.

The co-ordinates of a linear complex are the simultaneous invariants of the linear complex with each of six given linear complexes multiplied by certain constants. The six linear complexes can be chosen so that each one is in involution with the remaining five. The reader will easily perceive the equivalent theorems in the Theory of Screws. K and K' are the pitches, and the simultaneous invariant is merely double the virtual coefficient with its sign changed.

ZEUTHEN (H. G.)—*Notes sur un système de coordonnées linéaires dans l'espace.* Math. Ann.; Vol. i., pp. 432–454 (1869).

The co-ordinates of a line are the components of a unit force on the line decomposed along the six edges of a tetrahedron. These co-ordinates must satisfy one condition, which expresses that six forces along the edges of a tetrahedron have a single resultant force. The author makes applications to the theory of the linear complex.

Regarding the six edges as screws of zero pitch, they are not co-reciprocal. It may, however, be of interest to show how these co-ordinates may be used for a purpose different from that for which the author now quoted has used them. Let the virtual coefficients of the opposite pairs of edges be L, M, N. If the co-ordinates of a screw with respect to this system be $\theta_1 \ldots \theta_6$, then the pitch is

$$(L\theta_1\theta_2 + M\theta_3\theta_4 + N\theta_5\theta_6),$$

and the virtual coefficient of the two screws ϕ, θ is

$$\tfrac{1}{2} L (\theta_1\phi_2 + \theta_2\phi_1) + \tfrac{1}{2} M (\theta_3\phi_4 + \theta_4\phi_3) + \tfrac{1}{2} N (\theta_5\phi_6 + \theta_6\phi_5).$$

BATTAGLINI (G.)—*Sulle serie die sistemi di forze.* Napoli Rendiconto, viii., 1869, pp. 87–94. Giornale di Matemat, x., 1872, pp. 133–140.

This memoir deserves special notice in the history of the subject inasmuch as already remarked in § 13 it contains the earliest announcement of the *dynamical* significance of the cylindroid. Battaglini here shows that the cylindroid is the locus of the screws on which lie the wrenches produced by the composition of two variable forces on two fixed directions. See also p. 520.

* This equation expresses that the pitch of the screw denoted by the co-ordinates is zero.

BATTAGLINI (G.)—*Sulle dinami in involuzione.* Napoli Atti Accad. Sci., iv., 1869 (No. 14). Napoli Rendiconto, viii., 1869, pp. 166–167.

The co-ordinates of a dyname are the six forces which acting along the edges of a tetrahedron are equivalent to the dyname. This memoir investigates the properties of dynames of which the co-ordinates satisfy one or more linear equations. The author shows analytically the existence of two associated systems of dynames such that all the dynames of the first order are correlated to all the dynames of the second. These correspond to what we call two reciprocal screw complexes.

BALL (R. S.)—*A Problem in Mechanics. To determine the small oscillations of a particle on any surface acted upon by any forces* [1869]. Quart. Journ. Math., 1870, pp. 220–228.

With reference to this paper I may mention the following facts connected with the history of the present volume.

In the spring of 1869 I happened to attend a lecture at the Royal Dublin Society, given by my friend Dr G. Johnstone Stoney, F.R.S.

For one illustration he used a conical pendulum: he exhibited and explained the progression of the apse in the ellipse described by a heavy ball suspended from a long wire.

I was much interested by his exposition, and immediately began to work at the mathematical theory of the subject I was thus led to investigate some general problems relating to the small oscillations of a particle on a surface. Certain results, at which I arrived, seemed to me interesting and novel. They appeared in the paper now referred to. This paper was soon followed by another of a more general character and the subject presently began to develop into what was soon after called the "Theory of Screws."

BATTAGLINI (G.)—*Sul movimento geometrico infinitesimo di un sistema rigido.* Napoli Rendiconto, IX., 1870, pp. 89–100. Giornale di Matemat., x., 1872, pp. 207–216.

In this paper tetrahedral co-ordinates are employed in the analytical development of the statics of a rigid body, as well as the theory of small displacements. Besides the papers by this author to which I have specially referred there are several others (generally short) in Napoli Rendiconto, v.–x., both inclusive, which are of interest in connection with the fundamental notions involved in the theory of screws.

MANNHEIM (A.)—*Etude sur le déplacement d'une figure de forme invariable. Nouvelle méthode des normales; applications diverses.* Paris, Acad. Sci. Compt. Rend., lxvi., 1868, pp. 591–598. Paris, École Polytechn. Journ., cap. 43 (1870), pp. 57–121; Paris, Mém. Savants Étrang., xx., 1872, pp. 1–74.

This paper discusses the trajectories of the different points of a body when its movement takes place under prescribed conditions. It has been already cited (§ 121) for a theorem about the screws of zero pitch on a cylindroid. Another theorem of the same class is given by M. Mannheim. When a rigid body has freedom of the third order, then for *any* point on the surface of a certain quadric* the possible displacements are limited to a plane.

* The reader will easily see that this is the pitch quadric.

BALL (R. S.)—*On the small oscillations of a Rigid Body about a fixed point under the action of any forces, and, more particularly, when gravity is the only force acting.* Transactions of the Royal Irish Academy; Vol. xxiv., pp. 593–628 (January 24, 1870).

Certain dynamical problems which are here solved for the rotation of a body round a point were solved in subsequent papers for a body restricted in any manner whatever. Some of the chief results obtained are given in § 197.

This paper has its geometrical basis in the following theorem, due apparently to D'Alembert. *Recherches sur la Précession des Équinoxes*, Paris, 1749, p. 83.

Any small displacement of a rigid body rotating around a fixed point can be produced by the rotation around an axis passing through the point.

In 1776 Euler proved that the same law was true for displacements of finite magnitude. *Formulae generales pro translatione quacunque corporum Rigidorum.* Novi Commentarii Academiae Petropolitanae; Vol. xx., pp. 189–207.

KLEIN (F.)—*Notiz betreffend den Zusammenhang der Liniengeometrie mit der Mechanik starrer Körper.* Math. Ann.; Vol. iv., pp. 403–415 (June, 1871).

Among many interesting matters this paper contains the germ of the *physical* conception of reciprocal screws. We thus read on p. 413:—"Es lässt sich nun in der That ein physikalischer Zusammenhang zwischen Kräftesystemen und unendlich kleinen Bewegungen angeben, welcher es erklärt, wie so die beiden Dinge mathematisch co-ordinirt auftreten. Diese Beziehung ist nicht von der Art, dass sie jedem Kräftesystem eine einzelne unendlich kleine Bewegung zuordnet, sondern sie ist von anderer Art, sie ist eine *dualistiche.*

"Es sei ein Kräftesystem mit den Coordinaten Ξ, H, Z, Λ, M, N, und eine unendlich kleine Bewegung mit den Coordinaten Ξ', H', Z', Λ', M', N' gegeben, wobei man die Co-ordinaten in der im § 2 besprochenen Weise absolut bestimmt haben mag. *Dann repräsentirt*, wie hier nicht weiter nachgewiesen werden soll, *der Ausdruck*

$$\Lambda'\Xi + M'N + N'Z + \Xi'\Lambda + H'M + Z'N$$

das Quantum von Arbeit, welches das gegebene Kräftesystem bei Eintritt der gegebenen unendlich kleinen Bewegung leistet. Ist insbesondere

$$\Lambda'\Xi + M'H + N'Z + \Xi'\Lambda + H'M + Z'N = 0,$$

so leistet das gegebene Kräftesystem bei Eintritt der gegebenen unendlich kleinen Bewegung *keine* Arbeit. Diese Gleichung nun repräsentirt uns, indem wir einmal Ξ, H, Z, Λ, M, N, das andere Ξ', H', Z', Λ', M', N' als veränderlich betrachten, den Zusammenhang zwischen Kräftesystemen und unendlich kleinen Bewegungen."

BALL (R. S.)—*The Theory of Screws—a geometrical study of the kinematics, equilibrium, and small oscillations of a Rigid Body.* First memoir. Transactions of the Royal Irish Academy, Vol. xxv., pp. 137–217 (November 13, 1871).

This is the original paper on the Theory of Screws. At the time this paper was printed (1871) I had no suspicion that the Cylindroid had been ever studied by anyone besides myself. I subsequently learned that the same surface had been investigated by Plücker two or three years previously (1868–9) in connection with the linear complex (see pp. 20, 517). It also appeared that about the same time (1869) this surface presented itself in the Researches of Battaglini. Indeed, to this mathematician belongs, I believe, the distinction of having been the first to perceive that this particular conoid had a special dynamical significance (see pp. 20, 518). Plücker and Battaglini were certainly independent discoverers of the cylindroid, or rather rediscoverers, for neither of them was the earliest discoverer, for as shown in p. 510 the cylindroid was first introduced into science by Sir W. R. Hamilton so long ago as 1830. It is worthy of note that three investigators, and if I may add my own name a fourth also, following different lines of research, have each been independently led to perceive the importance of this particular surface in various theories of systems of lines.

In the paper now before us I had developed the doctrine of reciprocal screws which is of such fundamental importance in the theory. I had arrived at this doctrine independently, and not until after the paper was printed did I learn that the essential conception of Reciprocal Screws had been announced by Professor Klein a few months before my paper was read (pp. 17, 18, 520).

These facts have to be mentioned in explanation of the circumstance that this first paper contains no references to the names of either Plücker and Battaglini or Hamilton and Klein.

SOMOFF (J.)—*Sur les vitesses virtuelles d'une figure invariable, assujetties à des équations de conditions quelconques de forme linéaire.* St Pétersb. Acad. Sci. Bull., xviii., 1873, col. 162—184.

This paper is an important one in the history of the subject. Its scope may be realized from the paragraph here quoted.

"Dans le mémoire que j'ai l'honneur de présenter à l'Académie je donne un moyen analytique pour déterminer les vitesses virtuelles d'une figure invariable, en supposant que ces vitesses doivent satisfaire à des équations de condition de la forme générale que je vients de citer. Je prends en même temps en considération les propriétés des complexes linéaires de Plücker, auxquel les vitesses virtuelles d'une figure invariable sont intimement liées."

The analytical development of the Theory of the Constraints which follows is founded upon the conventions proposed by M. Résal in his "*Traité de Cinématique pure.*"

M. Somoff studies conditions of constraint which he has generalized from M. Mannheim's "*Étude sur le déplacement d'une figure de forme invariable*" (p. 519).

It is instructive to read M. Somoff's paper in the light of the Theory of Screws. For example on p. 179 he gives the theorem that every system of "virtual velocities" which satisfies three linear equations can be produced by two rotations around two rays common to the three corresponding linear complexes. In our language we express this by saying that any displacement of a body with three degrees of freedom can be produced by rotation around two screws of zero pitch belonging to the system. This is easily seen, for let θ be the screw about which the required displacement is a twist. Let ϕ be *any* other screw of the three-system, then the two screws of zero pitch on the cylindroid (θ, ϕ) are two axes of rotation that fulfil the required condition.

The cases of four and five degrees of freedom are also briefly discussed by Somoff, but without the conception of screw motion which he does not employ the results are somewhat complicated.

Reference may also be made to Somoff, "*Theoretische Mechanik,*" translated from the Russian by A. Ziwet, Leipzig, 1878–9.

CLIFFORD (W. K.)—*Preliminary Sketch of Biquaternions.* Proceedings of the London Mathematical Society, Nos. 64, 65, Vol. iv., pp. 381—395 (12th June, 1873).

This is one of the modern developments of that remarkable branch of mathematics with which the names of Lobachevsky and Bolyai are specially associated. A Biquaternion is defined to be the ratio of two twists or two wrenches or more generally of two Dynames in Plücker's sense or of two "motors" as Clifford prefers to call them. A "motor" may be said to bear the same relation to a screw which a vector bears to a ray. The calculus of Biquaternions is generalized from that of quaternions and belongs to the non-Euclidian geometry. See Klein's celebrated paper, "Ueber die sogenannte nicht Euclidische Geometrie." Math. Ann., Band IV., pp. 573–625. This paper of Clifford's has been the commencement of an extensive theory at which many mathematicians have since worked. Chap xxvi. discusses some of Clifford's theorems and in the course of these bibliographical notes there are several references to this theory. See under the names of Everett, Padeletti, Cox, Heath, Buchheim, Cayley, Burnside, Joly, Kotelnikof, and M'Aulay.

BALL (R. S.)—*Researches in the Dynamics of a Rigid Body by the aid of the Theory of Screws.* Second Memoir (June 19, 1873). Philosophical Transactions, pp. 15–40 (1874).

The chief advance in this paper is expressed by the theorem that a rigid body has just so many principal screws of inertia as it has degrees of freedom. This theorem is a generalization for all cases of a rigid system, no matter what be the nature and number of its constraints, of the well-known property of the principal axes of a rigid body rotating around a fixed point.

It is shown that if the screws on one cylindroid be regarded as impulsive screws, the system of corresponding instantaneous screws lie on another cylindroid. Any four screws on the one cylindroid, and their four correspondents on the others are equi-anharmonic. This theorem leads to many points of connexion between theoretical dynamics and modern geometry. It has been greatly developed subsequently.

A postscript to this paper gives a brief historical sketch which shows the relation of the theory of screws to the researches of Plücker and Klein on the Theory of the Linear Complex.

SKATOW.—*Zusammenstellung der Sätze von den übrigbleibenden Bewegungen eines Körpers, der in einigen Punkten seiner oberfläche durch normale Stützen unterstützt wird.* Schlömilch's Zeitschrift für Mathem. u. Physik, B. xviii., p. 224, 1873.

HALPHEN—*Sur le déplacement d'une solide invariable.* Bulletin de la Soc. Math., Vol. ii., pp. 56–62 (23 July, 1873).

The study of the displacements of a rigid body is distributed into six cases according to the number of degrees of freedom. This paper like so many others on the present subject has been suggested by the writings of M. Mannheim. It gives for instance a proof of Mannheim's theorem that all the displacements of a solid restrained by four conditions could be produced by two rotations around two determinate lines. These are of course in our language the two screws of zero pitch on the cylindroid expressing the freedom. Halphen considers in some cases conditions more general than those of Mannheim and adds some theorems of quite a new class. Thus still referring to the case of a body restrained by four conditions, *i.e.* with two degrees of freedom, he shows how the movements of every point are limited to a surface, and then calling the two screws of zero pitch the "axes" we have as follows. "Les projections, sur un plan donné, des éléments superficiels, décrit par les points du corps, sont proportionelles aux produits des segments interceptés, sur des sécantes parallèles issues de ces points, par un paraboloide passant par les deux axes, et ayant le plan donné pour plan directeur."

LINDEMANN (F.)—*Ueber unendlich kleine Bewegungen und über Kraftsysteme bei allgemeiner projectivischer Massbestimmung.* Math. Ann., Vol. vii., pp. 56–143 (July, 1873).

This is a memoir upon the statics and kinematics of a rigid body in elliptic or hyperbolic space. Among several results closely related to the Theory of Screws, we find that the cylindroid is only the degraded form in parabolic or common space of a surface of the fourth order, with two double lines. Lindemann both by this memoir and by that entitled "Projectivische Behandlung der Mechanik starrer Körper" in the same volume has become the pioneer of an immense and most attractive field of exploration. He has laid down the principles of Dynamics in Non-Euclidian space. One small part of this subject I have endeavoured to develop in Chap. XXVI.

WEILER (A.)—*Ueber die verschiedenen Gattungen der Complexe zweiten Grades.* Math. Ann., Vol. vii., pp. 145–207 (July, 1873).

In this elaborate memoir the author enumerates fifty-eight different species of linear complexes of the second order. The classification is based upon Kummer's surface, which defines the singularities of the complex. These investigations are of importance in the present subject because, to take a single instance, the screws of a system of the fourth order form a linear complex of the second order. This complex is of a special type included among the 58 species.

BALL (R. S.)—*Screw Co-ordinates and their applications to problems in the Dynamics of a Rigid Body.* Third memoir. Transactions of the Royal Irish Academy, Vol. xxv., pp. 259–327 (January 12, 1874).

The progress of the present theory was much facilitated by the introduction of screw co-ordinates. The origin and the use of such co-ordinates are here explained. It is, however, to be understood that screw co-ordinates, though no doubt arrived at independently, ought properly to be regarded as an adaptation for dynamical purposes of Klein's co-ordinates of a linear complex referred to six fundamental complexes, of which each pair are in involution or reciprocal, as we say in the terminology of this volume.

The pitch of a screw α as expressed in terms of its six co-ordinates $\alpha_1, \ldots \alpha_6$ is $\Sigma p_1\alpha_1^2$ where $p_1 \ldots p_6$, &c. are the pitches of the co-reciprocal screws of reference. The virtual coefficient of two screws α and β is $\Sigma p_1\alpha_1\beta_1$. In the dynamical part of the subject the chief result of this paper is the fundamental theorem that, when the six screws of reference are the six principal screws of inertia, then $p_1\alpha_1, p_2\alpha_2, \ldots p_6\alpha_6$ are the co-ordinates of the impulsive wrench which will make the body commence to move by twisting about the screw $\alpha_1 \ldots \alpha_6$.

This was, perhaps, all that could be desired in the way of a simple connexion between an impulsive screw and the corresponding instantaneous screw, so far as their co-ordinates were concerned. Long before this paper was published I had been trying to find a geometrical connexion between two such screws which would exhibit their relation in a graphic manner. But the search was not to be successful until the results in the Twelfth Memoir were arrived at.

EVERETT (J. D.)—*On a new method in Statics and Kinematics.* (Part I.) Messenger of Mathematics. New Series. No. 39 (1874), 45, 53 (1875).

The papers contain applications of quaternions. The operator $\varpi + V\sigma(\ \ )$ is a "motor," ϖ and σ being vectors, the former denoting a translation or couple, the latter a rotation or force. The pitch is $S\frac{\varpi}{\sigma}$. The equation to the central axis is $\rho = V\frac{\varpi}{\sigma} - x\sigma$. The work done in a small motion is $-S\varpi_1\sigma_2 - S\varpi_2\sigma_1$. The existence of k equations of the first degree between n motors is the condition of their belonging to a screw system of the first degree, and of order $n-k$. Several of the leading theorems in screws are directly deduced from motor equations by the methods of determinants.

STURM (Rudolf).—*Sulle forze in equilibrio.* Darmstadt, 1875.

This is an interesting geometrical memoir in which the beautiful methods of Möbius in his Lehrbuch der Statik have been followed up.

BALL (R. S.)—*The Theory of Screws.* A study in the Dynamics of a rigid body. Dublin, 8vo., 1876, pp. (1–194).

The substance of this volume (now out of print) has been incorporated in the present one. The necessity for a new work on the subject will be apparent from these bibliographical notes, from which it will be seen how much the subject has grown since 1876. It will be here sufficient to give an extract from the preface.

"The Theory presented in the following pages was first sketched by the author in a Paper communicated to the Royal Irish Academy on the 13th of November, 1871. This Paper was followed by others, in which the subject was more fully developed. The entire Theory has been re-written, and systematically arranged, in the present volume."

"References are made in the foot-notes, and more fully in the Appendix, to various authors whose writings are connected with the subject discussed in this book. I must, however, mention specially the name of my friend Professor Felix Klein, of Munich, whose private letters have afforded me much valuable information, in addition to that derived from his instructive memoirs in the pages of the Mathematische Annalen."

An abstract dated Nov. 1875 of the chief theorems in this book has been given in Math. Annalen, Vol. ix., pp. 541–553.

FIEDLER (W.)—*Geometrie und Geomechanik. Vierteljahrschrift der naturforschenden Gesellschaft in Zürich* (1876), xxi. 186, 228.

This valuable paper should be studied by any one desirous of becoming acquainted with the history of the subject. Dr Fiedler has presented a critical account of the manner in which the Theory of Screws has grown out of the works of the earlier mathematicians who had applied the higher geometry to Dynamics, especially Chasles, Poinsot, Möbius and Plücker. The paper contains an account of the chief results in the Theory so far as they were known in 1876. Many of the investigations are treated with much elegance, as might indeed have been expected from a mathematician so accomplished as the German translator of Dr Salmon's great works.

RITTERSHAUS (T.)—*Die Kinematische Kette, ihre Beweglichkeit und Zwangläufigkeit.* Der Civilingenieur, Vol. XXII. (1877).

This is the study of the kinematics of three rigid bodies whereof the first and second are hinged together, as are also the second and third. The cylindroid is employed to obtain many theorems. Of course it will be understood that the "Kinematische Kette" is a conception quite distinct from that of the Screw-chain discussed in the present volume (Chap. XXIV.). In a further paper (*loc. cit.* XXIV. 1878) the author develops cases in which the conditions are of increased generality.

CLIFFORD (W. K.)—*Elements of Dynamics.* 1878.

In this work, designed no doubt to be elementary but perhaps rather illustrating the breadth of view so characteristic of its gifted author, the fundamental theorem of the composition of twists and wrenches by the cylindroid is assigned an important position at the basis of mechanics.

SCHELL (W.).—*Theorie der Bewegung und der Kräfte.* 2nd edition. Leipzig, 1879. Vol. ii.

This is a comprehensive and valuable treatise on Theoretical Dynamics. It merits particular mention here because it contains an excellent exposition in the German language of many of the most important parts of the Theory of Screws. Part III., Chap. x., pp. 211–235, discusses the Cylindroid and reciprocal Screws, and includes a general account of the properties of the different screw-systems. Part IV., Chap. VIII., gives a general account of the Dynamical parts of the Theory, including the principal Screws of Inertia and Harmonic Screws.

BALL (R. S.)—*Note on the application of Lagrange's Equations of Motion to Problems in the Dynamics of a Rigid Body.* Proceedings of the Royal Irish Academy. 2nd Ser., Vol. iii., p. 213 (1879).

In this paper it is shown from Lagrange's well-known equations of motion in generalized co-ordinates that if T be the kinetic energy of a body twisting about a screw whose n screw co-ordinates referred to any co-reciprocal system with pitches $p_1, \ldots p_n$, are $\theta_1, \ldots \theta_n$, then the impulsive wrench which would have been capable of producing from rest the actual motion which the body possesses must have as its coordinates (§ 86)

$$\frac{1}{p_1}\frac{dT}{d\theta_1}, \ldots, \frac{1}{p_n}\frac{dT}{d\theta_n}.$$

BALL (R. S.)—*Extension of the Theory of Screws to the Dynamics of any Material System.* (Fourth Memoir.) Transactions of the Royal Irish Academy, Vol. xxviii., pp. 99–136 (1881).

The conception of a screw-chain is here introduced. The screw-chain is a geometrical entity which bears to an entire system, no matter how complex its parts or their connexions, the same relation which a screw bears to a single rigid body. One screw-chain can always be found which is reciprocal to $6\mu - 1$ screw-chains where μ is the number of material parts in the system.

One of the chief results obtained shows the extension of the notion of the principal screws of inertia of a single rigid body to a system of rigid bodies.

Chap. XXIV. of the present volume contains the essential parts of this memoir.

SCHELL (W.)—*Die sechs Grade der Beweglichkeit eines unveränderlichen Systems.* Central Zeitung für Optik und Mechanik, 1881.

Here is an interesting geometrical study of the degrees of freedom of a rigid body under the several conditions that 1, 2, 3, 4 or 5 of its points shall be constrained to lie on given surfaces.

If a force be applied along a normal to the surface at the point of the body which is constrained to lie on that surface then that force will be counteracted by the constraints. Every motion of the body which is possible must be a twist about a screw reciprocal to a screw of zero pitch on that normal.

Constraint of the most general nature cannot, however, be so produced. It is sufficient to mention that in the case of freedom of the fifth order the screw reciprocal to the system must have zero pitch, if the constraint is of the nature supposed by Schell, while in the general case the pitch may have any value.

BALL (R. S.)—*On Homographic Screw Systems.* Proceedings of the Royal Irish Academy, Ser. 2, Vol. iii. p. 435 (1881).

The theory of Homographic Screws shows the connection between certain geometrical theories of an abstract nature and Dynamics. The intimate alliance between geometry and the higher branches of Rigid Dynamics is illustrated in this paper. Invariant functions of eight screws are studied, and a generalized type of homographic ratio involving eight screws is considered. (See Chap. XIX.)

BALL (R. S.)—*On the Elucidation of a question in Kinematics by the aid of Non-Euclidian Space.* Report of British Association, York, 1881, p. 535.

Certain peculiarities which presented themselves in the geometrical representation of the screws of a three-system by points in a plane are here shown to be due to the conventions of Euclidian space. The screws of a three-system in non-Euclidian space can be arranged in equal pitch hyperboloids, which have eight common points and eight common tangent planes. In Euclidian space the corresponding quadrics are inscribed in a common tetrahedron and pass through four common points as explained in Chap. XV.

BALL (R. S.)—*Certain Problems in the Dynamics of a Rigid System moving in Elliptic Space.* (Fifth Memoir.) Transactions of the Royal Irish Academy, Vol. xxviii., pp. 159–184 (1881).

The chief theorem proved in this paper is, that though the virtual moment of two homonymous vectors is zero only when the two vectors are "rectangular," yet the virtual moment of two heteronymous vectors is always zero.

I may here mention another memoir which bears on the same subject. The title is, *On the Theory of the Content.*—Transactions of the Royal Irish Academy, Vol. xxix., pp. 123–181 (1887).

In this it is shown that the order in which two heteronymous vectors in elliptic space are applied to a rigid system may be inverted without affecting the result, which is, however, not a vector at all. On the other hand, when two homonymous vectors in elliptic space are applied to a rigid system, the result is, in every case, a homonymous vector; but then the order of application could not be inverted without changing the result.

These papers have contributed to Chap. XXVI. of the present volume.

PADELETTI (Dino)—*Osservazioni sulla teoria delle dinami (Theory of Screws).* Rendiconto della R. Accademia di Scienze Fis. e Nat. di Napoli, Fascicolo 2° Feb. 1882.

The author here gives a general account of the Theory of Screws so far as it had been developed up to 1876. The method he has employed for deducing the equations of the cylindroid is novel and instructive. The same author in the same journal for May 1882 has a paper entitled, *Su un Calcolo nella teoria delle dinami analogo a quello dei quaternioni.*

COX (Homersham)—*On the application of Quaternions and Grassmann's Ausdehnungslehre to different kinds of Uniform Space.* Cambridge Philosophical Transactions, Vol. xiii., Part II., pp. 69–143 (1882).

So far as the Theory of Screws is concerned the chief result in this paper is the demonstration that the homologue of the cylindroid in non-Euclidian space which Lindemann had already shown to be of the fourth degree may be represented by the equation

$$(p_\alpha - p_\beta)(w^2 + z^2)\, xy = (1 - p_\alpha p_\beta)(x^2 + y^2)\, wz.$$

The function known as the sexiant (§ 230) is here generalized into the corresponding function of six screws in non-Euclidian space. It is of course a fundamental theorem that a ray crossing two screws of equal pitch meets the cylindroid again in a third screw which it cuts perpendicularly (§ 22). This is here generalized into the theorem that a transversal across two screws of equal pitch on the cylindroid in elliptic space intersects that surface also in two other generators which are conjugate polars with respect to the absolute.

I may take this opportunity to observe that the function $\frac{p_\alpha - p_\beta}{1 - p_\alpha p_\beta}$ which enters into the above equation of the surface has an instructive property. If p_α and p_β be transformed into $\frac{p_\alpha + m}{1 + mp_\alpha}$ and $\frac{p_\beta + m}{1 + mp_\beta}$ respectively, where m is different from unity, then the above function is unaltered. Hence it follows that if the pitch p of every screw on a screw-system of the nth order in non-Euclidian space receive the transformation into $\frac{p + m}{1 + mp}$ then the screws so altered will still constitute an n-system. Thus we generalize that well-known feature of an n-system of screws in ordinary space which asserts that if the pitches of the screws in an n-system be augmented by a constant the screws so altered will remain an n-system. (See Proceedings of the Royal Irish Academy, 2nd Series, Vol. iv., p. 256 (1884).)

PADELETTI (Dino)—*Sulla più semplice forma dell' equazioni di equilibrio di un sistema rigido vincolato.* Rendiconto della R. Accademia Scienze Fis. e Mat. di Napoli, Fascicolo 1°, 1883.

In this short paper the author discusses separately two different cases of freedom and by the aid of the reciprocal screw-system gives in each case the equations of equilibrium.

HEATH (R. S.)—*On the Dynamics of a Rigid Body in Elliptic Space.* Phil. Trans. Part II., 1884, pp. 281–324.

"The special features of the method employed are the extensive use of the symmetrical and homogeneous system of coordinates given by a quadrantal tetrahedron, and the use of Professor Cayley's co-ordinates in preference to the 'Rotors' of Professor Clifford to represent the position of a line in space." The Theory of Screws is considered and the nature of the cylindroid in Elliptic Space discussed. The general equations of motion referred to any moving axes are then found, and in a particular case they reduce to a form corresponding to Euler's equations. When there are no acting forces these equations are solved in terms of the theta-functions. This paper has been already cited in §§ 412, 420.

BUCHHEIM (A.)—*On the Theory of Screws in Elliptic Space.* Proceedings of the London Math. Soc., Vol. xiv., p. 83; Vol. xvi., p. 15; Vol. xvii., p. 240; Vol. xviii., p. 88.

In these papers the methods of the Ausdehnungslehre of Grassmann have been applied to the Biquaternions of Clifford. Reference should also be made to another paper by the same author, "*A Memoir on Biquaternions.*" (American Journal of Mathematics, Vol. vii., No. 4, p. 23, 1884.) If A, B be two biquaternions they determine a linear singly infinite series of biquaternions $\lambda A + \mu B$ when λ, μ are scalars: this set is called a cylindroid, so that if C is any biquaternion of the cylindroid A, B we have $C = \lambda A + \mu B$. A remarkable investigation of the equation to this surface in elliptic space is given, and a generalization of the plane representation of the cylindroid is shown. In these writings it is the methods employed that are chiefly noticeable. We find however much more than is implied by the modest disclaimer of the lamented writer, who in the last letter I had from him says, "I have been but slaying the slain, i.e. discovering over again results obtained by you and Clifford."

SEGRE (C.)—*Sur une expression nouvelle du moment mutuel de deux complexes linéaires.* Kronecker's Journal, pp. 169–172 (1885).

A remarkable form for the expression of the virtual coefficient of two screws on a cylindroid is given in this paper. Translated into the terminology of the present volume we can investigate Segre's theorem as follows.

Let two screws on the cylindroid make angles θ, ϕ, with one of the principal screws, while the zero pitch screws make angles $+\alpha$, $-\alpha$. Let ρ be the anharmonic ratio of the pencil parallel to these four screws so that

$$\rho = \frac{\sin(\theta - \alpha)}{\sin(\theta + \alpha)} \cdot \frac{\sin(\phi + \alpha)}{\sin(\phi - \alpha)}.$$

Then as usual

$$p_\theta = p_0 + m \cos 2\theta,$$

$$0 = p_0 + m \cos 2\alpha;$$

whence

$$p_\theta = 2m \sin(\alpha - \theta) \sin(\alpha + \theta),$$

$$p_\phi = 2m \sin(\alpha - \phi) \sin(\alpha + \phi);$$

whence

$$4m^2 \sin^2(\theta - \alpha) \sin^2(\phi + \alpha) = \rho p_\theta p_\phi.$$

Thus

$$2m \sin(\theta - \alpha) \sin(\phi + \alpha) = \sqrt{\rho}\sqrt{p_\theta p_\phi},$$

$$2m \sin(\theta + \alpha) \sin(\phi - \alpha) = \sqrt{\rho^{-1}}\sqrt{p_\theta p_\phi};$$

adding, we easily obtain

$$\varpi_{\theta\phi} = \tfrac{1}{2}(\sqrt{\rho} + \sqrt{\rho^{-1}})\sqrt{p_\theta p_\phi}.$$

If therefore we make

$$\rho = e^{2i\epsilon},$$

we have as the result Segre's theorem that

$$\varpi_{\theta\phi} = \sqrt{p_\theta p_\phi} \cos \epsilon.$$

D'EMILIO (R).—*Gli assoidi nella statica e nella cinematica. Nota su la teoria delle dinami.* Atti del Reale Istituto Veneto di scienze, (6) iii. 1135–1154 (1885).

This is an account of the fundamental laws of the different screw-systems.

MINCHIN (G. M.)—*Treatise on Statics.* 3rd edition, Vol. ii. (1886).

In pp. 17–43 of this standard work the Theory of Screws is discussed. An instructive construction for the cylindroid is given on p. 20. We may also note the following theorem proved on p. 25, "If the wrench on any screw of the cylindroid is replaced by a force and a couple at the centre of the pitch-conic (centre of the cylindroid) the axis of this couple will lie along the perpendicular to the diameter of the pitch-conic which is conjugate to the direction of the force—or in other words, the plane of the couple will be that of the axis of the cylindroid and this conjugate diameter."

SCHÖNFLIES (Arthur)—*Geometrie der Bewegung in synthetischer Darstellung*, pp. 1–194, 8vo. Leipzig, 1886.

The third chapter of this work, pp. 79–192, is devoted to the geometrical study of the movement of a rigid system. The author uses the word *parameter* to express what we have designated as the *pitch*. As an illustration of the theorems given I cite the following from p. 92, "*Bewegt sich ein unveränderliches System beliebig im Raume, und ist in irgend einem Augenblick eine Gerade desselben senkrecht zur Tangente der Bahn eines ihrer Punkte, so ist sie es zu den Bahntangenten aller Punkte.*"

In the language of the present volume in which the dynamical and kinetical conceptions are so closely interwoven, this theorem appears as follows. Let two screws α and β be reciprocal and let the pitch of α be zero. A twist of a rigid body about β can do no work against a force on α. But α may be considered to act on the rigid body at any point in its line of application. Hence the displacements of every such point must be perpendicular to α.

The following suggestive theorems may be quoted from pp. 116, 117:

"Die sämmtlichen Punkte des Systems deren Bahnen nach einem festen Punkte D des Raumes gerichtet sind, liegen in jedem Augenblick auf einer Raumcurve dritter Ordnung C."

"Die Raumcurve C enthält die unendlich fernen imaginären Kreispunkte der zur Axe der Schraubenbewegung senkrechten Ebenen."

This work contains indeed much that it would be interesting to quote. I must however content myself with one more remark from p. 153, which I shall give in our own terminology. When a rigid body has freedom of the second order it can of course be twisted about any screw on a cylindroid. Such a twist can always be decomposed into two rotations around the two screws of zero pitch P and Q. The rotation around P does not alter P. Hence whatever be the small displacement of the system the movement of P can never be other than a rotation around Q, and the movement of Q can never be other than a rotation around P.

BALL (R. S.)—*Dynamics and Modern Geometry: a new chapter in the Theory of Screws.* Sixth Memoir. Cunningham Memoirs of the Royal Irish Academy, No. iv., pp. 1–44 (1886).

We represent the several screws on the cylindroid by points on the circumference of a circle. The angle between two screws is the angle which their corresponding points subtend at the circumference. The shortest distance of any two screws is the projection of the corresponding chord on a fixed ray in the plane of the circle. Any chord passing through the pole of this ray intersects the circle in points corresponding to reciprocal screws. The pitch of any screw is the distance of its corresponding point from this ray. A system of points representing instantaneous screws and the corresponding system representing the impulsive screws are homographic. The double points of the homography correspond to the two principal screws of inertia. This paper is the development of an earlier one. See Proceedings of the Royal Irish Academy, 2nd Series, Vol. iv. p. 29. The substance of it has been reproduced in Chaps v. and xii. of the present volume.

Ball (R. S.)—*On the Plane Sections of the Cylindroid.* Seventh Memoir. Trans. of the Royal Irish Academy, Vol. xxix., pp. 1–32 (1887).

This is a geometrical study of the cylindroid regarded as a conoidal cubic with one nodal line and three right lines in the plane at infinity. Plane sections of the cylindroid are shown in plates drawn to illustrate calculated cases. It is shown that the chord joining the points in which two reciprocal screws intersect a fixed plane envelops a hyperbola which has triple contact with the cubic curve in which the fixed plane cuts the cylindroid. See Chap. xiii. of the present volume.

I may take this opportunity of mentioning in addition to what has been said on the subject of models of the cylindroid in Chap. xiii. that very simple and effective models of this surface can now be obtained from Martin Schilling, Halle a Saale. See his catalogue for Feb. 1900.

Roberts (R. A.)—*Educational Times*, xlvi. 32–33 (1887).

In this it is shown that under the circumstances described the shadow of the cylindroid $z(x^2 + y^2) - 2mxy = 0$ on the plane $z = 0$ exhibits the hypocycloid with three cusps.

Ball (R. S.)—*A Dynamical Parable, being an Address to the Mathematical and Physical Section of the British Association.* Manchester, 1887.

This has been given in Appendix ii. p. 496. It may be added here that it has been translated into Hungarian by Dr A. Seydler, and into Italian by G. Vivanti.

Tarleton (F. A.)—*On a new method of obtaining the conditions fulfilled when the Harmonic Determinant has equal roots.* Proceedings of the Royal Irish Academy, 3rd Series, Vol. i. No. 1, p. 10 (1887).

This discusses the case of equal roots in the harmonic determinant so important in the Theory of Screws as in other parts of Dynamics. It should be studied in connection with § 85 of the present volume; also Note ii. p. 484. See also Zanchevsky, p. 531.

Ball (R. S.)—*How Plane Geometry illustrates general problems in the Dynamics of a Rigid Body with Three degrees of Freedom.* Eighth Memoir. Transactions of the Royal Irish Academy, Vol. xxix., pp. 247–284 (1888).

The system of the third order is of such special interest that it is desirable to have a concise method of representing the screws which constitute it. We here show that the screws of such a system correspond to the points in a plane. This is the development of an earlier paper communicated to the Royal Irish Academy in 1881. Proceedings, 2nd Series, Vol. iii., pp. 428–434.

In this method of representation the screws on a cylindroid belonging to the system are represented by the points on a straight line. The screws of any given pitch will have as their correspondents the points on a certain conic. A pair of points conjugate to the conic of zero pitch will correspond to a pair of reciprocal screws. The conic which represents the screws of zero pitch, and the conic which represents the screws of infinite pitch, will have a common conjugate triangle. The vertices of that triangle correspond to the principal screws of the system. It is proved that the pitch quadrics of a three-system are all inscribed in a common tetrahedron and have four common points on the plane at infinity.

The points which represent a series of impulsive screws and the points which represent the series of corresponding instantaneous screws are homographic. The three double points of the homography represent the three principal screws of inertia.

The three harmonic screws about any one of which the body would oscillate for ever in the vicinity of a position of stable equilibrium are determined as the vertices of the common conjugate triangle of two conics.

This memoir is the basis of Chap. XV. in the present volume.

HYDE (E. W.)—*Annals of Mathematics*, Vol. iv., No. 5, p. 137 (1888).

The author writes: "I shall define a screw to be the sum of a point-vector and a plane-vector perpendicular to it, the former being a directed and posited line, the latter the product of two vectors, hence a directed but not posited plane." Prof. Hyde proves by his calculus many of the fundamental theorems in the present theory in a very concise manner.

GRAVELIUS (Harry)—*Theoretische Mechanik starrer Systeme. Auf Grund der Methoden und Arbeiten und mit einem Vorworte von Sir Robert S. Ball.* Berlin, 1889. 8vo., p. 619.

The purport of this volume is expressed in the first paragraph of the preface: "Das vorliegende Werk stellt sich die Aufgabe, zusammenhängend und als Lehrbuch die in zahlreichen Arbeiten von Sir Robert Ball geschaffene Theorie der Mechanik starrer Systeme darzustellen. Es umfasst somit dem Inhalte nach sämmtliche Abhandlungen des Herrn Ball." Thus the work is mainly a translation of the Theory of Screws and of the subsequent memoirs up to the date 1889. Herr Gravelius has however added much, and his original contributions to the theory are specially found in Chap. XIX. "Projective Beziehungen räumlicher Schraubengebilde." I feel very grateful to Herr Gravelius for his labour in rendering an account of the subject into the German language.

ZANCHEVSKY (I.)—*Theory of Screws and its Application to Mechanics*, pp. I—XX., 1—131. Odessa, 1889.

I must first acknowledge the kindness with which my friend Mr G. Chawner, Fellow of King's College, has assisted me by translating the Russian in which this book is written. I here give some passages from the introduction.

Zanchevsky remarks that in the Theory of Screws I omitted to give a proof of the reality of all the roots of the equation of the nth degree which determines the principal Screws of Inertia, and then he gives a proof derived from a theorem of Kronecker. "*Zur Theorie der linearen und quadratischen Formen.*" Monatsberichte der Acad. der Wissenschaften zu Berlin, 1868, p. 339. The theorem is as follows. Let U and V be two homogeneous quadratic forms with n variables. If the discriminant of $\lambda U + \mu V$ when equated to zero gives a single imaginary root then no member of the system $\lambda U + \mu V$ can be expressed as the sum of n squares. We should, however, in this matter refer to the earlier paper of Weierstrass, p. 513. From this theorem Zanchevsky proves the reality of the roots of the Harmonic Determinant. (See § 85.) Then follows a discussion of the principal Screws of Inertia for a constrained system.

Chap. I. contains an exposition of Plücker's theory of the linear complex of the 1st order. Here will be found the conception of the screw, its co-ordinates, the virtual coefficient of two screws, and the connection between the systems of vectors which determine reciprocal screws. He remarks that this connection may be directly derived from the works of Lornoff.

Chap. II. is devoted to groups of screws. He discusses in detail groups containing three members, investigating some special cases not dwelt on before. He then gives the formulæ for what are termed "Oblique Co-ordinates."

In Chap. III. the application of the Theory of Screws to mechanics is discussed and the leading parts of the Theory of Screws in relation to dynamical problems with freedom of the nth order are set forth, and he adds, "The lack of books on the Theory of Screws both in Russia and abroad makes us hope that our work will be received with indulgence."

BALL (R. S.)—*The Theory of Permanent Screws.* Ninth Memoir. Transactions of the Royal Irish Academy, Vol. xxix., pp. 613—652. 1890.

Using Screw-chain co-ordinates an emanent (see Salmon's *Higher Algebra*, § 125, or Elliott's *Algebra of Quantics*) is here shown to vanish. This involves a general property of the function T which expresses the kinetic energy.

$$\dot{x}_1 \frac{dT}{dx_1} + \ldots + \dot{x}_n \frac{dT}{dx_n} = 0.$$

It is shown that for the permanent screw-chains,

$$\frac{dT}{dx_1} = 0, \ldots, \frac{dT}{dx_n} = 0.$$

The special cases for the different degrees of freedom of a single rigid body are considered in detail. If the rigid body has three degrees of freedom then there are three permanent screws, about any one of which the body will continue to twist if once set twisting.

In general, if the body be set twisting about a screw θ, a restraining wrench on some other screw η would be necessary if the motion were to continue as a twist about θ.

To find η we employ the plane representation. We construct first a system of points homographic with the points θ. The double points of the homography are representative of the three permanent screws. If we draw the ray connecting θ with its correspondent, then η is the pole of this ray with respect to the conic of zero pitch, while the pole of the same ray with respect to the conic of inertia gives the screw about which the acceleration is imparted to θ.

For freedom of the first and second orders there is only one permanent screw; for freedom of the third, fourth, and fifth, there are three permanent screws. When the body is quite free the permanent screws are triply infinite. The Theory of Permanent Screws is given in Chap. XXV.

HENRICI (O.)—*The Theory of Screws.* Nature, xlii. 127-132. London, 1890.

Under the form of a review of the work of Gravelius (see p. 531) we have here an original and suggestive discussion of the entire subject. Professor Henrici has pointed out several promising lines along which new departures might be taken in the further development of the present theory.

KÜPPER (C.)—*Die Schraubenbewegung, das Nullsystem und der lineare Complex.* Monatshefte für Mathematik und Physik. Vienna, 1890, pp. 95-104.

In this the theory of the linear complex has been developed from the Theory of Screws. The object appears to have been to introduce the study of the subject into the High Schools in Germany.

CAYLEY (A.)—*Non-Euclidian Geometry.* Transactions of the Cambridge Philosophical Society, Vol. xv., pp. 37–61 (1894). Read Jan. 27, 1890. See also Collected Papers, Vol. xiii., p. 480.

This is perhaps the best paper in the English language from which to obtain a general view of the Non-Euclidian Geometry. The development is here conducted mainly along geometrical lines. On this account a study of this paper is specially recommended in connection with Chap. XXVI. of the present volume.

BUDDE (E.)—*Allgemeine Mechanik der Punkte und starren Systeme.* 2 vols. 8vo. Berlin, 1891.

This comprehensive work may be cited in illustration of progress made in the use of the Theory of Screws in advanced text-books of Dynamics in Germany. There is an excellent account of the theory of the cylindroid in Vol. II., pp. 596–603. The only exception, and it is a very small one, which I feel inclined to take to this part of Professor Budde's work is that he speaks of the composition of *Screws.* It seems to me better to preserve the notion of a screw as simply a geometrical entity and to speak of the composition rather of twists or of wrenches on the screws than of composition of the screws themselves. Vol. II. pp. 639–644 gives an account of the fundamental parts of the theory of reciprocal screw systems. The geometrical construction for the cone of screws which can be drawn through any point reciprocal to a cylindroid (§ 22), and which was originally given in the *Theory of Screws*, 1876, p. 23, has been here reproduced. A good account is also given, Vol. II., pp. 905–908, of the geometrical theory of the restraints of the most general type. This subject is developed both by the elegant methods of Mannheim and also by those of the Theory of Screws.

ROUTH (E. J.)—*Treatise on Analytical Statics*, Vol. i., 2nd Edition, 1896.

In this standard work several of the fundamental Theorems of the Theory of Screws will be found. See pp. 202–208.

KLEIN (F.)—*Nicht-Euclidische Geometrie: Vorlesung.* 1889–90. Ausgearbeitet von Fr. Schilling. Göttingen, 1893.

This is a lithographed record of Klein's lectures. It is invaluable to any one who desires to become acquainted with the further developments of that remarkable Theory which is of such great importance in the subject of this volume as in so many other departments of Mathematics. The bearing of the Theory of Screws in its relation to the Non-Euclidian geometry is discussed by the author.

BURNSIDE (W.)—*On the Kinematics of Non-Euclidian Space.* London Math. Soc. Proceedings, xxvi. 33–56, Nov. 1894.

The paper consists of a number of applications of a construction for the resultant of two displacements (or motions), the construction being formally independent of the nature of the space, Euclidian, elliptic or hyperbolic, in which the motions are regarded as taking place.

§ II. of the paper gives the application to elliptic space. The main point in this case is to deduce synthetically, from the construction, the existence of finite motions which correspond to the velocity-systems that Clifford has called right- and left-vectors (the same words are here applied to the finite displacements themselves). This deduction is materially aided by considering the system of equidistant surfaces of a given pair of conjugate lines, the two sets of generators on which constitute respectively the right-parallels and the left-parallels of the given pair. In this way the existence of two sets of finite motions, each individual motion of which leaves unchanged each of a doubly-infinite number of straight lines is demonstrated (the right- and left-vectors). It is also shown that, when a right- (left-) vector is represented as the resultant of two rotations through two right angles, the axes of the two rotations are left- (right-) parallels. Hence from the original construction the resultant of two right- (left-) vectors is again a right- (left-) vector.

Lastly it is shown, still synthetically, that every right-vector is permutable with every left-vector, thus proving in a different way what is given in § 427 of this volume. See also p. 526. It is also shown that the laws according to which right- (left-) vectors combine together are the same as those by which finite rotations round a fixed point combine.

The remainder of § II. is concerned with the determination of all distinct types of "continuous groups" of motions in elliptic space.

§ III. gives the application of the construction to hyperbolic space. Here attention is first directed to a type of displacement which leaves no finite point or line undisplaced. It is also shown that no displacement in hyperbolic space can leave more than one real line unchanged. This fact, combined with the properties of the previously mentioned special type of displacements, is then used to determine all the distinct types of continuous groups of motion in hyperbolic space.

JOLY (C. J.)—*The Theory of Linear Vector Functions.* Transactions of the Royal Irish Academy, Vol. xxx., pp. 597–647 (1894).

In this memoir the close connexion between the quaternion theory of linear vector functions and the Theory of Screws is developed. "The axes of the screws of the resultants of any wrenches acting on three given screws belong therefore to one of the congruencies of lines treated of in the present paper, and every geometrical relation described in it may be applied to problems in Rational Mechanics." A remarkable quintic surface is discovered which under certain conditions degrades into the cylindroid. At the close of the memoir the linear vector functions expressing screw-systems of the third, fourth and fifth orders are discussed.

BALL (R. S.)—*The Theory of Pitch Invariants and the Theory of Chiastic Homography.* Tenth Memoir. Transactions of the Royal Irish Academy, Vol. xxx., pp. 559–586 (1894).

It is shown that if $\alpha_1 \ldots \alpha_6$ be the six co-ordinates of a screw α, while $h_1, \ldots h_6$ are the angles which α makes with the six co-reciprocal screws of reference, then expressions of the form

$$\alpha_1 \cos h_1 + \ldots + \alpha_6 \cos h_6$$

are invariants in the sense that they are unaltered for every screw on the same ray as α.

If $k_1, \ldots k_6$ be the similar angles for any other screw, then

$$\frac{\cos h_1 \cos k_1}{p_1} + \frac{\cos h_2 \cos k_2}{p_2} + \ldots + \frac{\cos h_6 \cos k_6}{p_6} = 0,$$

where $p_1 \ldots p_6$ are the pitches of the screws of reference.

If two instantaneous screws α and β and the corresponding impulsive screws η and ξ are so related that α is reciprocal to ξ, then β must be reciprocal to η. This clearly implied that there must in all cases be some relation between the virtual coefficients $\varpi_{\alpha\xi}$ and $\varpi_{\beta\eta}$. The relation is here shown to be

$$\frac{p_\alpha}{\cos(\alpha\eta)} \varpi_{\beta\eta} = \frac{p_\beta}{\cos(\beta\xi)} \varpi_{\alpha\xi}.$$

In this paper also the notion of chiastic homography is introduced. The characteristic feature of chiastic homography is, that every three pairs of correspondents α, η; β, ξ; γ, ζ, fulfil the relation

$$\varpi_{\alpha\xi}\varpi_{\beta\zeta}\varpi_{\gamma\eta} = \varpi_{\alpha\zeta}\varpi_{\beta\eta}\varpi_{\gamma\xi}.$$

The homography of impulsive and instantaneous systems is chiastic, and the relation has other physical applications. The substance of this paper has been reproduced in Chaps. XX. and XXI. of the present volume.

APPELL (B.)—*Sur le Cylindroïde.* Revue de mathématiques spéciales, 5th year, 1895, pp. 129, 130.

It had been shown in the *Theory of Screws*, 1876, that the projections of any point on the generators of a cylindroid form an ellipse. Appell has here shown conversely that if the projections of a point on the generators of a *conoidal* surface lie on a plane curve, then the conoid can be no other than a cylindroid. ROUBADI (C.), pp. 181–183 of the same volume, gives some further geometrical investigations about the cylindroid.

We may now enunciate a theorem still more general, that if the projections of every point on the generators of a *ruled surface* other than a cylinder are to form a plane curve then that curve must be an ellipse and the ruled surface must be the cylindroid (see p. 20).

BALL (R. S.)—*Further Development of the Relations between Impulsive Screws and Instantaneous Screws.* Eleventh Memoir. Transactions of the Royal Irish Academy, Vol. XXXI., pp. 99–144 (1896).

It is shown that when η is the impulsive screw and α the instantaneous screw, the kinetic energy of the mass M twisting about α with a twist velocity $\dot{\alpha}$ is

$$M\dot{\alpha}^2 \frac{p_\alpha}{\cos(\alpha\eta)} \varpi_{\alpha\eta}.$$

The twist velocity acquired by a given impulse is proportional to

$$\frac{\cos(\alpha\eta)}{p_\alpha}.$$

There is a second general relation, besides that proved in the last Memoir, between two pairs of impulsive screws η, ξ, and their corresponding instantaneous screws α, β for a free rigid body.

This relation is as follows

$$\frac{p_\alpha}{\cos(\alpha\eta)}\cos(\beta\eta) + \frac{p_\beta}{\cos(\beta\xi)}\cos(\alpha\xi) = 2\varpi_{\alpha\beta}.$$

The following theorem is also proved.

If two cylindroids be given there is, in general, one, and only one, possible correlation of the screws on the two surfaces, such that a rigid body could be constructed for which the screws on one cylindroid would be the impulsive screws, and their correspondents on the other cylindroid the instantaneous screws.

KOTELNIKOF (A. P.)—*Screws and Complex Numbers.* Address delivered 5th May, 1896. Printed (in Russian) by order of the Physical Mathematical Society in the University of Kazan.

After an introduction relating to the place of the Theory of Screws in

Dynamics the author introduces the complex numbers called Biquaternions by Clifford. Again, Mr Chawner translates:

"The more I studied these numbers the more clearly I grasped two properties in them to which I assign very great importance. First I found that I had only to have recourse to a little artifice to make the Theory of Biquaternions perfectly analogous, nay, perfectly identical, with the Theory of Quaternions. I found that I had only to introduce the idea of the functions of complex numbers of the form $a + \omega b$ where ω is a symbol with the property $\omega^2 = 0$ and at once all formulæ in the Theory of Quaternions could be regarded as formulæ in the Theory of Biquaternions. Second, I found that to the various operations in biquaternions there correspond various, more or less valuable, constructions of the Theory of Screws, and conversely that to the constructions of the Theory of Screws, which are so important to us, there correspond various operations with biquaternions. To these results I attach great importance. Thanks to biquaternions I can produce perfect parallelism between the constructions and theorems of the Theory of Vectors and those of the Theory of Screws. This I call the Theory of Transference and devote a great part of my book to it." (Theory of Vectors Kasan, 1899.)

He also mentions the Screw Integrals of certain differential equations and says, "If from two screw integrals corresponding to two given screws (we will call them α and β) we construct a third with the aid of Poisson's brackets, then the screw of the latter will be the vector product of the screws α and β of the given integrals. This circumstance allows us to use biquaternions in order to investigate the properties of screw integrals and their groups."

BALL (R. S.)—*The Twelfth and concluding Memoir on the Theory of Screws, with a Summary of the Twelve Memoirs.* Twelfth Memoir. Transactions of the Royal Irish Academy, Vol. xxxi., pp. 145–196 (1897).

At last I succeeded in accomplishing what I had attempted from the first. I could not develop the complete theory until I had obtained a geometrical method for finding the instantaneous screw from the impulsive screw. This has been set forth in this Memoir, and in Chap. XXII. of this volume.

RENÉ DE SAUSSURE.—*Principles of a new Line Geometry.* Catholic University Bulletin, Jan. 1897, Vol. iii. No. 1. Washington, D.C.

The distance and the angle between two rays are here represented as a single complex quantity known as the *Distangle*, $P + QI$, where I is a geometrical unit symbol like $\sqrt{-1}$. The quantity $(P + QI) \div I$ will be regarded as the angular measure of the same interval and will be known as the codistangle formed by the two lines. A *Codistangle* is a complete representation of a wrench, and the laws of the composition of wrenches are obtained.

M^c^AULAY (Alex.)—*Octonions, a development of Clifford's Bi-quaternions.* 8vo., pp. 1–253. Cambridge (1898).

"An octonion is a quantity which requires for its specification and is completely specified by a motor and two scalars of which one is called its ordinary scalar and the other its convert. The axis of the motor is called the axis of the octonion." In Chap. V. a large number of examples are given of the applications of Octonions to the Theory of Screws. Many of the well-known theorems in the subject are presented in an interesting manner. A discussion of Poinsot's theory of rotation is also given by the octonion methods. On p. 250 Mr M^c^Aulay has kindly pointed out that the "reduced wrench" is a conception which cannot have place in the special case when two reciprocal screw-systems have a screw in common. I had overlooked this exception (§ 96). The existence of n real principal screws of inertia of a rigid body with n degrees of freedom is proved also by Octonions, p. 248, and also the n harmonic screws, p. 248.

JOLY (C. J.)—*The Associative Algebra applicable to Hyperspace.* Proceedings of the Royal Irish Academy, 3rd Ser., Vol. v., No. 1, pp. 73–123 (1898).

The algebra considered in the present paper is that where units $i_1, i_2 \dots i_n$ satisfy equations of the type $i_s^2 = -1$ and $i_s i_t + i_t i_s = 0$. In this profound memoir there is a discussion of the Theory of Screws in a space of m dimensions. We learn that "when a system compounded from m screws is defined by a linear function (f), the reciprocal system is defined by the negative of the conjugate of that function ($-f'$)." The canonical representation of a screw in Hyperspace is given and the vector equation to the locus which is the analogue of the cylindroid. The following result, p. 106, is of much interest. "Thus in spaces of even order, the general displacement of a body may be effected by rotations of definite amounts in a number of definite hyper-perpendicular planes, one determinate point being held fixed; in spaces of odd order, a translational displacement must be added to the generalized rotation; but by proper choice of base-point this displacement may be made perpendicular to all the planes of rotation." This remark is illustrated by the well-known laws of the displacement of a body in two or three dimensions respectively.

JOLY (C. J.)—Bishop Law's Mathematical Prize Examination in the University of Dublin, Michaelmas, 1898.

Many of Sir William Hamilton's discoveries in quaternions were first announced in questions which he proposed from time to time at the Law Prize examination. This is, so far as I know, the only examination in which quaternion problems are still habitually proposed. Professor C. J. Joly in the Law Prize paper for 1898 has given the following questions containing applications of Quaternions to the Theory of Screws.

(*a*) The origin being taken as base-point, let μ and λ denote the couple and the force of any wrench, then the transformation

$$\mu = \frac{\mu}{\lambda} \cdot \lambda = S\frac{\mu}{\lambda} \cdot \lambda + V\frac{\mu}{\lambda} \cdot \lambda$$

contains Poinsot's theorem of the Central Moment.

(*b*)
$$\rho = V\frac{\mu_1 + s\mu_2}{\lambda_1 + s\lambda_2} + x(\lambda_1 + s\lambda_2)$$

is the vector equation of a ruled surface (the cylindroid) formed by the central axes of wrenches compounded from two given wrenches (μ_1, λ_1) and (μ_2, λ_2).

(*c*) The form of this equation shows that the locus of the feet of perpendiculars dropped from an arbitrary point on the generators of a cylindroid is a conic section.

(*d*) If (μ, λ) is any wrench compounded from three given wrenches (μ_1, λ_1), (μ_2, λ_2), and (μ_3, λ_3), the couple of this wrench is a determinate linear vector function of the force, or $\mu = \phi\lambda$, and the function ϕ adequately defines this 'three-system' of wrenches.

(*e*) Examine the scalar and vector parts of the quaternion $\phi\lambda \,.\, \lambda^{-1}$. Show that the pitch of any wrench of the system is inversely proportional to the square of that radius of a certain quadric which is parallel to its axis; also that the locus of feet of perpendiculars drawn from the origin to the central axes of the system is a surface (Steiner's Quartic) containing three double lines intersecting in the origin.

(*f*) The screws (μ, λ) and (μ', λ') being reciprocal if $S\mu\lambda' + S\lambda\mu' = 0$, show that the screws reciprocal to the system $\mu = \phi\lambda$ belong to the system $\mu' = -\phi'\lambda'$, or that a linear vector function and the negative of its conjugate determine, respectively, a 'three-system' of screws and its reciprocal 'three-system.'

Two other theorems communicated to me by Professor Joly may also find a place here.

If a body receive twists about four screws of a three-system and if the amplitude of each twist be proportional to the sine of the *solid angle* determined by the directions of the axes of the three non-corresponding screws, then the body after the last twist will have regained its original position.

If four wrenches equilibrate and if their axes are generators of the same system of a hyperboloid, their pitches must be equal.

WHITEHEAD (A. N.)—*Universal Algebra*, Vol. i., Cambridge (1898), pp. i–xxvi, 1–586.

It would be impossible here to describe the scope of this important work, the following parts of which may be specially mentioned in connection with our present subject.

Book v. Chap. I. treats of systems of forces, in which the inner multiplication and other methods of Grassmann are employed. Here as in many other writings we find the expression Null lines, and it may be remarked that in the language of the Theory of Screws a null line is a screw of zero pitch.

Chap. II. of the same book contains a valuable discussion on Groups of Systems of Forces. Here we find the great significance of anharmonic ratio in the higher branches of Dynamics well illustrated.

Chap. III. on Invariants of Groups continues the same theories and is of much interest in connection with the Theory of Screws.

Chap. IV. discusses among other things the transformation of a quadric into itself, and is thus in close connection with Chap. XXVI. of the present volume.

Whitehead's book should be specially consulted in the Theory of Metrics, Book VI. The Theory of Forces in Elliptic Space is given in Book VI. Ch. 3, in Hyperbolic Space in Book VI. Chap. 5, and the Kinematics of Non-Euclidian Space of all three kinds in Book VI. Chap. 6. There are also some passages of importance in Statics in Book VII. Chaps. 1 and 2, Book VIII. Chap. 4, and on Kinematics in Book VII. Chap. 2 and Book VIII. Chap. 4. The methods of Whitehead enable space of any number of dimensions to be dealt with almost as easily as that of 3 dimensions.

STUDY (E.)—*Eine neue Darstellung der Kräfte der Mechanik durch geometrische Figuren.* Berichte über die Verhandlungen der königlich-sächsischen Gesellschaft der Wissenschaften zu Leipzig. Mathematisch-physische Classe, Vol. li., Part II., pp. 29–67 (1899).

This paper is to develop a novel geometrical method of studying the problems referred to.

CARDINAAL (J.)—*The representation of the screws of Ball passing through a point or lying in a plane according to the method of Caporali.* Koninklijke Akademie van Wetenschappen te Amsterdam, 23rd Feb. 1899.

This is a development of a lecture before the Gesellschaft deutscher Naturforscher und Aerzte, Düsseldorf (Sept. 1898). The object of this memoir is to obtain from the principles of the Theory of Screws a representation of the rays of a certain quadratic complex treated of by R. Sturm and Caporali. See Sturm's *Liniengeometrie*, iii. pages 438–444. See also a paper by Cardinaal, "Über die Anwendung der Caporali'schen Abbildung des Strahlencomplexes zweiten Grades auf die Bewegung eines starren Körpers mit Freiheit vierten Grades," Jahresbericht der Deutschen Mathematiker-Vereinigung, vii. 1. This is the lecture above referred to, and it should be studied in connection with the geometrical theory of screw-systems of the fourth order.

JOLY (C. J.)—*Astatics and quaternion functions.* Proceedings of the Royal Irish Academy, 3rd Series, Vol. v. No. 3 (pp. 366–369), 1899.

STUDY (E.)—*Die Geometrie der Dynamen.* Deutsche Mathematiker-Vereinigung VIII. 1, 1900.

INDEX.

Accelerating Screw-chain 409
Accelerator in a 2-system 420
Address to British Association 496
Amplitude of a Twist defined 11
Angle between two Screws 276
Angle between two Screws in a 3-system 204
Anharmonic property, fundamental 113
Anharmonic property of Homographic Systems 266
Appell on Cylindroid 29, 535
Association, British, address to 496
Axioms of the Content 435, 448, 451
Axis, Homographic 127
Axis of Inertia 123
Axis, Principal properties of 260, 400

Battaglini 20, 34, 249, 518, 519, 520
Biquaternions 521
Bolyai 521
Bonola 439
British Association, address to 496
Buchheim 433, 522, 528
Budde 533
Burnside, Prof. W. 522, 533
Burnside, Prof. W. S. 281

Canonical, Co-reciprocals 38; Displacement of Rigid Body 7; System of Forces on Rigid Body, 10
Caporali 539
Cardinaal 539
Cartesian. Equation of Cylindroid 19, 224; Equations of Screw 39
Casey 30, 495
Cayley 20, 433, 509, 511, 514, 522, 527, 533
Chains, Screw-; Harmonic, 395; Homography of 373; Impulsive and Instantaneous 392; Principal of Inertia 394; Theory of 367
Chasles 1, 4, 510, 511, 512, 514, 524
Chawner 75, 531, 536
Chiastic Homography 306, 307
Circle, Representing Cylindroid 45
Clifford 433, 464, 509, 521, 525, 527, 528, 536
Coefficient, Virtual 17, 18
Complex, Linear 17, 20
Composition of Twists and Wrenches 16, 18; of Vectors 473
Cone, Reciprocal to a Cylindroid 27
Conic, The Pitch-, 24, 204
Conjugate Screws, of Inertia 71, 299; of Potential 89; of Quadratic n-system 240
Conjugate Screw-chains of Inertia 396
Conoid of Plücker 20
Constraints 63, 64, 119
Contact of Conic and Cubic 490
Content, Axioms of the 435, 448, 451
Content defined 433
Co-ordinates, of Impulsive Screw 356; of a Rigid Body 355; of Restraining Wrench-chain 407; of a Screw 31, 36; of a Twist or Wrench 34
Co-reciprocal Screws 33, 38, 335
Correspondence 383
Corresponding Screw-systems 271
Cox, Homersham 522, 527
Cubic equation in 3-system 492

Cylindroid, Chiastic Homography on 307; Constants necessary to define a 68; Defined by two Screws 20; Discovery of by Hamilton 510, 520; Drawing of Plane section 154; Dynamical property discovered by Battaglini 20; Equation of 19, 224; Equation of Central Section 166; Equation of Plane Section 152; Fundamental property 25; Geometrical investigation of 146, 150; Hyperbola in triple contact 164; Impulsive and Instantaneous 112; Models of 150, 530; Name suggested by Cayley 20; Nodal line upon 24; Normal 308; Parabola related to Plane Section 156; Permanent Screw on 422; Relation between two 108; Principal Screws on 21, 43; Reducing to a plane 23; Remarkable geometrical property, 535; Represented by a Circle 45; Represented by a Straight line 199; Restraining Screw on 419; Tangential section of 60; Three-system 180

D'Alembert 520
Degrees of certain Surfaces 242
D'Emilio 528
Departure defined 448
Differential equation for Kinetic Energy, 356
Displacement, canonical form of 1
Double Screws 264, 309
Dyname, correspondence of 383; defined 274; Parallel Projections of, 382
Dynamical parable, a 496

Elliott 532
Ellipse of Inertia 114; of Potential 116
Ellipsoid of Inertia 187
Emanants 274, 275, 408, 469
Enclosing Screw Systems defined 233
Energy, Kinetic, Emanant equation of 405; Expression for 296
Equations, Linear Systems of 67
Equilibrium of Four Forces 186; general theorem of 64
Euler 83, 231, 406, 520, 527
Everett 26, 522, 523
Extent defined 434

Fiedler 524
Forces in Theory of Screws 11; in Non-Euclidian Space 480
Freedom, various degrees of 63
Fundamental problem with free body 336

Geiser 513
Geometrical Theory of Impulsive and Instantaneous Screws 322
Goebel 151
Grace 495
Grassman 509, 515
Gravelius 531, 532
Grubb (Sir H.) 150, 151

Halphen 522
Halsted 439
Hamilton (Sir W. R.), 509, 510, 512, 515, 517, 521, 537
Harmonic Motion 99
Harmonic Screws 94, 106, 118, 133, 193, 397
Heath 452, 465, 522, 527
Henrici 532
Heteronymous Vectors 475, 480
Homography, Chiastic 306, 309; of Content 415; Double points 126, 127, 342; Screw-chains 370, 393; Screw Systems 262
Hyde, 531
Hyperbola in triple contact with Cylindroid, 164

Impulsive Screws, Chains 392; Cylindroid 112; Defined 71; Fundamental formulæ 298; Theory of two pairs 325
Inertia, Axis of 123; Conjugate Screws 71; Conjugate Screw-chains 395; Ellipse of 114; Ellipsoid of 187; Principal Screws of 69; Principal Screws on Cylindroid 114; Principal Screws on Three-system 188; Principal Screws on Four-system 230; Principal Screws on Five-system 252; Principal Screw-chains of 394
Infinite Pitch, Screws of 22, 39
Infinity, Screws at 291
Instantaneous Screws 71; Chains 392; Definition 10
Intensity of a Wrench defined 11
Intervene defined 434; Expression for 443
Invariants of Eight Screws 260
Invariants, Pitch 289
Involution, meaning of 249

Joly 30, 178, 182, 390, 492, 522, 534, 537, 538, 539

Kinetic Energy, Differential Equation for 356; Expression for 296; Property of 401
Klein 17, 18, 33, 34, 241, 249, 286, 433, 439, 517, 518, 520, 521, 533
Kotelnikof 522, 535
Kronecker 531
Kummer 517

Lagrange 76, 96, 97, 98, 395, 406, 489
Law Prize Examination 537
Lewis 46, 151
Lindemann 433, 523
Linear Complex 20
Linear Equations, note on 67
Lobachevsky 521
Lornoff 531

M'Aulay (Alex.) 37, 84, 522, 536
MacAulay (F. S.) 437
Manchester, address at 496
Mannheim 45, 513, 519, 521, 522, 533
Mass-chains 367
Minchin 151, 529
Möbius 31, 102, 248, 255, 511, 512, 524
Models of Cylindroid 150, 530
Motion, Equations of, in Screw-chain Co-ordinates 405
Motor defined 477

Non-Euclidian Space 433
Normal Cylindroids 308
Notation, fundamental 11

Object, Definition of 433
Octonions 536
Orthogonal Transformation 280, 465
Oscillations about a fixed point 194
Oscillations, Screw-chain 397

Padeletti 522, 526, 527
Panton 244
Parable, a Dynamical 496
Parabola in contact with Cylindroid 158
Parallel Projections, 379
Parameters of a Screw-System 65
Pascal line 307
Pectenoid 255
Permanent Screw-chains 410; Cylindroid 422; Definition 399; Different Degrees of Freedom 432; Equations of 421; Geometrical Construction for 427
Pitch, Axis of 46; Conic 24, 204, 210; Co-ordinates 36; Co-reciprocals 282; on Cylindroid 20; Definition of 7; Infinite 22, 39; Invariant 289, 294; Non-Euclidian Space 478, 527; Quadrics 227; Stationary 221
Planes, Four imaginary, in 3-system 202
Plane representative Cylindroid 45; Dynamical problems 120; Three-System 197
Plücker 17, 20, 34, 150, 177, 249, 274, 515, 516, 517, 520, 524, 531
Poinsot 4, 297, 510, 511, 513, 524, 536
Poisson 536
Polar Screw 238, 241
Potential 87; Conjugate Screws of 89; Ellipse of 116; Principal Screws of 90
Principal Axis of a Rigid Body 260, 400
Principal Screws on Cylindroid 21, 43
Principal Screws of Inertia 69; on a Cylindroid 114; Chains 394; Five-System 252; Four-System 230; Three-System 188, 207
Principal Screws of Potential 90
Projections, Parallel 379

Quadratic Screw-Systems 233
Quadric, The Pitch- 172
Quadric of Potential 192
Quadrics in Three-System 173
Quaternions 30, 512, 515

Range defined 434
Reaction of Constraints 64
Reality of the Principal Screws of Inertia 77
Reciprocal Screws, Cylindroid 37; Chains 288; Defined 26; Property of 347; System 63
Reduced Wrench 84
René de Saussure 536
Representation of Cylindroid in a Plane 45
Résal 521
Resolution of Twists and Wrenches 31
Restraining Screw-chain 407; Cylindroid 419; Co-ordinates 314; Locus 416
Restrictions on Forces 11; Movement 13

Ribaucour 513
Rittershaus 524
Roberts, R. A. 530
Rodrigues 17, 512
Roubadi 535
Routh 76, 98, 484, 533

Salmon 76, 495, 524, 532
Saussure, M. René de 10
Schell 525
Schilling 530
Schönemann 513
Schönflies 529
Screws, Angle between two 276; Anharmonic function of eight 267; on same axis 41; Cartesian Equation of 38; Conjugate 71; Co-ordinates 31; Co-ordinates, Transformation of 42; Co-reciprocal 33; Definition of 7; Harmonic 94; Imaginary on a three-system 201; at Infinity 41, 291; Intersecting in a three-system 212; Invariants of eight 266; Impulsive and Instantaneous, two pairs of 310; Instantaneous defined 10; on One Line 109; in Non-Euclidian Space 433; Parallel to a plane 277; Permanent 399; Permanent Equations of 421; Polar 238; Principal of Inertia 69; Reciprocal to a Cylindroid 26; Reciprocal defined 26; Reciprocal to five Screws 30, 246; Reciprocal to four Screws, locus of 29; Reciprocal of System 63; at Right Angles 276; Restraining on Cylindroid 419; Seven 31; Stationary pitch 221; Theory of, based on two principles 5
Screw-chains, Accelerating 409; of Conjugate Inertia 395; defined 369; Equations of motion of 405; of Fifth order 378, 384; of First order 369; of Fourth order 377; Harmonic 395; Homography in 370; Impulsive 392; Instantaneous 392; Intermediate Screws of 368; Lagrange's equations applied to 406; Oscillations on 397; Permanent 410; Principal of Inertia 394; Reciprocal 388; Second order 370; Third order 375
Screw Systems 62; Correspondence in 271; of Fifth order 246; of First order 101; of Fourth order 218; Homographic 262; Parameters of 65; Permanent Screws in 399; of Second order 107; of Sixth order 258; of Third order 170; two of Third order correlated 344
Segre 528
Sexiant 248
Sextic Surface, Joly on 493
Seydler 530
Signs, Theorem as to 285
Sixth order, Permanent Screws in 432
Skatow 522
Solid angle, Formulæ with 495
Somoff 521
Spottiswoode 516
Stationary pitch, Screws of 214
Steiner's quartic 179
Stoney, G. J., His lecture experiment 519
Study 538, 539
Sturm 524
Surfaces, Degrees of certain 242
Sylvester 249, 250, 514

Tangential Section of Cylindroid 60
Tarleton 76, 530
Three pairs of Impulsive and Instantaneous Screws 330
Three-System 170; Permanent Screws in 428; Plane representation of 197
Townsend 465
Transformation, Orthogonal 465; Screw Co-ordinates 42
Triangle of Twists 49
Triple Contact of Conic and Cubic 490
Twist, Amplitude of 11; Composition of 18; Definition of 7; Triangle of 49
Twist Velocity, Acquired by an impulse 81; Defined 11; Expression for 297; Theorem relating to 305

Vanishing Emanant 408
Vectors 464; Analogy with twists 10; Composition of 473; Fundamental laws of 475; Orthogonal Co-ordinates 470
Velocity, Twist defined 11; Expression for 297
Virtual Coefficient in Co-ordinates 36; Defined 17; as Emanant 278; Symmetry of 18
Vivanti 530

Weiler 523

Weierstrass 513
Whitehead 433, 538
Williamson 76, 281
Wrenches, Composition of 18; Definition of 10; by displacement 88; Intensity of 11; Reduced 84; Restraining 314
Wrench chain, Defined 388; Restraining 407

Zanchevsky 75, 530, 531
Zeuthen 34, 518
Ziwet 521

Lightning Source UK Ltd.
Milton Keynes UK
UKOW04f0318101214

242872UK00002B/320/P